Herbert Pöllmann
Cementitious Materials
De Gruyter Graduate

Also of Interest

Highlights in Mineralogical Crystallography
Thomas Armbruster, Rosa Micaela Danisi (Eds.), 2015
ISBN 978-3-11-041704-3, e-ISBN 978-3-11-041710-4

Rietveld Refinement. Practical Pattern Analysis using Topas 5.0
Robert E. Dinnebier, Andreas Leineweber, 2017
ISBN 978-3-11-045621-9, e-ISBN 978-3-11-046138-1

Multi-Component Crystals. Synthesis, Concepts, Function
Edward R. T. Tiekink, Julio Zukerman-Schpector (Eds.), 2017
ISBN 978-3-11-046365-1, e-ISBN 978-3-11-046495-5

Symmetry. Through the Eyes of Old Masters
Emil Makovicky, 2016
ISBN 978-3-11-041705-0, e-ISBN 978-3-11-041714-2

Pore Scale Geochemical Processes
Carl Steefel, Simon Emmanuel, Lawrence Anovitz (Eds.), 2016
ISBN 978-0-939950-96-6, e-ISBN 978-1-5015-0207-1

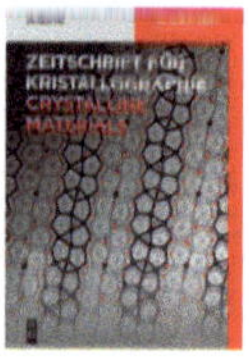

Zeitschrift für Kristallographie – Crystalline Materials
Pöttgen, Rainer (Editor-in-Chief)
ISSN 2194-4946, e-ISSN 2196-7105

Cementitious Materials

Composition, Properties, Application

Edited by
Herbert Pöllmann

DE GRUYTER

Physics and Astronomy Classification Scheme 2010
61 (Structure of solids and liquids; crystallography)

Editor
Prof. Dr. Dr. Herbert Pöllmann
Martin-Luther-Universität Halle
Institute for Geosciences
Von-Seckendorff-Platz 3
06120 Halle, Germany
herbert.poellmann@geo.uni-halle.de

ISBN 978-3-11-047373-5
e-ISBN (PDF) 978-3-11-047372-8
e-ISBN (EPUB) 978-3-11-047391-9
Set-ISBN 978-3-11-048673-5

Library of Congress Cataloging-in-Publication Data
A CIP catalog record for this book has been applied for at the Library of Congress.

Bibliographic information published by the Deutsche Nationalbibliothek
The Deutsche Nationalbibliothek lists this publication in the Deutsche Nationalbibliografie; detailed bibliographic data are available on the Internet at http://dnb.dnb.de.

Cover image: Herbert Pöllmann
Typesetting: PTP-Berlin, Protago-TEX-Production GmbH, Berlin
Printing and binding: Hubert & Co. GmbH und Co. KG
♾ Printed on acid-free paper
Printed in Germany

www.degruyter.com

Preface

Our current knowledge on cementitious and binder materials for building and construction is based on a long history and the use of many different natural materials in their production. When these construction materials were first used, their reactions were based on the carbonation of lime or hydration reactions to form hydrates during the hardening process. In the beginning, the properties of cements were improved when it became obvious that the hydraulic reactions of cement during the hardening process led to a better resistance against water and aggressive surroundings. Therefore, hydraulic cements are materials that react with water and are able to resist well afterwards to dissolution in water. The stability of cements against water is also used in immobilization applications of harmful and toxic materials to avoid the redissolution of the phases. Further developments have been made based on the increasing knowledge of people involved in the practical and direct usage of these materials. Only a very basic understanding of the relevant chemical and mineralogical background was available to them and changes were largely made after the failure of building materials. Nevertheless, in the past decades there has been a tremendous increase in knowledge on different natural materials, the processing, and the final products and its stability on the long term. New technologies and improved measurement possibilities have made it possible to increase our understanding of chemical reactions and the microstructural development of hardening cementitious systems. It was highly necessary to improve the properties of the materials being used and to avoid failures caused by using inappropriate available resources. The use and application of additives and admixtures has also come more into focus.

Nowadays, many different types of cement, composite cements, special cements, and other products like gypsum and lime exist. Special cements fulfill the requirements for special applications like ultrarapid hardening cements for repair applications or for special resistance at high temperatures or for special resistance against chemical attacks.

Our understanding of the compositions of the raw materials, the development of phases formed during the cement and binder production, and the newly formed microstructures during the hydration process are all based on the formation of crystalline and amorphous hydrated mineral phases. It seems obvious that the determination of physical parameters like strength development is easier and can be obtained with higher precision than the measurements and descriptions of the development of minerals and the microstructure during hydration.

Therefore, it is highly necessary that our knowledge of modern methods applied for the quantitative determination of phases and applied for the investigations of different binders and cements and their background are summarized and that knowledge about them is disseminated to as many people as possible.

DOI 10.1515/9783110473728-001

All these measurements are based on a deep understanding of atomic arrangements and crystallographic properties. Nowadays, a lot of knowledge is already available about reducing energy, reducing the CO_2 production of cement, improving the properties of the materials themselves, and avoiding failures. But, recently, the invention of completely new cements, cementitious materials, and binders and the use of different kinds of raw materials has also been progressing. Secondary raw materials, secondary fuels, different types of cements, and the application of grinding aids and admixtures and additives lead to many different mineral hydration phases. Among them, not all have yet been investigated and described. Therefore, the book focuses on several aspects where basic knowledge on chemistry and mineralogy is combined in the description of crystallographic parameters of crystalline and amorphous phases. Several applications and uses do show the possibilities of these new materials.

The book is based on 4 chapters dealing with cement composition and hydration, a selection of special cements and binder mineral phases, the use of special cementitious and binder materials, and some selected special measurements and properties. Overall, in this widespread area of cementitious materials, 15 different articles on special topics are included in the book to show and summarize the many different possibilities and challenges of inorganic binding materials.

The book can be of use for all those who are interested in the field of applied crystallography for materials and it can be used as a basis for understanding that there is still some work ahead to fully understand these systems and to improve the properties of these materials. Even the reuse of industrial residues and natural and artificial pozzolanes is very important.

Today, in light of environmental concerns, especially with respect to the need to reduce CO_2, different raw materials are replacing limestone. New cement types with different compositions, new phases, and properties must be determined for these materials.

It is obvious that a better understanding of the reactions and developing microstructures of the cement and binder materials can lead to more environmentally friendly binders and cements which can be developed with optimized properties obtained simultaneously. Knowledge on sustainability and environmental concerns related to cements must be included.

Understanding the control, the optimization, and the further development of cementitious materials is strongly related to applied mineralogy and crystallography.

These different aspects may impact young scientists to become interested in this wide field of investigation. All summarized data on raw materials, environment, properties, reuse of other industrial residues, waste fuels, and the development of better properties and handling using additives and admixtures are related to the crystalline phases, their structures, phase changes, and solid solutions. By using other raw materials, amorphous mineral phases are increasingly included. Some of these phases are well-known to show low crystallinity, others are even completely amorphous. With new cement types, even some completely new phases are present and must be investigated. The determination of crystal structures can be performed on pure phases synthesized in the laboratory.

For cements and binders, knowledge on practical experience must be combined with basic scientific knowledge in order to further develop and improve these materials, which are produced in very high quantities.

Thanks are due to all the contributors, who put in a lot of time to bring together their results in this book, and also to the editorial office for their continuous support and help.

Halle, November 2016
Herbert Pöllmann

Contents

Part IV: **Measurement and properties**

Part I: **Cement composition and hydration**

Ángeles G. De la Torre*, Isabel Santacruz, Laura León-Reina, Ana Cuesta, and Miguel A.G. Aranda

1 Diffraction and crystallography applied to anhydrous cements

Abstract: In this chapter, X-Ray powder diffraction (XRPD) combined with the Rietveld method is discussed to quantify both full crystalline and amorphous/crystalline non-quantified phases in anhydrous cements. For the latter, different approaches, such as internal and external standard methods, are addressed. In particular, the use of powder diffraction is shown in the characterization of Portland clinkers, Portland cements with supplementary cementitious materials, anhydrous cements for alkaline-activation, and ye'elimite-containing cements. In addition, the structural details of phases that are usually present in Portland clinkers are shown. Finally, the use of diffraction in the characterization of anhydrous cementitious materials under non-ambient conditions, viz. high temperature and high pressure, is also reviewed.

Keywords: clinker, Rietveld quantitative phase analysis, internal and external standard methods

1.1 Introduction

X-Ray powder diffraction (XRPD) combined with the Rietveld method is a powerful tool for characterizing materials and obtaining a quantitative phase analysis of them in general [1, 2] and of cementitious related systems in particular [3–6]. This chapter gives some insight into the Rietveld quantitative phase analysis (RQPA) of cement-related materials. The Rietveld method is known to be a standardless one, but the structural descriptions for each phase present in a mixture must be known and included in the Rietveld control file. Tab. 1.1 [7–27], which summarizes the findings of a previous publication [5], shows an updated compilation of the crystal structures of phases that are commonly present in anhydrous cements and could be used to perform RQPA. Here, we have made a selection of the "best" crystal structures since sometimes there is more than one structural description available in the literature for a given phase. The criteria for the selection of the "best" ones have been: (i) type of data, i.e. single crystal diffraction was preferred over powder diffraction; and among

***Corresponding author: Ángeles G. De la Torre,** Departamento de Química Inorgánica, Cristalografía y Mineralogía, Universidad de Málaga, Málaga, Spain, mgd@uma.es
Isabel Santacruz, Departamento de Química Inorgánica, Cristalografía y Mineralogía, Universidad de Málaga, Málaga, Spain
Laura León-Reina, Servicios Centrales de Investigación, Universidad de Málaga, Málaga, Spain
Ana Cuesta, Miguel A.G. Aranda, Alba Synchrotron Light Source, Barcelona, Spain
DOI 10.1515/9783110473728-002

powder diffraction, synchrotron or neutron were preferred over laboratory; and (ii) anisotropic atomic displacement parameters (ADPs) were preferred over isotropic ADPs. Moreover, if two or more suitable structural descriptions exist for a given phase, two (or more) RQPA of them should be carried out. The structure chosen should be that which gives the best fit based on a lower R_F factor, hopefully linked to a higher quantification value.

The powder Diffraction Database (www.icdd.com) is the appropriated tool that should be used to identify minor phases by searching the extra-peaks of the experimental pattern. Once the phase is identified, the crystal structure can be searched for in three structural data bases: (i) AMCSD 'American Mineralogist Crystal Structure Database' (http://rruff.geo.arizona.edu/AMS/amcsd.php); (ii) COD 'Crystallography Open Database' (www.crystallography.net); and (iii) ICSD 'Inorganic Crystal Structure Database' (www.fiz-karlsruhe.de/icsd.html). Special care should be taken when using the COD database as ADPs are not included in the downloaded files. Finally, site occupation factor(s) for a given crystal structure sometimes must be adapted to describe a given stoichiometry, mainly for solid solutions, but this should be done with caution and by experts. An example of this is the solid solution C_4AF, where the Al/Fe ratio may be different from 1.0, which affects the diffraction signals.

1.2 Rietveld quantitative phase analysis (RQPA) – full crystalline phase content, internal and external standard methods

The Rietveld method was developed in the late sixties [28] for the extensive characterization of polycrystalline compounds and uses the entire data of the measured powder pattern instead of only the reflection (peak) intensities; this method makes it possible to properly deal with strongly overlapping reflections. The Rietveld method carries out least-squares refinements to optimize a theoretical line profile until it fits (in the best possible way) the measured sample powder diffraction profile (whole-profile). For a successful RQPA, several steps must be fulfilled: (i) the sample has to be properly prepared; (ii) the diffractometer should be well aligned and maintained, and the optimal set-up should be used; (iii) every crystalline phase in the sample should be identified. When strong peak overlapping is present in the diffraction patterns, we compute the RQPA with the phases which are clearly present in the pattern and, from the net intensity in the difference curve, the remaining low-content phases are determined; and finally, (iv) when the main (all) phases are identified, the final RQPA is carried out. As mentioned before, the crystal structures of all crystalline components must be known. A list of Rietveld programs can be found elsewhere (www.ccp14.ac.uk/solution/rietveld_software/index.html), where GSAS [29, 30] and FULLPROF [31, 32] are the most widely used packages. Many other commercial packages can also be used, such as BGMN [33], SIROQUANT [34], TOPAS (Bruker AXS) and HighScore Plus (PANalytical BV).

Tab. 1.1: Structural details for phases that may be present in OPC clinkers.

Phase	Formula	Crystal system/notation	ICSD codes	Ref#
Alite	$(Ca_{2.93}Mg_{0.07})O_3(SiO_2)_{0.98}(Al_2O_3)_{0.01}$	Monoclinic/M_3	94742	[7]
	$Ca_{2.89}SiMg_{0.11}O_5$	Monoclinic/M_3	64759	[8]
	$Ca_{2.96}Mg_{0.03}Al_{0.01}(Si_{0.99}Al_{0.01})O_5$	Triclinic/T_3	162744	[9]
	Ca_3SiO_5	Triclinic/T_1	4331	[10]
Belite	Ca_2SiO_4	Monoclinic/β	81096	[11]
	Ca_2SiO_4	Orthorhombic/α′	81097	[11]
	Ca_2SiO_4	Orthorhombic/γ	81095	[11]
	$Ca_{1.85}Na_{0.15}(SiO_4)_{0.85}(BO_3)_{0.15}$	Orthorhombic/α'_H	431571	[12]
Aluminate	$Ca_3Al_2O_6$	Cubic	1841	[13]
	$Ca_{8.5}NaAl_6O_{18}$	Orthorhombic	100220	[14]
	$Ca_{8.25}Na_{1.5}Al_6O_{18}$	Monoclinic	100221	[14]
Ferrite	Ca_2AlFeO_5	Orthorhombic	9197	[15]
Lime	CaO	Cubic	52783	[16]
Periclase	MgO	Cubic	9863	[17]
Arcanite	K_2SO_4	Orthorhombic	79777	[18]
Aphthitalite	$K_3Na(SO_4)_2$	Rhombohedral	26018	[19]
Thenardite	Na_2SO_4	Orthorhombic	81506	[20]
Ca-Langbeinite	$Ca_2K_2(SO_4)_3$	Orthorhombic	40989	[21]
Ternesite	$Ca_5(SiO_4)_2(SO_4)$	Orthorhombic	85123	[22]
Ellestadite	$Ca_{10}(SiO_4)_3(SO_4)_3Cl_2$	Hexagonal	154205	[23]
Fluorellestadite	$Ca_{10}(SiO_4)_3(SO_4)_3F_2$	Hexagonal	97203	[24]
Ye'elimite	$Ca_4[Al_6O_{12}]SO_4$	Orthorhombic	237892	[25]
Ye'elimite	$Ca_{3.8}Na_{0.2}[Al_{5.6}Fe_{0.2}Si_{0.2}O_{12}]SO_4$	Pseudo-cubic	194482	[26]
Mayenite	$Ca_{12}Al_{14}O_{33}$	Cubic	241243	[27]

Apart from the raw data, any Rietveld program needs a control file to execute the refinements, where the crystal structures of the different components must be included. The fit is carried out by optimizing all appropriate variables such as: (i) the scale factor of every crystalline phase; (ii) the background parameters; (iii) the unit cell parameters for every crystalline phase; (iv) the peak shape parameters for every computed phase; (v) the correction parameters which may be phase-dependent (such as preferred orientation, extinction, and so on) or pattern-dependent (zero-shift, absorption correction when working in transmission geometry, etc.). The structural descriptions (atomic positional parameters, atomic displacement parameters and occupation factors) are not usually optimized for RQPA. The best available structural description should be used to extract the best possible scale factor value for every phase (see Tab. 1.1 and those included in [5]). This is evaluated by the flatness of the difference curve and also by obtaining low Rietveld-disagreement indices (R-factors) [35, 36]. However, low R-factors may also be obtained through false analyses. Hence, it is important to optimize those parameters that allow their minimization without excessive correlations, such as background coefficients.

There are alternative whole-pattern QPA methods for crystalline phases with unknown structures [37–39] that will not be discussed/reviewed here.

The application of RQPA to clinkers/cements/pastes is not straightforward for the following reasons: (i) the high number of phases, more than 6; (ii) each phase has its own mass absorption coefficient which may result in a microabsorption problem; (iii) poor particle statistics due to the small mean penetration depth of X-rays; (iv) the preferred orientation of some phases (e.g. [−1 0 1] for alite-M_3, [010] axis for gypsum and [104] for calcite) [3, 5, 6]; (v) phases can crystallize as several polymorphs that must be previously identified; (vi) peak shape anisotropy; (vii) scale factors are computed for ideal/stoichiometric phases but some can show atomic impurities inside.

Whole-pattern quantitative phase analysis approaches

The output of a RQPA is a set of m-crystalline phase scale factors ($\sum_m S_\alpha$ for a sample with m crystalline phases). The parameter to be calculated is the phase weight content (W_α), which is related to the scale factor (S_α) by [40, 41]:

$$S_\alpha = K_e \frac{W_\alpha}{\rho_\alpha V_\alpha^2 \mu_s}, \tag{1.1}$$

where K_e is a constant which depends on the diffractometer operation conditions, ρ_α is the crystallographic density of the α-phase, V_α is the unit cell volume of α phase, and μ_s is the sample mass absorption coefficient. K_e and μ_s are not known and they cannot be derived from the powder diffraction pattern of the studied sample.

There are three main ways to derive the phase content, W_α, from the Rietveld refined scale factor, S_α. These three methods are based on different mathematical approaches and they have different experimental complexities. They will be described in detail in the following.

(I) Normalization to full crystalline phase content method

The simplest approach is to make the approximation that the sample is composed only of crystalline phases with known structures, which will be incorporated into the control file. The weight fraction of an α-phase in a m-crystalline phase mixture is given by [40]:

$$W_\alpha = \frac{S_\alpha(ZMV)_\alpha}{\sum_{i=1}^{m} S_i(ZMV)_i}. \tag{1.2}$$

Instead of using ρ_α, the relation between S_α and W_α is based on the '*ZMV*' term with *Z* being the number of chemical units/formulas within the unit cell, *M* being the molecular mass of the chemical formula, and *V* the unit cell volume. Once the crystal structure is known, the '*ZMV*' term is known. The use of equation (1.2) in RQPA eliminates the need to measure the instrument calibration constant, K_e, and the sample mass absorption coefficient, μ_s. However, if the sample contains amorphous phases and/or misfitting problems of the analyzed phases, and/or some amounts of non-quantified crystalline phases (e.g. unknown crystal structure), the analyzed weight fractions will be overestimated. These contributions will be referred to here as amorphous and crystalline non-quantified (ACn). Thus, the resulting weight fractions are only accurate (close to the true value) if the amount of ACn is very small (almost negligible). However, although the amount of ACn in anhydrous cements is low, it is not negligible, and it is very relevant in the case of cement pastes.

(II) Internal standard method (spiking method)

A second approach consists of mixing the sample with a known amount of a crystalline standard (W_{st}). This standard must be free of amorphous content or at least its ACn content must be well-known. NIST standard reference material (SRM) 676a, corundum (α-Al_2O_3) powder, has been certified to have a phase purity of 99.02 ± 1.11 % (95 % confidence interval) by RQPA against a suitable primary standard (silicon powder carefully prepared from a single crystal). This novel certification method permits the quantification of the amorphous content for any sample of interest by the spiking method [42] using standard powder.

There are some drawbacks associated with the spiking method; for instance, the addition of the standard dilutes the crystalline phases in the sample, which may be quite problematic for low-content phases, and it might alter the hydration of cement pastes [43, 44]. RQPA will give an overestimated value of the internal standard content (R_{st}) due to the presence of ACn in the sample. The spiking method calculates the (overall) ACn content of the sample from the small overestimation of the internal crystalline standard, equation (1.3) [45, 46]:

$$\mathrm{ACn} = \frac{1 - W_{\mathrm{st}}/R_{\mathrm{st}}}{100 - W_{\mathrm{st}}} \times 10^4\ (\mathrm{wt\%}). \tag{1.3}$$

The errors associated with this approach and the optimum amount of the standard to be added have been discussed elsewhere [47]. The optimum amount of the internal standard strongly depends on the amount of ACn to be determined. Therefore, the higher the expected ACn value, the lower the addition of the standard.

One of the main issues with the internal standard method is the sample preparation. The artificial mixture (sample and internal standard) must be properly homogenized as the particles should be randomly arranged, and this is not a trivial task. Fig. 1.1 gives raw patterns of an ordinary Portland cement (OPC) type I mixed with ≈ 20 wt% of crystalline quartz and prepared following different methodologies: (i) an agate mortar for 20 minutes and (ii) McCrone micronizing mill [48] with zirconia grinding agents, isopropanol as grinding media and grinding times of 5, 10, and 15 minutes. The inset of the right side shows some peaks of alite (main phase in OPC, see Section 1.3.1).

It can be observed that the $[10\bar{1}]$ reflection is less intense when McCrone mill is used for 10 or 15 minutes. Tab. 1.2 gives RQPA results of these samples and the refined preferred orientation coefficients of alite along $[10\bar{1}]$ using the March-Dollase algo-

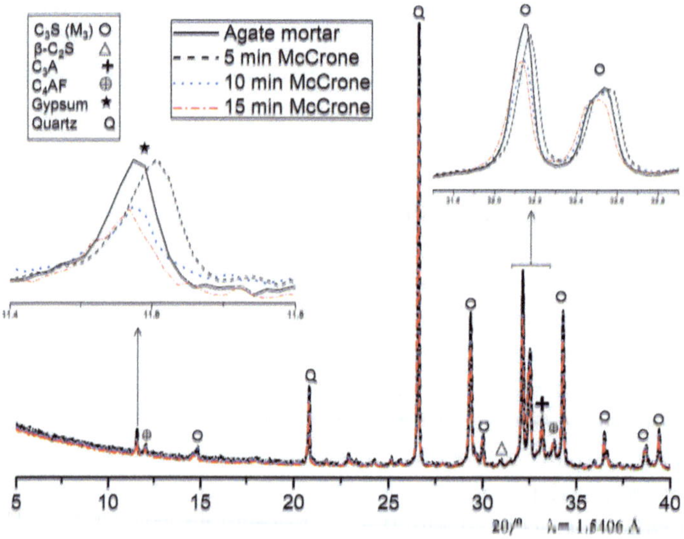

Fig. 1.1: Raw patterns of OPC type I mixed with Quartz as internal standard prepared with different methodologies: hand homogenized in an agate mortar for 20 minutes and using a McCrone micronizing mill for 5, 10, and 15 minutes. Insets display selected angular ranges detailing peaks of alite (right) and gypsum (left).

rithm [49]. The closer this value approaches 1.0, the less oriented the sample is, i.e. better homogenized. The use of the mill during at least 10 minutes minimized the preferred orientation of alite and the derived value of ACn is lower. Type I OPC contains clinker, gypsum (or other sulfate source) and may contain calcite up to 5 wt%. Consequently, the ACn content should comprise the subcooled phase (aluminate rich) of the clinker and misfitting of quantified crystalline phases. Moreover, the amorphization of the sample with grinding should be avoided. The main peak of gypsum (which is the softest phase in OPC) is displayed in the left inset of Fig. 1.1, and the full width at half maximums (FWHM) have been measured (see Tab. 1.2). No amorphization is observed when milling for 10 minutes, since the FWHM stays almost constant at around 0.055°, when compared to agate mortar. However, a slight increase in this value may indicate that some kind of amorphization is taking place. Consequently, the use of the micronizing mill for 10 minutes is highly recommended to prepare samples to perform the internal standard methodology to quantify ACn contents.

Tab. 1.2: RQPA including ACn contents derived by the internal standard method with Quartz of an OPC type I, prepared with different methodologies. Preferred orientation (P.O.) coefficient for alite and gypsum and FWHM of first reflection of gypsum.

	Agate mortar	McCrone		
wt%		5 min	10 min	15 min
ACn	19(1)	20(1)	13(1)	17(1)
Alite C_2S M_3	47.5(4)	46.8(4)	50.7(4)	48.5(4)
Belite β-C_2S	11.9(5)	12.8(5)	13.7(5)	12.9(5)
Aluminate C_3A cubic	7.5(2)	7.4(2)	7.9(2)	7.6(2)
Ferrite C_4AF	6.5(2)	6.5(2)	7.4(2)	7.2(2)
Aphthitalite	1.2(2)	1.0(2)	1.1(2)	1.5(2)
Calcite	4.7(2)	4.0(2)	4.3(2)	3.8(2)
Bassanite	1.0(1)	0.8(1)	0.9(1)	0.7(1)
Gypsum	0.7(1)	0.7(1)	1.0(1)	0.8(1)
P.O. C_3S $[10\bar{1}]$	0.885(3)	0.916(4)	0.956(4)	0.912(3)
P.O. gypsum [010]	0.56(2)	0.56(2)	0.75(3)	0.63(3)
FWHM*	0.053°	0.060°	0.052°	0.071°

* Gypsum reflection placed at 11.5° (2θ/Cu$K\alpha_1$)

(III) External standard method (*G*-factor approach)

The external standard method can be used to avoid complications that may arise from mixing an internal standard with the sample. This approach requires the recording of two patterns in identical diffractometer configuration/conditions for Bragg-Brentano $\theta/2\theta$ reflection geometry, and hence it is a more time demanding method. This method, proposed some time ago [50], was recently applied to anhydrous cements

[51] and organic mixtures [52]. This methodology consists of determining the diffractometer constant, K_e, with an appropriate standard (e.g. silicon powder from Si-single crystal). See equation (1.4), derived from equation (1.1), where the variables stand for the (external) standard rather than that for the phase(s):

$$G = K_e = S_{st}\frac{\rho_{st} V_{st}^2 \mu_{st}}{W_{st}}. \tag{1.4}$$

The calculated G-factor represents a calibration factor for the whole experimental setup and comprises the used diffractometer, radiation, optics, and all data acquisition conditions (e.g. detector configuration or integration time) [50]. This G-factor is used to determine the mass concentration of each phase of the sample under study by:

$$W_\alpha = S_\alpha \frac{\rho_\alpha V_\alpha^2 \mu_s}{G}. \tag{1.5}$$

This method allows one to determine the absolute weight fractions by using a diffractometer constant that must be previously determined. However, the mass attenuation coefficient of the sample is needed (μ_s) and must be independently determined. The most common way to obtain it in cements, is by X-ray fluorescence spectrometry [51]. This methodology has been applied to anhydrous OPC [51] and to OPC pastes [53–55]. From equation (1.5), it is clear that the weight fractions of all computed phases within the control file do not need to add up to 100 wt%. From the difference between 100 wt% and the sum of the crystalline phase contents, the overall ACn content can be derived.

Recent works [43, 44] have compared the internal and the external standard methods, highlighting the strengths and weaknesses of each approach. Moreover, both methodologies have been applied to calcium sulfoaluminate based materials [56]. This study confirmed that the increase of the ACn content in a cement related system could be followed using the external standard methodology. Fig. 1.2 shows a selected angular range of the Rietveld plots for a belite-ye'elemite clinker (see Section 1.3.4), where the main peaks are labeled.

The inset shows the ACn contents obtained by the external standard method for this clinker, as a function of the amount of added glass; the least-squares fit data (line, intercept, slope) are also included. The open symbol indicates the quantified ACn content in the sample without any added glass, which was 17.9 wt%. In addition, the ACn content of this sample obtained with the internal standard method was 16.2 wt% [56], which is in agreement with the value obtained by the external standard methodology.

The quantification of the ACn content in cement-related materials is a hot topic due to its inherent difficulty. The approaches mentioned above, internal and external standard methods, can determine the total amount of ACn but cannot distinguish among the individual contributions of different amorphous phases [56–60]. Moreover, the quantified amounts of ACn in clinkers or cements differ considerably, from negligible [60] to 20 wt% [56, 58, 59].

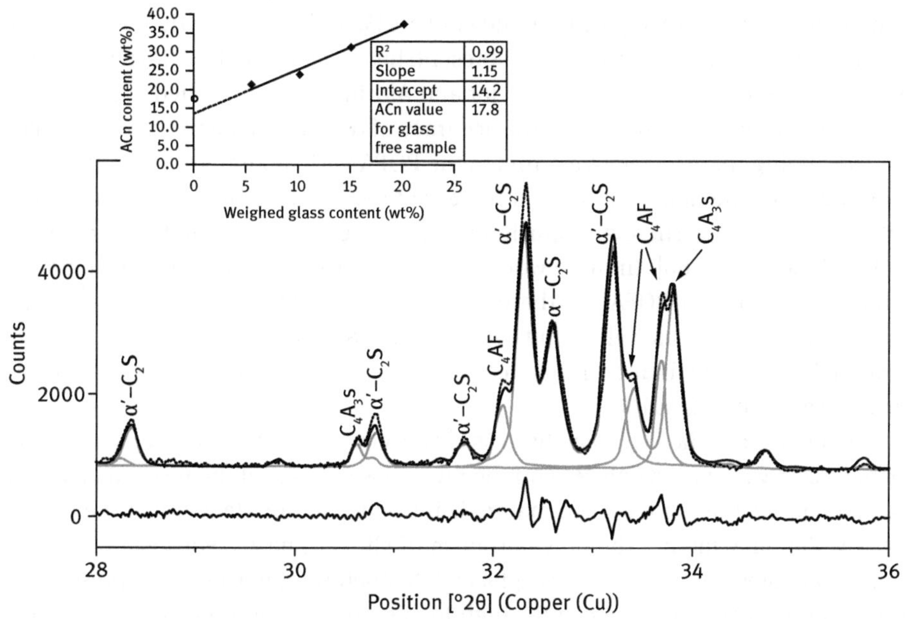

Fig. 1.2: Selected angular range of the Rietveld plots for a belite-ye'elimite-ferrite clinker. Dotted lines are the experimental pattern, black solid lines stand for the calculated pattern, solid line at the bottom stand for the difference curve and grey solid lines represent the individual phase patterns. Main peaks attributable to a given phase have been labeled. Inset: ACn contents, in weight percentage, as a function of the amount of added glass obtained by external standard method, least-square fit data. Open circle indicates the derived ACn content in the samples without any added glass (modified with permission of García-Maté et al. [56]).

An alternative approach for the quantification of ACn is the combination of the Rietveld with the profile summation method, the so called "Partial Or No Known Crystal Structure" (PONKCS) method [39]. This approach needs an appropriate calibration factor for each unknown or amorphous phase and has been applied to cement related systems [61–63].

1.3 Diffraction for quantifying phases in anhydrous clinkers and cements

1.3.1 Quantitative phase analysis of Portland clinkers

Hereafter, cement nomenclature will be used, i.e. C = CaO, S = SiO_2, A = Al_2O_3, F = Fe_2O_3, M = MgO, $\bar{S}$ = SO_3, $\bar{C}$ = CO_2, H = H_2O, K = K_2O, and N = Na_2O. The main OPC clinker phases are Ca_3SiO_5 (alite or C_3S), Ca_2SiO_4 (belite or C_2S) and $Ca_3Al_2O_6$ (tricalcium aluminate or C_3A) in white Portland clinkers and also $Ca_4Al_2Fe_2O_{10}$ (calcium aluminoferrite or ferrite) in grey Portland clinkers [5, and references therein]. In addition, several other minor phases may be present: lime (CaO), periclase (MgO), arcanite, aphtitalite, and others, see Tab. 1.1. Taylor and Aldridge [64] published the first work on RQPA of Portland clinker. Later, the accuracy of this type of analysis was checked [65], concluding that the relative errors in the RQPA with laboratory XRPD are of the order of 2 % for the main phases and increase to approximately 5–10 % for the low content components. There are four interesting reviews concerning the mineralogy of OPC clinkers [4–6, 66]. In these works, the polymorphism of alite is considered as an important issue for RQPA. Furthermore, the Fe/Al ratio in ferrite or the pseudo-polymorphism of C_3A, as well as the degree of crystallinity of these phases have important effects on the final performances of cements such as sulforesistance, which can be assessed by RQPA [67].

A general use of the RQPA can be to study the effects of using alternative fuels [68] or revalorized waste materials [69] on the phase assemblage of the resulting clinkers [70]. Commonly, these alternative materials introduce minor elements in the clinker matrix which may modify the polymorphism of clinker phases. For instance, tricalcium silicate presents seven polymorphs with temperature; however, the presence of foreign ions may stabilize some of the high-temperature pseudo-polymorphs at room temperature (RT) [64]. Mg and/or sulfate ions (the latter mainly coming from fuels) stabilize the T_3 and/or M_3 monoclinic forms [8, 71]. Fig. 1.3 displays a selected region of the Rietveld plot for laboratory prepared grey clinkers with (a) 0.4 wt% of MgO in the raw materials in which the high temperature triclinic T_3 polymorph of C_3S was stabilized and quantified and (b) shows the same selected region of a clinker with 2.5 wt% of MgO with the monoclinic M_3 polymorph [6]. It is highlighted that XRPD is an effective tool to distinguish among polymorphs, which is indicated in Fig. 1.3 as evidence of T_3 or M_3 [8]. These are clear examples of where foreign ions have formed solid solution with C_3S, which can be followed initially by the unit cell variations. Moreover, when the appropriate amount of stabilizers is present, high temperature polymorphs or even new phases can be stabilized and quantified [72].

1.3.2 Quantitative phase analysis of Portland cements with supplementary cementitious materials (SCM)

Portland clinker is mixed with a setting regulator to produce Portland cement, type I in the European Standard BS ENV 197-1. For this purpose, calcium sulfate phases are commonly added, such as gypsum ($C\bar{S}H_2$), bassanite ($C\bar{S}H_{0.5}$), and anhydrite ($C\bar{S}$). Natural gypsum usually contains other calcium sulfate phases and even impurities such as quartz or calcite [73]. RQPA is very useful for testing the phase assemblage of the setting regulator and to control any possible dehydration or amorphization process during the grinding procedure. Any gypsum dehydration process will lead to bassanite or even to soluble anhydrite-III phases which present different solubilities, and

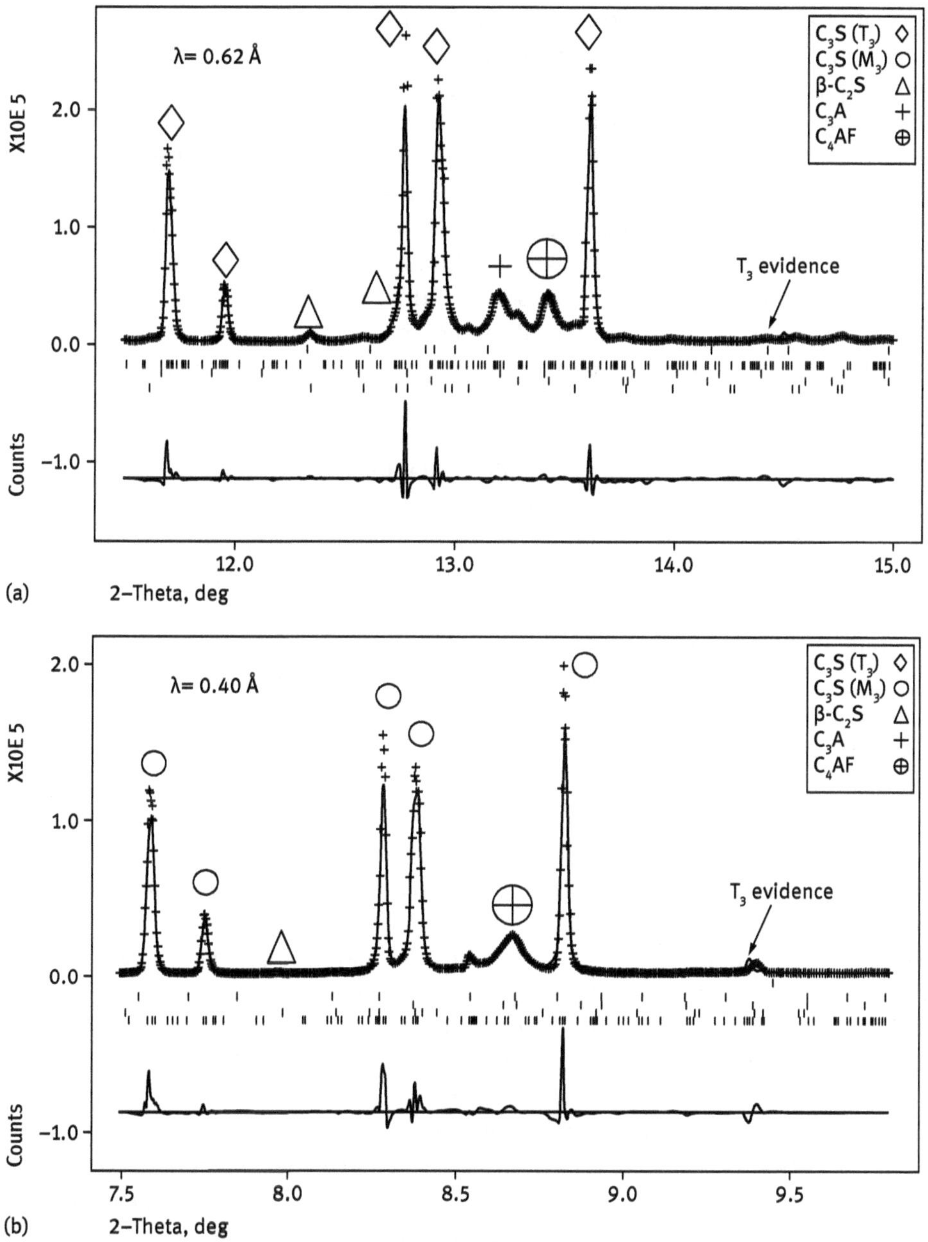

Fig. 1.3: Selected region of Rietveld plots for a (a) low MgO content laboratory grey Portland and (b) high MgO content laboratory grey Portland clinker with main peaks labeled.

this will have an impact on the setting [74]. Consequently, the determination of the composition of the setting regulator is key to controlling and optimizing the setting, viscosity, and hydration of mortars and concretes [75, 76].

Moreover, the use of supplementary cementitious materials (SCMs) [77] is widespread to produce cements with different properties and with the final aim of reducing the amount of clinker in cements, which is an eco-strategy for reducing carbon dioxide emissions and costs. The replacement of a part of the clinker may not imply negative effects on the mechanical performance or durability of the so-called blended cement [78], but this analysis is beyond the scope of this chapter. Limestone is added up to 35 wt% to produce CEM II/B-L cements and granulated blastfurnace slag (GBFS), silica fume, fly ash, burnt shale, or pozzolane minerals (such as calcined kaolinite) [79] are added to produce types II, III, IV, and V, respectively. RQPA is needed to control the mineralogy of these additions, including the quantification of amorphous contents [80–86]. While limestone is a highly-crystalline rock, the others are highly amorphous materials, but they may contain some crystalline phases (for instance mullite or hematite) that must be quantified in blended cements [5]. Fig. 1.4 shows a selected region of the Rietveld plot for a type II commercial Portland cement, as a representative example of a blended cement, where the main peaks are labeled. In that OPC, not only the clinker components have been quantified, but also the setting regulator (gypsum, anhydrite, and dolomite), and some phases coming from the addition of fly ash (mullite and quartz). The insets of the Fig. 1.4 include a table with the RQPA of the OPC (normalized to 100 wt% of crystalline phases), and two enlarged areas of the pattern.

1.3.3 Quantitative phase analysis of anhydrous cements for alkaline-activation

The production of alkaline cements requires two main components: a cementitious component and an alkaline activator [87]. For the latter caustic alkalis (e.g. NaOH or KOH) or alkaline salts (e.g. Na_2CO_3) are normally used. The cementitious compound is usually formed by industrial by-products or wastes, as well as some aluminosilicate raw materials. These materials include coal fly ash, granulated blast furnace or phosphorus slags, volcanic glass, zeolite, metakaolin, silica fume and non-ferrous slags [87]. The alkali activation consists of the reaction of the cementitious component (usually highly amorphous) with alkaline activators (alkaline solutions) at a certain temperature to produce a solid material. The amorphous/glassy constituent of the cementitious component transforms into a compact cement matrix in the process. A deep discussion about these processes is out of the scope of this chapter; however, some authors [88–90] have stated that the reactivity of the cementitious component mainly depends on the composition of the amorphous/vitreous phase. Consequently, the full characterization of these materials is of great importance in the development of new binders. Fig. 1.5 (top) shows a selected range of Rietveld plots of a fly ash, as an

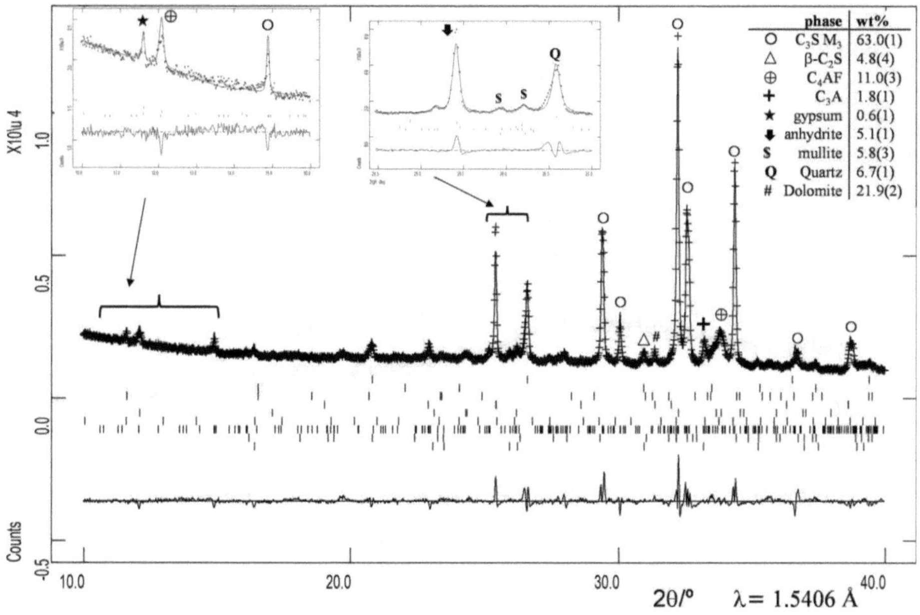

Fig. 1.4: Selected region of a Rietveld plot, $CuK\alpha_1$, for a commercial Portland cement (CEM II/BV 32.5R) with main peaks labeled. The insets show enlarged regions highlighting characteristic diffraction peaks for setting regulator (left inset) and crystalline phases from the addition (center inset). RQPA results are given in the right inset.

example of a cementitious component of alkaline cements, measured in reflection geometry with $CuK\alpha_1$ strictly monochromatic radiation where the ACn content has been determined by the external standard methodology (*G*-factor). Fig. 1.5 (bottom) gives the same fly ash mixed with ZnO as internal standard. The inset gives the RQPA, including the amorphous content as determined with external and internal standard methodologies (as explained before).

1.3.4 Quantitative phase analysis of ye'elimite-containing cements

Ye'elimite containing-cements can be considered because eco-cements as less CO_2 is released during their production compared to OPC. This is due to the presence of ye'elimite in their composition, which is a low calcite-demanding phase and consequently releases less CO_2 (0.22 t/t_{phase}) than other phases present in OPC, such as alite (0.58 t/t_{phase}), belite (0.51 t/t_{phase}), calcium aluminate (0.49 t/t_{phase}) or calcium aluminate ferrite (0.22 t/t_{phase}) [91]. For this reason, and due to their high performances, there has been an increasing amount of research on these cements. The mineralogy of these new binders is of great importance in understanding and predicting the final performance of the derived mortars and concretes.

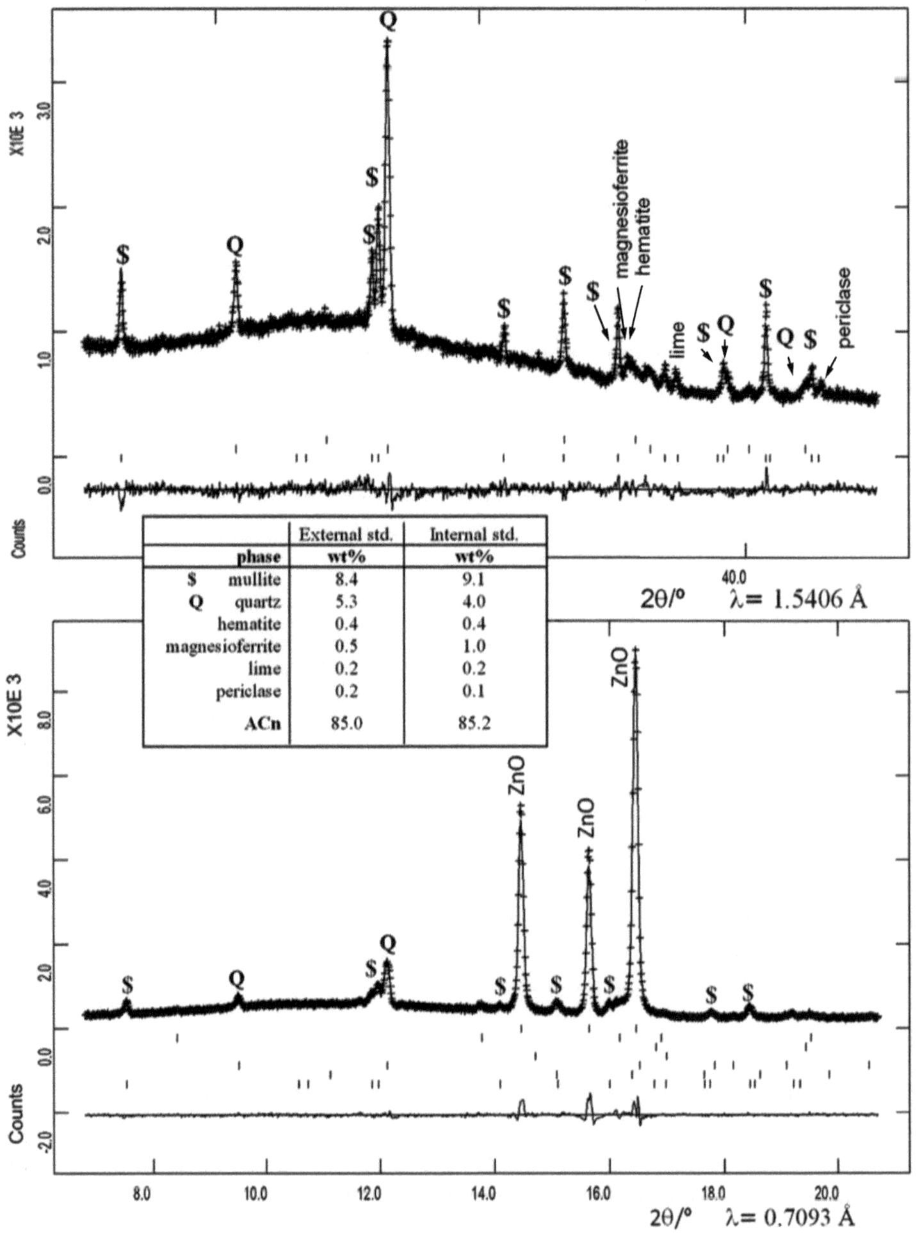

	phase	External std. wt%	Internal std. wt%
$	mullite	8.4	9.1
Q	quartz	5.3	4.0
	hematite	0.4	0.4
	magnesioferrite	0.5	1.0
	lime	0.2	0.2
	periclase	0.2	0.1
	ACn	85.0	85.2

Fig. 1.5: Selected region of Rietveld plots, (top) CuKα_1 of a fly ash usually used as cementitious component of alkaline cements and (bottom) MoKα_1 of the same fly ash mixed with ZnO. RQPA result using the internal and external standard methods are given in the inset. RQPA result using the internal and external standard methods are given in the inset.

Ye'elimite, also called Klein's salt or tetracalcium trialuminate sulfate $Ca_4[Al_6O_{12}]SO_4$, crystallizes with the sodalite type structure group with the following general composition $M_4[T_6O_{12}]X$ [92]. The crystal structure of stoichiometric ye'elimite at room temperature was not very clear since it has been reported to be cubic,[93, 94] tetragonal [95], and orthorhombic [96], the latter based on space group Pcc2. More recently, this structure has been revised using powder diffraction and atomistic calculations and both methodologies agree and confirm that the most stable structure at room temperature is the Pcc2 orthorhombic phase [24]. In addition, it has been reported [97] that it is improbable that stoichiometric ye'elimite presents a cubic structure at room temperature, since dopants are needed to stabilize it. The polymorphism of ye'elimite and its solid solutions in presence of Na^+ and Fe^{3+} have been investigated [98]. In 2014, the crystal structure of doped ye'elemite ($Ca_{3.8}Na_{0.2}Al_{5.6}Fe_{0.2}Si_{0.2}O_{12}SO_4$) was studied by high-resolution SXRPD, and a pseudo-cubic structural description based on space group $I\bar{4}3m$ was published [25]. Moreover, it was reported that the incorporation of Fe_2O_3 into the crystal structure of ye'elimite in a belite-calcium sulfoaluminate clinker leads to the formation of cubic ye'elemite [99].

Ye'elimite-containing cements can be gathered in four main groups [91]: (i) calcium sulfo-aluminate (CSA), (ii) belite calcium sulfo-aluminate or sulfobelite (BCSA or BYF, from belite-ye'elemite-ferrite cements), (iii) belite-alite-ye'elimite (BAY) cements, and (iv) ternesite-ye'elimite-belite (a.k.a. TYB).

(i) CSA cements are the most environmentally friendly cements of the group since they contain the highest amount of ye'elimite, and thus release the lowest CO_2 emissions (up to 40 % less CO_2 than OPC), where this reduction depends on the CSA composition [91]. Ye'elimite is their main phase, > 50 wt%, and other phases such as β-belite, ferrite, anhydrite, perovskite, and gehlenite can also be present [59, 100]; in addition, minor elements may also form solid solutions. Fig. 1.6 shows a selected region of two Rietveld plots for a commercial CSA cement, measured in reflection and transmission geometries, with $CuK\alpha_1$ and $MoK\alpha_1$ (monochromatic radiations), Fig. 1.6 (a) and (b), respectively. The main peaks are labeled and the effect of preferred orientation of gypsum is highlighted by arrows. A table with the obtained RQPA by using both patterns is also included in Fig. 1.6 as an inset.

This sample contains a well-known weighed amount of gypsum, i.e., 23.8 wt%. This is a layered compound and it usually displays a preferred orientation effect in powder diffraction patterns. The table of the inset shows the March–Dollase [47] preferred orientation coefficient, along the [010] axis, which behaved as expected. The refined value for a flat sample in reflection geometry was lower than 1, i.e. 0.73, while the value for a flat sample in transmission geometry was higher than 1, i.e. 1.08. This last value means that this phase showed very little preferred orientation when collecting the pattern in transmission configuration. Moreover, the RQPA obtained when working on transmission with strictly monochromatic $MoK\alpha_1$ radiation yielded more accurate gypsum values (i.e. gypsum: 26.8 wt%) than when the reflection geometry is used. This may be due to the higher irradiated volume that enhances

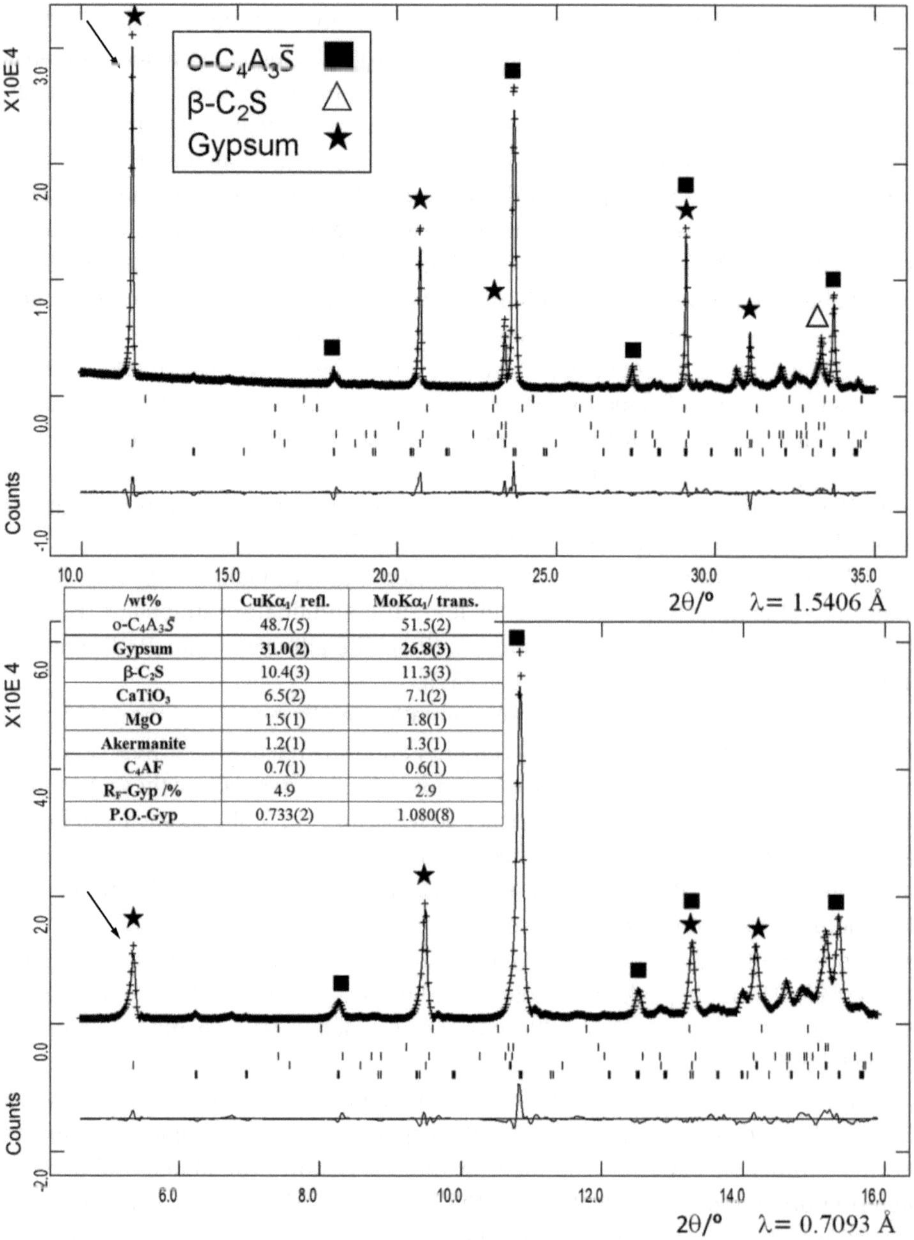

/wt%	$CuK\alpha_1$/ refl.	$MoK\alpha_1$/ trans.
o-$C_4A_3\bar{S}$	48.7(5)	51.5(2)
Gypsum	**31.0(2)**	**26.8(3)**
β-C_2S	10.4(3)	11.3(3)
$CaTiO_3$	6.5(2)	7.1(2)
MgO	1.5(1)	1.8(1)
Akermanite	1.2(1)	1.3(1)
C_4AF	0.7(1)	0.6(1)
R_F-Gyp /%	4.9	2.9
P.O.-Gyp	0.733(2)	1.080(8)

Fig. 1.6: Selected region of the Rietveld plots for a CSA clinker with 23.6 wt% of gypsum collected in the reflection mode in flat sample with strictly monochromatic $CuK\alpha_1$ radiation (top) and in transmission mode flat sample with strictly monochromatic $MoK\alpha_1$ radiation (bottom). Arrows highlight the preferred orientation effect on first gypsum peak. RQPA results are given in the inset.

particle statistics, which is also corroborated by lower R_F values, see table in Fig. 1.6 [46, 101]. These CSA cements can be used alone or in combination with other cements to provide improved performances, as they can show high mechanical strengths (> 80 MPa at 7 days) [102], low shrinkage, high impermeability, and a strong resistance to sulfate attack. Although CSAs are promising cements, they need high amounts of aluminum raw material(s) for their production, which usually come from expensive sources (e.g. bauxite); thus, they are not expected to replace OPC in massive applications. Furthermore, their use is strongly limited in Europe by the lack of standards concerning special cements derived from non-Portland clinkers. Given the present state of European standard regulations, binders based on CSA cements cannot be used in structural concrete according to the EN 206-1; only three formulations of CSA cements – produced by Buzzi Unicemin Trino (Italy) – obtained a CE mark based on an ETA (European Technical Approval) procedure in June 2013, released by DIBt (Deutsches Institut für Bautechnik), also allowing for their use for structural applications [103].

(ii) The term belite calcium sulfo-aluminate usually refers to clinkers containing belite as a phase (40–50 wt%) and intermediate ye'elimite contents (20–30 wt%). Some industrial trials of low energy belite-ye'elimite-ferrite (BYF) cements were prepared, and reported by [104–106]; for the latter, elemental sulfur and/or sulfur-rich fuels were used. However, the main technological disadvantage of these cements is related to their low mechanical strengths developed at early ages due to the slow hydration of belite. However, this problem is being overcome by the activation of belite and the presence of relatively high amounts of ye'elimite [59, 107–109]. The production of active BYF cements involves the stabilization of highly reactive C_2S polymorphs, i.e. a β-modified form and α-forms through the addition of minor elements. For instance, when borax is added to the raw materials, the high-temperature α'_H-C_2S polymorph is stabilized during clinkering. A revised crystal structure of boron-bearing α'_H-C_2S was recently published, see Tab. 1.1 [11]. Fig. 1.7 shows both Rietveld plots of (a) BYF without any activator and (b) BYF with 2.0 wt% of B_2O_3 added as borax [59, 110]. The former contains β-C_2S and orthorhombic ye'elimite, and the latter α'_H-C_2S and pseudo-cubic ye'elimite.

(iii) Developing belite-alite-ye'elimite (BAY) cements is an alternative way to solve the problem of low early age mechanical strengths [111–119], since alite is responsible for early mechanical strengths in OPC. Their manufacture may release 15 % less CO_2 than OPC, depending on their composition. The reaction of alite and ye'elimite with water will develop cements with high mechanical strengths at early ages, while belite will contribute to strength developments at later ages. However, clinkering this type of binder can be difficult due to the formation/decomposition temperature incompatibility between C_3S and $C_4A_3\bar{S}$; alite begins to be formed ≈ 1300 °C, and ye'elimite starts its decomposition at 1300–1350 °C [120–122]. Fortunately, the addition of minor quantities of fluorite or minor amounts of oxides such as those derived from Mg [113], Cu [114], Mn [117], Ti [118], or Zn [119] in the raw materials is able to overcome this limitation.

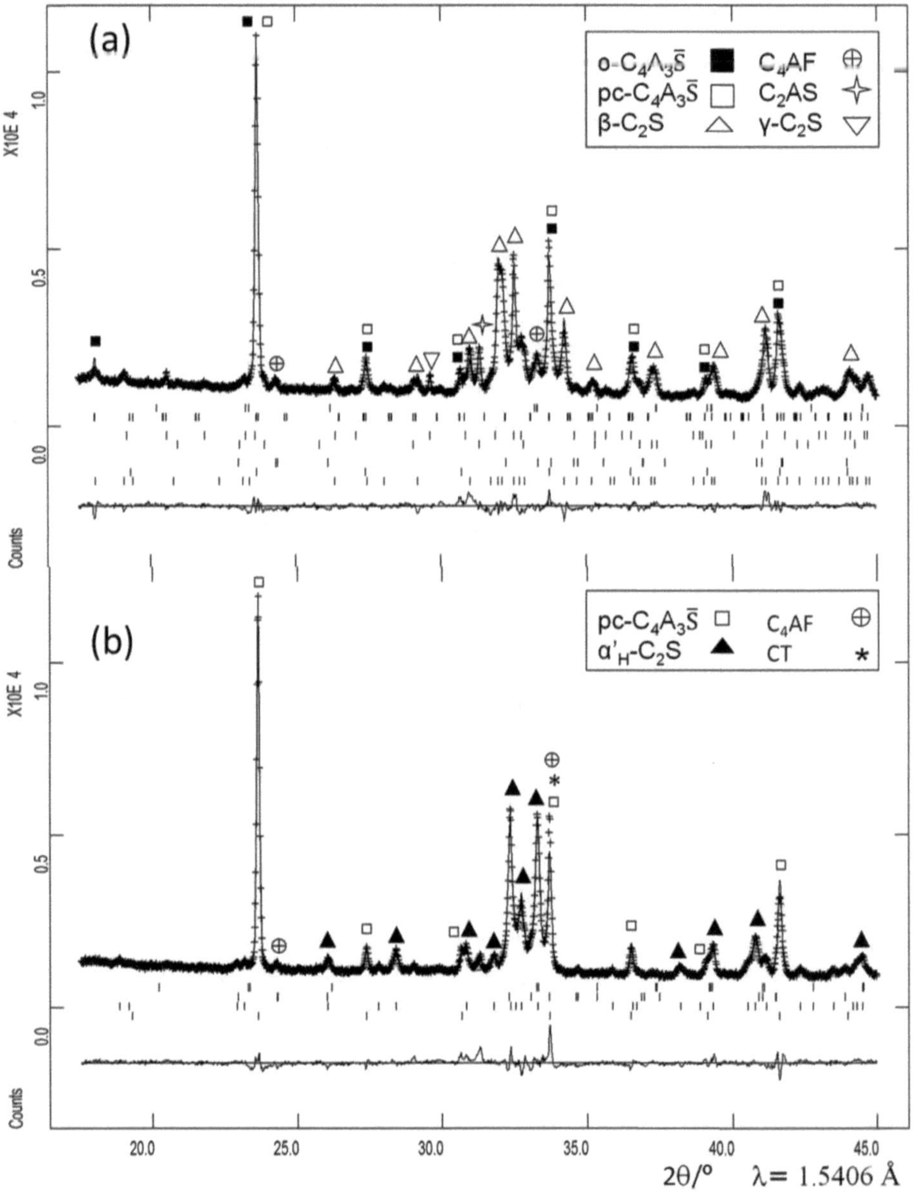

Fig. 1.7: Selected region of a Rietveld plot, CuKα_1, for (a) a laboratory-prepared belite-ye'elimite-ferrite clinker and (b) an active BYF clinker with main peaks labeled. o-$C_4A_3\bar{S}$ stands for orthorhombic ye'elimite and pc-$C_4A_3\bar{S}$ stands for pseudocubic ye'elimite.

(iv) Finally, ternesite-ye'elimite-belite cements have recently been proposed to be an environmentally friendly alternative to OPC [99, 123]; the contents of the three phases are approximately 50, 25, 20 wt%, respectively. Ternesite (calcium sulfospurrite or $Ca_5(SiO_4)_2SO_4$) is a phase which can be formed in belite sulfoaluminate clinkers from the reaction of anhydrite and belite between 1100 and 1200 °C [123]. The formation of this phase depends on the $SO_3/(Al_2O_3 + Fe_2O_3)$ ratio. Therefore, by increasing the SO_3 content in the raw meal, clinkers with high contents of ternesite and low contents of belite can be prepared [99]. Traditionally, ternesite was considered to be a hydraulically inert phase [124]; however, it has been recently reported [99] that this phase is more reactive than belite in BYF cements [125], and is therefore a very promising eco-material.

1.4 Diffraction for characterizing anhydrous cementitious materials in non-ambient conditions: high temperature and high pressure

XRPD is a very useful tool to follow phase transitions/reactions with temperature, pressure, degree of humidity, and so on [126].

On the one hand, the understanding of the phase formation mechanisms during the clinkering of new binders is mandatory in order to successfully and safely transfer this knowledge to the industrial system. Synchrotron XRPD, SXRPD, was used coupled to RQPA, to follow in situ clinkering reactions at high temperatures of up to 1500 °C of belite Portland [127] and BYF clinkers [120]. The preheated raw meals were loaded in platinum capillaries and very high-energy synchrotron X-rays ($\lambda \approx 0.30$ Å) were used to penetrate the holder and to obtain accurate intensities. Fig. 1.8 shows a selected range of the SXRPD raw patterns for an active BYF clinker collected on heating from 780 °C to 1375 °C (top), and a raw SXRPD pattern at a certain temperature where platinum peaks (from sample holder) dominate the scattering within the pattern (bottom). In the inset, the Rietveld quantitative plot of the appropriate region of the previous pattern is shown, where the diffraction peaks from the clinker phases are present.

These studies allowed the polymorphic transformation from β-C_2S to α'_H-C_2S to be followed with temperature and its dependence on the amount of activators. Moreover, the formation and decomposition of $C_4A_3\bar{S}$ and the melting of C_4AF were determined. Finally, the optimum clinkering temperatures were established as a function of the activators.

On the other hand, SXRPD is very well suited for high-pressure studies which is key for determining the bulk modulus values of relevant phases. This methodology has been extensively applied to hydrated phases (see the following chapter) and, recently, the bulk modulus of ye'elimite has also been determined [96].

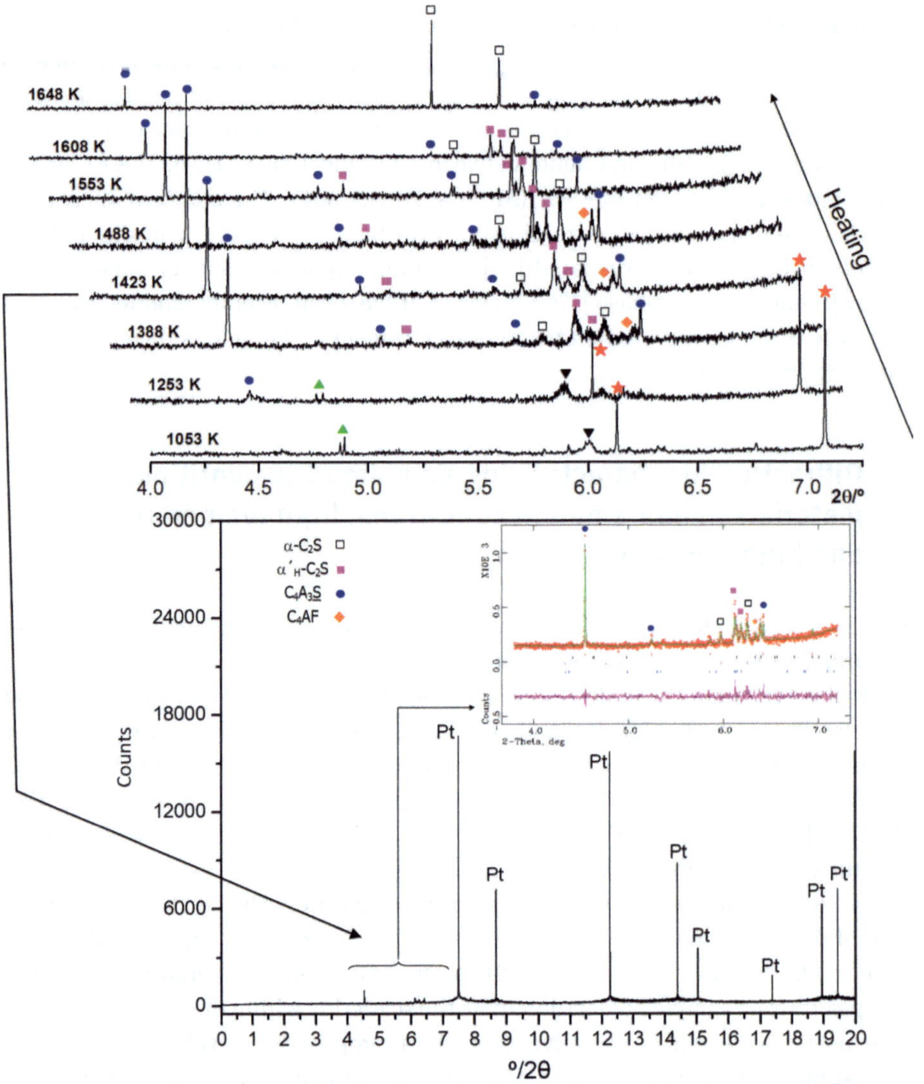

Fig. 1.8: Top: Selected range of the raw patterns for an active BYF where the symbols highlight CaO (red stars), β-SiO_2 (black triangles), $CaSO_4$ (green triangles), $C_4A_3\bar{S}$ (blue circles), α'_H-C_2S (pink squares), C_4AF (orange diamonds), and α-C_2S (open squares). Bottom: raw SXRPD pattern at a certain temperature where platinum peaks (from sample holder) are labeled. Inset: Rietveld plots of the appropriate region of the previous pattern with clinker phase diffraction peaks labeled (modified with permission of De la Torre et al. [120]).

Acknowledgment: We would like to acknowledge the funding we received from Spanish MINECO (BIA2014–57658-C2–2-R, which is co-funded by FEDER, and BIA2014–57658-C2–1-R research grants).

References

[1] Madsen IC, Scarlett NVY, Cranswick, LMD, Lwin, T. Outcomes of the International Union of Crystallography Commission on Powder Diffraction Round Robin on Quantitative Phase Analysis: samples 1a to 1h. J Appl Cryst. 2001; 34: 409–426.

[2] Scarlett NVY, Madsen IC, Cranswick LMD, et al. Outcomes of the International Union of Crystallography Commission on Powder Diffraction Round Robin on Quantitative Phase Analysis: samples 2, 3 and 4, synthetic bauxite, natural granodiorite and pharmaceuticals. J Appl Cryst. 2002; 35: 383–400.

[3] León-Reina L, De la Torre AG, Porras-Vázquez JM, et al. Round robin on Rietveld quantitative phase analysis of Portland cements. J Appl Cryst. 2009; 42: 906–916.

[4] Le Saout, Kocaba V, Scrivener K. Application of the Rietveld method to the analysis of anhydrous cements. Cem Concr Res. 2011; 41: 133–148.

[5] Aranda MAG, De la Torre AG, León-Reina L. Rietveld Quantitative Phase Analysis of OPC clinkers, cements and hydration products. Rev Miner Geochem. 2012; 74: 169–209.

[6] Aranda, MAG, De la Torre AG, León-Reina L. Powder diffraction characterization of cements. In: Gilmore, C, Kaduk J, Schenk, H, International Tables for Crystallography, ed. International Union for Crystallography. 2017; Vol H: In press.

[7] De la Torre AG, Bruque S, Campo J, Aranda MAG. The superstructure of C_3S from synchrotron and neutron powder diffraction and its role in quantitative phase analyses. Cem Concr Res. 2002; 32: 1347–1356.

[8] Nishi F, Takeuchi Y, Maki I. Tricalcium silicate $Ca_3O[SiO_4]$: The monoclinic superstructure. Z Kristallogr 1985; 172: 297–314.

[9] De la Torre AG, De Vera RN, Cuberos AJM, Aranda MAG. Crystal structure of low magnesium-content alite: Application to Rietveld quantitative phase analysis. Cem Concr Res. 2008; 38: 1261–1269.

[10] Golovastikov NI, Matveeva RG, Belov NV. Crystal structure of the tricalcium silicate $(CaOSi_2)_3$=C_3S. Kristallografiya. 1975; 20: 721–729.

[11] Mumme WG, Hill RJ, Bushnell-Wye G, Segnit ER. Rietveld crystal structure refinements, crystal chemistry and calculated powder diffraction data for the polymorphs of dicalcium silicate and related phases. N Jb Miner Abh. 1995; 169: 35–68.

[12] Cuesta A, Losilla ER, Aranda MAG, Sanz J, De la Torre AG. Reactive belite stabilization mechanisms by boron-bearing dopants. Cem Concr Res. 2012; 42: 598–606.

[13] Mondal P, Jeffery JW. The crystal structure of tricalcium aluminate, $Ca_3Al_2O_6$. Acta Cryst B. 1975; 31: 689–697.

[14] Takeuchi Y, Nishi F, Maki I. Crystal-chemical characterization of the $3(CaO)Al_2O_3$-Na_2O solid-solution series. Z Kristallogr. 1980; 152: 259–307.

[15] Colville AA, Géller S. The crystal structure of brownmillerite, Ca_2FeAlO_5. Acta Cryst B. 1997; 27: 2311–2315.

[16] Smith DK, Leider HR. Low-temperature thermal expansion of LiH, MgO and CaO. J App Cryst. 1968; 1: 246–249.

[17] Sasaki S, Fujino K, Takeuchi Y. X-ray determination of electron-density distributions in oxides, MgO, MnO, CoO, and NiO, and atomic scattering factors of their constituent atoms. Proc Japan Acad. 1979; 55: 43–48.

[18] Ojima K, Nishihata Y, Sawada A. Structure of potassium sulfate at temperatures from 296 K down to 15 K. Acta Cryst B. 1995; 51: 287–293.

[19] Okada K, Ossaka J. Structures of potassium sodium sulphate and tripotassium sodium disulphate. Acta Cryst B. 1980; 36: 919–921.

[20] Rasmussen SE, Jorgensen JE, Lundtoft B. Structures and phase transitions of Na_2SO_4. J Appl Cryst. 1996; 29: 42–47.

[21] Speer D, Salje E. Phase transitions in langbeinites I: Crystal chemistry and structures of K-double sulfates of the langbeinite type $M_2^{(2+)}K_2(SO_4)_3$, $M^{(2+)}$ = Mg, Ni, Co, Zn, Ca. Phys Chem Min. 1986; 13: 17–24.

[22] Irran E, Tillmanns E, Hentschel G. Ternesite, $Ca_5(SiO_4)_2SO_4$, a new mineral from the Ettringer Bellerberg/Eifel, Germany. Min Petr. 1997; 60: 121–132.

[23] Saint-Jean SJ, Hansen S. Nonstoichiometry in chlorellestadite. Solid State Sci. 2005; 7: 97–102.

[24] Pajares I, De la Torre AG, Martinez-Ramirez S, Puertas S, Blanco-Varela MT, Aranda MAG. Quantitative analysis of mineralized white Portland clinkers: The structure of Fluorellestadite. Powder Diffr. 2002; 17: 281–286.

[25] Cuesta A, De la Torre AG, Losilla ER, et al. Structure, Atomistic Simulations, and Phase Transition of Stoichiometric Yeelimite. Chem Mater. 2013; 25: 1680–1687.

[26] Cuesta A, De la Torre AG, Losilla ER, Santacruz I, Aranda MAG. Pseudocubic crystal structure and phase transition in doped ye'elimite. Cryst Growth Des. 2014; 14: 5158–5163.

[27] Palacios L, De la Torre AG, Bruque S, Garcia-Munoz JL, Garcia-Granda S, Sheptyakov D, Aranda MAG. Crystal structures and in-situ formation study of Mayenite electrides. Inorg Chem. 2007; 46: 4167–4176.

[28] Rietveld HM. A profile refinement method for nuclear and magnetic structures. J Appl Cryst. 1969; 2: 65–71.

[29] Larson AC, Von Dreele RB. Los Alamos National Lab. Rep. No. LA-UR-86–748. Los Alamos; 1994.

[30] Toby BH. EXPGUI, a graphical user interface for GSAS. J Appl Cryst. 2001; 34: 210–213.

[31] Rodríguez-Carvajal J. Recent advances in magnetic structure determination by neutron powder diffraction. Phys B. 1993; 192: 55–69.

[32] Roisnel T, Rodríguez-Carvajal J. WinPLOTR: a Windows tool for powder diffraction patterns analysis. Mat Sci For. 2001; 378–381: 118–123.

[33] Bergmann J, Friedel P, Kleeberg R. BGMN – a new fundamental parameter based Rietveld program for laboratory X-ray sources, its use in quantitative analysis and structure investigations CPD Newsletter, Commission of Powder Diffraction, International Union of Crystallography. 1998; 20: 5–8.

[34] Taylor JC. Computer programs for standardless quantitative analysis of minerals using the full powder diffraction profile. Powder Diffraction. 1991; 6: 2–9.

[35] Young RA, The Rietveld Method. International Union of Crystallography monographs on Crystallography. Oxford: Oxford University Press; 1993.

[36] McCusker LB, Von Dreele RB, Cox DE, Louer D, Scardi P. Rietveld Refinement Guidelines. J Appl Cryst. 1999; 32: 36–50.

[37] Smith DK, Johnson GG, Scheible A, Wims AM, Johnson JL, Ullrnann G. Quantitative X-ray powder diffraction method using the full diffraction pattern. Powder Diffr. 1987; 2: 73–77.

[38] Taylor JC, Zhu R. Simultaneous use of observed and calculated standard profiles in quantitative XRD analysis of minerals by the multiphase Rietveld method: the determination of pseudorutile in mineral sands products. Powder Diffr. 1992; 7: 152–161.

[39] Scarlett NVY, Madsen IC. Quantification of phases with partial or no known crystal structures. Powder Diffr. 2006; 21: 278–284.

[40] Hill RJ, Howard CJ. Quantitative phase analysis from neutron powder diffraction data using the Rietveld method. J Appl Cryst. 1987; 20: 467–474.

[41] Bish DL, Howard SA. Quantitative phase analysis using the Rietveld method. J Appl Cryst. 1998; 21: 86–91.

[42] Cline JP, Von Dree RB, Winburn R, Stephens PW, Filliben JJ. Addressing the amorphous content issue in quantitative phase analysis: the certification of NIST standard reference material 676a. Acta Cryst. Sect A. 2011; 67: 357–367.

[43] Gualtieri AF, Riva V, Bresciani A, Maretti S, Tamburini M, Viani A. Accuracy in quantitative phase analysis of mixtures with large amorphous contents. The case of stoneware ceramics and bricks. J Appl Cryst. 2014; 47: 835–846.

[44] Madsen IC, Scarlett NVY, Kern A. Description and survey of methodologies for the determination of amorphous content via X-ray powder diffraction. Z Kristallogr. 2011; 226: 944–955.

[45] De la Torre AG, Bruque S, Aranda MAG. Rietveld quantitative amorphous content analysis. J Appl Cryst. 2001; 34: 196–202.

[46] León-Reina L, García-Maté M, Álvarez-Pinazo G, et al. Accuracy in Rietveld quantitative phase analysis: a comparative study of strictly monochromatic Mo and Cu radiations. J Appl Cryst. 2016; 49: 722–735.

[47] Westphal T, Walenta G. Fuellmann T, Gimenez M, Bermejo E, Scrivener K, Poellmann H. Characterisation of cementitious materials – Part III. Int Cem Rev. July 2002: 47–51.

[48] O'Connor BH, Chang WJ. The amorphous character and particle size distribution of powders produced with the micronizing mill for quantitative X-ray powder diffractometry. X-Ray Spec. 1986; 15: 267–270.

[49] Dollase WA. Correction of intensities for preferred orientation in powder diffractometry: application of the March model. J Appl Cryst. 1986; 19: 267–272.

[50] O'Connor BH, Raven MD. Application of the Rietveld refinement procedure in assaying powdered mixtures. Powder Diffr. 1988; 3: 2–6.

[51] Jansen D, Stabler CH, Goetz-Neunhoeffer F, Dittrich S, Neubauer J. Does Ordinary Portland Cement contain amorphous phase? A quantitative study using an external standard method. Powder Diffr. 2011; 26: 31–38.

[52] Schreyer M, Guo L, Tjahjono M, Garland M. Three approaches to total quantitative phase analysis of organic mixtures using an external standard. J Appl Cryst. 2011; 44: 17–24.

[53] Jansen D, Bergold ST, Goetz-Neunhoeffer F, Neubauer J. The hydration of alite: a time-resolved quantitative X-ray diffraction approach using the G-factor method compared with heat release. J Appl Cryst. 2011; 44: 895–901.

[54] Jansen D, Goetz-Neunhoeffer F, Lothenbach B, Neubauer J. The early hydration of Ordinary Portland Cement (OPC): An approach comparing measured heat flow with calculated heat flow from QXRD. Cem Concr Res. 2012; 42: 134–138.

[55] Jansen D, Neubauer J, Goetz-Neunhoeffer F, Haerzschel R, Hergeth WD. Change in reaction kinetics of a Portland cement caused by a superplasticizer – Calculation of heat flow curves from XRD data. Cem Concr Res. 2012; 43: 327–332.

[56] García-Maté M, Santacruz I, Cuesta A, et al. Amorphous determination in calcium sulfoaluminate materials by external and internal methods. Adv Cem Res. 2015; 27: 417–423.

[57] Suherman PM, van Riessen A, O'Connor B, Li D, Bolton D, Fairhurst H. Determination of amorphous phase levels in Portland cement clinker. Powder Diffr. 2002; 17: 178–185.

[58] Whitfield PS, Mitchell LD. Quantitative Rietveld analysis of the amorphous content in cements and clinkers. J Mat Sci. 2003; 38: 4415–4421.
[59] Álvarez-Pinazo G, Cuesta A, García-Maté M, et al. Rietveld quantitative phase analysis of yeelimite-containing cements. Cem Concr Res. 2012; 42: 960–971.
[60] Snellings R, Bazzoni A, Scrivener K. The existence of amorphous phase in Portland cements: Physical factors affecting Rietveld quantitative phase analysis. Cem Concr Res. 2014; 59: 139–146.
[61] Williams RP, Hart D, van Riessen, A. Quantification of the extent of reaction of metakaolin-based geopolymers using X-ray diffraction, scanning electron microscopy, and energy-dispersive spectroscopy. J. Am Ceram Soc. 2011; 94: 2663–2670.
[62] Bergold ST, Goetz-Neunhoeffer F, Neubauer J. Quantitative analysis of C-S-H in hydrating alite pastes by in-situ XRD. Cem Concr Res. 2013; 53: 119–126.
[63] Snellings R, Salze A, Scrivener KL. Use of X-ray diffraction to quantify amorphous supplementary cementitious materials in anhydrous and hydrated blended cements. Cem Concr Res. 2014; 64: 89–98.
[64] Taylor JC, Aldridge LP. Full-profile Rietveld quantitative XRD analysis of Portland cement: Standard XRD profiles for the major phase tricalcium silicate (C_3S: $3CaO{\cdot}SiO_2$). Powder Diffr. 1993; 8: 138–144.
[65] De la Torre AG, Aranda MAG. Accuracy in Rietveld quantitative phase analysis of Portland cements. J Appl Cryst. 2003; 36: 1169–1176.
[66] Ludwig HM, Zhang W. Research review of cement clinker chemistry. Cem Concr Res. 2015; 78: 24–37.
[67] Robledo-Gutiérrez M, Blanco-Varela MT, Carmona-Quiroga PM. Mineralogical Composition of Clinker as an Indicator of Sulfate Resistance: A Rietveld XRD/Takashima Approach. J Am Ceram Soc. 2013; 96: 3637–3642.
[68] Husillos-Rodriguez N, Martinez-Ramirez S, Blanco-Varela MT. Activated carbon as an alternative fuel. Effect of carbon ash on cement clinkerization. J Cleaner Prod. 2016; 119: 50–58.
[69] Ferreira AV, Righi A, Araujo FGS, Espinosa DCR, Tenório JAS. Applications of the Rietveld method to quantify the crystalline phases of Portland cement clinker doped with nickel and chromium. Powder Diffr Suppl. 2008; 23: S42–S45.
[70] Herfort D, Moir GK, Johansen V, Sorrentino F, Bolio Arceo H. The chemistry of Portland cement clinker. Adv Cem Res. 2010; 22: 187–194.
[71] Li X, Xu W, Wang S, Tang M, Shen X. Effect of SO_3 and MgO on Portland cement clinker: Formation of clinker. Const Build Mat. 2014; 58: 182–192.
[72] Gineys N, Aouad G, Sorrentino F, Damidot D. Incorporation of trace elements in Portland cement clinker: Thresholds limits for Cu, Ni, Sn or Zn. Cem Concr Res. 2011; 41: 1177–1184.
[73] Walenta G, Fullmann T. Advances in quantitative XRD analysis for clinker, cements, and cementitious additions. Powder Diffr. 2004; 19: 40–44.
[74] Seufert S, Hesse C, Goetz-Neunhoeffer F, Neubauer J. Quantitative determination of anhydrite III from dehydrated gypsum by XRD. Cem Concr Res. 2009; 39: 936–941.
[75] Tang FJ, Gartner EM. Influence of sulphate source on Portland cement hydration. Adv Cem Res. 1988; 1: 67–74.
[76] Roessler C, Eberhardt A, Kucerova H, Moeser B. Influence of hydration on the fluidity of normal Portland cement pastes. Cem Concr Res. 2008; 38: 897–906.
[77] Snellings R, Mertens G, Elsen J. Supplementary Cementitious Materials. Rev Miner Geochem. 2012: 74; 211–278.
[78] Lothenbach B, Scrivener K, Hooton RD. Supplementary cementitious materials. Cem Concr Res. 2011; 41: 217–229.

[79] Taylor-Lange SC, Lamon EL, Riding KA, Juenger MCG. Calcined kaolinite–bentonite clay blends as supplementary cementitious materials. App Clay Sci. 2015; 108: 84–93.

[80] Kumar S, Kumar R, Bandopadhyay A, Alex TC, Ravi-Kumar B, Das SK, Mehrotra SP. Mechanical activation of granulated blast furnace slag and its effect on the properties and structure of portland slag cement. Cem Concr Res. 2008; 30: 679–685.

[81] Korpa A, Kowald T, Trettin R. Phase development in normal and ultra high performance cementitious systems by quantitative X-ray analysis and thermoanalytical methods. Cem Concr Res. 2009; 39: 69–76.

[82] Gonçalves JP, Tavares LM, Toledo-Filho RD, Fairbairn EMR. Performance evaluation of cement mortars modified with metakaolin or ground brick. Const Build Mat. 2009; 23: 1971–1979.

[83] De Weerdt K, Haha MB, Le Saout G, Kjellsen KO, Justnes H, Lothenbach B. Hydration mechanisms of ternary Portland cements containing limestone powder and fly ash. Cem Concr Res. 2011; 41: 279–91.

[84] Narmluk M, Nawa T. Effect of fly ash on the kinetics of Portland cement hydration at different curing temperatures. Cem Concr Res. 2011; 41: 579–589.

[85] Husillos-Rodriguez N, Martinez-Ramirez S, Blanco-Varela MT, et al. Re-use of drinking water treatment plant (DWTP) sludge: Characterization and technological behaviour of cement mortars with atomized sludge additions. Cem Concr Res. 2010; 40: 778–786.

[86] Snellings R, Mertens G, Cizer O, Elsen J. Early age hydration and pozzolanic reaction in natural zeolite blended cements: Reaction kinetics and products by in situ synchrotron X-ray powder diffraction. Cem Concr Res. 2010; 40: 1704–1713.

[87] Shi C, Fernández Jiménez A, Palomo A. New cements for the 21st century: The pursuit of an alternative to Portland cement. Cem Concr Res. 2011; 41: 750–763.

[88] Criado M, Fernandez-Jimenez A, De la Torre AG, Aranda MAG, Palomo A. An XRD study of the effect of the SiO_2/Na_2O ratio on the alkali activation of fly ash. Cem Concr Res. 2007; 37: 671–679.

[89] Fernández-Jiménez A, De la Torre AG, Palomo A, López-Olmo G, Alonso MM, Aranda MAG. Quantitative determination of phases in the alkali activation of fly ash. Part I. Potential ash reactivity, Fuel. 2006; 85: 625–634.

[90] Fernández-Jiménez A, De la Torre AG, Palomo A, López-Olmo G, Alonso MM, Aranda MAG. Quantitative determination of phases in the alkali activation of fly ash. Part II. Degree of reaction. Fuel. 2006; 85: 1960–1969.

[91] Aranda MAG, De la Torre AG. Sulfoaluminate cements, In: Pacheco-Torgal F, Jalali S, Labrincha J, editors. Eco-efficient concrete. Cambridge: Woodhead Publishing; 2013.

[92] Depmeier W. Aluminate sodalites – A family with strained structures and ferroic phase transitions. Phys Chem Miner. 1988; 15: 419–426.

[93] Saalfeld H, Depmeier W. Silicon-free compounds with sodalite structure. Krist Tech. 1972; 7: 229–233.

[94] Halstead PE, Moore AE. The composition and crystallography on an anhydrous calcium aluminosulphate occurring in expanding cement. J Appl Chem. 1962; 12: 413–417.

[95] Peixing Z, Yimin C, Piping S, Guanying Z, Wenmel H, Jiaguo W. The crystal structure of C_4A_3s. Proceedings of the 9th International Congress on the Chemistry of Cement, New Delhi, India; 1992.

[96] Calos NJ, Kennard CHL, Whittaker AK, Davis RL. Structure of calcium aluminate sulphate $Ca_4Al_6O_{16}S$. J Sol State Chem. 1995; 119: 1–7.

[97] Hargis CW, Moon J, Lothenbach B, Winnefeld F, Wenk HR, Monteiro PJM. Calcium sulfoaluminate sodalite ($Ca_4Al_6O_{12}SO_4$) crystal structure evaluation and bulk modulus determination. J Am Ceram Soc. 2014; 97: 892–898.

[98] Andac O, Glasser FP. Polymorphism of calcium sulphoaluminate ($Ca_4Al_6O_{16} \cdot SO_3$) and its solid solutions. Adv Cem Res. 1994; 6: 57–60.
[99] Bullerjahn F, Schmitt D, Ben Haha M. Effect of raw mix design and of clinkering process on the formation and mineralogical composition of (ternesite) belite calcium sulphoaluminate ferrite clinker. Cem Concr Res. 2014; 59: 87–95.
[100] García-Maté M, Santacruz I, De la Torre AG, León-Reina L, Aranda MAG. Rheological and hydration characterization of calcium sulfoaluminate cements. Cem Concr Comp. 2012; 34: 684–691.
[101] Cuesta A, Álvarez-Pinazo G, García-Maté M, Santacruz I, Aranda MAG, De la Torre AG, León-Reina L Rietveld quantitative phase analysis with molybdenum radiation. Powder Diff. 2015; 30: 25–35.
[102] García-Maté M, De la Torre AG, Cabeza A, Losilla ER, Aranda MAG, Santacruz I. Tailored setting times with high compressive strengths in bassanite calcium sulfoaluminate eco-cements. Cem Concr Comp. 2016; 72: 39–47.
[103] Paul G, Boccaleri E, Buzzi L, Canonico F, Gastaldi F. Friedel's salt formation in sulfoaluminate cements: A combined XRD and ^{27}Al MAS NMR study. Cem Concr Res. 2015; 67: 93–102.
[104] Popescu CD, Muntean M, Sharp JH. Industrial Trial Production of Low Energy Belite Cement. Cem Concr Comp. 2003; 25: 689–693.
[105] Walenta G and Saint-Antonin L. Aether® lower carbon cements [Internet]. [cited July 25, 2016], Available from: http://ec.europa.eu/environment/life/project/Projects/index.cfm?fuseaction=home.showFile&rep=file&fil=LIFE09_ENV_FR_000595_LAYMAN.pdf
[106] Hanein T, Imbabi MS, Glasser FP, Bannerman MN. Lowering the carbon footprint and energy consumption of cement production: A novel Calcium Sulfoaluminate cement production process. In: Proceedings in the 1st International Conference on Grand Challenges in Construction Materials, Los Angeles; 2016.
[107] Gartner EM, Li G. High belite sulfoaluminate clinker: fabrication process and binder preparation, World Patent Application WO2006/018569 A2. 2006.
[108] Cuberos AJM, De la Torre AG, Álvarez-Pinazo G, Martín-Sedeño MC, Schollbach K, Pöllmann H, Aranda MAG. Active Iron-Rich Belite Sulfoaluminate Cements: Clinkering and Hydration. Env Sci Tech. 2010; 44: 6855–62.
[109] Morin V, Walenta G, Gartner E, Termkhajornkit P, Baco I, Casabonne JM. Hydration of a Belite-Calcium Sulfoaluminate-Ferrite cement: Aether™. Proceedings of the 13th International Congress on the Chemistry of Cement. Madrid, Spain; 2011.
[110] Álvarez-Pinazo G, Santacruz I, León-Reina L, Aranda MAG, De la Torre AG. Hydration Reactions and Mechanical Strength Developments of Iron Rich Sulfobelite Eco-cements. Ind Eng Chem Res. 2013; 52: 16606–16614.
[111] Duvallet T, Zhou Y, Robl TL, Andrews R. Synthesis and Characterization of High-Iron Alite-Calcium Sulfoaluminate-Ferrite Cements produced from Industrial By-Products. Coal Combust Gasification Prod. 2014; 6: 29–34.
[112] Londono-Zuluaga D, Tobón J, Aranda MAG, Santacruz I, De la Torre AG. Clinkering and hydration of Belite-Alite-Ye'elimite cement. Cem Concr Comp. 2017; 80: 333–341.
[113] Liu X, Li Y. Effect of MgO on the composition and properties of alite-sulphoaluminate cement. Cem Concr Res. 2005; 35: 1685–1687.
[114] Ma S, Shen X, Gong X, Zhong B. Influence of CuO on the formation and coexistence of $CaO \cdot SiO_2$ and $3CaO \cdot 3Al_2O_3 \cdot CaSO_4$ minerals. Cem Concr Res. 2006; 36: 1784–1787.
[115] Ma S, Snellings R, Li X, Shen X, Scrivener KL. Alite-ye'elimite cement: Synthesis and mineralogical analysis. Cem Concr Res. 2013; 45: 15–20.
[116] Li J, Ma H, Zhao H. Preparation of Sulphoaluminate-Alite Composite Mineralogical Phase Cement Clinker From High Alumina Fly Ash. Key Eng Mater. 2007; 334/335: 421–424.

[117] Lili R, Xiaocun L, Tao Q, Lian L, Deli Z, Yanjun L. Influence of MnO_2 on the burnability and mineral formation of alite-sulphoaluminate cement clinker. Silics Ind. 2009; 74: 183–87.
[118] Liu XC, Li BL, Qi T, Liu XL, Li, YJ. Effect of TiO_2 on mineral formation and properties of alite-sulphoaluminate cement. Mater Res Innovations. 2009; 13: 92–97.
[119] Perez-Bravo R, Álvarez-Pinazo G, Compana JM, Santacruz I, Losilla ER, Bruque S, De la Torre AG. Alite sulfoaluminate clinker: Rietveld mineralogical and SEM-EDX analysis. Adv Cem Res. 2013; 26: 10–20.
[120] De la Torre AG, Cuberos AJM, Alvarez-Pinazo G, Cuesta A, Aranda MAG. In situ powder diffraction study of belite sulfoaluminate clinkering. J Synch Rad. 2011; 18: 506–514.
[121] Li X, Zhang Y, Shen X, Wang Q, Pan Z. Kinetics of calcium sulfoaluminate formation from tricalcium aluminate, calcium sulfate and calcium oxide. Cem Concr Res. 2014; 55: 79–87.
[122] Puertas F, Varela MTB, Molina SG. Kinetics of the thermal decomposition of $C_4A_3\$$ in air. Cem Concr Res. 1995; 25: 572–580.
[123] Shen Y, Qian J, Huang Y, Yang D. Synthesis of belite sulfoaluminate-ternesite cements with phosphogypsum. Cem Concr Comp. 2015; 63: 67–75.
[124] Sherman N, Beretka J, Santoro L, Valenti GL. Long-term behaviour of hydraulic binders based on calcium sulfoaluminate and calcium sulfosilicate. Cem Concr Res. 1995; 25: 113–126.
[125] Ben Haha M, Bullerjahn F, Zajac M. On the reactivity of ternesite. In: Proceedings of the 14th International Congress on the Chemistry of Cement. Beijing, China; 2015.
[126] Aranda MAG. Recent studies of cements and concretes by synchrotron radiation crystallographic and cognate methods. Crystallography Reviews. 2016; 22: 150–196.
[127] De la Torre AG, Morsli K, Zahir M, Aranda MAG. In situ synchrotron powder diffraction study of active belite clinkers. J Appl Cryst. 2007; 40: 999–1007.

Miguel A.G. Aranda*, Ana Cuesta, A.G. De la Torre,
Isabel Santacruz, and Laura León-Reina

2 Diffraction and crystallography applied to hydrating cements

Abstract: Diffraction techniques combined with crystallographic tools have been extensively used for the thorough characterization of the hydration processes in building materials. In this chapter, we summarize some works which employed crystallographic tools in wet cement chemistry. Initially, we briefly introduce the quantitative phase analysis of hydrating cements, mainly applying Rietveld methodology. In this context, the mineralogical phases that develop in hydrating Portland cements are reviewed. Then, special attention is dedicated to the quantification of phases in hydrating cementitious systems. We subdivide this section in three parts as a function of the increasing complexity: (i) model (reference) samples; (ii) neat Portland cements; and (iii) blended Portland cements with supplementary cementitious materials. Then, the use of the pair distribution function for characterizing the atomic order of nanocrystalline and amorphous phases in cements is reviewed. This is important as the main phase in Portland cement pastes is C-S-H (or related) gel(s). Finally, advanced crystallographic synchrotron-based characterization techniques are discussed. We pay close attention to spatially-resolved studies where imaging and mapping techniques allow for a better understanding of these heterogeneous materials.

Keywords: C-S-H gel, Rietveld quantitative phase analysis, pair distribution function, synchrotron radiation, spatially-resolved characterization

Initial information to readers: This chapter is based on four previous reviews of the authors [1–4]. On the one hand, the contents have been appropriately updated to the best knowledge and judgment of the authors, though it is not possible to report on every work dealing with powder diffraction and cement hydration. On the other hand, information concerning classic studies may be similar to that previously reported in the above-mentioned reviews.

***Corresponding author: Miguel A.G. Aranda,** Alba Synchrotron Light Source, Barcelona, Spain, g.aranda@cells.es
Ana Cuesta, Alba Synchrotron Light Source, Barcelona, Spain
A.G. De la Torre, Isabel Santacruz, Departamento de Química Inorgánica, Cristalografía y Mineralogía, Universidad de Málaga, Málaga, Spain
Laura León-Reina, Servicios Centrales de Investigación, Universidad de Málaga, Málaga, Spain

DOI 10.1515/9783110473728-003

2.1 Introduction to Rietveld quantitative phase analysis of hydrating cements

Quantitative phase analysis based on powder diffraction has already been discussed in this volume (see Chapter 1). An introduction to and details on Rietveld quantitative phase analysis (RQPA) are also included in that chapter. It is worth noting that Hugo Rietveld did not envisage this use for this analysis in his two seminal papers [5–7]. Two initial works established the basis of RQPA [8, 9]. RQPA was first applied to the analysis of a Portland clinker six years later [10]. The guidelines for Rietveld studies have been previously reported [11]. In 2012, a review article was devoted to the RQPA of blended Portland cements and their hydration products [1]. The analyses become very complex in hydration reactions of blended cements, therefore to carry out the best possible analysis a lot of care has to be exercised. This chapter is mainly based on the application of RQPA on cement pastes.

All RQPA procedures consist of:

(i) taking the best possible data sets (likely as a function of some variables like time, water content, pressure, and so on);
(ii) using the best possible control file;
(iii) following the most appropriate data analysis protocol.

Data collection

The powder pattern(s) must be recorded as free as possible of systematic errors. Therefore, synchrotron X-ray data are preferred over X-ray laboratory data but it is difficult to get beamtime at synchrotron beamlines [3]. Hence, most of the studies are carried out using laboratory sources with copper radiation ($\lambda = 1.54$ Å). For laboratory analyses, it is necessary to strictly use monochromatic radiation, as it minimizes the peak overlapping which is one of the main sources of systematic errors. A second source is particle statistics which are critical in cement pastes studies where the number of contributing phases is sometimes larger than 10. In order to apply this method, all phases must have enough randomly arranged microcrystals to yield smooth Debye–Scherrer diffraction cones that would result in accurate powder diffraction peak intensities when using point or linear detectors. Rotating the sample is essential to improve particle statistics but sometimes not enough. Reflection and transmission geometries have been used and both have advantages and drawbacks. Finally, strictly monochromatic molybdenum ($\lambda = 0.71$ Å) radiation with a set up working in transmission geometry has two key advantages, increasing particle statistics and decreasing preferred orientation maintaining small peak overlapping [12, 13]. This methodology has the potential to replace copper radiation analyses.

Crystal structures

Structural descriptions for each phase present in cement must be inserted in the Rietveld control file. A thorough compilation of cement phases was reported by Taylor [14]. Two updated compilations of crystal structures for the RQPA of Portland anhydrous cements and hydration products have been recently reported [1, 15]. This issue will be treated in detail in the next section. All the phases must be identified by searching for the peaks of the experimental pattern in the Powder Diffraction Database (www.icdd.com).

If the crystal structures are not known, quantification of these phases may be achieved using reference patterns, such as hkl phases, with the appropriate software [10, 16]. This is of particular importance for the analysis of nanocrystalline/amorphous phases like gels in cement chemistry as it is discussed below.

Data analysis

Rietveld analysis must be carried out with the appropriate program and the control file. The Rietveld procedure and quantitative phase analysis have already been explained in this volume and they are not treated further here. Nevertheless, it is important to have in mind that the strong peak overlapping in cement pastes makes the surveillance of phase peak shape parameters critical. Finally, rotating the sample in reflection geometry may increase particle statistics but it does not minimize preferred orientation. Some hydrated cement phases show this effect, which usually must be corrected with the March-Dollase algorithm by giving the appropriate axis [17]. The axes for some hydrated cement phases are [001] for portlandite; [001] for AFm family and [100] for ettringite. The order for modifying the appropriate parameters (scale factors, background coefficients, unit cell values, peak shape parameters, preferred orientation if needed, etc.) has been previously discussed [11].

Finally, a full list of the programs available for RQPA is not compiled here as they were given in Chapter 1.

2.2 Typical phases in hydrating cements

Hereafter, cement nomenclature will be used, i.e. C = CaO, S = SiO_2, A = Al_2O_3, F = Fe_2O_3, M = MgO, S = SO_3, C = CO_2, H = H_2O, K = K_2O, and N = Na_2O. Different phases can appear during the hydration of ordinary Portland cements (OPC). For instance, Tab. 2.1 lists the most typical hydrated phases that may be present in Portland cement pastes. The chemical formula, space group, unit cell parameters, and ICSD/PDF codes are detailed for all the phases [18–49]. It should be taken into account that the mineralogy of the hydrated compounds could depend upon the additions in blended cements [50].

Typical hydrated cement phases may be crystalline, such as portlandite: $Ca(OH)_2$ or CH, or amorphous/nanocrystalline phases such as calcium silicate hydrate gel, C-S-H gel, in cement nomenclature. It should be pointed out that a review of the main hydrated phase of OPC cements, calcium silicate hydrate, has been previously reported [51]. Several models for the nanostructure of the C-S-H gel have been described. Moreover, the crystal structures of tobermorite and jennite have also been discussed.

AFm and AFt set of phases should also be defined. AFm stands for the abbreviation for 'alumina ferric oxide mono-sulfate' or 'Al_2O_3–Fe_2O_3–mono', in the same way AFt stands for the related 'tri-sulfate/Al_2O_3–Fe_2O_3–tri'. By far the most common AFt phase is ettringite. Ettringite is a calcium sulfoaluminate hydrate and has the formula $[Ca_3(Al,Fe)(OH)_6 \cdot 12H_2O]_2 \cdot (SO_4)_3 \cdot nH_2O$, where normally $n \leq 2$. The term AFt refers to the three units of CX and there is an alternative way of writing the formula, $C_3(A,F) \cdot 3CX \cdot yH_2O$ [or $C_6(A,F)X_3 \cdot yH_2O$], where $y = n + 30$. The crystal structure can be described as compact columns of $[Ca_3Al(OH)_6 \cdot 12H_2O]^{3+}$ composition, running parallel to the c-axis, with the X anions and, usually, H_2O molecules in the intermediate channels [14]. The structural model for ettringite was proposed by Moore and Taylor by single-crystal X-ray diffraction analysis of a natural specimen [32]. However, no consideration was given to hydrogen atoms. For this reason, the crystalline structure of ettringite was revised by Hartman & Berliner [52] and Goetz-Neunhoeffer & Neubauer [31] to obtain a more complete structure model for this phase.

The term AFm includes a wide range of different compositions. These compounds have a layer structure with the general formula $[Ca_2Al(OH)_6]X \cdot nH_2O$ ($n = 8–14$) where X denotes an exchangeable singly charged (e.g. OH^- or chloride) or half a formula unit of a doubly charged anion (for instance sulfate, carbonate, and aluminosilicate) placed in the interlayer space jointly with water molecules. Some Fe(III) may also substitute aluminum. The AFm compounds crystallize in the form of hexagonal plates. The most ordinary AFm-type phases are kuzelite [35], stratlingite [44, 45], and C_2AH_8. There are other AFm-type phases such as monocarbonates [41] or monochlorides [36] that usually appear in chemically aggressive environments. The hydration states of AFm phases strongly depend upon relative humidity, temperature, and anion type in the interlayer space. A recent study combining thermodynamic calculations and powder diffraction has reported the different hydration states of the most important AFm phases [53].

Finally, hydrogarnet phases have a cubic structure related to that of grossular or garnet ($Ca_3Al_2Si_3O_{12}$) with the general formula $X_3Y_2(SiO_4)_3$. Hydrogarnet ($Ca_3(Al,Fe)_2(SiO_4)_y(OH)_{4(3-y)}$; $0 < y < 3$) includes a group of minerals where the $[SiO_4]^{4-}$ tetrahedra are partially or completely replaced by OH^- [29]. The Al-containing hydrogarnet includes hydrogrossular ($Ca_3Al_2(SiO_4)_y(OH)_{4(3-y)}$; $0 < y < 3$) with the end member katoite ($Ca_3Al_2(OH)_{12}$ or C_3AH_6).

It is important to mention that an excellent review dealing with the density of cement phases including those of AFm and other hydrates has been published [54].

2.3 Diffraction for quantifying phases in hydrating cements

2.3.1 Quantitative phase analysis of key reference phases

In situ X-ray powder diffraction coupled with the Rietveld method allows studying the hydration kinetics of single phases in cement chemistry. The hydration of alite, C_3S, which is the text-book example, has been extensively studied [55, 56]. A relevant recent example was the study of the hydration of alite and alite-gypsum mixture using the Rietveld method and mass balance calculations [57, 58]. Using this approach, the authors were able to disentangle the time-dependence contents of C-S-H gel and capillary water. The results were compared to those obtained by using the internal and external standards methods. Furthermore, the hydration kinetics and mechanism of alite has also been studied by Rietveld method and calorimetry [59]. Specifically, the role of mechanical activation was investigated [60], including the influence of the reactivity of the amorphous fraction of the alite sample [61]. The authors have shown that the reactivity of alite powders was strongly affected by the drying technique after milling. Fig. 2.1 displays the Rietveld plot (after one day of hydration) of fully-reacted mechanically-activated alite [60].

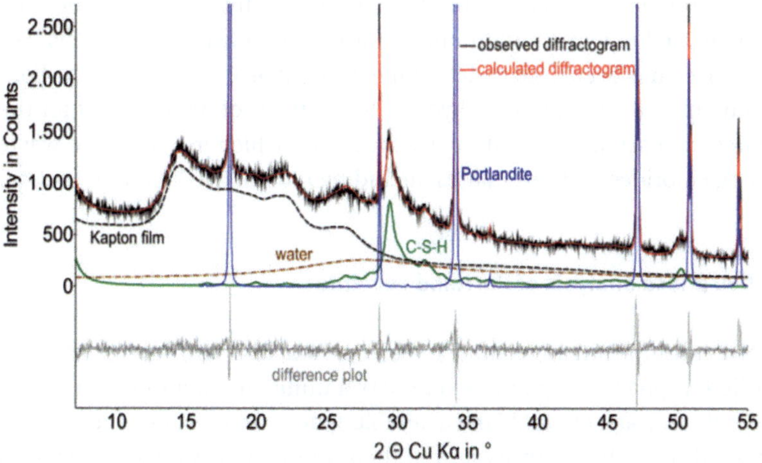

Fig. 2.1: Rietveld plot for mechanically-activated alite after 24 h of hydration (w/s = 1 and T = 23 °C). Alite is dissolved completely, while "long-range ordered" C-S-H, portlandite, and the remaining free water can be detected. Further details are given in the original publication. Reproduced under the Creative Commons BY license from http://dx.doi.org/10.1016/j.cemconres.2015.06.005 with permission [60].

The in situ hydration of C_3A and C_4AF at variable temperatures was studied by synchrotron X-ray powder diffraction, SXRPD. Phase quantifications were not carried out but the intermediate phases were identified [62]. Energy-dispersive SXRPD was also employed to characterize the hydration of C_4AF and CSH_2 at variable temperatures in order to identify the intermediate phases [63]. In the following examples, the Rietveld method has been employed for phase content determination in the hydrating systems. In a first set of studies, ettringite was quantified in the hydration of C_3A and $CSH_{0.5}$ samples with and without superplasticizer [64]. The hydration mechanisms of C_3A with variable gypsum contents were investigated in depth [65]. The hydration of C_4AF in a Portland cement environment with low sulfate content was studied by quantitative in situ X-ray powder diffraction, using the external standard method, and heat flow calorimetry. The different hydrated aluminate phases were identified and quantified and the mineralogical results were in agreement with the calorimetric data [66].

Very recently, a careful study of the hydration of CA_2 using heat flow calorimetry, in situ X-ray powder diffraction and in situ ^{1}H-time domain NMR explained the different observed hydration kinetics, which very likely arise from the different particle sizes and even the (possible) amorphous content [67].

In an additional set of works, the role of ye'elimite polymorphism in hydration reactions and kinetics was investigated by modifying the water-to-ye'elimite ratio and sulfate sources [68]. In a subsequent work, the hydration of C_4AF at room temperature was investigated under different conditions: (i) without gypsum; (ii) with gypsum; (iii) with gypsum and stoichiometric ye'elimite; and (iv) with gypsum and doped ye'elimite [26].

To end this section, it is worth underlining that powder diffraction and the Rietveld method can be used to extract additional information in hydrated systems. For instance, this approach has been used to determine the uptake of chloride and carbonate ions by calcium monosulfoaluminate hydrate [69]. This is of high interest for the safe-controlled decommissioning of old nuclear reactors, which may produce waste streams containing chlorides and carbonates, including radioactive $^{36}Cl^-$ and $^{14}CO_3^{2-}$.

2.3.2 Quantitative phase analysis of Portland cements

The hydration processes in OPC are complex: dissolutions, oversaturations, and precipitations of different phases (crystalline, nanocrystalline, and amorphous). These depend on temperature, water content (and available space), and several other parameters. An understanding of these processes and their interplays is slowly progressing [70]. During the hydration of OPC, several anhydrous phases (may) disappear and some hydrated phases appear. Powder diffraction has the ability to follow the dissolution and precipitation processes. Anhydrous phases were treated in a previous chapter. Here we focus on the hydrated phases, and so Tab. 2.1 shows a list of phases that may be present in cement pastes. For instance, Fig. 2.2 displays a typical Rietveld plot for a gray OPC hydrated paste for 3 days (water-to-cement ratio of 0.5) showing the phase assemblage. A related work employed a water-to-cement ratio of 0.4 and combined RQPA with thermal analysis data [71]. To end this introductory paragraph, it is important to discuss the status of the sample/paste where the study/analysis will be carried out. The paste may contain free water, and so hydration will continue, or alternatively, the hydration can be stopped by a number of procedures. In order to compare the results from diffraction with those from other techniques (for instance microstructural analysis), it is necessary to stop cement hydration and to dry the sample. There are good reasons to follow any of these two options but the possible sample modifications in the arresting/drying processes should be carefully taken into account [72, 73].

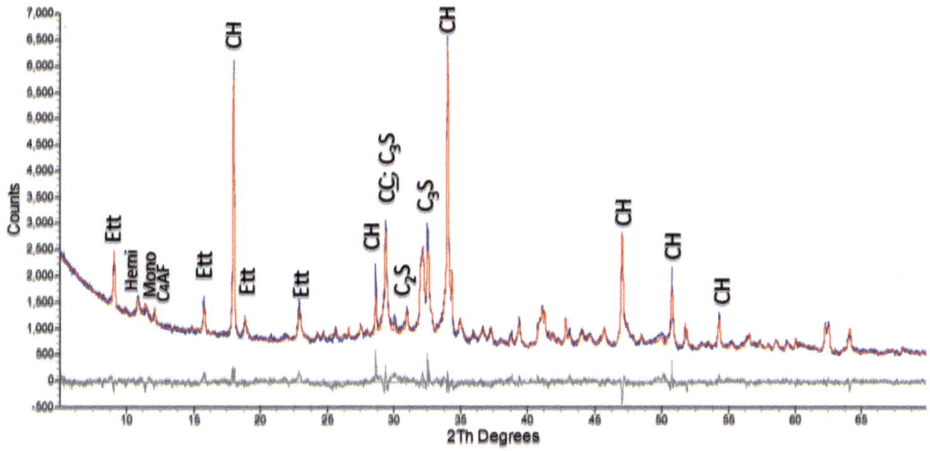

Fig. 2.2: Rietveld plot of a grey OPC paste cured for 3 days with the difference plot given at the bottom. The main peaks due to a given phase have been labelled. Ett stands for AFt, Mono stands for monocarbonate, and Hemi for hemicarbonate. Figure produced by the authors for this work.

Laboratory X-ray powder diffraction (LXRPD) has been widely used to study the hydration products of OPC in reflection [74] and transmission [75] geometries. Moreover, the results can be compared to those obtained from thermal analysis and electron microscopy. One study focused on the comparison of the results for the hydration of a CEM-I OPC cement (w/c = 0.5) in reflection and transmission geometries. The advantages and disadvantages of the employed configurations were discussed [76]. To choose the appropriate experimental powder diffraction configuration is important as there are systematic errors that can be avoided/minimized. Unfortunately, there is still not a single error-free protocol when using laboratory powder diffraction.

Tab. 2.1: Structural and crystallographic details for hydrated phases in cements.

Phase	Formula	s.g.	μ (cm^{-1})	*a* (Å)	*b* (Å)	*c* (Å)	α (°)	β (°)	γ (°)
Portlandite	$Ca(OH)_2$	P-3m	211.4	3.5853		4.895			
Gibbsite	$Al(OH)_3$	$P2_1/n$	57.1	8.684	5.078	9.736		94.54	
Clinotobermorite	$Ca_5(Si_6O_{17})5H_2O$	C1	168.6	11.274	7.344	11.468	99.18	97.19	90.02
Tobermorite-14Å	$Ca_5Si_6O_{16}(OH)_2 \cdot 7H_2O$	Bb	135.6	6.735	7.425	27.987			123.25
Tobermorite-11Å	$Ca_4Si_6O_{15}(OH)_2 \cdot 5H_2O$	B11m	144.3	6.735	7.385	22.487			123.25
Jennite	$Ca_9Si_6O_{18}(OH)_6 \cdot 8H_2O$	P-1	164.3	10.576	7.265	10.931	101.30	96.98	109.65
Dicalcium silicate hydrate	$Ca_2(SiO_4) \cdot H_2O$	$P2_12_12_1$	225.6	9.487	9.179	10.666			
Hydrogarnet/C_3AH_6	$Ca_3Al_2(OH)_{12}$	Ia-3d	163.7	12.565					
Hydrogarnet-Fe	$Ca_3Al_{1.69}Fe_{0.31}(OH)_{12}$	Ia-3d	189.4	12.603					
Katoite	$Ca_3Al_2(OH)_{7.6}(SiO_4)_{1.1}$	Ia-3d	187.8	12.270					
Katoite	$Ca_{2.93}Al_{1.97}(OH)_{9.44}(SiO_4)_{0.64}$	Ia-3d	185.5	12.38					
Hydroandradite	$Ca_3Fe_2(OH)_{8.64}(SiO_4)_{0.84}$	Ia-3d	385.5	12.5424					
Hydroandradite	$Ca_3Fe_2(OH)_{6.64}(SiO_4)_{1.34}$	Ia-3d	406.4	12.4297					
'CAH_{10}'	$CaAl_2(OH)_8(H_2O)_2 \cdot 1.84H_2O$	$P6_3/m$	58.5	16.387		8.279			
AFt									
Ettringite	$Ca_6Al_2(OH)_{12}(SO_4)_3 \cdot 26H_2O$	P31c	84.4	11.229		21.478			
Ettringite-CO_3	$Ca_6Al_2(OH)_{12}(CO_3)_3 \cdot 26H_2O$	*No structure*	*77.6*	*10.8344*		*21.250*			
Ettringite-BO_3	$Ca_6Al_2(OH)_{12}[B(OH)_4]_4(OH)_2 \cdot 22H_2O$	P3c	71.0	11.0296		10.6992			
Thaumasite	$Ca_6Si_2(OH)_{12}(CO_3)_2(SO_4)_2 \cdot 24H_2O$	$P6_3$	85.7	11.046		10.409			

Tab. 2.1: (continued)

Phase	*V* (Å³)	*Z*	*ρ* (g/cm³)	ICSD codes	PDF codes	Ref
Portlandite	54.49	1	2.26	15471	01-072-0156	[18]
Gibbsite	427.98	8	2.42	6162	01-070-2038	[19]
Clinotobermorite	929.78	2	2.61	90034	01-074-2595	[20]
Tobermorite-14Å	1170.43	2	2.23	152489	00-029-0331	[21]
Tobermorite-11Å	935.35	2	2.46	–	–	[22]
Jennite	759.50	1	2.32	151413	00-018-1206	[23]
Dicalcium silicate hydrate	928.81	8	2.72	73404	01-081-1987	[24]
Hydrogarnet/C_3AH_6	1983.75	8	2.53	202316	01-084-1354	[25]
Hydrogarnet-Fe	2001.88	8	2.49	431957	–	[26]
Katoite	1847.28	8	2.88	172077	00-038-0368	[27]
Katoite	1897.41	8	2.73	49772	01-077-1713	[28]
Hydroandradite	1973.07	8	3.03	–	–	[29]
Hydroandradite	1920.36	8	3.23	–	–	[29]
'CAH_{10}'	1925	6	1.55	407150	01-088-1410	[30]
AFt						
Ettringite	2345.34	2	1.78	155395	00-041-1451	[31]
Ettringite-CO_3	*2160.23*	*2*	*1.76*	–	00-036-1465	[32]
Ettringite-BO_3	1127.2	1	1.72	236592	–	[33]
Thaumasite	1099.90	2	1.88	–	–	[34]

Tab. 2.1: (continued)

Phase	Formula	s.g.	μ (cm^{-1})	*a* (Å)	*b* (Å)	*c* (Å)	α (°)	β (°)	γ (°)
AFm									
Kuzelite/C_4ASH_{12}	$Ca_2Al(OH)_6[(SO_4)_{0.5}\cdot 3H_2O]$	R-3	115.8	5.7586		26.7946			
Friedel's salt	$Ca_2Al(OH)_6[Cl\cdot 2H_2O]$	R-3c	146.7	5.724		46.689			
Friedel's salt	$Ca_2Al(OH)_6[Cl\cdot 2H_2O]$	R-3	141.7	5.7487		23.492			
Friedel's salt	$Ca_2Al(OH)_6[Cl\cdot 2H_2O]$	R-3	259.1	5.873		23.362			
Kuzel's salt	$Ca_2Al(OH)_6[Cl_{0.5}(SO_4)_{0.25}\cdot 2.5H_2O]$	R-3	126.9	5.7508		50.4185			
Hydrocalumite	$Ca_2Al(OH)_6[Cl_{0.5}(CO_3)_{0.25}]\cdot 2.4H_2O$	C 2/c	130.1	10.020	5.751	16.286		104.22	
Chloro-carboaluminate	$Ca_2Al(OH)_6[Cl_{0.45}(CO_3)_{0.27}]\cdot 2.27H_2O$	R-3c	129.0	5.7400		46.7402			
Monocarbo-aluminate	$Ca_2Al(OH)_6[(CO_3)_{0.5}\cdot 2.5H_2O]$	P1	124.7	5.7747	8.4689	9.923	64.77	82.75	81.43
Hemicarbo-aluminate	$Ca_2Al(OH)_6[(CO_3)_{0.25}OH_{0.5}\cdot 2H_2O]$	R-3c	113.9	5.7757		48.812			
Monocarbo-Ferrite	$Ca_2Fe(OH)_6[(CO_3)_{0.5}\cdot 3.1H_2O]$	R-3c	222.1	5.9196		47.8796			
Borate-AFm	$Ca_2Al(OH)_6[(HBO_3)_{0.5}\cdot 3.1H_2O]$	R-3c	106.4	5.7764		49.5499			
C_2AH_8	$Ca_2Al(OH)_6[Al(OH)_4\cdot 3H_2O]$	*No structure*	107.1	5.7907		64.6960			
Strätlingite/C_2ASH_8	$Ca_2Al(OH)_6[AlSiO_2(OH)_4\cdot 3H_2O]$	R-3m	98.8	5.745		37.770			
Strätlingite/C_2ASH_8	$Ca_2Al(OH)_6[(Al,Si)_2O_2(OH)_6\cdot 2.25H_2O]$	R-3m	91.0	5.7536		37.732			
Nitroaluminate	$Ca_2Al(OH)_6[NO_3\cdot 2H_2O]$	P-3c	111.2	5.7445		17.2350			
Bromoaluminate	$Ca_2Al(OH)_6[Br\cdot 2H_2O]$	R-3	163.8	5.758		24.4980			
Iodoaluminate	$Ca_2Al(OH)_6[I\cdot 2H_2O]$	R-3	353.1	5.772		26.5380			
Bromochloroaluminate	$Ca_2Al(OH)_6[Br_{0.478}Cl_{0.522}\cdot 2H_2O]$	R-3c	153.5	5.7537		48.1080			

Tab. 2.1: (continued)

Phase	*V* (Å^3)	*Z*	*ρ* (g/cm^3)	ICSD codes	PDF codes	Ref
Kuzelite/C_4ASH_{12}	769.5	3	2.02	100138	01-083-1289	[35]
Friedel's salt	1324.78	3	2.11	88617	01-089-8294	[36]
Friedel's salt	672.34	3	2.08	51890	01-072-4773	[37]
Friedel's salt	697.85	3	2.21	51892	01-072-4775	[37]
Kuzel's salt	1444.04	3	2.01	–	00-019-0203	[38]
Hydrocalumite	909.73	1	2.09	63250	01-078-2050	[39]
Chloro-carboaluminate	1333.7	6	2.054	–	–	[40]
Monocarbo-aluminate	1453.01	6	2.22	59327	01-087-0493	[41]
Hemicarbo-aluminate	1410.1	3	1.9	263124	–	[42]
Monocarbo-Ferrite	1453.01	6	2.22	–	–	[43]
Borate-AFm	1431.8	3	1.92	236591	–	[33]
C_2AH_8	1878.76	6	1.89	–	00-045-0564	–
Strätlingite/C_2ASH_8	1079.59	3	2.05	69413	01-080-1579	[44]
Strätlingite/C_2ASH_8	1081.70	3	1.95	431951	–	[45]
Nitroaluminate	492.55	1	2.07	280171	01-089-6723	[46]
Bromoaluminate	703.4	3	2.303	280087	–	[47]
Iodoaluminate	765.69	3	2.421	280150	–	[48]
Bromochloroaluminate	1379.25	6	2.182	280444	–	[49]

From the diffraction point of view, the process of the hydration of OPCs can be subdivided into two periods: (i) early hydration from minutes to about 1 day where the hydration reactions are commonly followed in situ; and (ii) hydration after about two days where the reactions are snapshotted ex situ at given time intervals. The best experimental approach to follow the period between about 12 hours and 2 days is sample dependent as, if the reaction kinetics are fast, the ex situ approach may yield more accurate results. The use of temperature and humidity chambers can be required to study the early hydration processes at the laboratories [74]. However, in this case, experimental conditions must be checked very carefully as some problems may develop including partial drying problems. Alternatively, early age in situ studies can be carried out by appropriately protecting/covering the paste [77], which likely results in less errors but does not allow for controlling the relative humidity. Moreover, the external and internal standard methodologies are important to study the RQPA of OPC cement in order to determine the developed amount of amorphous material [78]. The role of isopropanol to stop the hydration process in the pastes has also previously been studied [78]. The early age hydration of OPC cements has also been studied by LXRPD by focusing on the role of different water-to-cement ratios [79, 80]. The phase development on hydration after one day was followed by diffraction and thermogravimetry for an OPC and an ultra-high performance cementitious system [81]. RQPA results at early hydration using the external standard G-method were successfully compared with the results obtained from the heat flow measured by calorimetric experiments [59, 82]. A combination of X-ray diffraction, thermogravimetric, and SEM/EDS methods was also successfully applied to determine the phase distributions for OPC pastes as a function of hydration time [71]. RQPA results for OPC have also been compared with those obtained from different spectroscopic techniques such as Fourier transformed infrared (FTIR) and Mössbauer spectroscopy [83]. All the hydrated phases were quantified and the study confirmed the low reactivity of the C_4AF phase during early hydration.

The use of an intense and monochromatic synchrotron X-ray source coupled with a fast detection system permits the collection of powder diffraction data with short times enabling time-resolved analysis of rapid reaction sequences [84, 85]. For instance, the evolution of hydrated phases during the early hydration of Portland cements with a time resolution of 10 seconds was carried out [84]. Sub-second data collection and analysis have been also reported [86]. The early age phase development of active belite cement pastes has also been followed in situ by synchrotron X-ray powder diffraction and the results were compared with the hydration at later ages determined by laboratory X-ray powder diffraction [87]. Synchrotron powder diffraction data analyzed by the Rietveld method were also used to characterize the long-term leaching behavior of concretes by quantifying portlandite dissolution at different hydration times [88]. These results were successfully compared to those obtained by DTA/TGA.

The use of admixtures in the hydration of cements is key for the performances of the derived concretes and mortars [89, 90]. Hence, it is not a surprise that the influence of different additives, like superplasticizers or retarders on the hydration of cements has been extensively investigated. For instance, the effects of aliphatic sugar alcohols on the hydration of alite and OPC were investigated and compared with those of sucrose [91]. Superplasticizers, a very common type of admixture, change the reaction kinetics of some phases in OPC and this was analyzed by calorimetry and RQPA [92]. A follow-up study using the same methodology was focused on the influence of different types of cationic admixtures. The organic additives belong to the family of polydiallyldimethylammonium with different anionic counterions. The role of these admixtures on the hydration behavior of an OPC was disentangled [93].

2.3.3 Quantitative phase analysis of Portland cements with supplementary cementitious materials (SCM)

For a number of reasons, including decreasing both CO_2 emissions and price, it is allowed to add other materials at the grinding stage (in Europe) or at the concrete production stage (USA). The main SCMs are fly ash, ground granulated blast furnace slag, silica fume, calcined clays, natural/industrial pozzolans, and limestone [94, 95]. These materials are commonly blended with clinker to produce different Portland cement types or used as a replacement for a portion of clinker in concrete mixtures. Several other methods in addition to powder diffraction are required to properly asses the hydration degree of the SCMs [96, 97]. To the best of our knowledge, the first two studies applying Rietveld quantitative analysis to the hydration of blended cements took place just a decade ago. RQPA was used to study the hydration progress of cement pastes prepared by adding blast furnace slag (BFS) and limestone powder [98]. In that case, selective dissolution was also employed to distinguish between the amorphous content coming from BFS and the hydrated material, the C-S-H gel. In a second independent work, the effect of water curing conditions on the compressive strengths of cement pastes containing pulverized fly ash (PFA) was quantitatively investigated [99].

Firstly, we must underline a key study [100] where several SCMs were employed (blast furnace slag, metakaolin, limestone, and quartz) in addition to two Portland cements (gray and white). The content of amorphous materials in the initial (anhydrous) samples and in the hydrated blended cements was quantified by the "Partial Or No Known Crystal Structure" (PONKCS) method. The precision and accuracy of the method were assessed through a comparison to a series of mixes of known phase composition and of increasing complexity resulting in accuracies of 2 wt% in the best cases. To illustrate the complexity of these systems, Fig. 2.3 displays the powder diffraction plots of a blended Portland cement (30 wt% of siliceous fly ash) hydrated for 1 day [101].

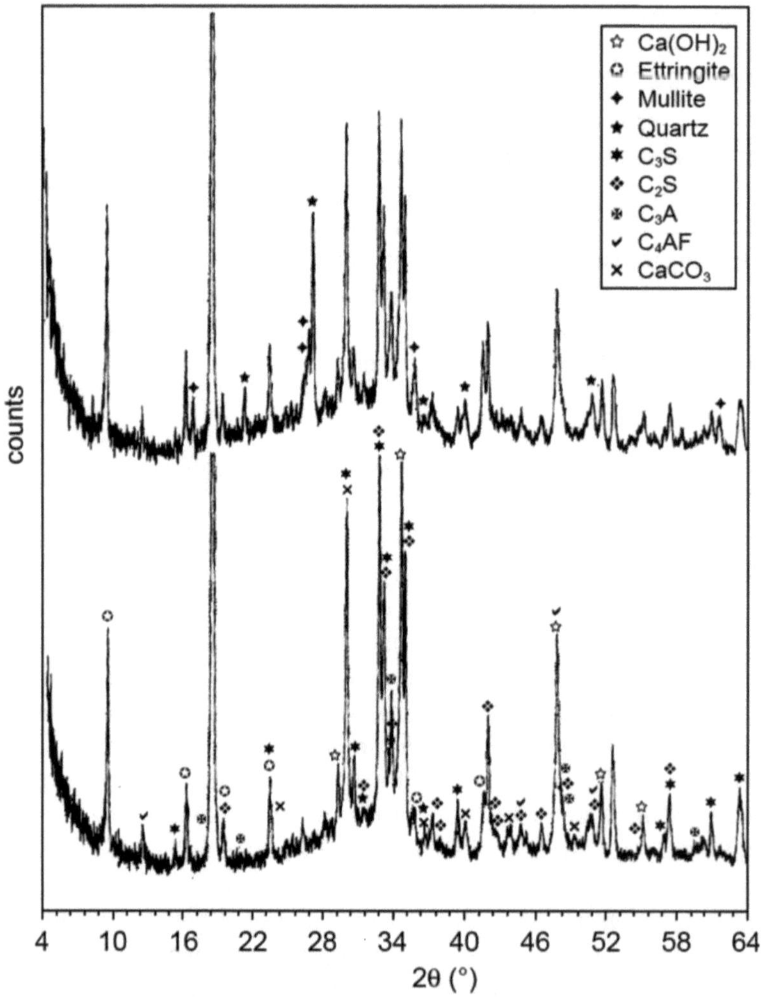

Fig. 2.3: Powder diffraction patterns for OPC without any addition (bottom) and OPC with 30 wt% of fly ash (top), after 1 day of hydration. Further details are given in the original publication. Reproduced with permission [101].

Concerning PFA blended Portland cements, selective dissolution and RQPA have been applied to understand the effect of fly ash on the hydration mechanism of OPC at different curing temperatures [102]. In an independent work, the effects of two different low calcium FA on the hydration of OPC pastes containing 50 wt% of FA were investigated up to 550 days of hydration. The results were compared with a reference blend of OPC containing 50 wt% of inert quartz powder which allowed for a distinction between the filler effect and pozzolanic reaction [103]. Moreover, the hydration mechanisms of ternary blended Portland cements containing supplementary cementitious materials such as fly ash and limestone powder [104] were widely studied using thermogravimetry, scanning electron microscopy and isothermal calorimetry together with RQPA. The publication was focused on studying the time-evolution of all phases during the hydration to understand the pozzolanic effect and the variations in chemical shrinkages.

The addition of silica fume to OPC has also been studied by RQPA jointly with ^{27}Al MAS NMR [105]. The study concluded that silica fume could accelerate the early age hydration of all the phases present in OPC, especially C_3S, although the hydration of C_2S might be delayed. In addition, the hydration of OPC in combination with silica fume and fly ash has been investigated [106]. This study confirmed that although silica fume accelerates cement hydration and fly ash retards this process, the combination of both additions can affect the hydration mechanism in different ways.

Recently, the hydration mechanisms of blended cements with various limestone sizes were studied in order to investigate the influence of the particle size distribution [107]. They concluded that the blended cements with a combination of different limestone sizes give a more efficient hydration process. Other less common SCMs have also been investigated. For instance, RQPA coupled with TGA were used to follow the reactivity on OPC blended with calcined kaolinite-bentonite clay [108]. Another thorough work was focused on the early-age hydration and pozzolanic reaction in OPC blended with natural zeolites [109]. One key conclusion of this study was that the addition of natural zeolites accelerates the onset of C_3S hydration and precipitation of CH and AFt. In a subsequent work, these authors enlarged the study by analyzing the early-age hydration of OPC blended with micronized zeolitite and quartzite powders [110].

2.4 Total diffraction pair distribution function studies of hydrating cements

Total scattering methods (often referred to as pair distribution function 'PDF' or radial distribution function) have been used for many decades to gain insight into the disordered structure of liquids and amorphous materials. The experimental set up essentially consists of a collimated, monochromatic beam impinging on a sample and the scattering data is nowadays recorded in a suitable 2D detector. Very short wavelength X-ray photons (or neutrons) allow one access to large momentum transfer values and reduce experimental artifacts. The PDF method involves the sine Fourier transform of the measured structure factor over the widest possible momentum transfer range, providing a direct measure of the probability, $G(r)$, of finding an atom surrounding a central atom at a radial distance. The weighting factor of each atomic species is scaled by their concentration and by the number of electrons (for X-rays). The advantage is that the average atomic structural information may be obtained when no (sharp) Bragg peaks are present in the measured diffraction pattern and all the scattering data is taken into account (including diffuse scattering), hence the name total scattering. A recent focused work has reviewed the use of total scattering methods in building materials to characterize amorphous and poorly crystalline phases [111]. Furthermore, quantitative phase analyses based on PDF investigations can also be obtained.

The characterization of nanocrystalline and amorphous materials is challenging but its analysis in samples containing large amounts of crystalline materials, f.i. cement pastes, is even more difficult. At this stage, it is important to clarify the differences between nanocrystalline and amorphous phases. In a nanocrystalline phase, the atomic structure can be described to a good approximation by using the crystal structure truncated in the real space by a nanosized shape function. Surface relaxation effects may nowadays be taken into account with suitable models and software. Conversely, the atomic structure of an amorphous compound cannot be properly described by using a truncated crystal structure [112].

PDF technique was used to study the local structure of synthetic C-S-H(I), showing nanocrystalline ordering with a particle diameter close to 3.5 nm, which is similar to a size-broadened 1.1 nm tobermorite crystal structure [113]. The C-S-H component in hydrated tricalcium silicate was found to be similar to C-S-H(I); only a slight bend and additional disorder within the CaO sheets was required to explain its nanocrystalline structure. In a subsequent work [114], these researchers studied different C-S-H samples with varying Ca/Si ratios (between 0.6 and 1.8). The PDF analysis results suggested that the C-S-H structure evolves from tobermorite-like to jennite-like as a function of increasing Ca/Si ratio. The evolution of these short- and medium-range order structural characteristics was associated with the alteration of the Ca-O layers and silicate depolymerization. In a comprehensive study [112], X-ray total scattering was used to determine the local atomic environment and nanostructure of the C-S-H gel present in hydrated tricalcium silicate (C_3S), and the C(-A)-S-H gels present in blended C_3S-slag and alkali-activated slag. The work revealed large intrinsic differences in the extent of nanoscale ordering between the gel derived from C_3S and that resulting from alkali-activated slag systems. As an example, Fig. 2.4 shows the PDF data for hydrated OPC that have been used to quantitatively study the deformation behavior of the C-S-H [115].

Total-scattering PDF and Rietveld methodology have been very recently employed to determine the local structure of a gel and to carry out quantitative phase analysis of $CaAl_2O_4$ hydrated at 50 °C which yielded crystalline hydrogarnet and nanocrystalline aluminum hydroxide [116]. Furthermore, five reference samples have also been studied for a good understanding of nanocrystalline aluminum hydroxide gels which had a gibbsite crystal structure with average particle size close to 5 nm [116]. PDF approach using synchrotron radiation has also been employed to study several other amorphous (or nanocrystalline) phases in cement research including alkali-silica-reaction gel [117], metakaolin [118], and geopolymers [119].

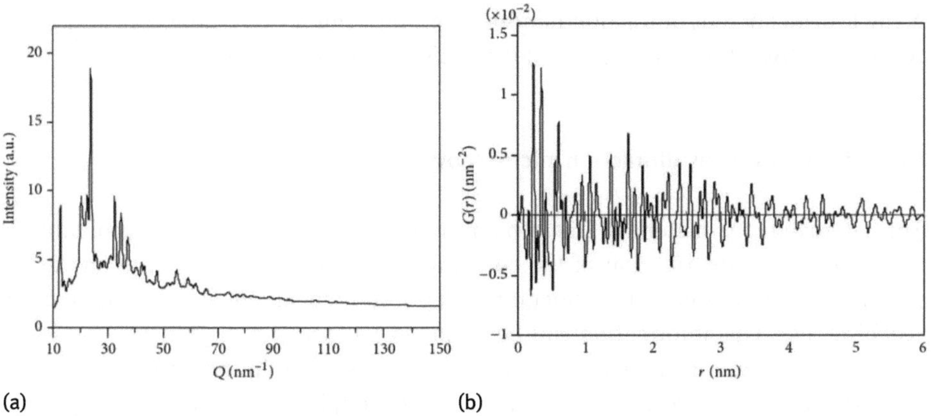

Fig. 2.4: (a) Total-scattering synchrotron X-ray diffraction pattern for the OPC paste; (b) PDF data derived from the diffraction pattern shown in (a). Reproduced under the Creative Commons BY license from http://dx.doi.org/10.1155/2016/8936084 [115].

2.5 Advanced crystallographic characterization of hydrating cements: spatially-resolved studies

In a previous section, we have reviewed the applications of diffraction to follow the phase assemblage and contents of different hydrating cements with time. However, for heterogeneous materials, like cement pastes and mortars, spatially-resolved phase information can be very valuable. This information can be extracted from synchrotron experiments where focused beams can be produced and diffraction/scattering patterns can be taken in a number of configurations. In this section, we review recent spatially-resolved crystallographic studies using synchrotron radiation. This part is based on a review of one of the authors which is summarized and updated here [3].

First, X-ray microscopic and tomographic approaches must be distinguished. In X-ray diffraction microscopy studies, 2D maps are produced where each pixel contains a powder pattern. The technique that provides this information is commonly named scanning X-ray diffraction microscopy. In X-ray tomography studies, 3D reconstructed volumes are produced where each voxel contains a powder pattern. The technique that provides this information is commonly named X-ray diffraction microtomography, although some authors has termed it 'pencil-beam' synchrotron X-ray diffraction microtomography to highlight the need of a focused tiny beam [120].

Secondly, there is an alternative to image-forming optics where the X-rays scattered by the sample are reconstructed by appropriate algorithms. This set of techniques is commonly named coherent diffraction imaging (CDI) techniques, and the two most common techniques have also been applied to hydrating cements. The techniques are X-ray ptychographic forward coherent diffraction nano-tomography and

X-ray Bragg coherent diffraction nano-tomography. Interested readers are addressed to Aranda MAG (2016) to learn about the basis of these techniques [3].

2.5.1 Scanning X-ray diffraction microscopy

For heterogeneous materials, like cement binders, in some cases it is invaluable to have the spatial distribution of the constituent phases. For instance, to characterize the alteration profile (as a function of depth) due to sulfate attack, carbonation, leaching, etc. This information can be obtained by focusing the synchrotron beam to a size of several micrometers and scanning the sample in the appropriate direction. This technique works in transmission, and sample preparation is as important as beam conditioning. For this technique to provide useful information, a relatively thin cross section must be prepared along the appropriate direction without modifying the phase assemblage of the pristine sample. Fig. 2.5 displays the procedure for this type of analysis [121]. Typical thicknesses for the flat slices range from 100 to 500 μm. This technique is also known simply as synchrotron microdiffraction.

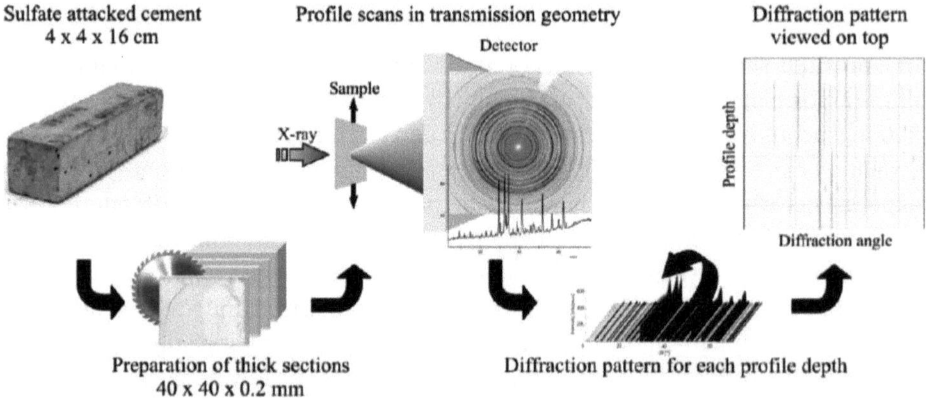

Fig. 2.5: Experimental setup and workflow for synchrotron X-ray powder micro-diffraction. A focused 'pencil' X-ray beam is raster scanned across the sample in many positions on a thin flat sample. Further details are given in the original publication. Reproduced with permission [120].

Synchrotron microdiffraction was initially employed to determine, with spatial resolution, the changes in the phase composition due to sulfate attack with the final aim of establishing the durability of cementitious materials under these conditions [121]. For this study, the sample thickness was 200 μm and the X-ray beam size and energy were 10 μm and 11.6 keV, respectively. Furthermore, the sulfate attacks are strongly affected by the presence of SCM. The microstructural profile analyses of concrete deterioration after sulfate attack of OPC blended with fly ash, natural pozzolana, and granulated blast furnace slag have been extensively investigated [122–124].

Chloride ingress (transport and binding) in cement matrices has also very recently been analyzed by synchrotron microdiffraction and complementary techniques. In addition to a pure OPC paste, other samples were prepared by replacing the clinker with limestone, class-F FA, or ground granulated BFS [125]. Combined sulfate and chloride attack of cement pastes has also been investigated using this methodology [126].

Finally, synchrotron microdiffraction has also been employed to study pyroclastic aggregate concrete of 1900 years old Trajan's Markets concretes as well as current reproductions [127]. The mortar reproduction gains fracture toughness over 180 days through coalescence of C-A-S-H gel binder and crystallization of strätlingite and katoite, after pozzolanic consumption was completed.

2.5.2 X-ray diffraction micro-tomography

This technique is a combination of diffraction (crystalline-phase selective) and imaging (through tomographic reconstruction) that allows determining the three-dimensional spatial distribution of hierarchically-arranged phases [128, 129]. The principles of X-ray diffraction micro-tomography are depicted in Fig. 2.6 where the procedure is compared with classical X-ray absorption micro-tomography [129]. Three dimensional imaging of the microstructure of materials is commonly attained through tomographic reconstruction. In this case, instead of acquiring the transmitted X-ray beam, the powder patterns are collected by translation and rotation of the specimen with the appropriate dimensions. In contrast to the previous technique, synchrotron microdiffraction, flat thin sections are not needed. Furthermore, the appropriate rescaling of the voxel intensity to the total intensity of sample scattering makes it possible to obtain the absolute quantification of the phase proportions in each voxel. This technique was originally developed for synchrotron radiation but it has been very recently expanded to the use of X-ray laboratory radiation [130].

X-ray diffraction micro-tomography was initially used for 3D monitoring the evolution of microstructure and phase formation non-invasively [120, 131]. This technique gave a good insight about the 3D phase assemblage at intermediate and later hydration ages, but the long acquisition times did not allow the phase mapping at early stages of hydration. In an interesting subsequent approach, fast X-ray absorption micro-tomography, for early hydration ages, and X-ray diffraction micro-tomography, for later ages (after 7 days of hydration), were combined to get a better insight about the phase development in hydrating pastes [132].

This technique has been recently used to image the phases developed in OPC cement pastes under different hydrating conditions. In a first work, the hydration of a neat OPC paste was compared to that of a seed-nucleated sample [133]. The quantitative description of the phase spatial distribution by radial distribution func-

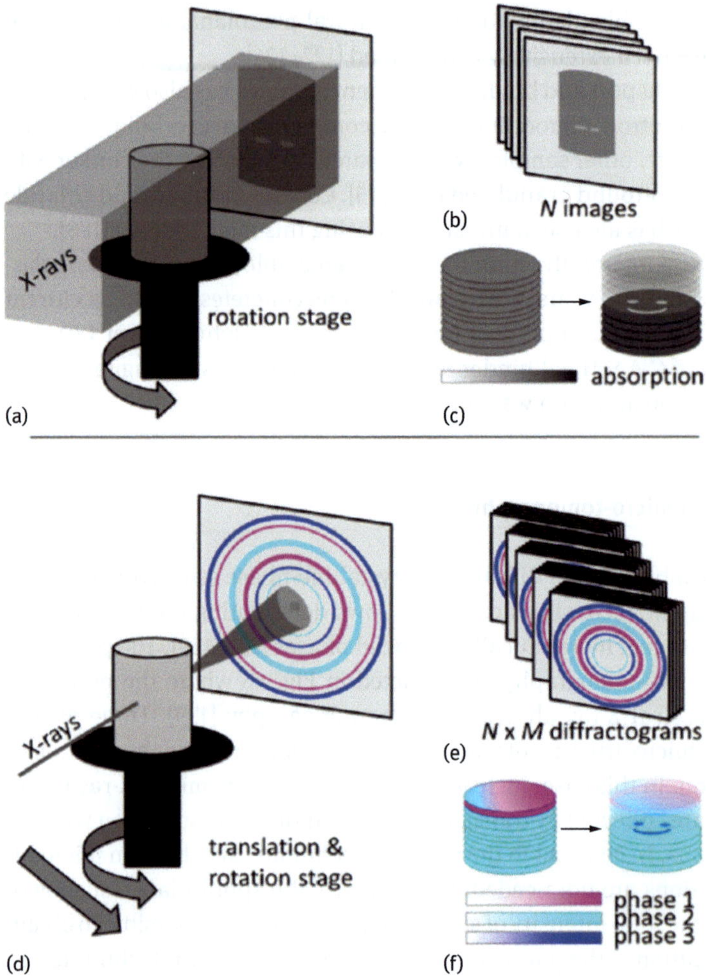

Fig. 2.6: Experimental setup and workflow for absorption tomography (a–c) and diffraction tomography (d–f). In traditional absorption tomography, the attenuation sample is recorded for a set of *N* viewing angles (a, b). Through tomographic reconstruction software, 3D images of the sample absorption density are obtained that reveal the interior microstructure (c). In diffraction tomography, a focused 'pencil' X-ray beam is raster scanned across the samples in *M* positions for a set of *N* viewing angles (d) resulting in $N \times M$ individual diffraction patterns (e). The diffraction information is then reconstructed using the same reconstruction algorithms yielding images based on diffraction contrast (f) that allow distinguishing materials with similar linear attenuation coefficients such as polymorphs. Further details are given in the original publication. Reproduced with permission [129].

tions allowed for the discrimination of different nucleation mechanisms. In a second work from the same research group, X-ray diffraction micro-tomography was used to map the C-S-H precipitation in absence and presence of superplasticizer [134]. The observed spatial correlation between C-S-H and unhydrated cement particle surfaces indicated that, in the absence of superplasticizers, C-S-H forms by a process of heterogeneous nucleation on the surface of the dissolving cement particles. In addition, when the superplasticizer is added to the system, the lack of significant spatial correlation between C-S-H and the surface of unhydrated particles indicated that C-S-H precipitates randomly throughout the available space in the paste.

2.5.3 X-ray ptychographic forward coherent diffraction nano-tomography

X-ray ptychographic forward coherent diffraction nano-tomography is a non-invasive imaging technique based on the coherent properties of synchrotron radiation and it allows recording the 3D mapping of the electron density in the studied sample. This variant of CDI does not require crystalline ordering and its quantitativeness combined with a resolution close to 100 nm (for a field of view of about 60–100 μm) makes it very appropriate for studying hierarchical microstructures in complex materials including cement pastes. Instead of fully illuminating a small sample with a featureless plane wave, a small X-ray beam size is used to raster scan an extended sample. The deconvolution of the effects due to the scattering of the sample from those due to the structured illumination can be ensured if the sample is scanned in sufficiently fine, overlapping steps. The data from multiple known scan positions are inverted to yield a 2D image, whose resolution is limited by the maximum scattering angle where there is a measurable signal and by the positioning accuracy of the sample. When combined with a rotation stage/strategy, a 3D tomographic image can be obtained. Furthermore, the high accuracy in measuring the electron density allows accurate segmentation of the data. Details about the data collection procedure were already reported [3]. Ptychography became practical only by combining it with iterative phase retrieval algorithms which reduced the sampling requirements drastically.

Ptychographic nano-tomography was applied to image a resin-impregnated hardened cement paste [135]. The 3D maps of six constituent phases were successfully segmented based on the electron density values. In a recent work [136], this technique was used to characterize the microstructure of C-S-H formed by hydrating C_3S. The 3D spatial resolution of the phase contrast images was close to 130 nm, whereas the resolution of the absorption images was poorer $\approx$ 250 nm. It was observed that the C-S-H density can depend on the particles' states of hydration. For fully hydrated particles, the estimated density of the outer-product C-S-H was larger than that of the inner-product C-S-H, whereas for the partially hydrated particles, the densities of the apparent outer and the inner products were very similar. This technique has also been used to determine the 3D mass density and attenuation coefficient distributions of pastes derived from ye'elimite-containing eco-cements. Ettringite and aluminum hydroxide gel volume distributions were mapped out in the segmented tomograms and the densities of the gels were determined [137].

Finally, soft X-ray ptychographic microscopy (2D imaging) has also been employed for studying C-S-H gel [138]. The use of low energy photons, 750 eV, enforces the use of high vacuum during imagining that may cause artefacts. In any case, the outer and inner C-S-H morphology was measured and distinguished, and the fibrillar microstructures reported.

2.5.4 X-ray Bragg coherent diffraction nano-tomography

Bragg coherent diffraction imaging (BCDI) is another variant of CDI technique. This non-invasive imaging tool requires individual crystals and its experimental setup has been previously described [3]. BCDI yields three-dimensional images of the crystals through inversion of the diffraction data by computational methods and it is also highly sensitive to crystal defects and strain fields inside crystals, seen as phase evolution. In a recent study [139], the early hydration of microcrystals of calcium monoaluminate was investigated by following the 3D Bragg diffraction electron density and strain evolution. The first Bragg density change during the hydration process was due to a loss of Bragg density and was seen as a removal of density but not phase. This work provided additional evidence supporting the through-solution reaction mechanism of calcium monoaluminate.

2.6 Conclusions and outlook

Powder diffraction coupled to RQPA is being increasingly used for characterizing cementitious materials. Initially, it was mainly devoted to measure anhydrous Portland clinkers and cements but it is nowadays being used for analyzing many types of hydrated products. Furthermore, RQPA is being increasingly used for studying the hydration reactions of blended cements. The number of by-products that can be blended with Portland cements is large and expanding. In addition, research on new types of eco-cements is skyrocketing. This is not covered in the present chapter but the use of RQPA in ye'elimite-containing eco-cements as wells as in alkaline-activated eco-cements is large and increasing.

As complexity of the studied cementitious materials growths (some samples contain more than nine crystalline phases) the powder averaging for all phases becomes critical. A possible way out is to use highly penetrating X-rays, compared to Cu or Co radiations. As the access to synchrotron radiation is limited, we therefore expect Mo-radiation diffractometers to be increasingly used for the RQPA of complex samples. However, this cannot be achieved at the cost of resolution, so diffractometers equipped with an incident beam Mo-Kα monochromator are needed [13]. Mo-Kα incident monochromator(s) coupled with fast detectors will likely make a breakthrough in the routine RQPA of cement pastes.

Most cement binders are based on nanocrystalline/amorphous gels. We all know that the characterization of nanocrystalline/amorphous materials (in presence of crystalline phases) is always complicated due to the lack of long range order and periodicity, and also because of their large chemical variability. Advances in synchrotron tools will tackle these issues starting with a better characterization of the local atomic order all the way up to the determination of chemical compositions and density values of these gels with very high spatial resolution. Here, tomographic pair-distribution-function methodology [140], still not applied to cements, will play a role.

The quantitative study of cement paste microstructures is very important to understand and predict their mechanical behavior as well as their chemical durability. In this area, synchrotron diffraction tools are very well suited as they do not require special sample preparation that may alter the microstructures. Here, the challenge is to continue developing spatially-resolved techniques, and sample preparation procedures to enter into a resolution range lower than 100 nm. High resolution, well below 100 nm, without a tradeoff of field-of-view, is important to properly characterize key pore microstructure details including connectivity and tortuosity. It will also be important to quantify the changes in the microstructure provoked by the uses of SCM, which can vary quite a lot (fly ashes, slags, partially burned clays, etc.). For this type of application, hard X-ray ptychographic forward coherent diffraction nano-tomography will become a suitable technique.

Acknowledgment: This work has been supported by Spanish MINECO through BIA 2014-57658-C2-1-R and BIA2014-57658-C2-2-R, which is co-funded by FEDER, research grants. Funding from the Junta de Andalucía, P11-FQM-07517, is also acknowledged.

References

[1] Aranda MAG, De la Torre AG, León-Reina L. Rietveld quantitative phase analysis of OPC clinkers, cements and hydration products. Rev Miner Geochem. 2012; 74: 169–209.

[2] Aranda MAG, De la Torre AG. Sulfoaluminate cements, In: Pacheco-Torgal F, Jalali S, Labrincha J. Eco-efficient concrete. Cambridge: Woodhead Publishing; 2013.

[3] Aranda MAG. Recent studies of cements and concretes by synchrotron radiation crystallographic and cognate methods. Crystallogr Rev. 2016; 22: 150–196.

[4] Aranda MAG, De la Torre AG, León-Reina L. Powder diffraction characterization of cements. In: Gilmore, C, Kaduk J, Schenk, H, International Tables for Crystallography, ed. International Union for Crystallography. 2017; Vol H. In press.

[5] Rietveld HM. Personal communication; 2010.

[6] Rietveld HM. Line profiles of neutron powder-diffraction peaks for structure refinement. Acta Crystallogr. 1967; 22: 151–152.

[7] Rietveld HM. A profile refinement method for nuclear and magnetic structures. J Appl Crystallogr. 1969; 2: 65–71.
[8] Hill RJ, Howard CJ. Quantitative phase analysis from neutron powder diffraction data using the Rietveld method. J Appl Crystallogr. 1987; 20: 467–474.
[9] Bish DL, Howard SA. Quantitative phase analysis using the Rietveld method. J Appl Crystallogr. 1988; 21: 86–91.
[10] Taylor JC, Aldridge LP. Full-profile Rietveld quantitative XRD analysis of Portland cement: Standard XRD profiles for the major phase tricalcium silicate (C_3S: $3CaO.SiO_2$). Powder Diffr. 1993; 8: 138–144.
[11] McCusker LB, Von Dreele RB, Cox DE, Louer D, Scardi P. Rietveld refinement guidelines. J Appl Crystallogr. 1999; 32: 36–50.
[12] Cuesta A, Álvarez-Pinazo G, García-Maté M, et al. Rietveld quantitative phase analysis with molybdenum radiation. Powder Diffr. 2015; 30: 25–35.
[13] León-Reina L, García-Maté M, Álvarez-Pinazo G, et al. Accuracy in Rietveld quantitative phase analysis: a comparative study of strictly monochromatic Mo and Cu radiations. J Appl Crystallogr. 2016; 49: 722–735.
[14] Taylor HFW. Cement Chemistry. London: Telford; 1997.
[15] Le Saoût G, Kocaba V, Scrivener K. Application of the Rietveld method to the analysis of anhydrous cements. Cem Concr Res. 2011; 41: 133–148.
[16] Scarlett NVY, Madsen IC. Quantification of phases with partial or no known crystal structure. Powder Diffr. 2006; 21: 278–284.
[17] Dollase WA. Correction of intensities for preferred orientation in powder diffractometry: application of the March model. J Appl Crystallogr. 1986; 19: 267–272.
[18] Petch HE. The hydrogen positions in portlandite, $Ca(OH)_2$, as indicated by the electron distribution. Acta Crystallogr. 1961; 14: 950–957.
[19] Saalfeld H, Wedde M. Refinement of the crystal structure of gibbsite, $Al(OH)_3$. Z Kristallogr. 1974; 139: 129–135.
[20] Merlino S, Bonaccorsi E, Armbruster T. The real structure of clinotobermorite and tobermorite 9 Å: OD character, polytypes, and structural relationships. Eur J Mineral. 2000; 12: 411–429.
[21] Bonaccorsi E, Merlino S, Kampf AR. The crystal structure of tobermorite 14 A (plombierite), a C-S-H phase. J Am Ceram Soc. 2005; 88: 505–512.
[22] Merlino S, Bonaccorsi E, Armbruster T. The real structure of Tobermorite 11 Å: normal and anomalous forms, OD character and polytypic modifications. Eur J Mineral. 2001; 13: 577–590.
[23] Bonaccorsi E, Merlino S, Taylor HFW. The crystal structure of jennite, $Ca_9Si_6O_{18}(OH)_6 \cdot 8(H_2O)$. Cem Concr Res. 2004; 34: 1481–1488.
[24] Yano T, Urabe K, Ikawa H, Teraushi T, Ishizawa N, Udagawa S. Structure of alpha-dicalcium silicate hydrate. Acta Crystallogr C. 1993; 49: 1555–1559.
[25] Lager GA, Armbruster T, Faber, J-jr. Neutron and X-ray diffraction study of hydrogarnet $Ca_3Al_2(O_4H_4)_3$. Am Mineral. 1987; 72: 756–765.
[26] Cuesta A, Santacruz I, Sanfelix SG, Fauth F, Aranda MAG, De La Torre AG. Hydration of C_4AF in the presence of other phases: a synchrotron X-ray powder diffraction study. Const Build Mater. 2015; 101: 818–827.
[27] Ferro O, Galli E, Papp G, Quartieri S, Szakall S, Vezzalini G. A new occurrence of katoite and re-examination of the hydrogrossular group. Eur J Mineral. 2003; 15: 419–426.
[28] Sacerdoti M, Passaglia E. The crystal structure of katoite and implications within the hydrogrossular group of minerals. Bull Mineral. 1985; 108: 1–8.
[29] Dilnesa BZ, Lothenbach B, Renaudin G, Wichser A, Kulik, D. Synthesis and characterization of hydrogarnet $Ca_3(Al_xFe_{1-x})_2(SiO_4)_y(OH)_{4(3-y)}$. Cem Concr Res. 2014; 59: 96–111.

[30] Guirado F, Galis S, Chinchón S, Rius J. Crystal structure solution of hydrated high-alumina cement from X-ray powder diffraction data. Angew Chem Int Ed. 1998; 37: 72–75.

[31] Goetz-Neunhoeffer F, Neubauer J. Refined ettringite ($Ca_6Al_2(SO_4)_3(OH)_{12} \cdot 26H_2O$) structure for quantitative X-ray diffraction analysis. Powder Diffr. 2006; 21: 4–11.

[32] Moore AE, Taylor HFW. Crystal structure of ettringite. Acta Crystallogr B. 1970; 26: 386–393.

[33] Champenois J-B, Mesbah A, Cau-dit-Coumes C, et al. Crystal structures of Boro-AFm and Boro-AFt phases. Cem Concr Res. 2012; 42: 1362–1370.

[34] Scholtzová E, Kucková L, Kožíšek J, Pálková H, Tunega D. Experimental and computational study of thaumasite structure. Cem Concr Res. 2014; 59: 66–72.

[35] Allmann R. Refinement of the hybrid layer structure $[Ca_2Al(OH)_6] \cdot 0.5(SO_4) \cdot 3H_2O$. Neues Jb Miner Monat. 1977: 136–144.

[36] Renaudin G, Kubel F, Rivera JP, François M. Structural phase transition and high temperature phase structure of Friedels salt, $3(CaO)(Al_2O_3)(CaCl_2) \cdot 10(H_2O)$. Cem Concr Res. 1999; 29: 1937–1942.

[37] Rousselot I, Taviot-Gueho C, Leroux F, Leone P, Palvadeau P, Besse JP. Insights on the structural chemistry of hydrocalumite and hydrotalcite-like materials: investigation of the series $Ca_2\ M^{(3+)}\ (OH)_6Cl\ 2(H_2O)$ ($M^{(3+)}$: Al, Ga, Fe, and Sc) by x-ray powder diffraction. J Solid State Chem. 2002; 167: 137–144.

[38] Mesbah A, François M, Cau-dit-Coumes C, Frizon F, Filinchuk Y. Crystal structure of Kuzel'salt $3CaO \cdot Al_2O_3 \cdot 1/2CaSO_4 \cdot 1/2CaCl_2 \cdot 11H_2O$ determined by synchrotron powder diffraction. Cem Concr Res. 2011; 41: 504–509.

[39] Sacerdoti M, Passaglia E. Hydrocalumite from Latium, Italy: its crystal structure and relationship with related systhetic phases. Neues Jahrbuch fuer Mineralogie. 1988: 462–475.

[40] Mesbah A, Rapin J-P, François M, et al. Crystal structures and phase transition of cementitious bi-anionic AFm-(Cl^-, CO_3^{2-}) compounds. J Am Ceram Soc. 2011; 94: 261–268.

[41] François M, Renaudin G, Evrad O. A cementitious compound with composition $3CaO \cdot Al_2O_3 \cdot CaCO_3 \cdot 11H_2O$. Acta Crystallogr C. 1998; 54: 1214–1217.

[42] Runcevski T, Dinnebier RE, Magdysyuk OV, Pöllmann H. Crystal structures of calcium hemicarboaluminate and carbonated calcium hemicarboaluminate from synchrotron powder diffraction data. Acta Crystallogr B. 2012; 68: 493–500.

[43] Dilnesa BZ, Lothenbach B, Le Saoût G, et al. Iron in carbonate containing AFm phases. Cem Concr Res. 2011; 41: 311–323.

[44] Rinaldi R, Sacerdoti M, Passaglia E. Straetlingite: crystal structure, chemistry and a reexamination of its polytype vertumnite. Eur J Mineral. 1990; 2: 841–849.

[45] Santacruz I, De la Torre AG, Álvarez-Pinazo G, et al. Structure of stratlingite and effect of hydration methodology on microstructure. Adv Cem Res. 2016; 28: 13–22.

[46] Renaudin G, François M. The lamellar Double-hydroxide (LDH) compound with composition $3CaO \cdot Al_2O_3 \cdot Ca(NO_3)_2 \cdot 10H_2O$. Acta Crystallogr C. 1999; 55: 835–838.

[47] Rapin JP, Noor NM, François M. The double layered hydroxide $3CaO \cdot Al_2O_3 \cdot CaBr_2 \cdot 10H_2O$. Acta Cryst C. 1999: 55 IUC9900092.

[48] Rapin JP, Walcarius A, Lefevre G, François M. A double-layered hydroxide, $3CaO \cdot Al_2O_3 \cdot CaI_2 \cdot 10H_2O$. Acta Crystallogr C. 1999; 55: 1957–1959.

[49] Rapin JP, François M. The double-layered hydroxide $3CaO \cdot Al_2O_3 \cdot 0.5CaBr_2 \cdot 0.5CaCl_2 \cdot 10H_2O$. Acta Crystallogr C. 2001; 57: 137–138.

[50] Matschei T, Glasser FP. Temperature dependence, 0 to 40 °C, of the mineralogy of Portland cement paste in the presence of calcium carbonate. Cem Concr Res. 2010; 40: 763–777.

[51] Richardson IG. The calcium silicate hydrates. Cem Concr Res. 2008; 38: 137–158.

[52] Hartman MR, Berliner. Investigation of the structure of ettringite by time-of-flight neutron powder diffraction techniques. Cem Concr Res. 2006; 36: 364–370.

[53] Baquerizo LG, Matschei T, Scrivener KL, Saeidpou M, Wadsö L. Hydration states of AFm cement phases. Cem Concr Res. 2015; 73: 143–157.
[54] Balonis M, Glasser, FP. The density of cement phases. Cem Concr Res. 2009; 39: 733–739.
[55] Bellmann F, Damidot D, Moser B, Skibsted J. Improved evidence for the existence of an intermediate phase during hydration of tricalcium silicate. Cem Concr Res. 2010; 40: 875–884.
[56] Bergold ST, Goetz-Neunhoeffer F, Neubauer J. Quantitative analysis of C-S-H in hydrating alite pastes by in-situ XRD. Cem Concr Res. 2013; 53: 119–126.
[57] Valentini L. RieCalc: quantitative phase analysis of hydrating cement pastes. J Appl Crystallogr. 2013; 43: 1899–1902.
[58] Valentini L, Dalconi MC, Favero M, Artioli G, Ferrari G. In-Situ XRD Measurement and Quantitative Analysis of Hydrating Cement: Implications for Sulfate Incorporation in C-S-H. J Am Ceram Soc. 2015; 98: 1259–1264.
[59] Jansen D, Bergold ST, Goetz-Neunhoeffer F, Neubauer J. The hydration of alite: a time-resolved quantitative X-ray diffraction approach using the G-factor method compared with heat release. J Appl Crystallogr. 2011; 44: 895–901.
[60] Bergold ST, Goetz-Neunhoeffer F, Neubauer J. Mechanically activated alite: new insights into alite hydration. Cem Concr Res. 2015; 76: 202–211.
[61] Bergold ST, Goetz-Neunhoeffer F, Neubauer J. Influence of the reactivity of the amorphous part of mechanically activated alite on its hydration kinetics. Cem Concr Res. 2016; 88: 73–81.
[62] Christensen AN, Jensen TR, Scarlett NVY, Madsen IC, Hanson JC. Hydrolysis of pure and sodium substituted calcium aluminates and cement clinker components investigated by in situ synchrotron X-ray powder diffraction. J Am Ceram Soc. 2004; 87: 1488–1493.
[63] Meller N, Hall C, Jupe, AC, et al. The paste hydration of brownmillerite with and without gypsum: a time resolved synchrotron diffraction study at 30, 70, 100 and 150 °C. J Mater Chem. 2004; 14: 428–435.
[64] Merlini M, Artioli G, Cerulli T, Cella F, Bravo A. Tricalcium aluminate hydration in additivated systems. A crystallographic study by SR-XRPD. Cem Concr Res. 2008; 38: 477–486.
[65] Quennoz A, Scrivener KL. Hydration of C_3A–gypsum systems. Cem Concr Res. 2012; 42: 1032–1041.
[66] Ectors D, Neubauer J, Goetz-Neunhoeffer F. The hydration of synthetic brownmillerite in presence of low Ca-sulfate content and calcite monitored by quantitative in-situ-XRD and heat flow calorimetry. Cem Concr Res. 2013; 54: 61–68.
[67] Hueller F, Ectors D, Neubauer J, Goetz-Neunhoeffer F. Influence of crystallinity and surface area on the hydration kinetics of CA_2. Cem Concr Res. 2016; 89: 136–144.
[68] Cuesta A, Alvarez-Pinazo G, Santacruz I, et al. Hydration mechanisms of two polymorphs of synthetic ye'elimite. Cem Concr Res. 2014; 63: 127–136.
[69] Mesbah A, Cau-dit-Coumes C, Renaudin G, Frizon F, Leroux F. Uptake of chloride and carbonate ions by calcium monosulfoaluminate hydrate. Cem Concr Res. 2012; 42: 1157–1165.
[70] Scrivener KL, Juilland P, Monteiro PJM. Advances in understanding hydration of Portland cement. Cem Concr Res. 2015; 78: 38–56.
[71] Soin AV, Catalan LJJ, Kinrade SD. A combined QXRD/TG method to quantify the phase composition of hydrated Portland cement. Cem Concr Res. 2013; 48: 17–24.
[72] Zhang J, Scherer GW. Comparison of methods for arresting hydration of cement. Cem Concr Res. 2011; 41: 1024–1036.
[73] Galan I, Beltagui H, Garcia-Mate M, Glasser FP, Imbabi MS. Impact of drying on pore structures in ettringite-rich cements. Cem Concr Res. 2016; 84: 85–94.

[74] Scrivener KL, Fullmann T, Gallucci E, Walenta G, Bermejo E. Quantitative study of Portland cement hydration by X-ray diffraction/Rietveld analysis and independent methods. Cem Concr Res. 2004; 34: 1541–1547.

[75] Mitchell LD, Margeson JC, Whitfield PS. Quantitative Rietveld analysis of hydrated cementitious systems. Powder Diffr. 2006; 21: 111–113.

[76] Dalconi MC, Favero M, Artioli G. In-situ XRPD of hydrating cement with lab instrument: reflection vs. transmission measurements. Z Kristallogr Proc. 2011; 1: 155–161.

[77] Jansen D, Goetz-Neunhoeffer F, Lothenbach B, Neubauer J. The early hydration of Ordinary Portland Cement (OPC): An approach comparing measured heat flow with calculated heat flow from QXRD. Cem Concr Res. 2012; 42: 134–138.

[78] Le Saoût G, Fullmann T, Kocaba V, Scrivener K. Quantitative study of cementitious materials by X-ray diffraction/Rietveld analysis using an external standard. Proceedings of 12th International Congress on the Chemistry of Cement; 2007.

[79] Hesse C, Degenkolb M, Gaeberlein P, Goetz-Neunhoeffer F, Neubauer J, Schwarz V. Investigations into the influence of temperature and w/c ratio on the early hydration of white cement. Cem Int. 2008; 6: 68–78.

[80] Hesse C, Goetz-Neunhoeffer F, Neubauer J, Braeu M, Gaeberlein P. Quantitative in situ X-ray diffraction analysis of early hydration of Portland cement at defined temperatures. Powder Diffr. 2009; 24: 112–115.

[81] Korpa A, Kowald T, Trettin R. Phase development in normal and ultra high performance cementitious systems by quantitative X-ray analysis and thermoanalytical methods. Cem Concr Res. 2009; 39: 69–76.

[82] Jansen D, Goetz-Neunhoeffer F, Stabler C, Neubauer J. A remastered external standard method applied to the quantification of early OPC hydration. Cem Concr Res. 2011; 41: 602–608.

[83] Choudhary HK, Anupama AV, Kumar R, et al. Observation of phase transformations in cement during hydration. Constr Build Mater. 2015; 101: 122–129.

[84] Weyer HJ, Muller I, Scmitt B, Bosbach D, Putnis A. Time-resolved monitoring of cement hydration: Influence of cellulose ethers on hydration kinetics. Nucl Instr Meth Phys Res B. 2005; 238: 102–106.

[85] Merlini M, Artioli G, Meneghini C, Cerulli T, Bravo A, Cella F. The early hydration and the set of Portland cements: In situ X-ray powder diffraction studies. Powder Diffr. 2007; 22: 201–208.

[86] Schlegel MC, Sarfraz A, Muller A, Panne U, Emmerling F. First Seconds in a Building's Life-In Situ Synchrotron X-Ray Diffraction Study of Cement Hydration on the Millisecond Timescale. Angew Chem Int Ed. 2012; 51: 4993–4996.

[87] Cuberos AJM, De la Torre AG, Martin-Sedeño MC, et al. Phase development in conventional and active belite cement pastes by Rietveld analysis and chemical constraints. Cem Concr Res. 2009; 39: 833–842.

[88] Marinoni N, Pavese A, Voltolini M, Merlini M. Long-term leaching test in concretes: An X-ray powder diffraction study. Cem Concr Comp. 2008; 30: 700–705.

[89] Cheung J, Jeknavorian A, Roberts L, Silva D. Impact of admixtures on the hydration kinetics of Portland cement. Cem Concr Res. 2011; 41: 1289–1309.

[90] Plank J, Sakai E, Miao CW, Yud C, Hong JX. Chemical admixtures — Chemistry, applications and their impact on concrete microstructure and durability. Cem Concr Res. 2015; 78: 81–99.

[91] Zhang L, Catalan LJJ, Balec RJ, Larsen AC, Esmaeili HH, Kinrade SD. Effects of saccharide set retarders on the hydration of ordinary Portland cement and pure tricalcium silicate. J Am Ceram Soc. 2010; 93: 279–287.

[92] Jansen D, Neubauer J, Goetz-Neunhoeffer F, Haerzschel R, Hergeth WD. Change in reaction kinetics of a Portland cement caused by a superplasticizer – Calculation of heat flow curves from XRD data. Cem Concr Res. 2012; 42: 327–332.
[93] Jansen D, Goetz-Neunhoeffer F, Neubauer J, Haerzschel R, Hergeth WD. Effect of polymers on cement hydration: A case study using substituted PDADMA. Cem Concr Comp. 2013; 35: 71–77.
[94] Snellings R, Mertens G, Elsen J. Supplementary cementitious materials. Rev Mineral Geochem. 2012; 74: 211–278.
[95] Juenger MCG, Siddique R. Recent advances in understanding the role of supplementary cementitious materials in concrete. Cem Concr Res. 2015; 78: 71–80.
[96] Kocaba V, Gallucci E, Scrivener KL. Methods for determination of degree of reaction of slag in blended cement pastes. Cem Concr Res. 2012; 42: 511–525.
[97] Scrivener KL, Lothenbach B, De Belie N, Gruyaert E, Skibsted J, Snellings R, Vollpracht A. TC 238-SCM: hydration and microstructure of concrete with SCMs. Mat and Struct. 2015; 48: 835–862.
[98] Hoshino S, Yamada K, Hirao H. XRD/Rietveld Analysis of the Hydration and Strength Development of Slag and Limestone Blended Cement. J Adv Concr Tech. 2006; 4: 357–367.
[99] Termkhajornkit P, Nawa T, Kurumisawa, K. Effect of water curing conditions on the hydration degree and compressive strengths of fly ash–cement paste. Cem Concr Comp. 2006; 28: 781–789.
[100] Snellings R, Salze A, Scrivener KL. Use of X-ray diffraction to quantify amorphous supplementary cementitious materials in anhydrous and hydrated blended cements. Cem Concr Res. 2014; 64: 89–98.
[101] Jozic D, Jelica J. The effect of fly ash on cement hydration in aqueous suspensions. Ceramics-Silikáty. 2006; 50: 98–105.
[102] Narmluk M, Nawa T. Effect of fly ash on the kinetics of Portland cement hydration at different curing temperatures. Cem Concr Res. 2011; 41: 579–589.
[103] Deschner F, Winnefeld F, Lothenbach B, et al. Hydration of Portland cement with high replacement by siliceous fly ash. Cem Concr Res. 2012; 42: 1389–1400.
[104] De Weerdt K, Haha MB, Le Saoût G, Kjellsen KO, Justnes H, Lothenbach B. Hydration mechanisms of ternary Portland cements containing limestone powder and fly ash. Cem Concr Res. 2011; 41: 279–291.
[105] Wei Y, Yao W, Xing X, Wu M. Quantitative evaluation of hydrated cement modified by silica fume using QXRD, ^{27}Al MAS NMR, TG–DSC and selective dissolution techniques. Constr Build Mater. 2012; 36: 925–932.
[106] Rong ZD, Sun W, Xiao HJ, Wang W. Effect of silica fume and fly ash on hydration and microstructure evolution of cement based composites at low water–binder ratios. Constr Build Mater. 2014; 51: 446–450.
[107] Knop Y, Peled A. Setting behavior of blended cement with limestone: influence of particle size and content. Mater Struct. 2016; 49: 439–452.
[108] Taylor-Lange SC, Lamon EL, Riding KA, Juenger MCG. Calcined kaolinite–bentonite clay blends as supplementary cementitious materials. Appl Clay Sci. 2015; 108: 84–93.
[109] Snellings R, Mertens G, Cizer O, Elsen J. Early age hydration and pozzolanic reaction in natural zeolite blended cements: Reaction kinetics and products by in situ synchrotron X-ray powder diffraction. Cem Concr Res. 2010; 40: 1704–1713.
[110] Snellings R, Mertens G, Adriaens R, Elsen J. In situ synchrotron X-ray powder diffraction study of the early age hydration of cements blended with zeolitite and quartzite fines and water-reducing agent. Appl Clay Sci. 2013; 72: 124–131.

[111] Meral C, Benmore CJ, Monteiro PJM. The study of disorder and nanocrystallinity in C-S-H, supplementary cementitious materials and geopolymers using pair distribution function analysis. Cem Concr Res. 2011: 41: 696–710.

[112] White CE, Daemen LL, Hartl M, Page K. Intrinsic differences in atomic ordering of calcium (alumino)silicate hydrates in conventional and alkali-activated cements. Cem Concr Res. 2015; 67: 66–73.

[113] Skinner LB, Chae SR, Benmore CJ, Wenk HR, Monteiro PJM. Nanostructure of Calcium Silicate Hydrates in Cements. Phys Rev Lett. 2010; 104: 195502.

[114] Soyer-Uzun S, Chae SR, Benmore CJ, Wenk HR, Monteiro PJM. Compositional evolution of calcium silicate hydrate (C-S-H) structures by total X-Ray scattering. J Am Ceram Soc. 2012; 95: 793–798.

[115] Suzuki H, Bae S, Kanematsu M. Nanostructural Deformation Analysis of Calcium Silicate Hydrate in Portland Cement Paste by Atomic Pair Distribution Function. Advances in Materials Science and Engineering. 2016: 8936084.

[116] Cuesta A, Ichikawa RU, Londono-Zuluaga D, et al. Aluminium Hydroxide Gel Characterization within a Calcium Aluminate Cement Paste by Combined Pair Distribution Function and Rietveld Analyses. Cem Concr Res. 2017; 96: 1–12.

[117] Benmore CJ, Monteiro, PJM. The structure of alkali silicate gel by total scattering methods. Cem Concr Res. 2010; 40: 892–897.

[118] White CE, Page K, Henson NJ, Provis JL. In situ synchrotron X-ray pair distribution function analysis of the early stages of gel formation in metakaolin-based geopolymers. Appl Clay Sci. 2013; 73: 17–25.

[119] White CE, Provis JL, Bloomer B, Henson NJ, Page K. In situ X-ray pair distribution function analysis of geopolymer gel nanostructure formation kinetics. Phys Chem Chem Phys. 2013; 15: 8573–8582.

[120] Voltolini M, Dalconi MC, Artioli G, Tucoulou R. Understanding cement hydration at the microscale: new opportunities from 'pencil-beam' synchrotron X-ray diffraction tomography. J Appl Crystallogr. 2013; 46: 142–152.

[121] Schlegel MC, Muller U, Panne U, Emmerling F. Deciphering the sulfate attack of cementitious materials by high-resolution micro-X-ray diffraction. Anal Chem. 2011; 83: 3744–3749.

[122] Stroh J, Meng B, Emmerling F. Monitoring of sulphate attack on hardened cement paste studied by synchrotron XRD. Solid State Sci. 2015; 48: 278–285.

[123] Schlegel MC, Muller U, Malaga K, Panne U, Emmerling F. Spatially resolved investigation of complex multi-phase systems using μXRF, SEM-EDX and high resolution SyXRD. Cem Concr Comp. 2013; 37: 241–245.

[124] Stroh J, Schlegel MC, Irassar EF, Meng B, Emmerling F. Applying high resolution SyXRD analysis on sulfate attacked concrete field samples. Cem Concr Res. 2014; 66: 19–26.

[125] Schlegel MC, Stroh J, Malaga K, Meng B, Panne U, Emmerling F. Pathway of a damaging mechanism – Analyzing chloride attack by synchrotron based X-ray diffraction. Cem Concr Comp. Solid State Sci. 2015; 44: 45–54.

[126] Stroh J, Meng B, Emmerling F. Deterioration of hardened cement paste under combined sulphate-chloride attack investigated by synchrotron XRD. Solid State Sci. 2016; 56: 29–44.

[127] Jackson MD, Landis EN, Brune PF, et al. Mechanical resilience and cementitious processes in imperial roman architectural mortar. PNAS. 2014; 111: 18484–18489.

[128] Bleuet P, Welcomme E, Dooryhee E, Susini J, Hodeau JL, Walter P. Probing the structure of heterogeneous diluted materials by diffraction tomography. Nat Mater. 2008; 7: 468–472.

[129] Birkbak ME, Leemreize H, Frolich S, Stock SR, Birkedal H. Diffraction scattering computed tomography: a window into the structures of complex nanomaterials. Nanoscale. 2015; 7: 18402–18410.

[130] Cersoy S, Leynaud O, Alvarez-Murga M, et al. Laboratory implementation of X-ray diffraction/scattering computed tomography. J Appl Crystallogr. 2015; 48: 159–165.
[131] Valentini L, Dalconi MC, Parisatto M, Cruciani G, Artioli G. Towards three-dimensional quantitative reconstruction of cement microstructure by X-ray diffraction microtomography. J Appl Crystallogr. 2011; 44: 272–280.
[132] Artioli G, Dalconi MC, Parisatto M, Valentini L, Voltolini M, Ferrari G. 3D imaging of complex materials: the case of cements. Int J Mater Res. 2012; 103: 145–150.
[133] Artioli G, Valentini L, Dalconi MC, et al. Imaging of nano-seeded nucleation in cement pastes by X-ray diffraction tomography. Int J Mater Res. 2014; 105: 628–631.
[134] Artioli G, Valentini L, Voltolini M, Dalconi MC, Ferrari G, Russo V. Direct imaging of nucleation mechanisms by synchrotron diffraction micro-tomography: superplasticizer-induced change of C-S-H nucleation in cement. Cryst Growth Des. 2015; 15: 20–23.
[135] Trtik P, Diaz A, Guizar-Sicairos M, Menzel A, Bunk O. Density mapping of hardened cement paste using ptychographic X-ray computed tomography. Cem Concr Comp. 2013; 36: 71–77.
[136] Da Silva JC, Trtik P, Diaz A, et al. Mass density and water content of saturated never-dried calcium silicate hydrates. Langmuir. 2015; 31: 3779–3783.
[137] Cuesta A, De la Torre AG, Santacruz I, et al. Chemistry and Mass Density of Aluminum Hydroxide Gel in Eco-Cements by Ptychographic X-ray Computed Tomography. J Phys Chem C. 2017; 121: 3044–3054.
[138] Bae S, Taylor R, Shapiro D, et al. Soft X-ray ptychographic imaging and morphological quantification of calcium silicate hydrates (C-S-H). J Am Ceram Soc. 2015; 98: 4090–4095.
[139] Liu X, Aranda MAG, Chen B, Wang P, Harder R, Robinson I. In situ Bragg coherent diffraction imaging study of a cement phase microcrystal during hydration. Cryst Growth Des. 2015; 15: 3087–3091.
[140] Jacques SDM, Di Michiel M, Kimber SAJ, et al. Pair distribution function computed tomography. Nature Comm. 2013; 4: 2536.

Bastian Raab* and Herbert Pöllmann

3 Synthesis of highly reactive pure cement phases

Abstract: Pure phases of ordinary Portland cements (OPC), belite cements, and calcium aluminate cements (CAC) are often synthesized to understand the formation of the phases during the sintering process and to investigate their hydration behavior in simplified conditions. Due to the need to reduce CO_2 emissions in cement production, the usage of by-products from other industries with a better homogenization of the chemical components is often discussed. In addition to the traditional solid state reaction, other low temperature synthesis methods such as self-propagating combustion synthesis, sol gel synthesis, and polymeric precursor processes (Pechini process) were used to understand phase formation by using better homogenized raw materials. By using these synthesis routes, cement phases with a much higher specific surface area and reactivity can also be achieved. The hydration process, which starts on the surface area of the particles, can be much better characterized using these kinds of synthesis methods. The target of this chapter is therefore to describe different synthesis methods for pure cement phases and the phase formation of the phases obtained by different synthesis routes.

Keywords: synthesis, cement phase, Portland cement, calcium aluminate cement, nano, sol-gel, polymer precursor process, GNP

3.1 Introduction

As a consequence of the high sintering temperature of around 1450 °C required for OPC production and the use of limestone, enormous amounts of CO_2 (0.63 t_{CO_2}/t_{Cement} in Germany) are produced [1]. To reduce CO_2 emissions, process parameters were optimized, secondary cementitious materials (SCMs) were added, and the usage of alternative raw materials is discussed. At the moment, intensive research and development activity is also taking place to find new binder materials. Pöllmann et al. give a summary of the different cement phases that occur in industrial cements [2–4]. In the last years, production of belite cements, belite sulfo aluminate cements, belite ferrate cements, and belite sulfo aluminate ferrate cements with lower lime contents and lower production temperatures between 1100 and 1350 °C took place on an industrial scale [5]. Belite sulfo aluminate cements [6, 7] or belite sulfo aluminate ferrate

***Corresponding author: Bastian Raab,** Technische Hochschule Nürnberg, Georg Simon Ohm, Fakultät Werkstofftechnik, Nürnberg, Germany, bastian.raab@th-nuernberg.de
Herbert Pöllmann, Department of Mineralogy and Geochemistry, Martin Luther University Halle-Wittenberg, Halle (Saale), Germany

DOI 10.1515/9783110473728-004

cements [5] contain the sulfate-containing phase ye'elimite $4CaO \cdot 3Al_2O_3 \cdot SO_3$ (C_4A_3s). In C_2S rich cements, higher reactive α-, α'_L-, or β C_2S modifications can be stabilized by NaF [8], B_2O_3 [9], or Na_2O [10]. Reactivity and early strength (till 28 days) are lower in belite-rich cements than in OPCs despite a stabilization of α'_L-C_2S and β-C_2S due to lower or no C_3S content [11, 12]. Martín-Sedeño et al. showed a lower hydration speed of C_2S despite the stabilization of the α-C_2S modification [12]. In further studies, old concrete was dehydrated and ground in order to use the dehydrated product again as a binder material in construction applications. High strength during hydration of these materials can be obtained by heating concrete at 800 °C [13, 14]. Some other research projects show the production of a belite rich binder by the dehydration of hydrated cement or concrete [13–17]. A decomposition of portlandite $Ca(OH)_2$ between 500 and 600 °C [17] and a C_2S formation at around 600 °C was measured [16]. Shui et al. showed a brownmillerite formation at a sintering temperature of about 800 °C [14]. The dehydration of cement phases was described in detail by Splittgerber & Mueller [18]. Shui et al. mentioned the higher water demand of binders produced by the dehydration of hydrated cement and this is the main reason for the lower strength [13, 14]. In a patent of Stemmermann et al., a binder material is also produced by a hydrothermal process at lower production temperatures [19]. In other projects, ceramic waste materials are used for the production of hydraulic binders [20]. These materials contain no carbonates and make the production of hydraulic binders with a lower energy demand possible. Summarized from all these publications, the following phases could be identified as the most important hydraulic phases:
Calcium silicates: $3CaO \cdot SiO_2$ (C_3S), $2CaO \cdot SiO_2$ (C_2S)
Calcium aluminates: $CaO \cdot Al_2O_3$ (C_3A), $12CaO \cdot 7Al_2O_3$ ($C_{12}A_7$), $CaO \cdot Al_2O_3$ (CA)
Calcium aluminum silicate: $2CaO \cdot Al_2O_3 \cdot SiO_2$ (C_2AS)
Calcium aluminum ferrates: $Ca_2(Al_xFe_{2-x})O_5$ ($C_4A_xF_{2-x}$)
Calcium aluminum sulfate: $4CaO \cdot 3Al_2O_3 \cdot SO_3$ (C_4A_4s)

Yoshioka has reported that a metastable solid solution in the system CA–CA_2–CAS_2 exists with hydraulic properties [21].

3.2 Synthesis methods

As precursor materials for the solid state reaction, stoichiometric amounts of solid $CaCO_3$ (Cc), SiO_2 (S), Fe_2O_3 (F), Al_2O_3 (A), and $CaSO_4$ (Cs) are used. A homogenization of the raw materials in a gas phase is possible but this application is most of the time limited to the precipitation on substrates [22, 23]. The production of powders from the liquid phase is more practical and common [22, 24, 25]. According to the literature, different sol gel, self combustion, and polymeric precursor methods were used in this case. In addition, the synthesis of oxides by decomposing their metals salts is described. The precursor powders synthesized by the synthesis routes are sintered at different temperatures. The results can be used to understand and optimize the production of hydraulic binder by using well homogenized raw materials or waste materials with very fine particle sizes or dehydrated concrete.

3.2.1 Solid state reaction

To achieve pure phases without any additional phases, long sintering times with a few intermediate grinding steps are needed. The formation of pure phases can be accelerated by packing the precursor powder very close and by pressing powder pellets. An unfortunate negative aspect of sintering pellets is that the pellets are present with a lower surface than a normal clinker and therefore more intensive grinding is needed.

In practice, the oxides CaO (produced by the decarbonation of $CaCO_3$), Al_2O_3, SiO_2, Fe_2O_3, and $CaSO_4$ were mixed in the corresponding ratios to achieve pure phases. The mixed oxides were ground together for some minutes to homogenize the powders and reduce the particle size. By using conventional grinding procedures, a minimum particle size of several micrometers can be realized in an adequate grinding time. The powders were mixed and sintered with the maximum temperatures shown in Tab. 3.1. Intermediate grinding steps were applied after a few hours until the pure phase was obtained. The procedure of this method is summarized in detail by Wesselsky & Jensen [26].

Tab. 3.1: Maximum sintering temperature by using the solid state reaction.

Phase	C_3S	C_2S	C_3A	C_4AF	CA	$C_{12}A_7$	C_4A_3s	C_5A_3MS
T_{max} (°C)	1500	1400	1400	1250	1500	1300	1300	1300

3.2.2 Sol gel method

“Sol” is defined as a dispersion with some colloids with a size ranging from 1 nm to 500 nm. The production of a sol is based on the hydrolysis of a silan (e.g. tetraethyl orthosilicate = TEOS). The basic work to understand the hydrolysis and condensation reactions from silans in order to produce silicate containing phases is done by Bergna [27] and Livage [28]. Bergna also describes the stability of silica sols in an aqueous solution [27]. The “traditional” sol gel process is done by changing the reaction parameters (pH) in the solution to destabilize the sol and promote gelation. The aqueous gel (hydrogel) is slowly dried to a xerogel. This xerogel is sintered to produce the silicate containing phases. The principle procedure is shown in Fig. 3.1 [29].

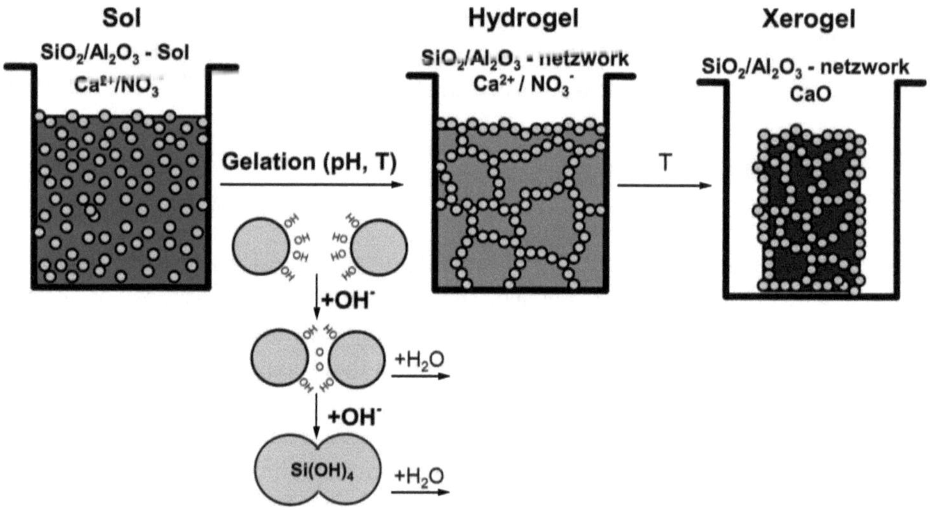

Fig. 3.1: Principle procedure of the sol gel process according to Brinkner et al. [29].

This method is already industrially used to produce layers for ceramics and glasses. Calcium silicates can be produced easily by using silica sol and a calcium salt. Fu & Lin [30], Saravanapavan & Hench [31], and Martınez et al. [32]. produced calcium silicate glasses from calcium nitrate und tetraethyl orthosilicate (TEOS). Chrysafi et al. showed the formation of crystalline C_2S at a sintering temperature of 600 °C and stabilized β-C_2S at room temperature (RT) [33]. A difference in phase formation by using water and ethanol cannot be seen [33]. Catauro & Laudisio [34] synthesized calcium silicates from calcium acetate und tetramethyl orthosilicate (TMOS) by using dimethyl sulfoxide. Stephan & Wilhelm [35] and Stephan & Plank [36] used commercially available SiO_2 sol and Al_2O_3 sol for the production of calcium silicates and calcium aluminates. Calcium silicates were also synthesized by using silica sols from Wang & Thomson [37] and Fujimori et al. [38]. The synthesis of calcium aluminates is described by using aluminum alkoxides or Al_2O_3 sol and calcium salts. Amorphous calcium aluminates were synthesized by using the sol gel method with the raw materials calcium nitrate und aluminum-*sec*-butylate ($Al(OC_4H_9)_3$) by Goktas & Weinberg [39] and Kerns et al. [40]. Ghosh & Pramanik used aluminum formate to produce ceramic phases in order to replace the more expensive aluminum alkoxides [41–43]. Page et al. directly produced white cements by using a sol gel method [44]. A general disadvantage by using silans and aluminum alkoxides compared to sols is the long gelation process up to six days and a more complex production control. The usage of commercial sols is recommended for a fast synthesis procedure. In our own sol gel synthesis, Al_2O_3 sol Aerodisp W630 (Degussa Aerosil) and SiO_2 sol Ludox™ (Grace Davison) were used as colloidal sources. The sol was mixed with metal nitrates because their solubility in water is higher than the solubility of metal chlorides or metal carboxylates. Since less water has to be evaporated, the gelation time is shorter. Additionally, the gelation time depends on temperature and pH value. The reaction therefore takes place at a temperature of 70 °C and a pH value of 5.5 for the synthesis of calcium silicates and a pH value of 6.5 for calcium aluminates. These parameters were optimized by Stephan & Wilhelm [35] and Stephan & Plank [36].

3.2.3 Self propagating combustion synthesis (SPCS)

Self-propagating combustion synthesis (SPCS) is another synthesis method for the synthesis of metal oxides. In this case, glycine is used as a fuel; the method is also called glycine nitrate process (GNP). By this method, oxidant (e.g. metal nitrates) and fuel (e.g. urea, glycine, citric acid) are used as reducing agents. These reducing agents were solved in water. After gelation and thermal activation, the combustion reaction takes place. The method was used by Kingsley & Patil to produce doped alumina [45]. Cueneyt-Tas synthesized different calcium aluminates with urea as fuel [46]. Hwang et al. described the influences of different fuels and the influence of fuel-to-oxidant ratio on the synthesis of zinc oxide [47, 48]. A combustion reaction in which the energy is set free by the reaction of Ti with B_2O_3 is described by Yi et al. but TiB_2 unfortunately occurs in the final product [49]. Calcium silicates were synthesized by Chandran et al. [50], Huang & Chang [51], and Sreekanth-Chakradhar et al. [52]. who applied SiO_2 sol and calcium nitrate as precursors. Stievano et al. described a strong exothermic reaction between colloidal silica and glycine at a starting temperature of 170 °C [53]. Different authors have described the synthesis of other ceramic phases by using the combustion process [54–58]. In some of our own syntheses, starting powders of $Ca(NO_3)_2 \cdot 4H_2O$, $Al(NO_3)_3 \cdot 9H_2O$, $Fe(NO_3)_3 \cdot 9H_2O$, and SiO_2 sol Ludox™ 50 (Grace Davison) were mixed in corresponding amounts and dissolved in distilled water at room temperature to yield one molar solutions. Glycine was added to the solution (glycine-to-oxidant ratio of 1.5) and stirred for one hour at room temperature. Hwang & Wu achieved a maximum combustion temperatures using a glycine-to-oxidant ratio of 1.5 and just a small amount of nitrous gases occur at this ratio [47]. After drying the mixture at 170 °C, it was heated up to 250 °C. The combustion reaction took place within a few minutes and a foamy and voluminous precursor powder was formed. These powders can be heated up to higher temperatures to achieve pure crystalline phases. In some publications, a stoichiometric amount of glycine ($C_2H_5N_2O$) in order to avoid theoretically nitrous gases is used. The stoichiometric equation for the production of CA by using the combustion process is given in the following formula:

$$
\begin{aligned}
&9Ca(NO_3)_2 \cdot 4H_2O + 18Al(NO_3)_3 \cdot 9H_2O + 40C_2H_5NO_2 + 36O_2 \\
&\qquad \xrightarrow{170\text{–}250\,°C} 9(CaO \cdot Al_2O_3) + 112N_2 + 80CO_2 + 298H_2O.
\end{aligned}
$$

3.2.4 Polymer precursor synthesis ("Pechini method", "citrate gel method", or "polymeric precursor process")

Instead of the sol gel method and SPCS, the polymer precursor synthesis method is well known for the synthesis of calcium aluminates and calcium silicates. Pechini patented the synthesis method by using citric acid and ethylene glycol as starting materials for the polymerization process [59]. Some authors [60–62] have used these starting materials to form a polymeric resin [63, 64] and called the synthesis route the "Pechini method" or "citrate gel method". In the first step, citric acid is dissolved in deionized water. Afterwards, metal salts are added in corresponding amounts. In most cases, metal nitrates are used because of their high solubilities. Afterwards, a glycol (e.g. ethylene glycol) is added and, by increasing the temperature, an ester is formed (Fig. 3.2). The polymerization reaction and the influence of polymer-to-cation ratio was described in detail by Nishio et al. [65] and Petrykin et al. [66]. Nishio & Tsuchiya have showed that the vacancy of the metal ions influence the polymerization process [65]. The type and amount of the polymer influences the decomposition of the polymer and this enables one to influence the phase formation and the particle size of the inorganic compound [67, 68]. Gülgün et al. [69], Hong & Young [61], Lee et al. [70], Pati et al. [71], Anjos et al. [72], Lee & Kriven [73], Kakali et al. [62], and Gaki et al. [63, 64] produced calcium aluminates and calcium silicates using the Pechini method with the starting materials citric acid and ethylene glycol for the polymerization process. In some of our own syntheses, an aqueous solution of citric acid ($C_6H_8O_7$), metal nitrate hydrates ($Ca(NO_3)_2 \cdot 4H_2O$, $Al(NO_3)_3 \cdot 9H_2O$, and $Fe(NO_3)_3 \cdot 9H_2O$), and silica sol was stirred and heated up to 60 °C (molar ratio: citric acid/metal cations = 1). In the next step, ethylene glycol ($C_2H_6O_2$) was added (molar ratio: ethylene glycol/citric acid = 2) and the solution was stirred at 80 °C until a viscous gel was achieved (Fig. 3.3). This gel is dried at 150 °C in a drier and a very fulminous foam was achieved (Fig. 3.3). This foam was crushed and used as a precursor material for the subsequent sintering process.

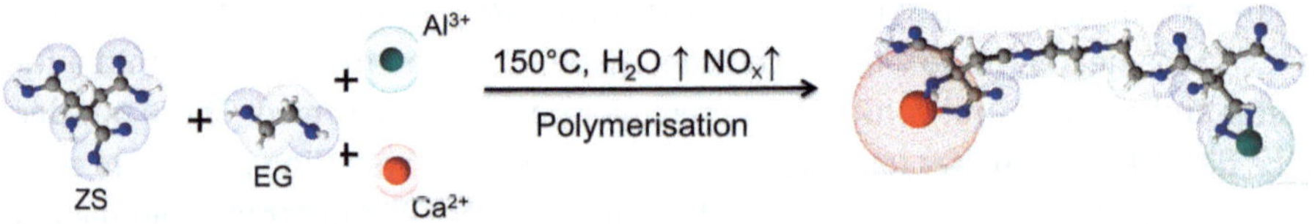

Fig. 3.2: Schematic process of the polymer precursor synthesis by using of citric acid, ethylene glycol, and metal ions (Ca^{2+}, Al^{3+}) to an ester, orbital structure: C-atom = black, H-atom = white, O-atom = blue, Ca-atom = red, and Al-atom = turquoise.

Fig. 3.3: Dried aqueous solution of metal nitrates, citric acid, and ethylene glycol at 80 °C before (1) and after (2) the polymerization process at 150 °C.

3.2.5 Spray method

In comparison to other methods, pure phases were often synthesized just by the homogenization of metal salts in an aqueous solution. After the drying and decomposition process, pure oxide phases were obtained [74]. The drying process was often performed by using a spray drier to achieve a fast drying process and to avoid segregation to receive well-homogenized precursor powders. The reason for the decomposition of the liquid jet to get spherical particles in a spray dryer are the vibration movements that result when the liquid jet exits the spray nozzle. Spherical particles with sizes of 1 μm up to 2 mm were thereby obtained [75]. Douy & Gervais [76] successfully synthesized different calcium aluminates by spray drying and decomposing aqueous calcium aluminum nitrate solutions. Due to the solubility, the decomposition temperature, and the decomposition products, the salts of organic acids (e.g. metal formate) are also very suitable for the drying process. Metal nitrates and metal chlorides have a very high solubility and hygroscopic behavior and these phases are therefore impracticable using the spray method [77]. As a consequence, these powders have the strong tendency to agglomerate after the spray drying process [78]. Spray dried metal salts have to be decomposed afterwards at ambient temperatures [79–81]. Block & Dolhert spray dried formate solutions by using the stoichiometric composition of $YBa_2Cu_3O_{7-\delta}$ [82]. Peshev & Pecheva already synthesized lithium ferrite spinel from a spray-dried solution of lithium formate monohydrate and iron(III)formate [83]. Thereby a direct formation of the spinel phase is described after the decomposition of the formates; also, a spherical shape of the particles was achieved. Kim et al. also showed the formation of spherical agglomerates by the spray drying process of a silica sol [84].

3.3 Phase formation of pure cement phases

3.3.1 Calcium aluminates (System $CaO–Al_2O_3$)

Different publications on the phase diagram of the binary system $CaO–Al_2O_3$ are summarized from Hallstedt [85].

3.3.1.1 $CaO\cdot Al_2O_3$ (CA)

Until now, two modifications of CA (orthorhombic and monoclinic) are known. The unit cell of the orthorhombic modification of CA (ICDD: 00-034-0440) was described for the first time by Ito et al. [74]. The metastable orthorhombic phase was later synthesized by Douy & Gervais [76] and Gaki et al. [64]. This metastable phase is formed only by well homogenized raw materials at a sintering temperature of around 900 °C. Gülgün et al. described the synthesis of orthorhombic CA by the polymer synthesis method [69]. Further results of calcium aluminates from amorphous precursors are presented by Uberoi & Risbud [86] and Vallino [87]. Janáková et al. determined the structure of metastable CA doped with Eu^{3+} [88]. The stable monoclinic modification of CA is very well described because it is the main phase in calcium aluminate cements (CAC) [89]. The synthesis of a pure phase of CA by using the solid state reaction (2 × 8 h at 1500 °C) and the sol gel method (2 × 2 h at 1500 °C) is only possible at higher sintering temperatures of 1500 °C [35].

During the heating process of an amorphous precursor powder (Fig. 3.4, 3.5) synthesized by the polymer method, the polymer and in-between formed calcite are completely decomposed at 650 °C. The oxidizing conditions for the decomposition of the polymer have to be high enough; otherwise, carbon is present up to higher temperatures (Fig. 3.11). After a sintering step of 2 h at 900 °C, the crystalline orthorhombic modification of CA occurs. Crystalline products by the polymer process show agglomerated particles with a diameter of about 50 nm (Fig. 3.6–3.8). The shape of the agglomerates have the same morphology of the polymer with the metal cations prior to pyrolysis of the polymer [90]. Sintering of the orthorhombic CA powders occurs in four further sintering steps from 1000 °C to 1300 °C, every step after a further two hours, and the orthorhombic modification converts into the stable monoclinic modification (Fig. 3.10). This can be shown by the appearance of the peaks at 19° 2θ and 24° 2θ and by the splitting of the peak at 35.7° 2θ ($Cu_{K\alpha}$) in the XRD diagram (Fig. 3.10). The particles agglomerated during this sintering step show the same typical morphology of CA sintered by solid state reaction (Fig. 3.9). Refined unit cell parameters of the orthorhombic and monoclinic modification are summarized in Tab. 3.2. Investigations by using different polymer contents show that there is no influence on the phase formation [91].

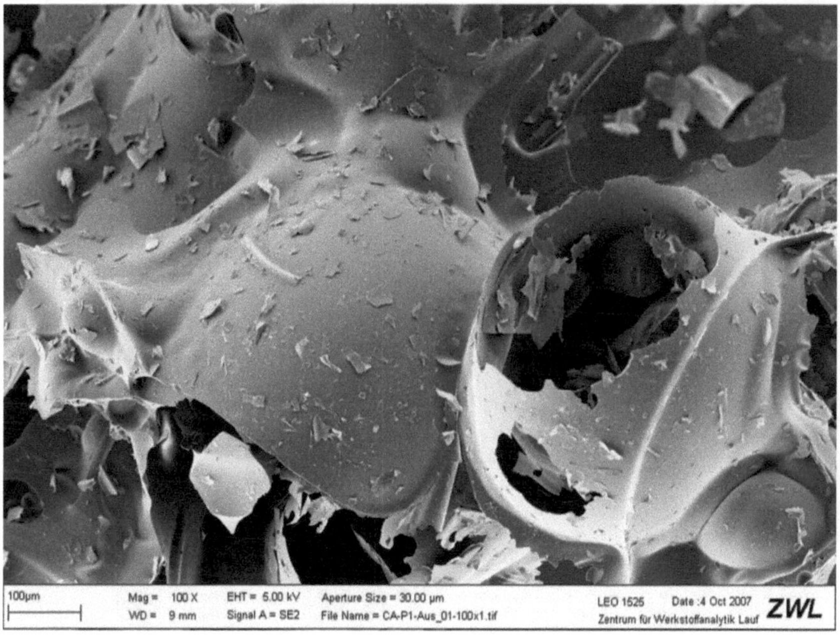

Fig. 3.4: SEM image of CA precursor synthesized by the polymer precursor method.

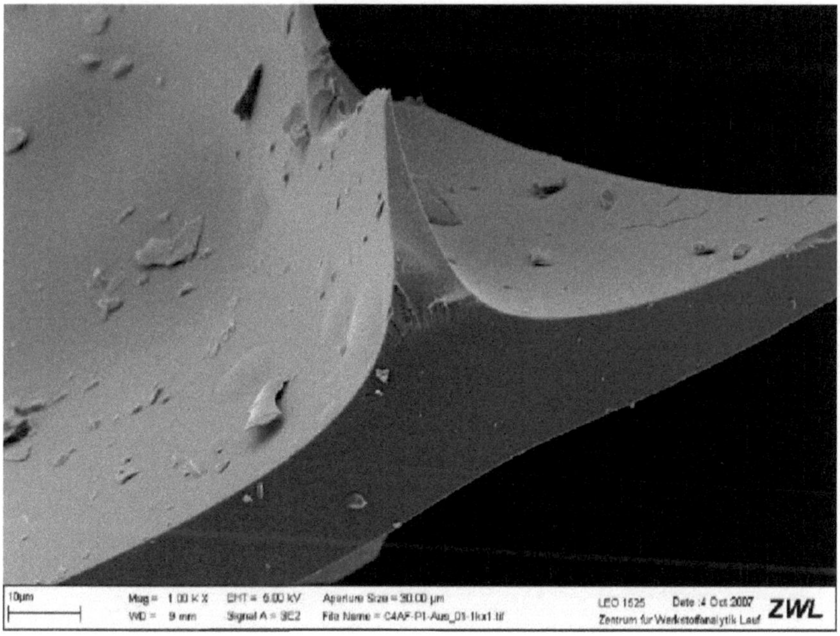

Fig. 3.5: SEM image of CA precursor synthesized by the polymer precursor method.

Fig. 3.6: SEM image of metastable CA synthesized by polymer precursor method at 900 °C (2 h).

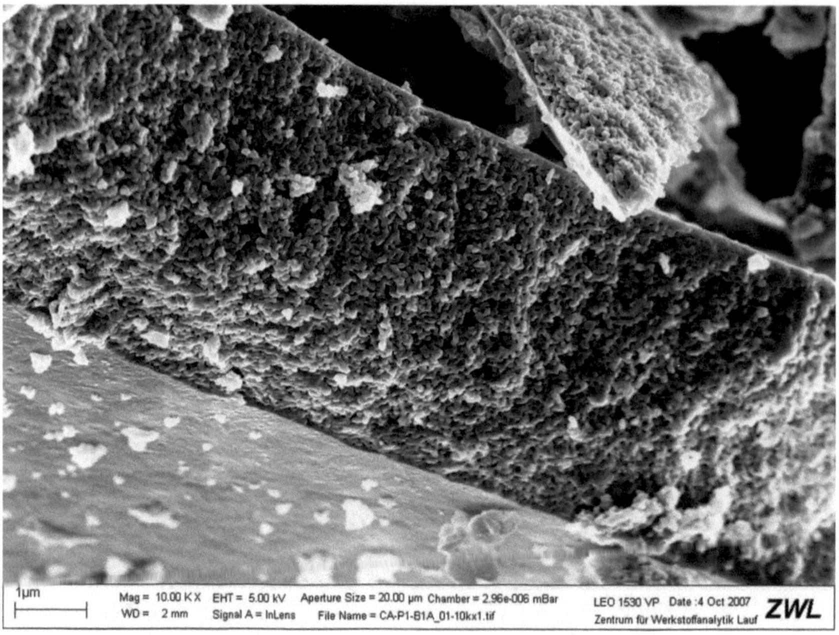

Fig. 3.7: SEM image of metastable CA synthesized by polymer precursor method 900 °C (2 h).

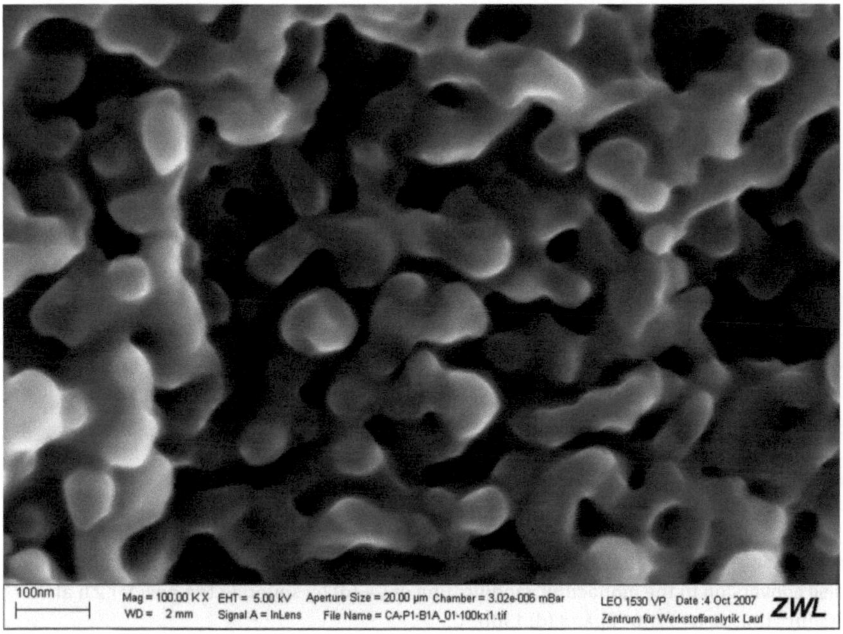

Fig. 3.8: SEM image of metastable synthesized by polymer precursor method 900 °C (2 h).

Fig. 3.9: SEM image of stable monoclinic CA synthesized by polymer precursor method 1500 °C (2 h).

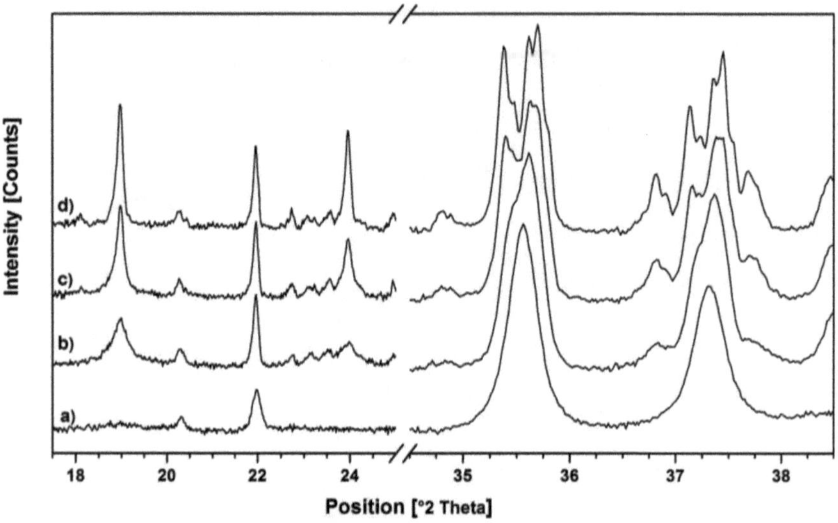

Fig. 3.10: Metastable orthorhombic CA (a) synthesized by the polymer precursor method and the stepwise conversion to the monoclinic modification of CA (d) by sintering the powder 2 h at (a) 1000 °C, (b) 1100 °C, (c) 1200 °C, and (d) 1300 °C.

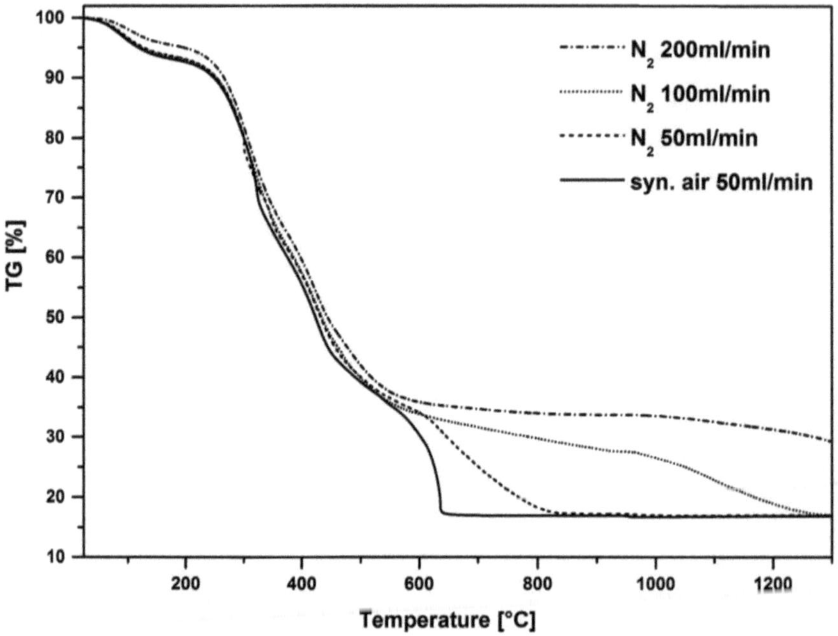

Fig. 3.11: TG curve of the precursor powder synthesized by the polymer precursor method at different atmospheres controlled by the flushing gas and rate, heating rate: 10 °C/min.

The oxidizing conditions during the decomposition of the precursors synthesized by the polymer precursor method and the spray methods were sufficient although a small loss of ignition was detected at 951 °C. Gülgün et al. showed also that carbon was lost at a temperature of 950 °C by the synthesis of CA using the polymer precursor method (Fig. 3.12) [69]. Hernandez & Gonzalez produced alumina by the polymer method and also detected an evaporation of carbon between 900 and 1100 °C [60]. The presence of carbonates can be excluded by own samples sintered at 700 °C at different atmospheres by using the IR spectroscopy. Lin et al. showed that by the decomposition of aluminum organic compounds –O–Al–C–O–Al–O– bonds were formed in the surface of amorphous Al_2O_3 [92]. These bonds are stable up to a temperature of about 900 °C. It can be assumed that these –O–Al–C–O–Al–O– bonds were also formed by using the polymer and spray methods for the synthesis of CA.

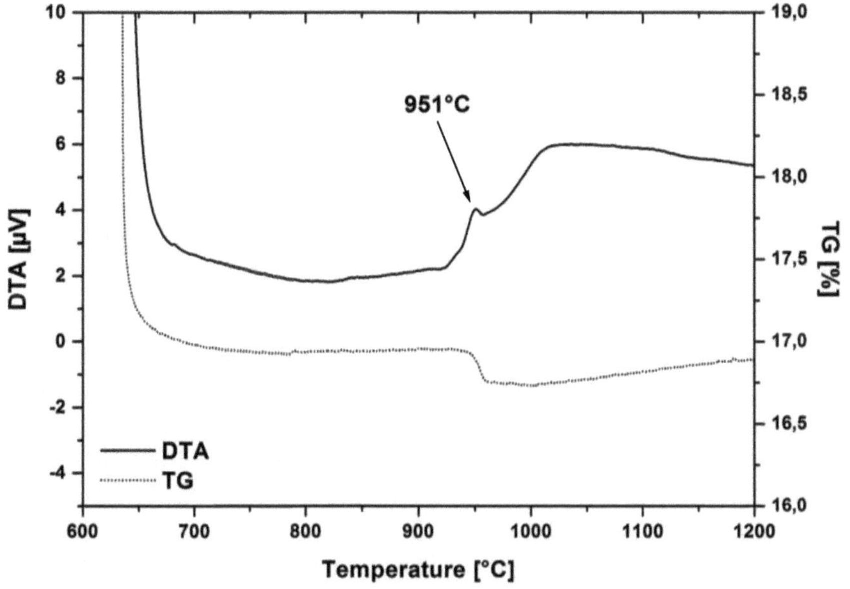

Fig. 3.12: Enlargement of the TG/DTA curve from Fig. 3.11 by using the precursor powder synthesized by the polymer precursor method, heating rate: 10 °C/min, flushing gas: synthetic air, flushing rate: 50ml/min.

Tab. 3.2: Refined unit cell parameters of orthorhombic and monoclinic CA.

modification	SG	*a* (Å)	*b* (Å)	*c* (Å)	β (°)
orthorhombic CA	–	8.737	8.076	15.125	90.00
monoclinic CA	P21/n	8.700	8.092	15.191	90.19

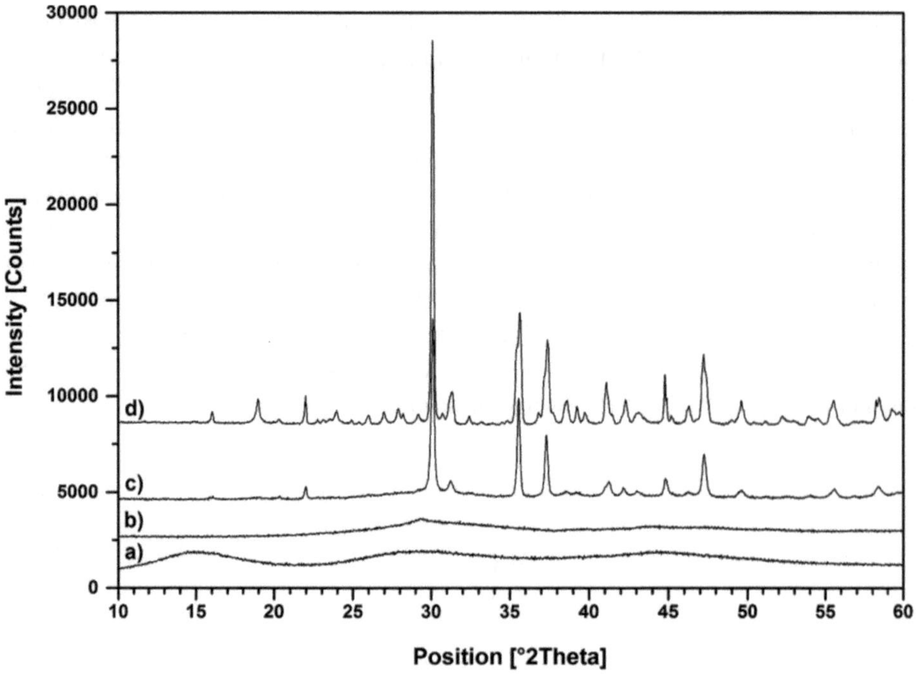

Fig. 3.13: XRD diagram of spray dried calcium aluminate format solution heated up to the following sintering steps: (a) RT, (b) 700 °C (2 h), (c) 900 °C (2 h), and (d) 1300 °C (2 h).

The phase formation of the calcium aluminate formate precursor powders synthesized by spray drying and the glycine nitrate process is similar to the polymer precursor method. Particle sizes are also similar to the polymer precursor method. After the decomposition of the formates, X-ray amorphous phases were formed and the crystallization of the orthorhombic CA started after a sintering step at 900 °C (2 h) (Fig. 3.13).

The crystallization process followed by the infrared (IR) analysis is shown in Fig. 3.14. Between 800 and 900 °C, the crystallization of the orthorhombic CA starts. A further increase of the sintering temperature between 900 and 1300 °C show some additional absorption bands in the IR diagram. A detailed identification of the absorption bands is shown by Bachiorrini [93].

High temperature X-ray diffraction measurements show that there is a conversion from the orthorhombic to the monoclinic modification of CA at 1100 °C. It could be proved by this test that the orthorhombic modification is a metastable phase at low sintering temperatures and not a high temperature modification (Fig. 3.15).

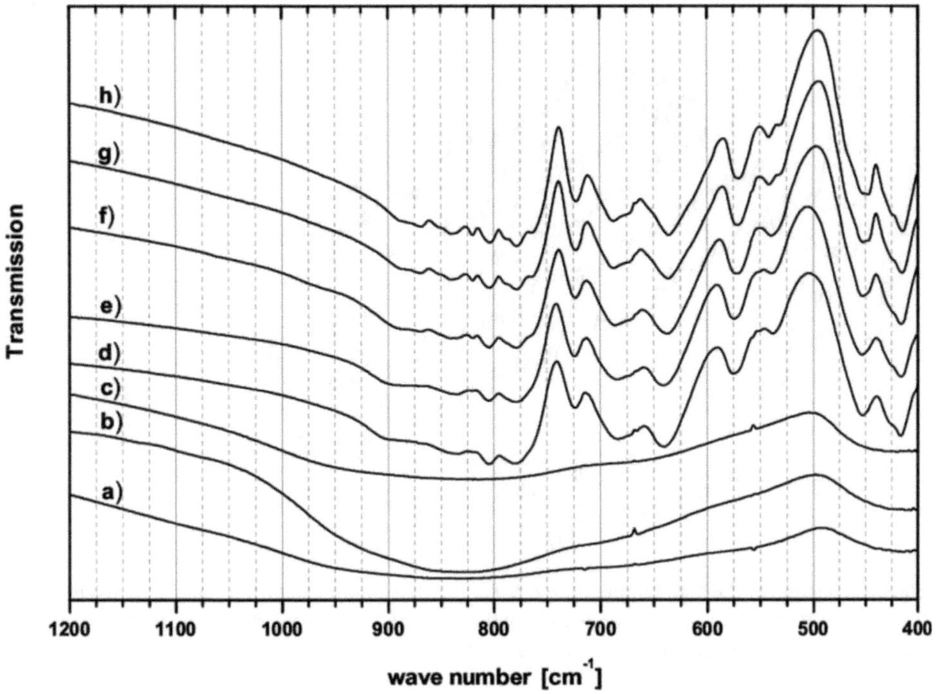

Fig. 3.14: Infrared diagrams of the precursor (CA) synthesized by the polymer precursor method sintered at the following temperatures: (a) 600 °C, (b) 700 °C, (c) 800 °C, (d) 900 °C, (e) 1000 °C, (f) 1100 °C, (g) 1200 °C, and (h) 1300 °C (2 h).

3.3.1.2 $12CaO{\cdot}7Al_2O_3$ ($C_{12}A_7$)

Mayenite $12CaO{\cdot}7Al_2O_3$ ($C_{12}A_7$; dodecacalcium heptaaluminate) is the most reactive phase in calcium aluminate cements. The general formula of $C_{12}A_7$ can also be named as $11CaO{\cdot}7Al_2O_3{\cdot}CaX_2$ ($X = \frac{1}{2}O^{2-}$, OH^-, Cl^-, F^-) [94]. Additionally, the metastable phase $5CaO{\cdot}3Al_2O_3$ (C_5A_3; pentacalcium trialuminate) with a similar chemical composition is described [95]. The unit cell and structure of C_5A_3 is described by Aruja [96] and Vincent & Jeffery [97]. The stability and phase transformation of both phases was investigated by Zhmoidin & Chatterjee [94] and Brisi et al. [98]. Using dry conditions and/or an exclusion of oxygen leads to the formation of C_5A_3 below a sintering temperature of 1280 °C. Brisi et al. already mentioned that the use of well homogenized starting material promote the formation of C_5A_3 in a water and oxygen containing atmosphere at a lower sintering temperature between 900 and 1000 °C [98].

The starting powders CaO and Al_2O_3 using the solid state reaction were mixed and sintered five times with intermediate grinding steps and a maximum temperature of 1300 °C (4 h) to obtain pure $C_{12}A_7$. As minor phases, CA and C_3A occurred during this sintering process. The formation of the metastable phase C_5A_3 cannot be reached by using this synthesis method and sintering conditions. The precursor powder syn-

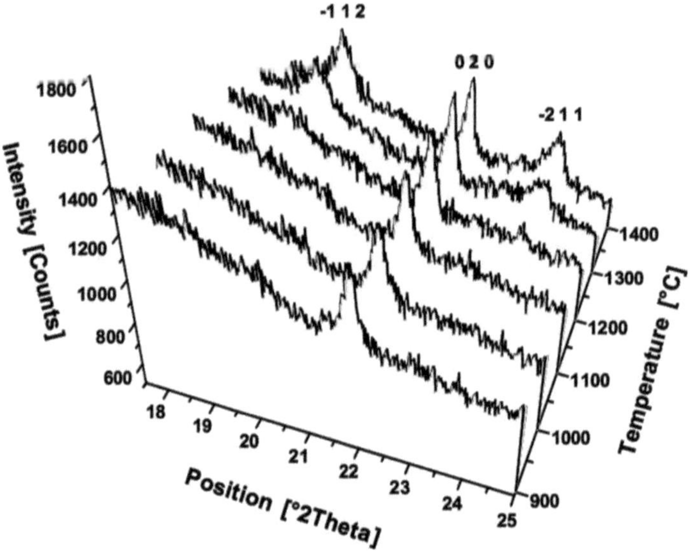

Fig. 3.15: Section of the high temperature X-ray diffraction diagram of orthorhombic CA in a temperature range of 900 °C to 1400 °C; appearing of the −1 1 2 and −2 1 1 peak due to the transformation of the orthorhombic to the monoclinic modification of CA .

thesized by GNP is X-ray amorphous (Fig. 3.16 (a)). In several sintering processes, the powder was heated up step by step. At the sintering step at 800 °C (2 h), the foamy precursor starts to crystallize. In the X-ray pattern a wide amorphous peak at about 31° 2θ occurs (Fig. 3.16 (b)) at this sintering step. The peak has the same position as the two main peaks of the metastable phase $5CaO \cdot 3Al_2O_3$ (C_5A_3). In Fig. 3.17, the SEM images shows that at this sintering step small particles of about 50 nm are formed out of the amorphous foam. Sintering the precursor powder at 900 °C (2 h), C_5A_3 plus small amounts of $3CaO \cdot Al_2O_3$ (C_3A) crystallize (Fig. 3.16 (c)). From the X-ray pattern, the crystallite size was measured by using the full width at half maximum (FWHM) [99]. The calculation was done using the fundamental parameter approach [100]. By assuming spherical crystallites, an average crystallite size of 43 nm was calculated and this fits quite well to the optical measured particle size by SEM. By sintering the powder at 1000 °C (2 h), the amounts of C_5A_3 and C_3A decrease and the stable phase $C_{12}A_7$ occurs (Fig. 3.16 (d)). Using a temperature of 1200 °C (2 h), the reaction is complete and pure $C_{12}A_7$ was obtained (Fig. 3.16 (e)). The maximum grain size is below 3 µm and a foamy structure is obtained after the final sintering step. The phase formation of precursor powders synthesized by polymer precursor method is similar – also C_5A_3 occurs at 900 °C. For the complete conversion of C_5A_3 to $C_{12}A_7$, a sintering step at 1200 °C (2 h) is necessary. The particle size of this $C_{12}A_7$ is slightly smaller but the particles are more agglomerated.

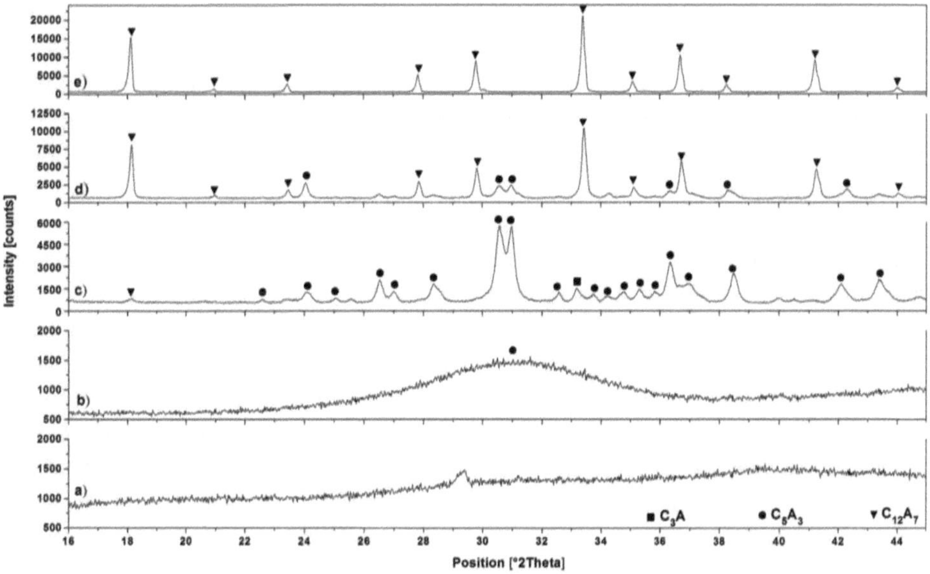

Fig. 3.16: Phase development of X-ray amorphous precursor powders synthesized by the GNP and sintered at the following steps: (a) 25 °C, (b) 800 °C (2 h), (c) 900 °C (2 h), (d) 1000 °C (2 h), and (e) 1200 °C (2 h).

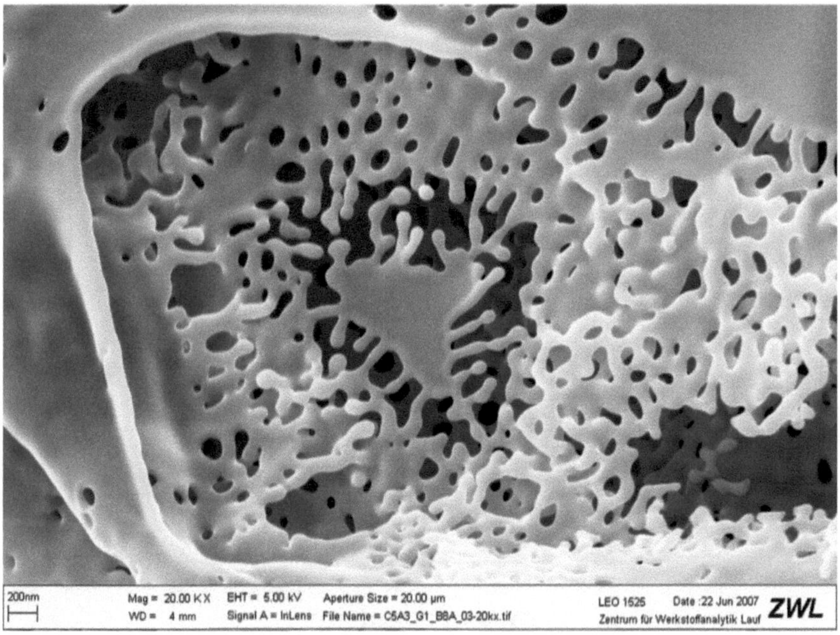

Fig. 3.17: Precursor powder of the GNP sintered 800 °C (2 h), Corresponding X-ray pattern given in Fig. 3.16 (b).

3.3.1.3 $3CaO \cdot Al_2O_3$ (C_3A)

Pure C_3A crystallizes in a cubic crystal structure. Due to the replacement of CaO by Na_2O and K_2O in the crystal structure of commercial cements, the formation of an orthorhombic and monoclinic phase was also described by Lee & Glasser [101] und Wistuba [102]. The synthesis of C_3A by using different synthesis methods is explained by Lazau et al. [103].

Pure C_3A without additional phases can be synthesized by sintering CaO and Al_2O_3 at 1400 °C (20 h) using the solid state reaction. Using the precursor powder from the sol gel synthesis a sintering at 1400 °C (2 h) is needed to synthesize pure C_3A. At 1300 °C (2 h), the additional phases CaO and $C_{12}A_7$ are present. When using better homogenized precursor powders by the glycine nitrate process or the polymer synthesis method, a sintering step at 1200 °C (2 h) is sufficient. This high temperature is in comparison to the synthesis of CA and $C_{12}A_7$ necessary because at lower sintering temperatures there is a preferred crystallization of CA (orthorhombic modification) metastable C_5A_3 and the stable phase $C_{12}A_7$. In addition, free lime (CaO) is formed at lower sintering temperatures. Higher sintering rates (compared to 10 °C/min) reduce the formation of these phases, although they are still present in small quantities. Increasing the sintering temperatures from 1000 °C (2 h) to 1200 °C (2 h) leads to an enormous increase of the intensity (Fig. 3.18) and a parallel reduction of the FWHM value due to the crystallization process.

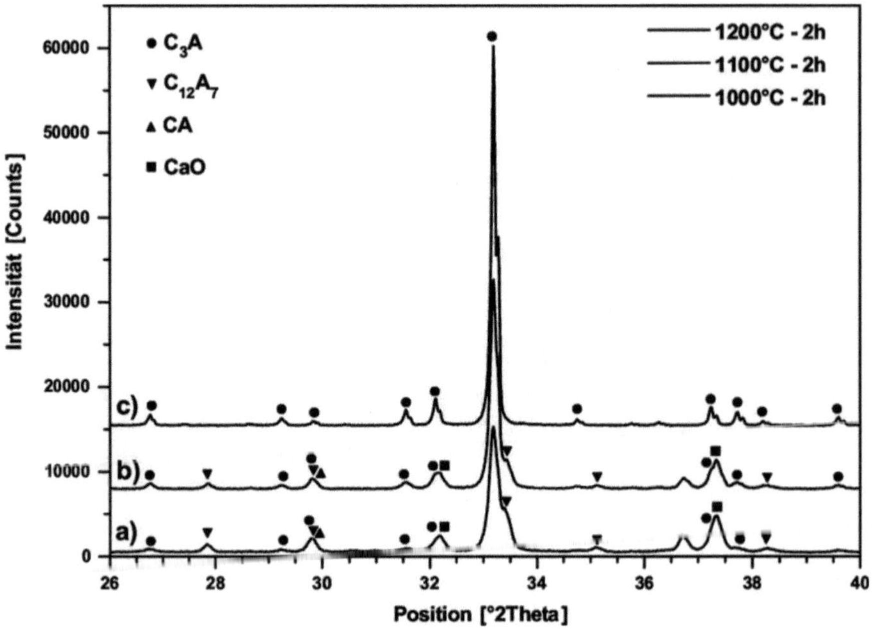

Fig. 3.18: X-ray diffraction diagrams of C_3A using the precursor produced by the polymer synthesis at the following sintering steps: (a) 1000 °C, (b) 1100 °C, and (c) 1200 °C (2 h).

The grain sizes using the solid state reaction determined by the SEM show very large sizes of more than 10 µm. The grain size of the powder using the sol gel method is in a smaller range of 1–2 µm. The grain sizes by the well homogenized precursor powders (glycine nitrate process or polymer precursor method) show grain sizes of 0.5–2 µm.

3.3.2 Calcium silicate (system $CaO–SiO_2$)

3.3.2.1 $2CaO \cdot SiO_2$ (C_2S)

C_2S is normally not very reactive in OPCs but the phase plays an important role in belite cements and calcium sulfo aluminate cements, which can be produced with less energy and accelerated by admixtures and impurities. The different modifications of C_2S are summarized very precisely by Mumme et al. [104], Remy et al. [105], and Taylor [95]. At room temperature, the less reactive γ-C_2S is stable and the more reactive β-C_2S can be stabilized with low particle sizes, high cooling rates, or foreign ions. Chan et al. have proved that a conversion to β-C_2S on RT can be avoided by particle sizes smaller than 10 µm [106]. Hong et al [61]. and Nettleship et al. [107] showed by using the Pechini process, that α'_L-C_2S can be stabilized at RT using a sintering temperature of 650 °C. Some other authors have also showed that C_2S can be produced at low temperatures of around 600 °C by using sol gel methods [33, 51, 108]. Dovál et al. synthesized β-C_2S from calcium nitrate hydrate and silica sol at 600 °C but small amounts of the side phase CaO were present [108].

Comparing the different synthesis methods, pure C_2S can be synthesized by the polymer method after a sintering step of 2 h at 650 °C. By using the polymer method, calcite and amorphous silica are stable after the decomposition of the polymer at 550 °C. At 550 °C, the decarbonation of calcite starts and after a sintering step at 650 °C (2 h) α'_L-C_2S was stabilized. The decarbonation of calcite occurs at a significantly lower temperature because the particle sizes are smaller than 50 nm directly after the decomposition of the polymer precursor. By increasing the sintering temperature up to 1100 °C, the conversion to β-C_2S takes place and the specific surface area and porosity, determined by BLAINE, decreases (Tab. 3.3). A decrease of the specific surface using this synthesis method was also published by Hong et al. using BET measurements [61]. All reflections of the α'_L-C_2S (without foreign ions) could be indexed by using the orthorhombic unit cell (a, 3b, c) described by Saalfeld [109] and Jelenić & Bezjak [110]. The β-C_2S was refined with the monoclinic space group $Pna2_1$ from Mumme et al. [104].

By sintering the precursor powder synthesized by the sol gel method at 650 °C (2 h) it's also possible to achieve α'_L-C_2S. In case of the combustion synthesis, after the exothermic reaction, without a sintering process a mixture of α'_L-C_2S and β-C_2S is stable. It is also possible to synthesize the phase C_2S by the combustion method despite a strong exothermic reaction which is very difficult to control and handle.

Tab. 3.3: Refined unit cell parameters, crystallite size, and measured specific surface by BLAINE of C_2S synthesized by the polymer precursor method at different sintering steps.

Sintering step	Phase	SG	*a* (Å)	*b* (Å)	*c* (Å)	β (°)	Crystallite size (nm)	Specific surface (BLAINE) (cm^2/g)
650 °C (2 h)	α'_L-C_2S	Pna21	20.292	9.567	5.639	90	14	17035
700 °C (2 h)	Mixture of α'_L-C_2S and β-C_2S						14	16207
800 °C (2 h)	(They can't clearly separated						16	9492
900 °C (2 h)	due to strong peak interfering)						38	9168
1000 °C (2 h)							60	4214
1100 °C (2 h)	β-C_2S	P21/n	5.512	6.757	9.315	94.50	89	1571

For the solid state reaction, quartz and lime were used as precursor. The sintering process took place at 1400 °C (24 h) in order to obtain a pure C_2S. After sintering and cooling in air, a mixture of β-C_2S and γ-C_2S is stable. The result shows that a part of the C_2S converts into γ-C_2S during cooling, which has a higher specific volume (volume increase of 12 %) [106]. The phase was also determined by XRD and can be seen by the cracks in SEM images.

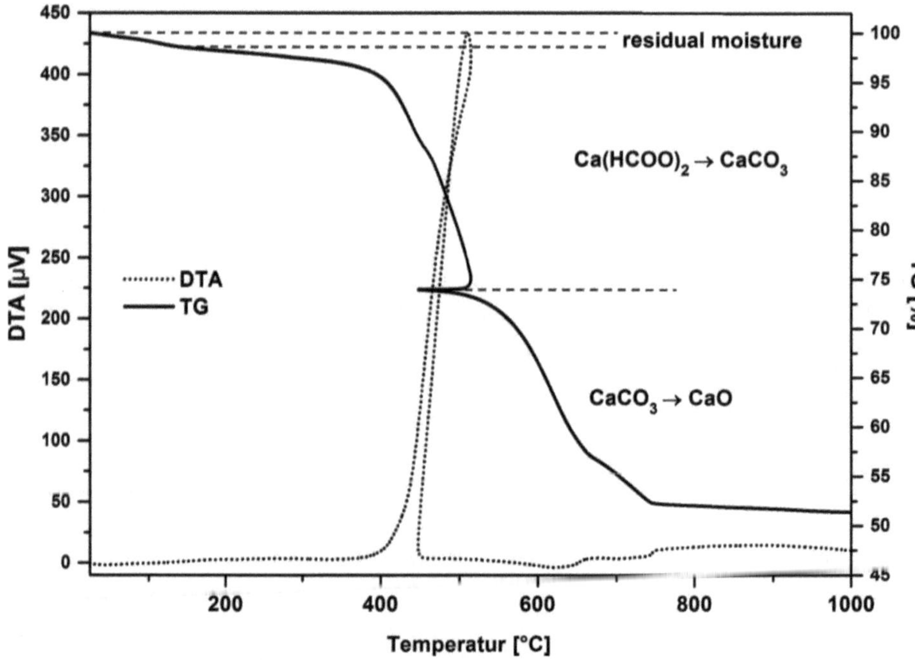

Fig. 3.19: TG/DTA diagram of spray dried calcium formate und silica sol for the production of C_2S; flushing gas: synthetic air; flushing rate: 200ml/min; heating rate: 10 °C/min.

For the spray method, a 0.25 molar solution of calcium formate and SiO_2 sol was used. After the spray process, a mixture of amorphous SiO_2, β-, and γ-$Ca(HCOO)_2$ was identified. The thermogravimetric analysis showed that the calcium formate decomposes at the temperature of 407 °C to $CaCO_3$. The decarbonation reaction occurs at a temperature of 532 °C (Fig. 3.19). Directly after the decarbonation, the formation of α'_L-C_2S starts. Fig. 3.20 shows that the typical spherical shapes of spray-dried particles also remain after a sintering step of 700 °C (2 h). At a higher resolution, the porous and nano scaled structure of the spherical particles can be seen (Fig. 3.21). The calculated crystallite size of 18 nm (Tab. 3.3) is in this case similar to the measured grain size by SEM of around 20 nm (Fig. 3.20).

In summary, the results of the investigations show that the formation of C_2S is possible by using different low temperature synthesis methods with well homogenized precursors. The results of the stabilized modifications α'_L-C_2S and β-C_2S are different which could be the consequence of the existence of both very similar phases at sintering temperatures used in these investigations. Investigations on the phase formation of the dehydration between 600 and 1000 °C of hydrated cements [16] or pure calcium silicate hydrates [111] show that a formation of C_2S at 600 °C is also possible. The stabilization of α'_L-C_2S and β-C_2S was also observed during the production of belite cements [112]. Up to 900 °C, α'_L-C_2S is the main phase, at higher temperatures than 1000 °C β-C_2S is present in higher quantities. Puertas & Trivino showed that α'_L-C_2S and β-C_2S can also be distinguished by IR spectroscopy, and that small amounts of calcite can be clearly identified by ν_2 (CO_3^{2-}) vibration at 874 cm^{-1} [113]. The formation of β-C_2S can be identified by ν_3 (SiO_4^{4-}) vibrations at 995 cm^{-1} [113]. Due to the lower symmetry attributed to the conversion from α'_L-C_2S to β-C_2S, an increasing amount of absorption band can be observed (Fig. 3.22).

In further investigations, the phase formation of C_2S was also investigated by HT-XRD. The precursor powder of the polymer synthesis method was heated up to different temperatures. The temperature was also detected during the cooling process. By heating, only the decarbonation of calcite and the formation of α'_L-C_2S could be identified. The decarbonation starts at 500 °C and the phase formation of α'_L-C_2S is finished at 650 °C. If the powder is cooled down from 700 °C, pure α'_L-C_2S remains. By using a sintering temperature of 1100 °C, the transformation of α'_L-C_2S (grey) to β-C_2S (light grey) occurs between 500 and 600 °C (Fig. 3.23). By increasing the sintering temperature to 1300 °C, the transformation to β-C_2S is already finished at 600 °C. Between 500 and 400 °C, a phase transition to γ-C_2S (white) took place (Fig. 3.24). This implies that the sintering temperature and the particle as well as the crystallite size influence the transition temperature.

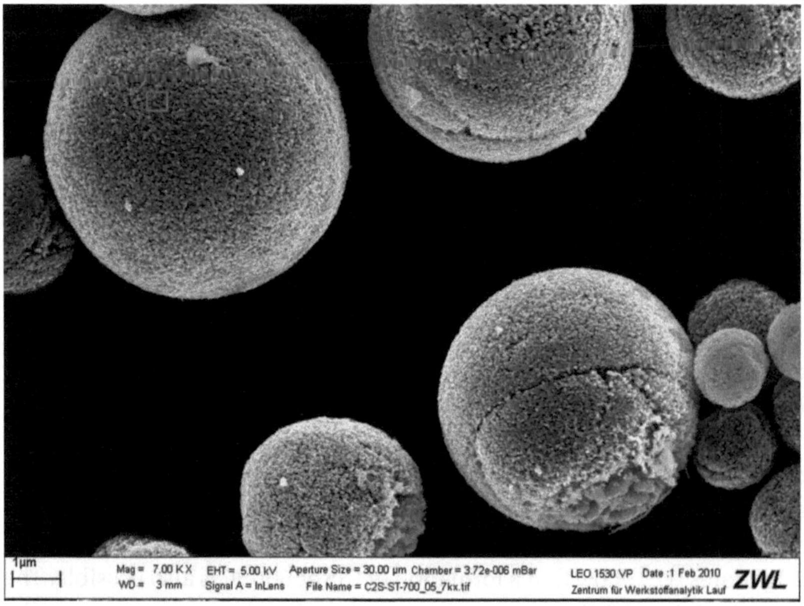

Fig. 3.20: SEM image of α'_L-C_2S; produced by spray drying of a calcium formate und silica sol solution at a sintering step of 700 °C (2 h).

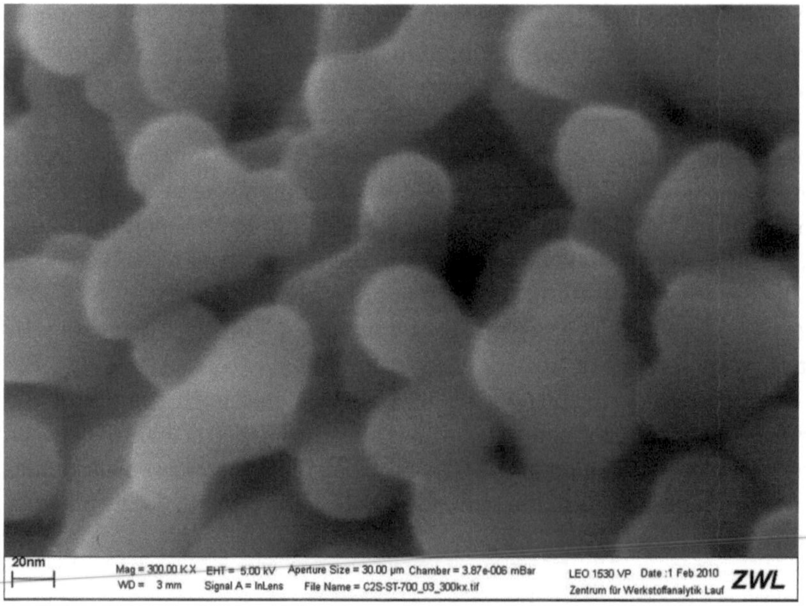

Fig. 3.21: SEM image of α'_L-C_2S; produced by spray drying of a calcium formate und silica sol solution and a following sintering step at 700 °C (2 h), red marked image section from Fig. 3.20.

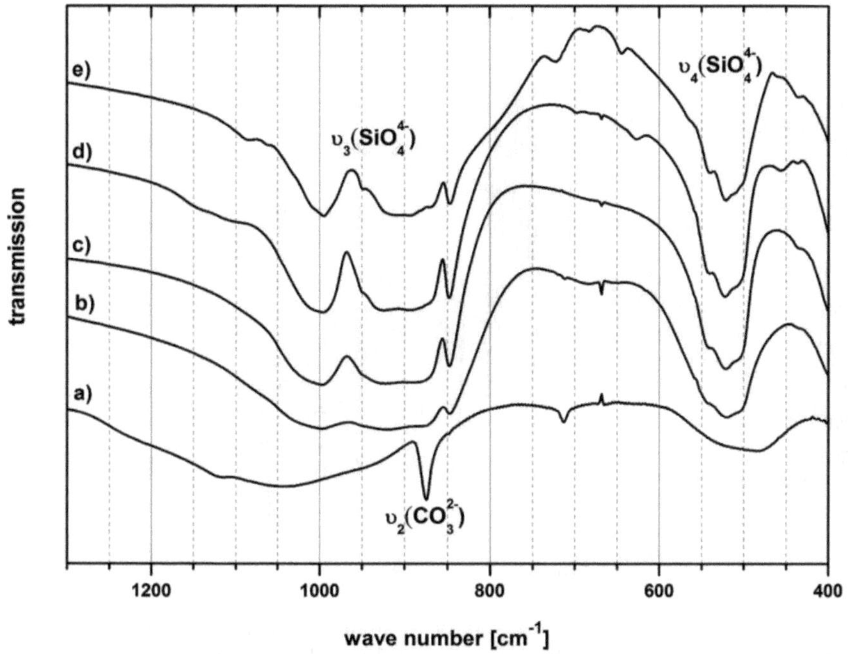

Fig. 3.22: IR diagram of C_2S, synthesized by the polymer precursor method and sintered at the following temperatures: (a) 550 °C, (b) 650 °C, (c) 800 °C, (d) 900 °C, and (e) 1100 °C (2 h); (*) calcite.

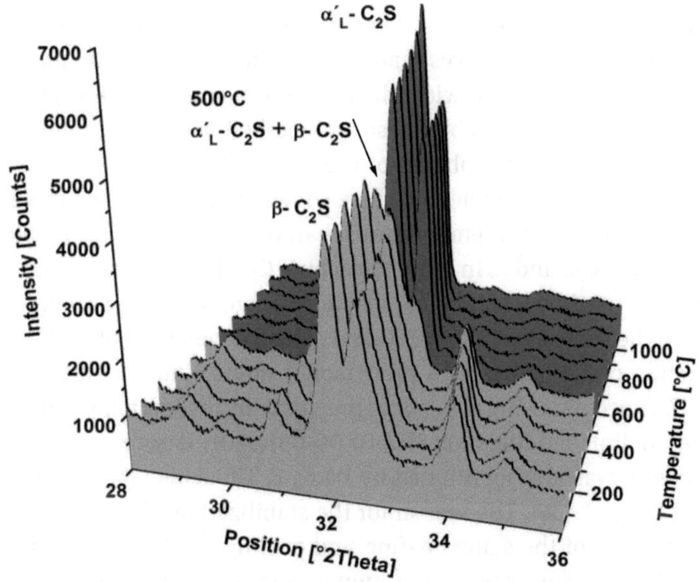

Fig. 3.23: HT-XRD diagram by cooling down α'_L-C_2S synthesized by the polymer precursor method and using a sintering temperature of 1100 °C; formation of β-C_2S (light grey) occurring at a temperature between 500 and 600 °C.

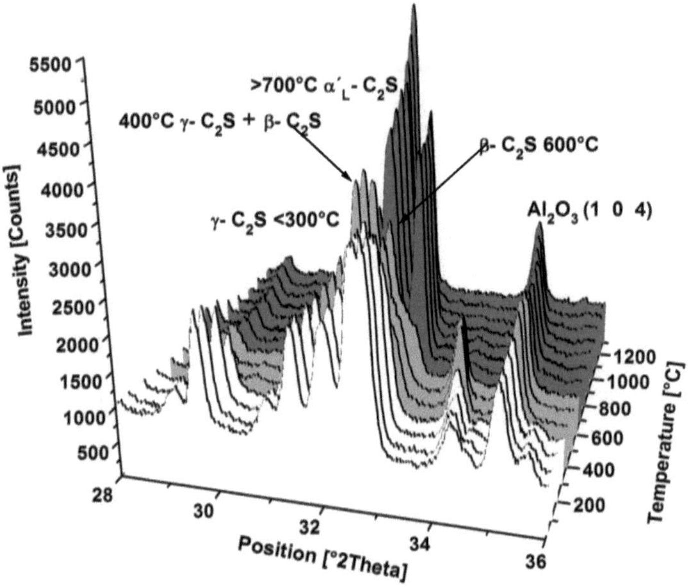

Fig. 3.24: HT-XRD diagram by cooling down α'_L-C_2S synthesized by the polymer precursor method and using a sintering temperature of 1300 °C; formation of β-C_2S (light grey) and γ-C_2S (white).

3.3.2.2 $3CaO\cdot SiO_2$ (C_3S)

C_3S is the main phase in OPCs and is formed above 1250 °C by a reaction of C_2S and C [114]. The polymorphs of C_3S were summarized and discussed by Regourd [115] and Taylor [95]. Wang & Thomson did some investigations into the kinetics of the phase formation of C_3S by using a sol gel method with silica sol and calcium nitrate as starting materials [37]. To synthesize pure C_3S by the sol gel method, very long sintering times are necessary to get no additional phases of C_2S and C [35, 36]. For the three alternative synthesis methods used, commercial silica sol was used as starting material. C_3S cannot be synthesized by lower temperatures than 1250 °C because the stable phases at this temperature are C_2S and C. In order to get pure C_3S, higher temperatures are necessary, as they were used by a solid state reaction. Therefore all C_3S samples were sintered together in the same oven at 1500 °C and cooled down in air at 20 °C.

By solid state reaction using quartz as a precursor, a monoclinic modification (ICDD: 01-086-0402) of C_3S occurs. Using SiO_2 sol as precursor by the low temperature synthesis methods, the triclinic modification (ICDD: 00-031-0301) is stable. A stabilization of this modification was also mentioned by Wang & Thomson [37], Fujimori et al. [38], and Stephan & Plank [36]. The reason for the stabilization of the different modifications, despite the use of the same heating and cooling conditions, is obviously the different particle sizes. For all three low temperature synthesis methods, the particle looked similar with the same particle size and degree of agglomeration.

3.3.3 Calcium aluminum silicate (system $CaO–Al_2O_3–SiO_2$)

In the system $CaO–Al_2O_3–SiO_2$, gehlenite ($2CaO \cdot Al_2O_3 \cdot SiO_2$) and anorthite are known as stable phases ($CaO \cdot Al_2O_3 \cdot 2SiO_2$). Additional metastable modifications of $CaO \cdot Al_2O_3 \cdot 2SiO_2$ and the metastable phase yoshiokaite ($Ca_{8-X/2}\square_{X/2}Al_{16-X}Si_XO_{32}$) are described. Yoshioka has also reported on the hydraulic properties of the kalsilite-type in the solid solution of yoshiokaite [21].

3.3.3.1 $CaO \cdot Al_2O_3 \cdot 2SiO_2$ (CAS_2)

Calcium aluminum disilicate crystallizes in nature in the triclinic modification anorthite. Takéuchi & Donnay determined the structure of the hexagonal modification of CAS_2 [116]. Ito published that the metastable phase has a monoclinic (pseudo hexagonal) structure [117]. Abe et al. [118], Iyatomi & Aoki [119], Abe & Sunagawa [120], Dimitrijević et al. [121], and Daniel et al. [122] investigated the phase formation of calcium aluminum silicates and confirmed the results of Ito [117]. A further orthorhombic modification of CAS_2 is described in the ICDD data set (ICDD: 00-005-0528). Takéuchi et al. have postulated that this modification have piezoelectric properties and has to be also monoclinic (pseudo-orthorhombic) [123]. Daniel et al. described a further monoclinic phase using Raman spectroscopy, but there is no published structural information about this phase [122]. All calcium aluminum disilicate modifications are summarized in Tab. 3.4.

Lee & Kim [124] and Lee & Lee [67] synthetized anorthite via the polymer precursor method and figured out that the crystallization of the metastable phases of CAS_2 starts at around 900 °C, and the transformation to anorthite starts at 1000 °C. Additionally, they showed that the polymer to cation ratio influences phase formation. The correct mineral name is "Dmisteinbergite". In many publications, a false name for the mineral is mentioned, "Omisteinbergite". A further triclinic high pressure modification of CAS_2 is mentioned by Angel [125].

By using the solid state reaction with the starting materials CaO, SiO_2, and Al_2O_3, only the stable triclinic phase could be reached at a sintering temperature of 1200 °C. By using the polymer precursor method with the starting materials silica sol and metal nitrates, the metastable phases occur at a temperature of 900 °C (2 h). An increasing polymer to cation ratio leads to higher crystallite sizes (Tab. 3.5). This higher crystallite size has the consequence that, instead of the pseudo-hexagonal, the pseudo orthorhombic or triclinic modification is present [124].

Tab. 3.4: Metastable and stable modifications of $CaO\cdot Al_2O_3\cdot 2SiO_2$.

Crystal system		Hexagonal	Monoclinic (pseudo-hexagonal)	Orthorhombic	Monoclinic (pseudo-orthorhombic)	Triclinic
Mineral name		Dmisteinbergite				Anorthite
RG		P6/mcm	C2	$P2_12_12_1$	$P2_1$	$P\bar{1}$
Lattice param- eters	*a* (Å)	5.10	10.24	8.22	8.228	8.1820
	b (Å)	5.10	17.74	8.61	8.621	12.8690
	c (Å)	14.72	14.99	4.84	4.827	14.1690
	α (°)	90	90	90	90	93.24
	β (°)	90	92.05	90	90	115.77
	γ (°)	90	90	90	90	91.21
ICDD		00-051-0064 00-031-0247 01-074-0814		00-005-0528		01-089-1472
ICSD		026486				086330
Literature		Takéuchi & Donnay [116]	Ito [117]		Takéuchi et al. [123]	Angel et al. [124]

Tab. 3.5: Crystalline modifications of CAS_2 by using the polymer precursor method and a sintering step of 950 °C (2 h) at different polymer contents; the amount of the different modifications was measured according to the intensities in the XRD diagram; crystallite size (L_{VOL}) measured according to the full width at half maximum (FWHM) by using the fundamental parameter approach.

$CaO\cdot Al_2O_3\cdot 2SiO_2$	Crystalline modification at a sintering step at 950 °C (2 h)		
Polymer content	**Monoclinic (pseudo-hexagonal)**	**Monoclinic (pseudo-orthorhombic)**	**Triclinic**
0.25	++++ (36 nm)	–	–
0.5	++++ (40 nm)	–	–
1.0	+++ (42 nm)	+	+
2.0	+	++ (53 nm)	++ (54 nm)

++++ single phase, +++ much, ++ medium, + little, – not present

3.3.3.2 Yoshiokaite

Yoshioka [21, 126] synthesized the phase and proved the existence of a solid solution series between the phases CA, CA_2, and CAS_2. He described also a nepheline type [126] and kalsilite type [21] of yoshiokaite with the general formula $Ca_{8-X/2}\square_{X/2}Al_{16-X}Si_XO_{32}$. The nepheline type was later synthesized from Kirkpatrick & Steele [127]. The structure of yoshiokaite with the kalsilite type was published by Steele & Pluth with the formula $Ca_{5.3}\square_{2.7}Al_{10.7}Si_{5.3}O_{32}$ [128]. Natural yoshiokaite were investigated by Vaniman & Bish [129]. Yoshiokaite was synthesized by heating a glass of the corresponding composition between 900 and 1100 °C [126]. The yoshiokaite with the kalsilite type composition was never published afterwards. The peaks in the XRD patterns described by Yoshioka [21] could be indexed by the metastable orthorhombic modification of CA which was later published by Ito et al. [74].

By using the polymer precursor method and the combustion method, the formula described by Steele & Pluth was used [128]. At a sintering temperature of 900 °C, the phase is still amorphous at a sintering step of 950 °C with a very small peak at the position of the gehlenite main peak. Using a sintering temperature of 1000 °C (2 h), yoshiokaite with additional very small amounts of gehlenite can be achieved. At higher temperatures of 1100 °C (2 h), the phase transformation of yoshiokaite to the stable phases C_2AS, CAS_2, CA_2, and small amounts of wollastonit-1A (CS) occur. At a sintering step of 1200 °C (2 h), the metastable phase completely disappeared. High temperature XRD measurements showed that by increasing the temperature from 1000 °C to 1100 °C the amount of yoshiokaite decreases and the stable phases occur. It can be confirmed that yoshiokaite is a metastable phase at lower sintering temperature and not at high temperatures.

3.3.3.3 $2CaO \cdot Al_2O_3 \cdot SiO_2$ (C_2AS)

Gehlenite is a main phase in many calcium aluminate cements [130]. Dovál et al. tried to synthesize gehlenite by using well homogenized raw materials boehmite, calcium nitrate hydrate, and silica sol in an aqueous solution [108]. They achieved a pure phase by using a sintering temperature of 1250 °C.

Using the polymer precursor method, gehlenite could be reached at a sintering temperature of 900 °C. Additional small amounts of β-C_2S und orthorhombic CA occur. These small amounts of additional phases could not be avoided because the crystallization temperature is slightly lower than the formation of gehlenite. At higher sintering temperatures up to 1300 °C, the additional phase decreases and the crystallinity of gehlenite increases.

3.3.4 Calcium aluminum ferrate – $Ca_2(Al_xFe_{2-x})O_5$

The crystal chemistry of brownmillerite, solid solutions $Ca_2(Fe_{2-X}Al_X)O_5$ is described by Redhammer et al. [131]. Iron rich brownmillerites ($0 < X < 0.6$) are stable in the modification with the space group (SG) Pnma at RT, alumina rich phases ($0.6 < X < 1.3$) are stable in the modification with the SG I2mb. In the case of the iron-rich brownmillerites, a phase transformation from the SG Pnma to I2mb occur at about 700 °C [132]. At higher temperatures, the iron rich brownmillerites show a phase transition from the SG Pnma to I2mb [131–133]. Producing pure brownmillerite phases by solid state reaction sintering temperatures of 1250 °C were used [134]. Lee et al. produced brownmillerites by using a polymeric precursor method [135]. Synthesis using the solid state reaction and hydration behavior of pure brownmillerite was studied by Negro & Safferi [136] and Neubauer et al. [134]. The results of the synthesis done by Lee et al. [135]. using the polymer precursor method showed that the peak positions for $Ca_4Al_2Fe_3O_{10}$ at 700 °C disagree with the results from Fukuda & Ando [132] and Redhammer et al. [131].

The phase forming of $Ca_2(Fe_{2-X}Al_X)O_5$ by the polymeric precursor process was studied with the following Al^{3+}/Fe^{3+} ratios: $Ca_2(Fe_{2-X}Al_X)O_5$ with $X = 0, 0.2, 0.4, 0.6, 0.8, 1.0$, and 1.2. After the decomposition of the polymer at 600 °C, calcite occurs. At a sintering step of 700 °C, the decarbonation is finished and brownmillerite is present. In Fig. 3.25, the refined unit cell volume V_0 dependent on sintering temperature and the Al^{3+}/Fe^{3+} ratio is shown. Comparing the measured unit cell parameters with the unit cell parameter determined by Redhammer et al. [131]. demonstrated that, by sintering alumina rich brownmillerites at 700 °C, iron rich brownmillerite phases were formed. Higher sintering temperatures (800 °C to 1000 °C) are needed for the incorporation of alumina into the brownmillerite structure. At 1000 °C, pure phases of brownmillerite occurred according to the starting composition. Additional crystallite sizes of brownmillerites were refined at different temperatures and Al^{3+}/Fe^{3+} ratios. The crystallite sizes increase by using higher iron contents and higher sintering temperatures (Fig. 3.26). Pure brownmillerites with crystallite sizes of minimum 40 nm are colored in grey. Phases with lower crystallite sizes are colored in light grey. Additional amorphous alumina is stable. The crystallite size of C_2F increases extensively at a sintering temperature of 1200 °C due to its lower melting temperature. Iron rich brownmillerites ($0 < X < 0.6$) are normally stable in the modification of SG Pnma. Characteristic for this modification are the $h + k + l = 2n + 1$ reflexes (111, 131, 151). Iron rich brownmillerites sintered at 700 °C (2 h) do not exhibit these reflexes because the brownmillerites are stable in the high temperature modification with the SG I2mb. At higher sintering temperatures of 800 °C (2 h; $X = 0$), 900 °C (2 h; $X = 0.2$), and 1000 °C (2 h; $X = 0.4$), a transformation into the stable modification (SG Pnma) has been observed. During this transformation of the SG an increase of the lattice parameter b_0 and decrease of c_0 could be observed. From 1000 °C (2 h) to 1200 °C (2 h), the b_0/c_0 ratio decreases again. The reason for this is the varying b_0/c_0 ratio concluding in the change of the Fe^{2+}/Fe^{3+} ratio.

3.3.5 Calcium aluminum sulfate – $4CaO{\cdot}3Al_2O_3{\cdot}SO_3$ ($C_4A_3\bar{s}$)

The polymorphs of ye'elimite $4CaO{\cdot}3Al_2O_3{\cdot}SO_3$ ($C_4A_3\bar{s}$) and the substitution by sodium and iron was published by Andac & Glasser [137]. Puertas et al. investigated the decomposition of ye'elimite synthesized by the solid state reaction [138]. They explained the evaporation of SO_3 and the formation of CA und $C_{12}A_7$ at 1300 °C. Palou et al. analyzed the reactivity of sulfo aluminate belite cements produced by a sol gel method [139].

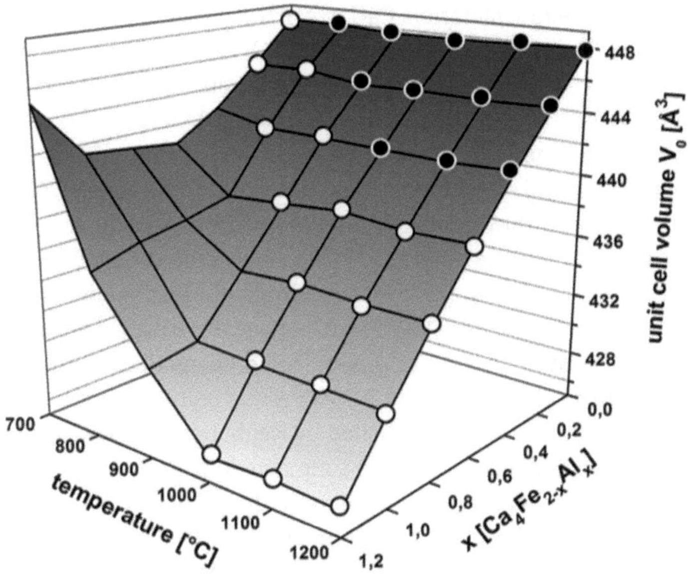

Fig. 3.25: Refined unit cell volume V_0 dependent on sintering temperature (700–1200 °C) and Al^{3+}/Fe^{3+} ratio ($0 < X < 1.2$), black dots = SG I2mb, white dots = SG Pnma.

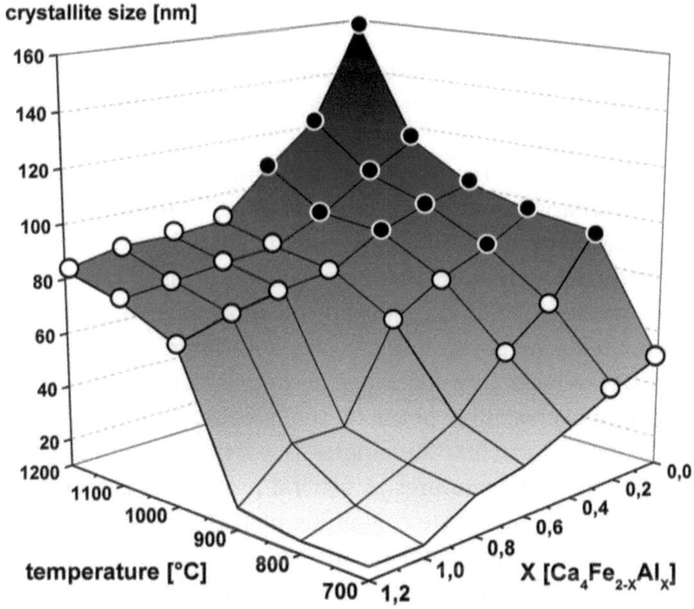

Fig. 3.26: Refined crystallite sizes dependent on sintering temperature (700 °C to 1200 °C) and Al^{3+}/Fe^{3+} ratio ($0 < X < 1.2$) by assuming spherical crystallite form, white dots = SG I2mb, black dots = SG Pnma.

Using the solid state reaction, the raw materials CaO, γ-Al_2O_3, and $CaSO_4$ were taken and heated up in the following sintering procedure: 1000 °C (2 h), 1100 °C (2 h), 1200 °C (2 h), and 1300 °C (2 h). After a sintering step at 1300 °C (2 h), pure ye'elimite in an orthorhombic modification described by Calos et al. occurred [140].

Using the polymer precursor method, the starting materials calcium nitrate hydrate, aluminum nitrate hydrate, and stoichiometric amounts of sulfuric acid (H_2SO_4) were used. The crystalline phases orthorhombic CA und anhydrite (Cs) occur at a sintering step of 1000 °C (2 h). At a sintering step of 1100 °C, the formation of C_4A_3s starts. By using sintering temperatures of 1200 °C (2 h), the additional phases CA and Cs are still present. Higher sintering temperatures (1300 °C) or longer sintering times leads to the decomposition of $CaSO_4$ and C_4A_3s. Pure phases could be synthesized by an increase of the content of sulfuric acid (H_2SO_4) by 70–100 % and using sintering temperatures of 1200 °C (4 h). Similar results are published by Neubauer [141]. He tried to substitute SO_4^{2-} by other anions and using the solid state reaction. The results show that by using low temperature synthesis methods the formation of orthorhombic CA und $CaSO_4$ occurs at low temperatures and the formation of ye'elimite can be avoided. The formation is also hindered because the decomposition temperature is decreased due to the low particle size.

3.4 Summary and Discussion

In this chapter, the results of the phase formation using different low temperature synthesis methods are concluded. Pure phases were synthesized by using the sintering temperatures and times in Tab. 3.6.

The temperatures, normally needed for the phase formation, can be decreased by several hundred degrees in comparison to the traditional solid state reaction. The most useful and practicable low temperature synthesis method is the polymer precursor method. In this case, the starting materials are very homogeneously incorporated into the polymer precursor. An important condition of the polymer precursor method is the oxidizing condition during the decomposition of the polymer. Otherwise, carbon is still in the powdery synthesis products. By a controlled decomposition of the polymer precursor, the phase formation of crystalline phases is very precise.

In particular, the formation of calcium aluminates is possible using the combustion method at lower sintering temperatures. A general problem of combustion method occurs during the synthesis of silicate containing phases because there is a strong exothermic reaction between the silica sol and the glycine. This reaction is very difficult to control and the targeted formation of the phase is nearly impossible.

The classical sol gel method is less practicable because of the long gelation time. Furthermore, very often small amounts of side phases occur when using the sol gel method and slightly higher sintering temperatures are necessary. The same results have been published by Stephan & Wilhelm [35] and Stephan & Plank [36].

Tab. 3.6: Sintering steps to obtain pure phases by different synthesis methods.

	Solid state reaction	Polymer precursor synthesis	Glycine nitrate process	Sol gel method
CA	1500 °C (16 h)	900 °C (2 h)	900 °C (2 h)	1500 °C (16 h)
C_5A_3	–	900 °C (2 h)	900 °C (2 h)	–
$C_{12}A_7$	1300 °C (50 h)	1200 °C (2 h)	1200 °C (2 h)	1300 °C (30 h)*
C_3A	1400 °C (20 h)	1200 °C (2 h)	1200 °C (2 h)	1400 °C (2 h)
Amorphous CaO and A_2O_3	–	700 °C (2 h)	–	–
C_2S	1400 °C (2 h)	650 °C (2 h)	1000 °C (2 h)	1000 °C (2 h)
C_3S	1500 °C (12 h)	1500 °C (12 h)	1500 °C (12 h)	1500 °C (12 h)
C_2AS	–	1300 °C (2 h)	1300 °C (2 h)	–
CAS_2	1200 °C (2 h)	900 °C (2 h)	900 °C (2 h)	–
CAS	–	950 °C (2 h)	–	–
$C_4A_{1.0}F_{1.0}$	1250 °C (60 h)	1000 °C (2 h)	1000 °C (2 h)	1250 °C (20 h)*
C_2F	–	700 °C (2 h)	–	–
C_4A_3s	1300 °C (60 h)	1250 °C (60 h)*	–	–

* very small amounts of additional phases

The formation of the pure phases by using the spray method is equivalent to the polymer precursor method. The advantage of the spray method is that the size of the particles can be controlled by the concentration of the solution. The crystallite size and grain size of these particles can be controlled by the sintering process afterwards. Another advantage of this method is that spray dryers are available in an all industrial scales.

Due to the lower sintering temperature and better homogenized starting materials, the following metastable phases could be reached in comparison to the solid state reaction:
- CA (orthorhombic modification), C_5A_3,
- α'_L-C_2S, β-C_2S,
- iron-rich brownmillerites with the space group I2mb,
- yoshiokaite and two monoclinic modifications of CAS_2.

In a direct comparison of the polymer precursor process and the solid state reaction, the phases summarized in Tab. 3.7 could be synthesized at significantly lower sintering temperatures.

Reasons for the formation of the phases at lower sintering temperatures in comparison to the solid state reaction are:
- a much better homogenization of the raw materials,
- a decrease of the decarbonation temperature of $CaCO_3$ from 900 to 550 °C,
- a faster diffusion on grain boundaries in comparison to the crystal lattice [142].

Tab. 3.7: Temperature to synthesized pure phases at significantly lower sintering temperatures between 700 and 950 °C.

Phase	Sintering temperature by using the polymer precursor method	Sintering temperature by using the solid state reaction
C_2S	650 °C (α'_L-C_2S)	1400 °C (γ-C_2S)
$Ca_2(Al_xFe_{2-x})O_5$ with $0 < x < 0.4$	700 °C (SG: I2mb)	1250 °C (SG: Pnma)
CA	900 °C (orthorhombic)	1500 °C (monoclinic)
C_5A_3	900 °C*	–
C_2AS	900 °C*	1250 °C
CAS_2	900 °C (monoclinic)	900 °C (triclinic)
$Ca_{5.3}\square_{2.7}Al_{10.7}Si_{5.3}O_{32}$	950 °C*	–

* contain very small amount of additional phases

Tab. 3.8: Phases that can be formed at temperatures higher than 900 °C despite using low temperature synthesis methods.

Phase	T_U (°C)	Phases below the sintering temperature T_U
C_3S	1250	C_2S, CaO
$C_{12}A_7$	1000	C_5A_3
C_3A	900	C_5A_3, $C_{12}A_7$, C_3A
C_2AS	900	C_2AS, C_2S, CA (ortho.)
$Ca_2(Al_xFe_{2-x})O_5$ with $0.6 < x < 1.2$	1000	$Ca_2(Al_xFe_{2-x})O_5$ with $0 < x < 0.4$
C_4A_3s	1000	CA (ortho.), Cs
$Ca_{20}Al_{32-2x}Mg_xSi_xO_{68}$	1100	CA (ortho.), CAS_2

The reason for the decrease in the decarbonation temperature is the much lower particle size between 10 and 50 nm after the decomposition of the precursor phases of the different synthesis methods. The nanoscale particles have a lower vapor pressure because more atoms are present on the surface with less contact points to the atoms beside. The mobility is much higher and the decarbonation temperature decrease. Ries et al. measured a similar lower decarbonation temperature also during the decomposition of strontium carbonate and barium carbonate [143].

By using the polymer precursor method, the same metastable phases occur at lower sintering temperatures T_U (Tab. 3.8, right column).

The usage of well homogenized raw materials leads to the formation of metastable phases (Tab. 3.9, left column) at lower sintering temperatures. The corresponding stable phase formed by the usage of higher sintering temperatures or a longer sintering time are summarized in the right column (Tab. 3.9).

Tab. 3.9: Metastable and corresponding stable phases by using low temperature synthesis methods

Metastable phase	Stable phase
α'_L-C_2S	γ-C_2S
β-C_2S	γ-C_2S
$Ca_2(Al_xFe_{2-x})O_5$ with $0 < x < 0.4$ (SG: I2mb)	$Ca_2(Al_xFe_{2-x})O_5$ with $0 < x < 0.4$ (SG: Pnma)
CA (orthorhombic)	CA (monoclinic)
$Ca_{5.3}\square_{2.7}Al_{10.7}Si_{5.3}O_{32}$	CA, CA_2, CAS_2, CS
CAS_2 (monoclinic – pseudo-hexagonal)	CAS_2 (triclinic)
CAS_2 (monoclinic – pseudo-orthorhombic)	CAS_2 (triclinic)
C_5A_3 (+ small amounts of C_3A)	$C_{12}A_7$

Tab. 3.10: Modifications of C_2S by using the polymer precursor method at different sintering steps.

Modification	Sintering step	D_{XRD} (g/cm³)
α'_L-C_2S	650 °C (2 h)	3.14
β-C_2S	1100 °C (2 h)	3.31
γ-C_2S	1300 °C (2 h)	2.96

The synthesis of C_2S by using the low temperature synthesis methods leads to the formation of the high temperature modifications α'_L-C_2S and β-C_2S. Higher sintering temperatures increase the crystallite and particle size and a conversion to the stable γ-C_2S modification (Tab. 3.10) occurs.

Chan et al. described how small particle sizes stabilize β-C_2S and a transformation to γ-C_2S is avoided due to the small particles which have more atoms with a higher surface energy [106]. This leads to a higher packing density and the stabilized β-C_2S with a higher X-ray density in comparison to γ-C_2S [23]. With the same theory, the stabilization of α'_L-C_2S cannot be explained. For the stabilization of the higher symmetric α'_L-C_2S, a higher nucleation probability at a lower sintering temperature (600 °C) has to be taken into account.

In one line of C_2S, iron rich brownmillerites $Ca_2(Al_xFe_{2-x})O_5$ with $0 < x < 0.4$ also show a stabilization of the high temperature modification (SG: I2mb) at low sintering temperatures. An increasing aluminum (x) content from 0 to 0.4 shows that the stabilization increases by using higher sintering temperatures of 700–900 °C. Also in this case, the higher symmetric high temperature modification is stabilized at lower sintering temperatures. The calculation of the crystallite size shows that the conversion of the metastable high temperature modifications into the stable modification occurs if the crystallite size is higher than 90 nm. The crystallite size could be seen as an indicator of the phase conversion or the crystallite size has a direct influence to the stabilization of the high temperature modification.

In the case of the monoclinic modifications of CAS_2, yoshiokaite, and orthorhombic CA, the formation can be concluded to have a higher nucleation probability at a lower sintering temperature. The high temperature modifications are not known and cannot be identified using our own high temperature X-ray diffraction measurements. By using the low temperature synthesis methods, the preferred formation of the monoclinic (pseudo-hexagonal) modification of CAS_2 was obtained and not the monoclinic (pseudo-orthorhombic) modification of CAS_2. One reason for this is that the sintering process of the precursor powder started at room temperature [118]. In a cooling process from 1250 to 950 °C, the preferred formation of the monoclinic (pseudo-orthorhombic) modification of CAS_2 was measured by Iyatomi & Aoki [119].

Calculated crystallite sizes using XRD of the phases C_2S, C_2F, C_5A_3, and orthorhombic CA directly after the decomposition of $CaCO_3$ formed in the meanwhile correlate with the grain sizes determined by SEM. It can be assumed that nanoscaled single crystals are present directly after the phase formation. At higher sintering temperatures, a very fast increase of the grain and crystallite size occur.

The synthesis of aluminum containing phases by using the polymer precursor method leads to the formation of an amorphous precursor at a sintering temperature up to 800 °C. Crystalline aluminum containing phases occur at sintering temperatures above 900 °C. Thermogravimetric measurements show a weight loss of about 0.2 mass-% up to 2 mass-% at around 950 °C. Gülgün et al. [69] and Hernández & González [60] measured the evaporation of carbon at these temperatures by using the polymer precursor method. Gülgün et al. discussed how carbon is absorbed on the surface as a consequence of decomposition with not enough oxygen [69]. Our own investigations with very high flushing rates of synthetic air have shown that weight loss is always present. It could therefore be excluded that carbon is present on the surface of the particles [91]. The presence of $CaCO_3$ is excluded by IR spectroscopy. Hernández & González also measured the weight loss by the synthesis of pure alumina [60]. A more reasonable explanation could be found by Lin et al. [92]. They identified amorphous –O–Al–C–O–Al–O– bond as a product of the decomposition of aluminum organic phases. The effect could be measured in our own investigations by the decomposition of $Al(OH)(HCOO)_2 \cdot xH_2O$. This could be the reason there were no crystalline calcium aluminates at lower sintering temperatures than 900 °C. Despite the investigations of Douy & Gervais [76], no formation of CA, $C_{12}A_7$, and C_5A_3 by the decomposition of spray-dried calcium nitrate hydrates and aluminum nitrate hydrates and using sintering temperatures less than 900 °C was detected either. A prevention of the crystallization because of the presence of the –O–Al–C–O–Al–O– bonds could be excluded. Tas has demonstrated the crystallization of calcium rich phases C_3A, $C_{12}A_7$, or C_5A_3 below 900 °C by using sintering times of several days using the glycine nitrate process [46]. Experiments from Turrillas et al. [144] and Splittgerber & Mueller [18] have also shown the formation of C_3A and $C_{12}A_7$ at around 600 °C by using C_3AH_6 as starting material. Formation of C_5A_3 was not observed. The reason is obviously the high humidity by the dehydration of hydrates that promote the direct formation of $C_{12}A_7$ [94, 98].

The high temperature phase C_3S cannot be synthesized at lower temperatures by well homogenized raw materials. The reason is that C_2S und CaO crystallize with the use of lower sintering temperatures than 1250 °C. In this case, only very high heating rates minimized the formation of stable phases at lower sintering temperatures. The same phenomenon occurs during the synthesis of ye'elimite (C_4A_3s). Anhydrite and the metastable orthorhombic modification of CA occur below 900 °C. Afterwards, the formation of C_4A_3s is hindered above 1000 °C. During the synthesis of C_4A_3s, a considerable decrease of the decomposition temperature of Cs and C_4A_3s was also estimated, which makes the synthesis of this pure phase very difficult.

Summarizing the results of the phase formation, it could be demonstrated that by using well homogenized raw materials the formation of metastable phases in Tab. 3.9 have to be considered by using sintering temperatures below 950 °C. Based on the phase diagram of $CaO–Al_2O_3–SiO_2$ and the use of low sintering temperatures, the formation of the marked metastable phases in Fig. 3.27 must be considered.

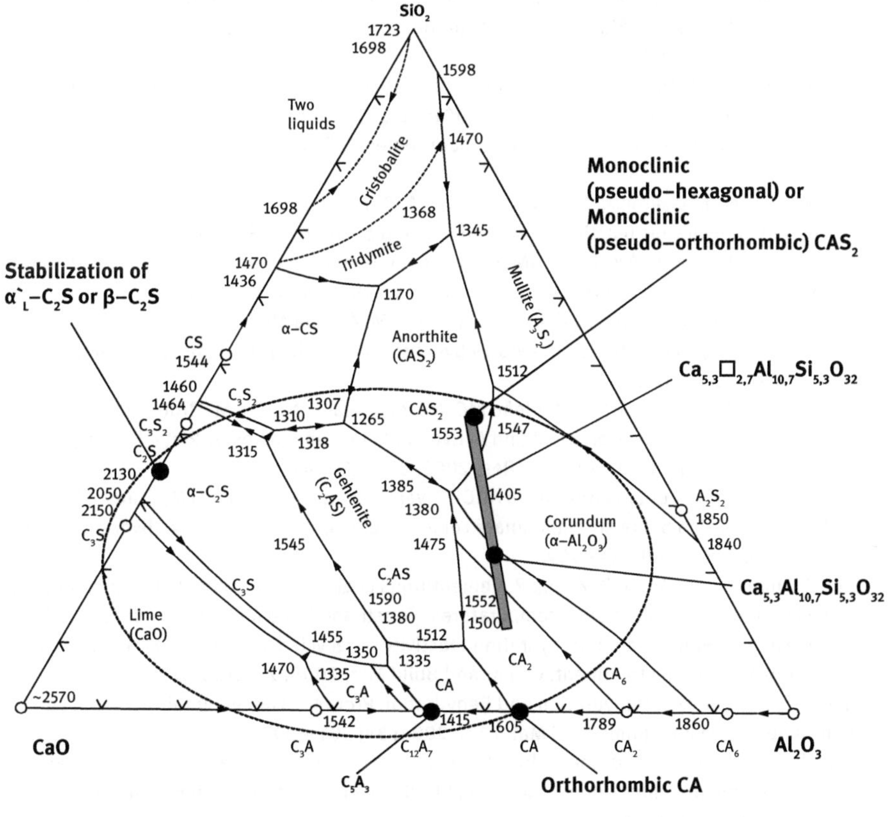

Fig. 3.27: Possible metastable phases in the system $CaO–Al_2O_3–SiO_2$, by using well homogenized raw materials and the usage of low sintering temperatures.

Acknowledgment: The authors would like to thank the Zentrum für Werkstoffanalytik Lauf GmbH for providing the possibility of SEM investigations and Prof. Ebbinghaus (Institute of Inorganic Chemistry, Martin-Luther-University Halle-Wittenberg) for providing the possibility to use the spray dryer.

References

[1] Walenta G. Reduction of Cement specific CO_2 emissions using XRD Rietveld. Building and Material Analysis Workshop. Halle, Germany; 2009.

[2] Pöllmann H. Mineralogy and Crystal Chemistry of Calcium Aluminate Cement. Calcium Aluminate Cements. Edinburgh, UK: IOM Communications Ltd.; 2001.

[3] Pöllmann H. Composition of cement phases. In: Bensted J, Barnes P, editors. Structure and Performance of Cements. 2nd edition. London: Spon Press; 2002: 25–56.

[4] Pöllmann H, Winkler N, Oberste-Padtberg R, Meyer R, Goeske J, Raab B. Quantitative mineralogical, chemical and application investigations on some High Alumina Cements from different sources. 28th International Conference on Cement Microscopy. Denver, USA; 2006.

[5] Popescu CD, Muntean M, Sharp JH. Industrial trial production of low energy belite cement. Cem & Concr Comp. 2003; 25: 689–693.

[6] Glasser FP, Zhang L. High-Performance Cement Matrices Based on Calcium Sulfoaluminate-Belite Compositions. Cem and Concr Res. 2001: 31; 1881–1886.

[7] Quillin K. Performance of belite-sulfoaluminate cements. Cem and Concr Res. 2001; 31: 1341–1349.

[8] Kacimi L, Simon-Masseron A, Salem S, Ghomari A, Derriche Z. Synthesis of belite cement clinker of high hydraulic reactivity. Cem and Concr Res. 2009; 39: 559–565.

[9] El-Didamony H, Sharara AM, Helmy IM, El-Aleem SA. Hydration Characteristics of β-C_2S in the Presence of some Accelerators. Cem and Concr Res. 1996; 26(8): 1179–1187.

[10] Morsli K, laTorre ÁGd, Stöber S, Cuberos AJM, Zahir M, Aranda MAG. Quantitative Phase Analysis of Laboratory-Active Belite Clinkers by Synchrotron Powder Diffraction. J Am Ceram Soc. 2007; 90(10): 3205–3212.

[11] Cuberos AJM, Torre ÁGDl, Martín-Sedeño MC, Moreno-Real L, Merlini M, Ordónez LM, et al. Phase Development in Conventional and Active Belite Cement Pastes by Rietveld Analysis and Chemical Constraints. Cem and Concr Res. 2009; 39: 833–842.

[12] Martín-Sedeño MC, Cuberos AJM, Torre ÁGDl, Álvarez-Pinazo G, Ordónez LM, Gateshki M, et al. Aluminum-rich belite sulfoaluminate cements: Clinkering and early age hydration. Cem and Concr Res. 2010; 40: 359–369.

[13] Shui Z, Xuan D, Chen W, Yu R, Zhang R. Cementitious characteristics of hydrated cement paste subjected to various dehydration temperatures. Constr and Build Mat. 2009; 23: 531–537.

[14] Shui Z, Xuan D, Wan H, Cao B. Rehydration reactivity of recycled mortar from concrete waste experienced to thermal treatment. Constr and Build Mat. 2008; 22: 1723–1729.

[15] Alonso C, Fernandez L. Dehydration and Rehydration Processes of Cement Paste Exposed to High Temperature Environments. J Mater Sci. 2004; 39: 3015–3024.

[16] Castellote M, Alonso C, Andrade C, Turrillas X, Campo J. Composition and Microstructural Changes of Cement Pastes upon Heating, as Studied by Neutron Diffraction. Cem and Concr Res. 2004; 34: 1633–1644.

[17] Handoo SK, Agarwal S, Agarwal SK. Physicochemical, Mineralogical and Morphological Characteristics of Concrete Exposed to Elevated Temperatures. Cement and Concrete Research 2002; 32: 1009–1018.

[18] Splittgerber F, Mueller A. Inversion of the Cement Hydration as a new method for Identification and/or Recycling? 11th International Congress on the Chemistry of Cement. Durban, South Africa; 2003.
[19] Stemmermann P, Garbev K, Schweike U, Bäuchle G, inventors. Belite-Based Binder Production Method; 2007.
[20] Puertas F, García-Díaz I, Barba A, Gazulla MF, Palacios M, Gómez MP, et al. Ceramic wastes as alternative raw materials for Portland cement clinker production. Cem & Concr Comp, 2008; 30: 798–805.
[21] Yoshioka T. A New Crystal with Kalsilite-type Structure on the $CaAl_2O_4$-SiO_2 Join. Bull Chem Soc Jap. 1970; 43: 2317–2321.
[22] Schubert U, Hüsing N. Synthesis of Inorganic Materials. Series. Weinheim: WILEY-VCH; 2000.
[23] Wegner K, Pratsinis SE, Köhler M. Chemische Technik: Prozesse und Produkte. In: Dittmeyer R, Keim W, Kreysa G, Oberholz A, editors. Nanomaterialien und Nanotechnologie, Band 2: Neue Technologie, Weinheim: WILEY-VCH Verlag GmbH & Co.KGaA; 2004.
[24] Kozuka H. Volume I: Sol-Gel Porcessing. In: Kozuka H, editor. Handbook of Sol-Gel Science and Technology: Processing Characterization and Applications – Volume I: Sol-Gel Processing. Norwell: Kluwer Academic Publishers; 2004.
[25] Rahaman MN. Ceramic Processing. Series, Boca Raton: CRC Press, 2007.
[26] Wesselsky A, Jensen OM. Synthesis of pure Portland cement phases. Cem and Concr Res. 2009; 39: 973–980.
[27] Bergna HE. The Colloid Chemistry of Silica. In: Bergna HE, editor. The Colloid Chemistry of Silica – An Overview, Washington: American Chemical Society. 1994: 1–51.
[28] Livage J. Basic Principles of Sol-Gel Chemistry. In: Aegerter MA, Menning M, editors. Sol-Gel Technologies for the Producers and Users. Boston: Kluwer Academic Publishers. 2004: 4–14.
[29] Brinker J, Scherer GW. Sol-gel Science: The Physics and Chemistry of Sol-gel Processing. Series. Boston: Academic Press; 1990.
[30] Fu Y-P, Lin C-H. Synthesis and microwave characterization of 2(MgO,CaO)-$2Al_2O_3$-$5SiO_2$ glass ceramics from the sol-gel process. J Mater Sci Lett. 2003; 38: 3081–3084.
[31] Saravanapavan P, Hench LL. Mesoporous calcium silicate glasses. I. Synthesis. J Non-Crystall Sol. 2003; 318: 1–13.
[32] Martınez A, Izquierdo-Barba I, Vallet-Regi M. Bioactivity of a CaO-SiO_2 Binary Glasses System. Chemistr Mater. 2000; 12: 3080–3088.
[33] Chrysafi R, Perraki T, Kakali G. Sol-gel Preparation of $2CaO{\cdot}SiO_2$. J Eur Ceram Soc. 2007; 27: 1707–1710.
[34] Catauro M, Laudisio G. Sol-gel Processing and Crystallization of Calcium Silicate Glasses. J Mater Sci Lett. 1998; 17: 311–312.
[35] Stephan D, Wilhelm P. Synthesis of Pure Cementitious Phases by Sol-Gel Process as Precursor. Zeitschrift für Anorganische und Allgemeine Chemie. 2004; 630: 1477–1483.
[36] Stephan D, Plank J. Influence of Foreign Oxides on Lattice Parameters and Reactivity of Pure Cement Phases of OPC. 12th International Congress on the Chemistry of Cement. Montreal, Canada; 2007.
[37] Wang Y, Thomson WY. Kinetic studies of tricalcium silicate formation from sol-gel precursors. J Mater Sci. 1996; 31: 1319–1325.
[38] Fujimori H, Yahata D, Yamaguchi N, Ikeda D, Ioku K, Goto S. Synthesis of Calcium Silicate by a Chelate Route with Aqueous Solution of Citric Acid. J Ceram Soc Jap. 2001; 109(5): 391–395.
[39] Goktas AA, Weinberg MC. Preparation and Crystallization of Sol-Gel Calcia-Alumina Compositions. J Am Ceram Soc. 1991; 74(5): 1066–1070.
[40] Kerns L, Weinberg MC, Myers S, Assink R. Al coordination in sol-gel and conventional calcium aluminate. J Non-Crystall Sol. 1998: 232–234, 86–92.

[41] Ghosh NN, Pramanik P. Synthesis of Nano-Sized Ceramic Powders using Precipitated Silica in Aqueous Sol-Gel Method. Nano Structured Materials. 1997; 8(8): 1041–1055.
[42] Ghosh NN, Pramanik P. Synthesis of Mullite Powders Using Precipitated Silica and Tetraethoxy Silane and Aluminium-Formate as Precursor Materials in Aqueous Medium. Bull Mater Sci, 1997; 20(2): 283–286.
[43] Ghosh NN, Pramanik P. Synthesis of Spodumene and Spodumene-Zirconia Composite Powders Using Aqueous Sol-gel Method. Bull Mater Sci. 1997; 20(2): 247–257.
[44] Page CH, Thombare CH, Kamat RD, Chartterjee AK. Development of Sol-gel Technology for Cement Manufacture. British ceramic: transactions and journal. 1991; 16: 643–660.
[45] Kingsley JJ, Patil KC. A Novel Combustion Process for the Synthesis of Fine Particles α-Alumina and Related Oxide Materials. Mat Lett. 1988; 6(11/12): 427–432.
[46] Tas AC. Chemical Preparation of the Binary Compounds in the Calcia-Alumina System by Self-Propagating Combustion Synthesis. J Am Cer Soc. 1998; 81(11): 2853–2863.
[47] Hwang C-C, Wu T-Y. Synthesis and characterization of nanocrystalline ZnO powders by a novel combustion synthesis method. Mat Sci and Eng. 2004; B111: 197–206.
[48] Hwang C-C, Wu T-Y, Wan J. Design and modify the combustion synthesis method to synthesize ceramic oxide powders. J Mat Sci. 2004; 39: 4687–4691.
[49] Yi HC, Guigne JY, Moore JJ, Schowengerdt FD, Robinson LA, Manerbino AR. Preparation of calcium aluminate matrix composites by combustion synthesis. J Mat Sci. 2002; 37: 4537–4543.
[50] Chandran RG, Patil KC, Candrappa GT. Combustion Synthesis, Characterization, Sintering and Microstructure of Mullite-Cordierite Composites. J Mat Sci Lett. 1995; 14: 548–551.
[51] Huang X-H, Chang J. Low-temperature synthesis of nanocrystalline beta-dicalcium silicate with high specific surface area. J Nano Res. 2007; 9: 1195–1200.
[52] Sreekanth-Chakradhar RP, Nagabhushana BM, Chandrappa GT, Ramesh KP, Rao JL. Solution combustion derived nanocrystalline macroporous wollastonite ceramics. Mat Chem and Phys. 2006; 95(1): 169–175.
[53] Stievano L, Piao LY, Lopes I, Meng M, Costa D, Lambert J-F. Glycine and lysine and reactivity on the surface of amorphous silica. Eur J Miner. 2007; 19: 321–331.
[54] Bosković SB, Matovic BZ, Vlajić MD, Krisitić VD. Modified Glycine Nitrate Procedure (MGNP) for the Synthesis of SOFC Nanopowders. Cer Int. 2007; 33: 89–93.
[55] Kim WJ, Park JY, Oh SJ, Kim YS, Hong GW, Kuk IH. Characteristics and sintering behavior of yttria powders synthesized by the combustion process. J Mat Sci Lett. 1999; 18: 411–413.
[56] Peng T, Liu X, Dai K, Xiao J, Song H. Effect of acidity on the glycine-nitrate combustion synthesis of nanocrystalline alumina powder. Mat Res Bull. 2006; 41: 1638–1645.
[57] Yang YJ, Wen TL, Tu H, Wang DQ, Yang J. Characteristics of lanthanum strontium chromite prepared by glycine nitrate process. Solid State Ionics. 2000; 135: 475–479.
[58] Chick LA, Pederson LR, Maupin GD, Bates JL, Thomas LE, Exarhos GJ. Clycine-Nitrate Combustion Synthesis of Oxide Ceramic Powders. Mat Lett. 1990; 10(1): 6–12.
[59] Pechini MP, inventor. US-Patent. USA; 1967.
[60] Hernández MT, González M. Synthesis of resins as alpha-alumina precursors by the Pechini method using microwave and infrared heating. J Eur Cer Soc. 2002; 22: 2861–2868.
[61] Hong S-H, Young JF. Hydration Kinetics and Phase Stability of Dicalcium Silicate Synthesized by the Pechini Process. J Am Cer Society. 1999; 87(7): 1681–1686.
[62] Kakali G, Chrysafi R, Gaki A, Perraki T, Tsitouras A. Wet Chemical synthesis of hydraulic compounds and materials. 12th International Congress on the Chemistry of Cement. Montreal, Canada; 2007.
[63] Gaki A, Chrysafi R, Kakali G. Chemical Synthesis of Hydraulic Calcium Aluminate Compounds Using the Pechini Technique. J Eur Cer Soc. 2007; 27: 1781–1784.

[64] Gaki A, Perraki T, Kakali G. Wet Chemical Synthesis of Monocalcium Aluminate. J Eur Cer Soc. 2007; 27: 1785–1789.

[65] Nishio K, Tsuchiya T. Sol-Gel Processing of Thin Films with Metal Salts. In: Kozuka H, editor. Handbook of Sol-Gel Science and Technology: Processing Characterization and Applications – Volume I: Sol-Gel Processing. Norwell, USA: Kluwer Academic Publishers; 2004.

[66] Petrykin V, Kakihana M. Chemistry and Applications of Polymeric Gel Precursors. In: Sakka S, editor. Handbook of Sol-Gel Science and Technology: Processing Characterisation and Applications – Volume I: Sol-Gel Processing. Norwell, USA: Kluwer Academic Publishers; 2004.

[67] Lee S-J, Lee M-H. Effect of degree of polymerization on synthesis of anorthite-type powder by wet-chemical method employing PVA. J Mat Sci Lett. 2003; 22: 1291–1293.

[68] Chai YL, Ray DT, Chen GJ, Chang YH. Synthesis of $LaSr_{0,8}Co_{0,2}Ni_{0,5}O_{3-\delta}$ Thin Films for High Sensitivity Sensing Material Using the Pechini Process. J All and Comp. 2002; 333: 147–153.

[69] Gülgün MA, Popoola OO, Kriven WM. Chemical Synthesis and Characterization of Caclium Aluminate Powders. J Am Cer Soc. 1994; 77(2): 531–539.

[70] Lee S-J, Lee C-H, Kriven WM. Synthesis of oxide ceramic powders by polymerized organic inorganic complex route. J Cer Proc Res. 2000; 1(2): 92–95.

[71] Pati RK, Panda AB, Pramanik P. Preparation of Nanocrystalline Calcium Aluminate Powders. J Mat Synth and Proc. 2002; 10(4): 157–161.

[72] Anjos PNM, Pereira EC, Gobato YG. Study of the structure and optical properties of rare-earth-doped aluminate particles prepared by an amorphous citrate sol–gel process. J All and Comp. 2005; 391: 277–283.

[73] Lee S-J, Kriven WM. Synthesis and hydration study of Portland cement components prepared by the organic steric entrapment method. Mat and Struc. 2005; 38: 87–92.

[74] Ito S, Ikai K, Suzuki K, Inagaki M. Metastable Orthorhombic $CaO{\cdot}Al_2O_3$. Journal Am Cer Soc. 1975; 58(1/2): 79–80.

[75] Gablenz S. Sprühtrocknung und Sprühhydrolyse – Neue Methoden zur Herstellung von Metalloxiden und zur Modifizierung von $BaTiO_3$. Halle (Saale): Martin-Luther-Universität Halle-Wittenberg; 2001.

[76] Douy A, Gervais M. Crystallization of Amorphous Precursors in the Calcia–Alumina System: A Differential Scanning Calorimetry Study. J Am Cer Soc. 2000; 83(1): 70–76.

[77] Severin JW, With GD, Hal HAMV. Low-Temperature Synthesis Route for $YBa_2Cu_3O_x$ Powder. Pysica. 1988; C152: 159–160.

[78] Lenggoro IW, Itoh Y, Iida N, Okuyama K. Control of size and morphology in NiO particles prepared by a low-pressure spray pyrolysis. Mat Res Bull. 2003; 38: 1819–1827.

[79] Messing GL, Zhang SC, Jayanthi GV. Ceramic Powder Synthesis by Spray Pyrolysis. J Am Cer Soc. 1993; 76(11): 2707–2726.

[80] Stopic S, Dovrak P, Friedrich B. Synthesis of Spherical Nanosized Copper Powder by Ultrasonic Spray Pyrolysis. World of Metallurgy – Erzmetalle. 2005; 58(4): 191–197.

[81] Stopic S, Dvorak P, Friedrich B. Synthesis of spherical nanosized silver powder by ultrasonic spray pyrolysis. Metall. 2006; 60: 299–304.

[82] Block J, Dolhert LE. Preparation of Orthorhombic $YBa_2Cu_3O_{7-\delta}$ by Spray Drying Formate Solutions. Mat Lett. 1991; 11(10–12): 334–338.

[83] Peshev P, Pecheva M. Preparation of Spinel Lithium Ferrite by Thermal Treatment of Spray-Dried Formates. Material Research Bulletin. 1978; 13(11): 1167–1174.

[84] Kim CS, Ahn KW, Rah SC, Kim SG. Preparation of Silica Nanostructured Spheres by Sol Spray Drying. Drying Technology. 2008; 26: 260–264.

[85] Hallstedt B. Assessment of the CaO-Al_2O_3 System. J Am Cer Soc. 1990; 73(1): 15–23.

[86] Uberoi M, Risbud SH. Processing of Amorphous Calcium Aluminate Powders at $< 900\,°C$. J Am Cer Soc. 1990; 73(6): 1768–1770.

[87] Vallino M. Kinetic Study of the Crystallization of Amorphous-Derived $CaO \cdot 2Al_2O_3$. Ceram Int. 1984; 10(1): 30–32.
[88] Janáková S, Salavcová L, Renaudin G, Filinchuk Y, Boyer D, Boutinaud P. Preparation and structural investigations of sol-gel derived Eu^{3+}-doped $CaAl_2O_4$. J Phys and Chem of Solids. 2007; 68: 1147–1151.
[89] Hoerkner W, Mueller-Buschbaum H. Zur Kristallstruktur von $CaAl_2O_4$. J Inorg and Nucl Chem. 1976; 38: 983–984.
[90] Raab B, Poellmann H. Investigation of the Hydration Behaviour of High Reactive Pure Cement Phases. 32nd International Conference on Cement Microscopy. New Orleans, USA: 2010.
[91] Raab B. Synthese und Charakterisierung nanoskaliger hydraulisch hochreaktiver Phasen des Portland- und Tonerdezements. Martin-Luther-Universität Halle-Wittenberg; 2010.
[92] Lin C, Cheng MYZ, Zhang C, Meng Q, Lin J. Bluish-White Emission from Radical Carbonyl Impurities in Amorphous Al_2O_3 Prepared via the Pechini-Type Sol-Gel Process. Inorg Chem. 2008; 47: 49–55.
[93] Bachiorrini A. Polymorphism of Monocalcium Aluminate. Study by Infrared Spectroscopy. Cem and Concr Res. 1985; 15(1): 167–73.
[94] Zhmoidin GI, Chatterjee AK. Conditions and Mechanism of Interconvertibility of Compounds $12CaO \cdot 7Al_2O_3$ and $5CaO \cdot 3Al_2O_3$. Cem and Concr Res. 1984; 14: 386–396.
[95] Taylor HFW. Cement Chemistry. Series. London: Academic Press; 1997.
[96] Aruja E. The Unit Cell of Orthorhombic Pentacalcium Trialuminate, $5CaO \cdot 3Al_2O_3$. Acta Crystallographica. 1956; 10: 337–339.
[97] Vincent MG, Jeffery JW. The Crystal Structure of Pentacalcium Trialuminate, $5CaO \cdot 3Al_2O_3$. Acta Crystallographica. 1978; B34: 1422–1428.
[98] Brisi C, Lucco-Borlera M, Montanaro L, Negro A. Hydration of $5CaO \cdot 3Al_2O_3$. Cem and Concr Res. 1986; 16: 156–160.
[99] Stokes AR, Wilson AJC. A method of calculating the integral breadths of Debye-Scherrer lines. Mathematical Proceedings of the Cambridge Philosophical Society. Cambridge University Press; 1942.
[100] Cheary RW, Coelho AA, Cline JP. Fundamental Parameters Line Profile Fitting in Laboratory Diffractometers. J Res Natl Inst Stand and Technology. 2004; 109: 1–25.
[101] Lee FC, Glasser FP. Powder diffraction data for compounds in the series $Na_x(Ca_{3-x}Na_x)Al_2O_6$. J Appl Crystallogr. 1979; 12: 407–410.
[102] Wistuba SP. Einfluss der Na_2O-Dotierung von Tricalciumaluminat und Portlandzementklinker auf die Wechselwirkung mit Polycarboxylat-basierten Fließmitteln. Münschen: Technische Universität München; 2008.
[103] Lazau I, Pacurariu C, Escedi Z. Tricalcium Aluminate Synthesis by Unconventional Methods. Romanian International Conference on Chemistry and Chemical Engineering. Bucharest, Romania; 2005.
[104] Mumme WG, Hill RJ, Bushnell-Wye G, Segnit ER. Rietveld crystal structure refinements, crystal chemistry and calculated powder diffraction data for the polymorphs of dicalcium silicate and related phases. Neues Jahrbuch für Mineralogie – Abhandlungen. 1995; 169(1): 35–68.
[105] Remy C, Guyot F, Madon M. High Pressure Polymorphism of Dicalcium Silicate Ca_2SiO_4. A Transmission Electron Microscopy Study. Physics and Chemistry of Minerals. 1995; 22: 419–427.
[106] Chan CJ, Kriven WM, Young JF. Physical Stabilization of the $\beta \rightarrow \gamma$ Transformation in Dicalcium Silicate. Journal of the American Ceramic Society. 1992; 75(6): 1621–1627.
[107] Nettleship I, Shull JL, Kriven WM. Chemical Preparation and Phase Stability of Ca_2SiO_4 and Sr_2SiO_4 Powders. J Eur Cer Soc. 1993; 11: 291–298.

[108] Dovál M, Palou M, Mojumdar SC. Hydration Behaviour of C_2S and C_2AS Nanomaterials, Synthesized by Sol-Gel Method. J Ther Anal and Calorimetry. 2006; 86(3): 595–599.

[109] Saalfeld H. X-Ray Investigation of Single Crystals of β-Ca_2SiO_4 (Larnite) at High Temperatures. Am Mineralogist. 1975; 60: 824–827.

[110] Jelenić I, Bezjak A. Electron Diffraction Evidence for Superstructures in α′-Modification of Dicalcium Silicate. Cem and Concr Res. 1982; 12: 785–788.

[111] Shaw S, Henderson CMB, Komanschek BU. Dehydration/recrystallization mechanisms, energetics, and kinetics of hydrated calcium silicate minerals: an in situ TG/DSC and synchrotron radiation SAX/WAXS study. Chem Geology. 2000; 167: 141–159.

[112] Hughes DC, Jaglin D, Kozłowski R, Mucha D. Roman cements – Belite cements calcined at low temperature. Cem and Concr Res. 2009; 39: 77–89.

[113] Puertas F, Trivino F. Examination by Infra-Red Spectroscopy for the Polymorphs of Dicalcium Silicate. Cem and Concr Res. 1985; 15: 127–133.

[114] Phillips B, Muan A. Phase Equilibria in the System CaO-Iron Oxide-SiO_2 in Air. J Am Cer Society. 1959; 42(9): 413–423.

[115] Regourd M. Détermination des réseaux de cristaux microscopiques. Application aux différentes formes du silicate tricalcique. Bulletin de la Société Française de Minéralogie et de Cristallographie. 1964; 87: 241–272.

[116] Takéuchi Y, Donnay G. The Crystal Structure of Hexagonal $CaAl_2Si_2O_8$. Acta Crystallographica. 1959; 12: 465–470.

[117] Ito J. High Temperature Solvent Growth of Anorthite on the Join $CaAl_2Si_2O_8$-SiO_2. Contributions to Mineralogy and Petrology. 1976; 59: 187–194.

[118] Abe T, Tsukamoto K, Sunagawa I. Nucleation, Growth and Stability of $CaAl_2Si_2O_8$ Polymorphs. Physics and Chemistry of Minerals. 1991; 17: 473–484.

[119] Iyatomi N, Aoki Y. Crystallization behaviour of $P2_1$ and C2 phases of anorthite composition. Neues Jahrbuch für Mineralogie – Monatshefte. 1992; 4: 177–192.

[120] Abe T, Sunagawa I. Hexagonal $CaAl_2Si_2O_8$ in a high temperature solution; metastable crystallization and transformation to anorthite. Min Journal. 1995; 17(6): 257–281.

[121] Dimitrijević R, Dondur V, Kremenović A. Thermally Induced Phase Transformations of Ca-exchanged LTA and FAU Zeolite Frameworks: Rietveld Refinement of the Hexagonal $CaAl_2Si_2O_8$ Diphyllosilicate Structure. Zeolites. 1996; 16: 294–300.

[122] Daniel P, Gillet ID, Gillet P, McMillan PF, Richet P. An In-Situ High-Temperature Structural Study of Stable and Metastable $CaAl_2Si_2O_8$ Polymorphs. Mineralogical Magazine. 1995; 59: 25–33.

[123] Takéuchi Y, Haga N, Ito J. The crystal structure of monoclinic $CaAl_2Si_2O_8$: a case of monoclinic structure closely simulating orthorhombic symmetry. Zeitschrift für Kristallographie. 1973; 137: 380–398.

[124] Lee S-J, Kim G-S. Characteristics and densification behavior of anorthite powder synthesized by a solution process employing a polymer carrier. J Cer Processing Research. 2002; 3(3): 136–140.

[125] Angel RJ. Order-disorder and the high-pressure $P\bar{1}$-$I\bar{1}$ transition in anorthite. Am Mineralogist. 1992; 77: 923–929.

[126] Yoshioka T. Metastable Solid Solution with Nepheline-type Structure in the CaO-Al_2O_3-SiO_2 System. Bul of the Chem Soc of Japan. 1970; 43: 1981–1987.

[127] Kirkpatrick RJ, Steele IM. Hexagonal $CaAl_2SiO_6$: A new synthetic phase. Am Mineralogist. 1973; 58: 945–946.

[128] Steele IM, Pluth JJ. Crystal structure of synthetic yoshiokaite, a stuffed derivative of the tridymite structure. Am Mineralogist. 1990; 75: 1186–1191.

[129] Vaniman DT, Bish DL. Yoshiokaite, a new Ca,Al-silicate mineral from the Moon. Am Mineralogist. 1990; 75: 676–686.

[130] Poellmann H, Winkler N, Oberste-Padtberg R, Meyer R, Goeske J, Raab B. Quantitative mineralogical, chemical and application investigations on some High Alumina Cements from different sources. 28th International Conference on Cement Microscopy. Denver, USA; 2006.
[131] Redhammer GJ, Trippelt G, Roth G, Amthauer G. Structural variations in the brownmillerite series $Ca_2(Fe_{2-x}Al_x)O_5$: Single-crystal X-ray diffraction at 25 °C and high-temperature X-ray powder diffraction (25 °C ≤ T ≤ 1000 °C). Am Mineralogist. 2004; 89: 405–420.
[132] Fukuda K, Ando H. Determination of the Pcmn/Ibm2 Phase Boundary at High Temperatures in the System $Ca_2Fe_2O_5$–$Ca_2Al_2O_5$. J of the Am Cer Society. 2002; 85(4): 1300–1302.
[133] Berastegui P, Eriksson SG, Hull S. A Neutron Diffraction Study of the Temperature Dependence of $Ca_2Fe_2O_5$. Mat Research Bulletin. 1999; 34(2): 303–314.
[134] Neubauer J, Goetz-Neunhoeffer F, Lindner I. Investigation of Hydration Behaviour of Ferrite Phase $C_6A_XF_{3-X}$ with Different Al^{3+}-content in Mixes with C_3A and Gypsum using a Revised Highly Efficient Isothermal Calorimeter. 11th International Congress on the Chemistry of Cement. Durban/South Africa; 2003.
[135] Lee S-J, Benson EA, Kriven WM. Preparation of Portland Cement Components by Poly(vinyl alcohol) Solution Polymerization. J of the Am Cer Society. 1999; 82(8): 2049–2055.
[136] Negro A, Stafferi L. Über die Hydratation der Calcium-Ferrite und Calciumaluminat-Ferrite. Zement-Kalk-Gips. 1979; 2: 83–88.
[137] Andac O, Glasser FP. Polymorphism of Calcium Sulphoaluminate ($Ca_4Al_6O_{16}\cdot SO_3$) and its Solid Solutions. Advances in Cement Research. 1994; 6(22): 57–60.
[138] Puertas F, Varela MTB, Molina SG. Kinetics of the Thermal Decomposition of C_4A_3s in Air. Cement and Concrete Research. 1995; 25(3): 572–580.
[139] Palou M, Dovál M, Drábik M. Applications of sol-gel technique to synthesize inorganic binder materials with upgraded hydration characteristics. 12th International Congress on the Chemistry of Cement. Montreal, Canada; 2007.
[140] Calos NJ, Kennard CHL, Whittaker AK, Davis RL. Structure of Calcium Aluminate Sulfate $Ca_4Al_6O_{16}S$. J Sol State Chemistry. 1995; 119: 1–7.
[141] Neubauer J. Realisierung des Dpeoniekonzeptes der "Inneren Barriere" für Rauchgasreinigungsrückstände aus Müllverbrennungsanlagen. Erlangen: Friedrich-Alexander-Universität Erlangen-Nürnberg; 1992.
[142] Kodas TT, Hampden-Smith MJ. Aerosol Processing of Materials. Series. New York: Wiley-VCH; 1999.
[143] Ries A, Simões AZ, Cilense M, Zaghete MA, Varela JA. Barium strontium titanate powder obtained by polymeric precursor method. Materials Characterization. 2003; 50: 217–221.
[144] Turrillas X, Convert P, Hansen. T, Aza AHd, Pena P, Rodriguez MA, et al. The Dehydration of Calcium Aluminate Hydrates Investigated by Neutron Thermodiffractometry. Calcium Aluminate Cements. Edinburgh, UK; 2001.

Barbara Lothenbach* and Frank Winnefeld

4 Thermodynamic modelling of cement hydration: Portland cements – blended cements – calcium sulfoaluminate cements

Abstract: This chapter gives a short introduction to thermodynamic calculations applied to different cementitious systems and shows possible applications: hydration of Portland and calcium sulfoaluminate cements as well as the effect of limestone and fly ash on the composition of hydrated PC and the effect of calcium sulfate, belite, and limestone on hydrated calcium sulfoaluminate cements. The results of thermodynamic modelling are summarized in binary and ternary diagrams in order to visualize the main effects of the clinker composition on hydrate assemblage and volume.

Keywords: blended cements, calcium sulfoaluminate cements, hydration, thermodynamic modelling, phase diagram

4.1 Introduction

Once cement is in contact with water, its constituents start to dissolve and hydration products such as C-S-H (calcium silicate hydrate), portlandite, ettringite, monosulfate, or monocarbonate will form. The kind and amount of solids hydrates depends on the composition of the cement, the temperature, and the reaction time [1–3]. Thermodynamic modelling can be used to calculate the stable phase assemblages during hydration [3–6] or to model the influence of the cement composition on the hydrates formed [7–9]. Thermodynamic modelling of such multicomponent-multiphase systems enables us to understand better the impact of different factors such as composition, hydration time, leaching, or temperature. In addition, thermodynamic modelling allows one to easily vary different parameters and to predict the composition of hydrate assemblages under different conditions, thus reducing the amount of experiments needed.

The kind and amount of hydrates can also simply be obtained by mass balance calculations if the relative stability of the hydrate phases is well known or by using ternary phase diagrams for hydrated Portland cements [10], which have been constructed based on the knowledge of the relative stability of the hydrates as obtained from thermodynamic modelling and from experimental data. Based on such

***Corresponding author: Barbara Lothenbach,** Empa, Swiss Federal Laboratories for Materials Science and Technology, Dübendorf, Switzerland, barbara.lothenbach@empa.ch
Frank Winnefeld, Empa, Swiss Federal Laboratories for Materials Science and Technology, Dübendorf, Switzerland

DOI 10.1515/9783110473728-005

pre-constructed ternary phase diagrams [10], a general overview of the effect of differences in the main oxides (CaO, SiO_2, Al_2O_3, SO_3, and CO_2) in Portland cements can be obtained easily from the cement composition. Such ternary phase diagrams are compatible with thermodynamic modelling if primarily the main oxides are of interest, as discussed in detail in Herfort & Lothenbach [10].

For more complex systems, if other oxides are important, in the absence of CaO and C-S-H, at elevated temperatures, or if the composition of the pore solution is of importance, thermodynamic modelling using geochemical software is more adequate. Thermodynamic modelling allows one to predict which phases are stable and to take into account the full complexity of cementitious systems [11, 12]. To obtain meaningful results from thermodynamic modelling, a complete set of thermodynamic data, some knowledge of geochemistry (e.g. which phases can reasonably precipitate or dissolve under the conditions of the experiment), as well as practical knowledge of the modelling software are needed.

This chapter illustrates the changes of Portland cement and calcium sulfoaluminate cements during hydration. Thermodynamic modelling and ternary phase diagrams are used to visualize the effect of limestone and fly ash on hydrated Portland cements and of calcium sulfate, belite, and limestone content on hydrated calcium sulfoaluminate cements.

4.2 Methods

4.2.1 Thermodynamic modelling and database for cementitious systems

Thermodynamic equilibrium calculations predict which solids are stable under the specific conditions of the experiment, as discussed in detail in previous works [11, 12], using generic data such as the solubility products, K_{SO}, of the different solids and complex formation constants of aqueous complexes. Thermodynamic modelling is thus based on the knowledge of the thermodynamic data (solubility, complex formation constants, ...) of all the solids, aqueous, and gaseous species which can form in the system. Thermodynamic data are valid for all geochemical systems and compiled in different thermodynamic databases. Several geochemical modelling codes (e.g. PHREEQC [13], CHESS [14], or GEMS [15, 16]) can be used to compute the equilibrium phase assemblage and speciation and they will yield comparable results if the same thermodynamic database is used.

The quality of the results of thermodynamic modelling depends directly on the quality and the completeness of the underlying thermodynamic database. Incomplete datasets or false data can falsify the results. Thermodynamic data for cements have been collected in different thermodynamic databases, the most recent critical databases available are the thermoddem dataset [17, 18] or the Cemdata07 dataset [19–22] and its updated version Cemdata14, which contains additionally very recent thermodynamic data [23–30]. In general, comparable results will be obtained for both databases, although some differences exist mainly for the relative stability of katoite and monosulfate (as discussed in detail elsewhere [12]) and for the stability of calcium aluminate hydrates and aluminium hydroxide at different temperatures [24]. Both cement databases have been obtained based on critical reviews of the available experimental data, are valid in the temperature range 0–100 °C and at near ambient pressure, and contain solubility product as well as the Gibbs free energy, enthalpy, entropy, and heat capacity used to calculate the stability at different temperatures. A more detailed discussion on the temperature dependence of cement hydrates can be found elsewhere [12, 17, 20, 24]. The database Cemdata14 [19, 20, 23–30] is summarized in Tab. 4.1 and covers the hydrates commonly encountered in Portland cement systems in the temperature range 0–100 °C, including C-S-H, hydrogarnet, hydrotalcite, AFm and AFt phases and their solid solutions.

In the last years, many thermodynamic data for hydrates have been measured, but some gaps still exist. Modelling of potassium, sodium, aluminium, and sulfate uptake by the C-S-H phase is partially an unsolved issue although some first models for C-S-H which take into account aluminium and sodium uptake in C-S-H have been published recently [43, 44]. Such models are especially of importance for low Ca/Si C-S-H (present in blended or alkali activated slag or fly ash systems) where a high uptake of alkali [45, 46] and of aluminium is observed [47–49]. The advantages and shortcomings of the available models for aluminium uptake are critically discussed elsewhere [46]; the development of further models is in progress. In this chapter, a general Al/Si in C-S-H of 0.05 has been considered in the thermodynamic calculations to simplify the complexity of the calculations and to allow comparison with the ternary phase diagrams.

A further challenge for thermodynamic modelling, as well as for the ternary phase diagrams discussed in the next section, is the slow kinetics of some precipitation and dissolution processes such that thermodynamic equilibrium may not be reached, although most precipitation and dissolution processes are sufficiently fast that thermodynamic equilibrium can be assumed. A notable exception is the dissolution of the clinker phases, which depends on the composition of the solution and is relatively slow in Portland cements. Precipitation of quartz, dolomite, and siliceous hydrogarnet also often does not occur under ambient conditions and in the time frame of the experiments. Generally, the formation of such solids is thus prevented during modelling even if they would be thermodynamically stable.

In this chapter, the results of series of thermodynamic calculations where the amount of the different minerals has varied systematically are presented in binary and ternary plots. Note that ternary plots of hydrated phases used to illustrate the amount of phases formed under different condition do not follow the same strict rules as the construction of the ternary phase diagrams discussed in the next section.

Tab. 4.1: Solubility products and thermodynamic data of Cemdata2014 at standard conditions (298 K, 1 atm). The database is also available at www.empa.ch/cemdata.

	$\log K_{S0}$	$\Delta_f G^0$ (kJ/mol)	$\Delta_f H^0$ (kJ/mol)	S^0 (J/K/mol)	a_0 (J/K/mol)	a_1	a_2	a_3	V^0 (cm³/mol)	Ref
(Al-)ettringite[a,b]	−44.9	−15205.94	−17535	1900	1939	0.789			707	[19, 20]
Tricarboaluminate[a]	−46.5	−14565.64	−16792	1858	2042	0.559	$-7.78 \cdot 10^6$		650	[19, 20]
Fe-ettringite[b]	−44.0	−14282.36	−16600	1937	1922	0.855	$2.02 \cdot 10^6$		717	[19, 22]
Thaumasite	−24.75	−7564.52	−8700	897.1	1031	0.263	$-3.40 \cdot 10^6$		330	[30]
C_3AH_6[c]	−20.50	−5008.2	−5537.3	422	290	0.644	$-3.25 \cdot 10^6$		150	[23, 24]
$C_3AS_{0.41}H_{5.18}$[c*]	*−25.35*	*−5192.9*	*−5699*	*399*	*310*	*0.566*	*$-4.37 \cdot 10^6$*		*146*	[23]
$C_3AS_{0.84}H_{4.32}$[e*]	*−26.70*	*−5365.2*	*−5847*	*375*	*331*	*0.484*	*$-5.55 \cdot 10^6$*		*142*	[23]
C_3FH_6[d**]	*−26.30*	*−4122.8*	*−4518*	*870*	*330*	*1.237*	*$-4.74 \cdot 10^6$*		*155*	[23]
$C_3FS_{0.84}H_{4.32}$[d,e]	−32.50	−4479.9	−4823	840	371	0.478	$-7.03 \cdot 10^6$		149	[23]
$C_3FS_{1.34}H_{3.32}$	−34.20	−4681.1	−4994	820	395	0.383	$-8.39 \cdot 10^6$		145	[23]
C_4AH_{19}[f]	−25.45	−8749.9	−10017.9	1120	1163	1.047		−1600	371	[24]
C_4AH_{13}	−25.00	−7324.3	−8300.2	700	711	1.047		−1600	274	[24]
$C_2AH_{7.5}$	−13.80	−4695.5	−5277.6	450	323	0.728			180	[24]
CAH_{10}	−7.60	−4623.0	−5288.2	610	151	1.113		3200	193	[24]
$C_4A\bar{S}H_{12}$[f,g]	−29.26	−7778.50	−8750	821	594	1.168			309	[19, 20]
$C_4A\bar{C}H_{11}$	−31.47	−7337.46	−8250	657	618	0.982	$-2.59 \cdot 10^6$		262	[19, 20]
$C_4A\bar{C}_{0.5}H_{12}$	−29.13	−7335.97	−8270	713	664	1.014	$-1.30 \cdot 10^6$	−800	285	[19, 20]
C_2ASH_8	−19.70	−5705.15	−6360	546	438	0.749	$-1.13 \cdot 10^6$	−800	216	[19, 20]
$C_4ACl_2H_{10}$	−27.27	−6810.90	−7604	731	498	0.895	$-2.04 \cdot 10^6$	1503	272	[29, 31]
$C_4A\bar{S}_{0.5}ClH_{12}$	−28.53	−7533.97	−8472***	820	557	1.141	$-1.02 \cdot 10^6$	751	289	[29, 31]
*C_4FH_{13}***	*−30.75*	*−6438.6*	*−7435*	*630*	*694*	*1.113*	*$2.02 \cdot 10^6$*	*−1600*	*286*	[23]
$C_4F\bar{S}H_{12}$[g]	−31.57	−6873.2	−7663	1430	577	1.234	$2.02 \cdot 10^6$		321	[25]
$C_4F\bar{C}H_{12}$	−34.59	−6674.0	−7485	1230	612	1.157	$-5.73 \cdot 10^5$		292	[26]
$C_4F\bar{C}_{0.5}H_{10}$	−30.83	−5952.9	−6581	1270	308	1.201	$-9.08 \cdot 10^5$	3200	273	[26]

Tab. 4.1: (continued)

	log K_{S0}	$\Delta_f G^0$ (kJ/mol)	$\Delta_f H^0$ (kJ/mol)	S^0 (J/K/mol)	a_0 (J/K/mol)	a_1	a_2	a_3	V^0 (cm^3/mol)	Ref
*M_4AH_{10}***	*−56.02*	*−6394.56*	*−7196*	*549*	*−364*	*4.21*	*$3.75 \cdot 10^6$*	*629*	*220*	[3, 19]
$1/2M_6A\bar{C}H_{13}$[h]	−33.29[v]	−4339.85	−4875.89	411	512.6				115	[28]
$1/2M_6F\bar{C}H_{13}$[h]	−33.64[v]	−3882.60	−4415.09	423	521.7				119	[28]
$C\bar{S}$ (anhydrite)	−4.357	−1322.12	−1434.60	106.7	70.2	−0.099			46	[32, 33]
$C\bar{S}H_2$ (gypsum)	−4.581	−1797.76	−2023.36	193.8	91.4	−0.318			75	[32, 33]
β-$C\bar{S}H_{0.5}$ (hemihyd)	−3.59****	−1436.34****	−1575.3****	134.3	124.1				62	[34]
$CK\bar{S}_2H$ (syngenite)	−7.20	−2884.91	−3172	326	201	0.308	$-1.78 \cdot 10^6$		128[k]	[3]
$Al(OH)_3$ (gibbsite)	−1.12	−1151.0	−1288.7	70	36	0.191			32	[32, 33]
$Al(OH)_3$ (mic)	−0.67	−1148.4	−1265.3	140	36	0.191			32	[24]
FeOOH (mic)	−5.6	−480.14	−509.3	200	101	−0.008	$-2.12 \cdot 10^6$		21	[23]
CH (portlandite)	−5.2	−897	−985	83	187	−0.022		−1600	33	[32, 33]
$SiO_{2,am}$	1.476	−848.90	−903	41	47	0.034	$-1.13 \cdot 10^6$		29	[20]
C-S-H (quaternary solid solution)										
TobH Ca/Si = 0.67										
$C_{2/3}SH_{1.5}$[i]	−6.19*****	−1668.56	−1841.51	89.9	141.6				55	[27]
TobD Ca/Si = 1.25										
$C_{5/6}S_{2/3}H_{1.83}$[i]	−6.90*****	−1570.89	−1742.42	121.8	166.9				48	[27]
JenH Ca/Si = 1.33										
$C_{1.33}SH_{2.17}$[i]	−10.96*****	−2273.99	−2506.27	142.5	207.9				76	[27]
JenD Ca/Si = 2.25										
$C_{1.5}S_{0.67}H_{2.5}$[i]	−10.47*****	−2169.56	−2400.72	173.4	232.8				81	[27]
C_3S		−2784.33	−2931	169	209	0.036	$-4.25 \cdot 10^6$		73	[35]
C_2S		−2193.21	−2308	128	152	0.037	$-3.03 \cdot 10^6$		52	[35]
C_3A		−3382.35	−3561	205	261	0.019	$-5.06 \cdot 10^6$		89	[35]
$C_{12}A_7$		−18451.44	−19414	1145	1263	0.274	$-2.31 \cdot 10^7$		518[l]	[35]

Tab. 4.1: (continued)

	log K_{S0}	$\Delta_f G^0$ (kJ/mol)	$\Delta_f H^0$ (kJ/mol)	S^0 (J/K/mol)	a_0 (J/K/mol)	a_1	a_2	a_3	V^0 (cm^3/mol)	Ref
CA		−2207.90	−2327	114	151	0.042	$-3.33 \cdot 10^6$		54[m]	[35]
CA_2		−3795.31	−4004	178	277	0.023	$-7.45 \cdot 10^6$		89[n]	[35]
C_4AF		−4786.50	−5080	326	374	0.073			130	[35]
$K\bar{S}$ (K_2SO_4 arcanite)		−1319.60	−1438	176	120	0.100	$-1.78 \cdot 10^6$		66	[36]
K (K_2O)		−322.40	−363	94	77	0.036	$-3.68 \cdot 10^5$		40	[37]
$N\bar{S}$ (Na_2SO_4 thenardite)		−1269.80	−1387	150	58	0.023			53	[36]
N (Na_2O)		−376.07	−415	75	76	0.020	$-1.21 \cdot 10^6$		25	[37]

a_0, a_1, a_2, a_3 are the empirical coefficients of the heat capacity equation: $C_p^0 = a_0 + a_1 T + a_2 T^{-2} + a_3 T^{-0.5}$; empty = 0.

All solubility products refer to the solubility with respect to the species $Al(OH)_4^-$, $Fe(OH)_4^-$, $SiO(OH)_3^-$, OH^-, H_2O, Ca^{2+}, K^+, Mg^{2+}, CO_3^{2-}, or SO_4^{2-}; Cement shorthand notation is used: A = Al_2O_3; C = CaO; F = Fe_2O_3; H = H_2O; K = K_2O; M = MgO; N = Na_2O; S = SiO_2; $\bar{C} = CO_2$; $\bar{S} = SO_3$.

* precipitates very slowly at 20 °C, generally not included in calculations; ** tentative values; *** typing error in [29], recalculated using G_f^0 and S from [29]; **** recalculated from ΔG_r^0 of −20 500 J/mol [19]; ***** recalculated from ΔG_f^0 values.

a, c, d, e, h, i ideal solid solutions cf. [23, 27, 28]; **b, f, g** non-ideal solid solutions, for details see [20, 25, 38]; **k** calculated from density data from [39], **l** [40], **m** [41], **n** [42].

4.2.2 Ternary phase diagrams

Ternary phase diagrams are a technique to predict and visualize the phase assemblage and the relative phase contents from 3 intensive variables. A detailed description of how to construct such ternary phase diagrams for hydrated Portland cements and blended cements can be found in Herfort & Lothenbach [10]. An Excel file for performing simple phase assemblage calculations based on two pre-defined ternary phase diagrams is available at www.empa.ch/cemdata, which can be used to plot cement compositions automatically onto two sub-ternary phase diagrams. For hydrated Portland cements the two $CaO–Al_2O_3–SiO_2$ and $C_3A–CaSO_4–CaCO_3$ ternary phase diagrams are sufficient to define the composition of the main hydrates formed in Portland cement; the minor amounts of MgO and Fe_2O_3 present are assumed to form hydrotalcite and FeOOH. In short, the $CaO–Al_2O_3–SiO_2$ ternary phase sub-diagram gives the phase relationships between the C(-A)-S-H phase, portlandite (or strätlingite in systems undersaturated in $Ca(OH)_2$) and the AFm and AFt phases, while the $C_3A–CaSO_4–CaCO_3$ ternary phase sub-diagrams plots the distribution of aluminium, sulfate, and carbonate to ettringite, monosulfate, monocarbonate, hemicarbonate, calcite, and calcium sulfate.

4.3 Portland cements

4.3.1 Hydration

Thermodynamic modelling can be used to calculate the kind and amount of solids present during the hydration of Portland cement. Cement clinkers in contact with water react, continuously releasing Ca, Si, Al, Fe, and hydroxide into the solution, which then precipitate as C-S-H, ettringite, and other hydrate phases. Experimental data or empirical description of clinker dissolution can be used to describe the dissolution of the anhydrous phases with time [50]. By combining such a dissolution model for the clinker phases with thermodynamic modelling, the amount and volume of the hydrates formed can be described as a function of time as shown in Fig. 4.1 and described in several papers [3, 4, 6, 19, 51]. Such a combination of kinetic and thermodynamic models reproduces the depletion of gypsum and/or anhydrite within the first day of hydration (Fig. 4.1), while ettringite precipitates. Once gypsum and anhydrite are depleted, AFm[1] phases are calculated to form: monocarbonate in calcite-containing cements, monosulfate in the absence of calcite [6]. C-S-H, portlandite, ettringite, monocarbonate, hydrotalcite, calcite, and unhydrated clinker are the main phases formed.

1 The expression AFm (aluminium-iron-mono)-phase is used as a generic term for the different AFm-type phases such as monosulfate, monocarbonate, hemicarbonate, hydroxy-AFm, as well as strätlingite.

The MgO present in the cement or in SCMs is calculated to react together with aluminium to hydrotalcite under the high pH conditions in Portland cement. With the exception of ettringite, which is limited by the amount of sulfate, the quantity of the other hydrates continues to increase with time as long as the clinker phases continue to react. At the same time, the amount of pore solution and the capillary porosity decrease as the volume of the solid phases increase. The total of the volume of the solid and liquid phase decreases during hydration as the water incorporated into the hydrates has a higher density than "free" water, resulting in chemical shrinkage (Fig. 4.1) of approximately 4 cm^3/100 g dry cement after 28 days, which agrees well with chemical shrinkage measurements in Portland cements [52]. The modelled changes of the composition of the solids and liquid phase, of the porosity, the amount of pore solution, and the heat evolved during hydration have been found to agree well with the experimental data determined as a function of time, as has been shown elsewhere [3, 4, 6, 19, 51, 53].

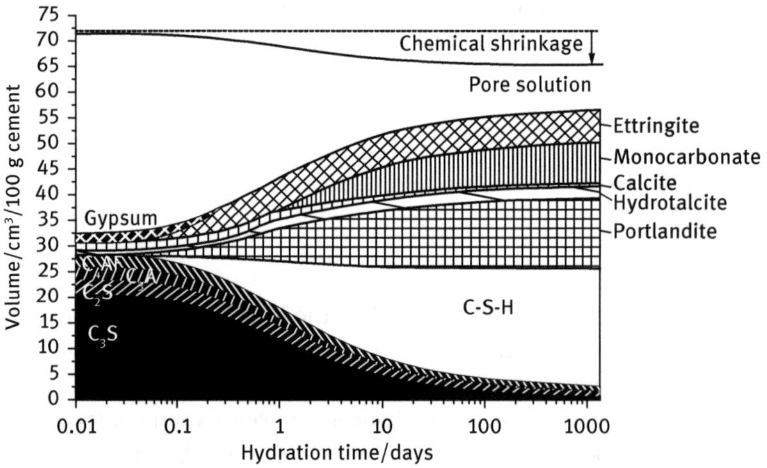

Fig. 4.1: Calculated composition of a Portland cement containing 4 mass-% of calcite as a function of hydration time. Adapted from Lothenbach (2010) [11].

Alternative options to visualize the changes during hydration are
(i) to plot the component ratios of the reacted cement fraction as done in Fig. 4.2 or
(ii) to use ternary phase diagrams (Fig. 4.3).

The plot of the component ratios shows the stable hydrate assemblage in a PC system as a function of the carbonate to aluminium and sulfate to aluminium ratio in the presence of excess C-S-H and portlandite. The stable hydrate assemblage depends on the molar fraction of carbonate and sulfate to aluminium. In hydrated Portland cements with no or very little carbonate, the formation of monosulfate and hemicarbonate can be expected. In modern Portland cements, which generally contain limestone, the stable phase assemblage includes ettringite, monocarbonate, and calcite [6–8, 54]. The fraction of carbonate and sulfate to aluminium changes during hydration, as the sulfate and carbonate containing phases (alkali sulfates, gypsum, anhydrite, and calcite) react relatively fast while aluminium is bound in the slower reacting clinker phases and thus becomes available more slowly. Thus, in a hydrating OPC, initially a high ratio of carbonate and sulfate to aluminium is present, which decreases during hydration as the clinkers react (Fig. 4.2).

The same information shown in the binary plot of the carbonate to aluminium and sulfate to aluminium ratio in Fig. 4.2 can also be obtained from ternary phase diagrams, as shown in Fig. 4.3. Ternary phase diagrams for Portland cements determine the phase assemblages and relative phase contents based on the cement composition and have been constructed from 3 intensive variables. A detailed description of how to construct such ternary phase diagrams for hydrated Portland cements and blended cements is given in Herfort and Lothenbach [10]. The $CaO–Al_2O_3–SiO_2$ ternary phase sub-diagram shown in Fig. 4.3 gives the phase relationships between the C(-A)-S-H, portlandite (or strätlingite in systems undersaturated in $Ca(OH)_2$), and AFm and AFt phases. Hydrated Portland cement plots within the region saturated with respect to $Ca(OH)_2$ and the presence of the following phases is predicted: portlandite, C(-A)-S-H, monosulfate, ettringite, and hemicarbonate for a Portland cement with only

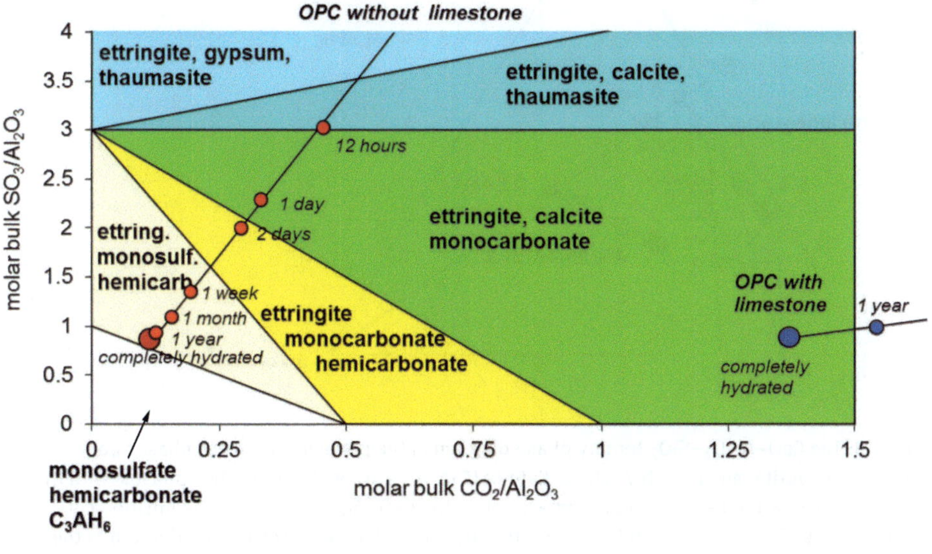

Fig. 4.2: Calculated phase assemblage of hydrating Portland cements plotted as a function of the carbonate to aluminium and the sulfate to aluminium ratio in the presence of excess C-(A-)S-H and portlandite. The changes during the hydration of an OPC with 4 % calcite and an OPC where no additional calcite has been added (0.3 % CO_2, data from [6]) are indicated by dots. From [11].

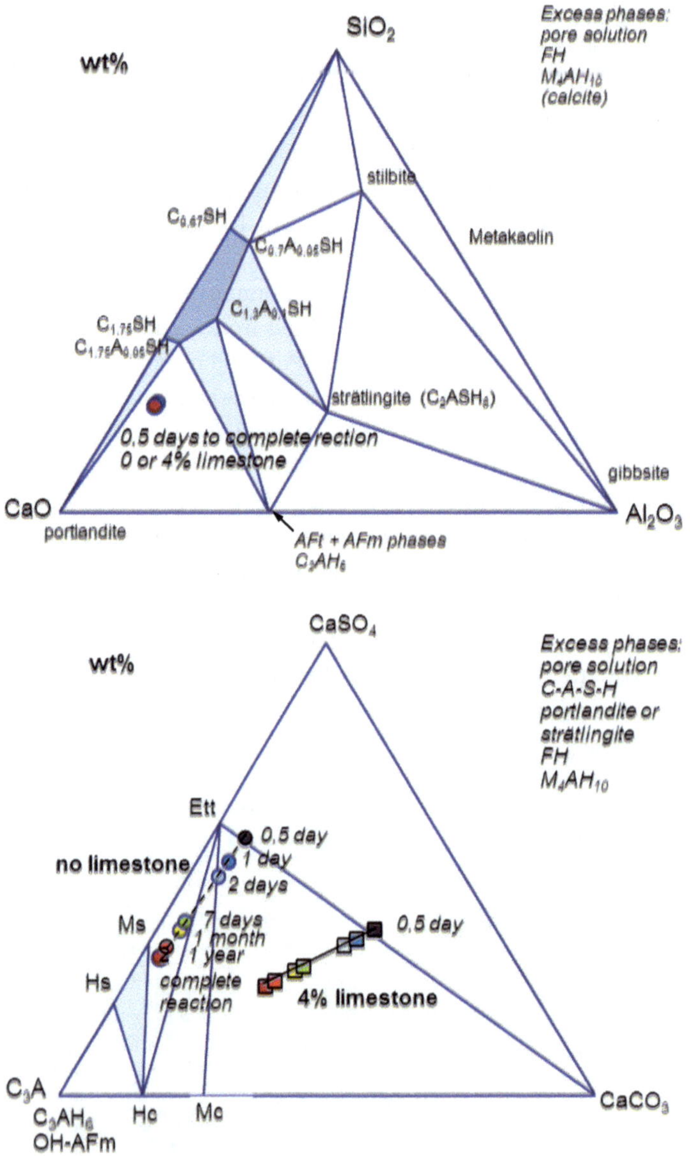

Fig. 4.3: The $CaO–Al_2O_3–SiO_2$ ternary phase diagram in the presence of excess phases pore solution, ferrihydrite, and possibly calcite. Katoite (C_3AH_6), ettringite and the AFm phases (with the exception of strätlingite) plot on the same point on the $CaO–Al_2O_3$ tie line on the position corresponding to C_3A. The $C_3A–CaCO_3–CaSO_4$ ternary phase diagram gives the distribution of the AFm and AFt phases in the presence of excess phases (C-S-H, portlandite or strätlingite, pore solution, ferrihydrite, and hydrotalcite). Ms = monosulfate, Hs = hemisulfate, Hc = hemicarbonate, Mc = monocarbonate, Ett = ettringite. The changes during the hydration of a PC with 4 % calcite (squares) and a PC where no additional calcite has been added (0.3 % CO_2, circles) are indicated. Cement composition and degree of reaction are taken from Lothenbach et al. (2008) [6].

minor carbonates. The distribution of AFm and AFt phases depends on the amount of carbonate and sulfate relative to aluminium as shown in the C_3A–$CaSO_4$–$CaCO_3$ ternary phase sub-diagram (Fig. 4.3). Fig. 4.3 also visualizes the increase of the relative amount of C_3A to sulfate and carbonate during hydration, while the fraction of portlandite and the calculated C(-A)-S-H composition vary only marginally during hydration of PC independent of the absence or presence of limestone.

4.3.2 Effect of limestone on the hydrate assemblage

As discussed, ettringite ($C_6A\bar{S}_3H_{32}$) and monosulfate ($C_4A\bar{S}_3H_{12}$) are stable in Portland cements in the absence of limestone. The presence of limestone leads to the stabilization of monocarbonate and ettringite instead of monosulfate [6, 8, 12, 48, 55], and to an increase of the volume of the solid phases according to equation (4.1):

$$\begin{gathered} 3C_4A\bar{S}H_{12} + 2C\bar{C} + 18H \Leftrightarrow C_6A\bar{S}_3H_{32} + 2C_4A\bar{C}H_{11} \\ (3 \times 309 + 2 \times 37 \Leftrightarrow 707 + 2 \times 262\,\text{cm}^3/\text{mol}) \\ (1001 \Leftrightarrow 1231\,\text{cm}^3/\text{mol}) \end{gathered} \tag{4.1}$$

The replacement of a low fraction of PC by limestone thus results in an increase of the volume of the hydrates. The effect of blending PC with limestone can be visualized in the two ternary phase sub-diagrams for Portland cements as shown in Fig. 4.4 [10]. While the addition of limestone has no effect in the CaO–Al_2O_3–SiO_2 ternary phase sub-diagram (the cement with and without limestone is indicated by overlapping circles), as the relative amount of CaO to SiO_2 and Al_2O_3 available to form AFt and AFm phases remains constant, clear differences are observed in the C_3A–$CaCO_3$–$CaSO_4$ ternary phase sub-diagram, as is visible in the position of the circles indicating PC blended with 0 to 6 mass-% limestone plotted in Fig. 4.4. If limestone is present, the composition moves from the monosulfate–ettringite axis into the ettringite–hemicarbonate–monocarbonate field and at more than 2 mass-% of limestone the composition plots within the ettringite–monocarbonate–calcite field. Any further addition of limestone will lead to no further changes in the kind of hydrates formed but to the presence of more limestone and to a dilution of the PC.

The effect of different amounts of limestone on the volume of the hydrates can best be visualized in binary plots based on thermodynamic calculations (or based on the ternary phase diagrams), as shown in Fig. 4.5 (a). A maximum in solid volume is predicted between 1–2 mass-% of limestone, where all monosulfate is destabilized. In agreement with the maximum in solid volume a maximum in compressive strength of PC is also observed at low limestone additions (Fig. 4.5 (b)). The presence of calcite in excess of that needed to form ettringite and monocarbonate dilutes the other hydrate, increases the porosity in the hydrated pastes, and reduces the compressive strength. For different cement compositions, a different optimum limestone addition exists, depending on its sulfate and alumina contents and the degree of reaction.

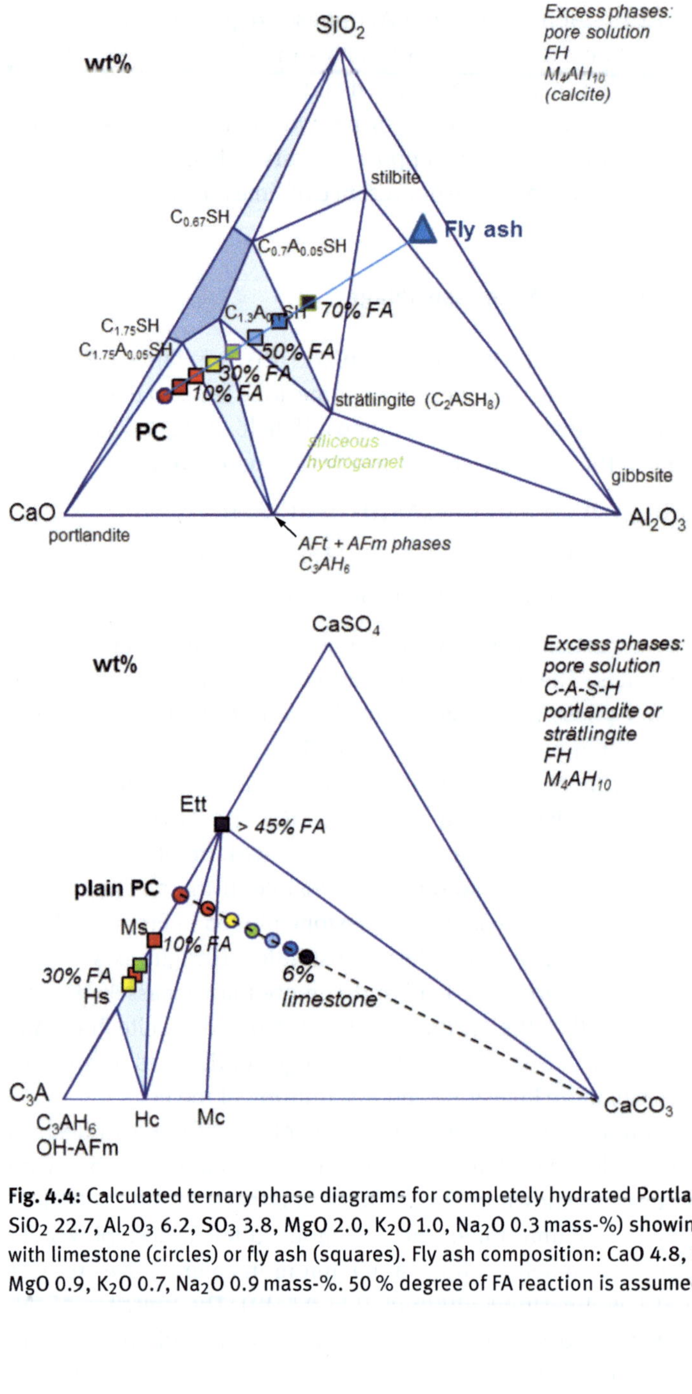

Fig. 4.4: Calculated ternary phase diagrams for completely hydrated Portland cement (CaO 64.0, SiO_2 22.7, Al_2O_3 6.2, SO_3 3.8, MgO 2.0, K_2O 1.0, Na_2O 0.3 mass-%) showing the effect of blending with limestone (circles) or fly ash (squares). Fly ash composition: CaO 4.8, SiO_2 59.1, Al_2O_3 33.7, MgO 0.9, K_2O 0.7, Na_2O 0.9 mass-%. 50 % degree of FA reaction is assumed.

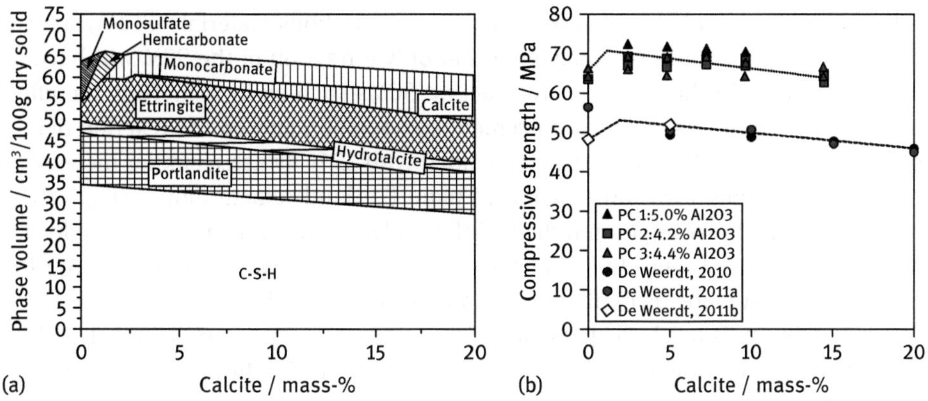

Fig. 4.5: (a) Calculated volume changes as a function of the amount of limestone in hydrated Portland cement (CaO 64.0, SiO_2 22.7, Al_2O_3 6.2, SO_3 3.8, MgO 2.0, K_2O 1.0, Na_2O 0.3 mass-%). (b) Influence of the addition of limestone on the compressive strength of Portland cement mortars after 28 days (PC 1, PC 2, PC3; cement composition in Herfort et al. 1999 [58] and De Weerdt 2010 [59]) and 90 days of hydration (De Weerdt 2011a [56], 2011b [48]); adapted from Damidot et al. (2011) [12]. The lines are intended as visual guides only.

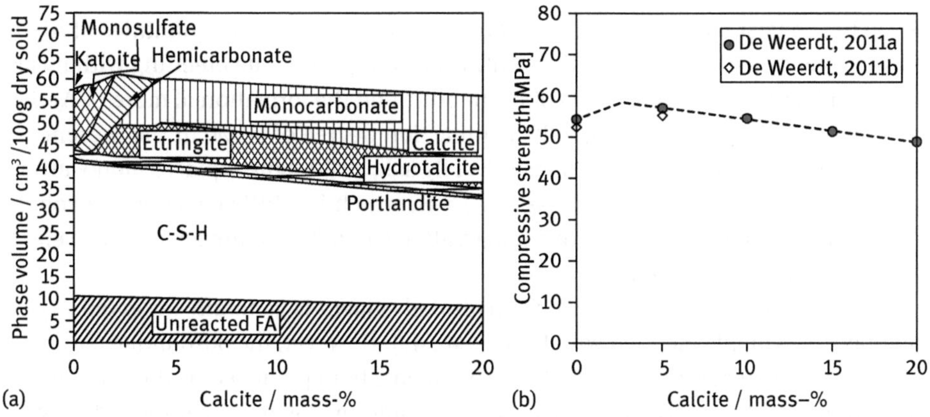

Fig. 4.6: (a) Calculated volume changes as a function of the amount of limestone in hydrated 65 % Portland cement + 35 % fly ash (25 % reaction degree; fly ash: CaO 4.8, SiO_2 59, Al_2O_3 34 mass-%). (b) Influence of the addition of limestone on the compressive strength of 65 % Portland cement – 35 % fly ash mortars (De Weerdt 2011a [56], 2011b [48]); reproduced from Damidot et al. (2011) [12]. The lines are intended as visual guides only.

If the cement is blended with supplementary cementitious materials (SCMs) such as fly ash or slag, it should be considered that only a fraction of the SCM will have reacted especially in the case of slow-reacting SCMs such as fly ash. Thus the effect of limestone additions depends on cement composition and on the reaction degree of the SCM used. In Portland cement blended with 35 % fly ash a similar increase in compressive strength was observed as in plain Portland cement mortar (Fig. 4.6 (b)) [48, 56]. Detailed investigation indicated that only roughly one quarter of the fly ash had reacted [48, 56], and thus a similar amount of Al_2O_3 was available for AFt formation in the FA-blend as in plain PC. If more reactive metakaolin (46 mass-% Al_2O_3) instead of fly ash is used, the presence of limestone has a more pronounced positive effect on the compressive strength [47, 57], in agreement with the higher volume maximum obtained from thermodynamic calculations [12, 47], as the influence of limestone is amplified by high Al_2O_3. The amount of SO_3 also plays an important role. A high SO_3/Al_2O_3 molar ratio in the hydrated cements results in relatively more ettringite and less monosulfate, and the influence of limestone is less pronounced.

4.3.3 Effect of fly ash on the hydrate assemblage

Fly ashes consist mainly of SiO_2, but can also contain significant quantities of Al_2O_3. In siliceous fly ashes, relatively little CaO is present [1, 60]. The blending with fly ash thus results in the destabilization of portlandite and the formation of more C(-A)-S-H, as shown in Fig. 4.7, assuming a fly ash reaction degree of 50 %. In the presence of more than approximately 30–40 mass-% of fly ash, strätlingite replaces monocarbonate or monosulfate, resulting in less C(-A)-S-H, which has little effect on the volume and on the measured compressive strength after 3 months and longer as shown e.g. in De Weerdt et al. (2011) [56].

The decrease of portlandite in the hydrated cement is generally observed experimentally in cements blended with fly ash [48, 49]. However, the portlandite content is observed to decrease much less in the experiments than predicted by thermodynamic modelling and the complete depletion of portlandite is generally observed only for blends with more than 60 % fly ash [61, 62]. Although portlandite is still present, the Ca/Si in C-S-H is observed to decrease from 1.5–1.8 in plain PC to Ca/Si = 1.3–1.4 in fly ash blends [48, 49, 63–65], while thermodynamic modelling calculates a constant Ca/Si = 1.6 as long as portlandite is present. The difference between the absence of portlandite in the model and its persistence in experiments with blended cements is probably related to the limited availability of water in hydrated systems, the slow kinetic, and the inhomogeneity in the microstructure.

At a higher addition of fly ash, the formation of strätlingite is calculated (Fig. 4.4, Fig. 4.7), which has been observed experimentally only in studies where high quantities of fly ash have been used [64, 66]. In contrast to modelling, which indicates that strätlingite is only stable in the absence of portlandite (Fig. 4.4, Fig. 4.7), the presence

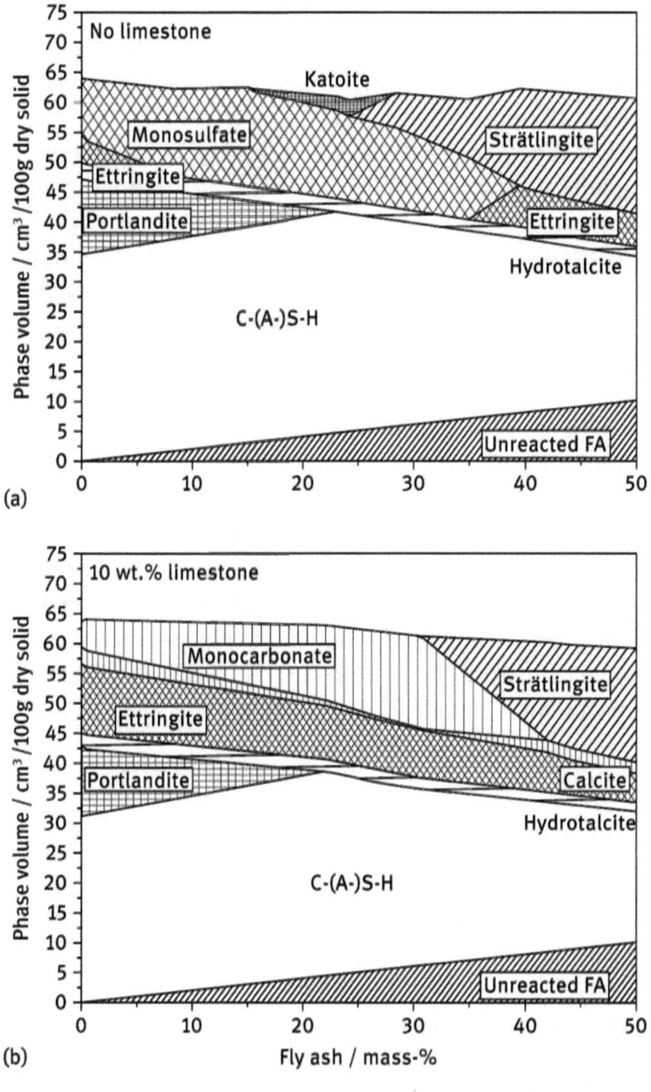

Fig. 4.7: Modelled effect of fly ash on the composition of hydrated Portland cement containing (a) no limestone and (b) 90 mass-% PC and 10 mass-% limestone. A complete reaction of the Portland cement (CaO 63, SiO_2 20, Al_2O_3 5.6, MgO 1.8, Na_2O 0.3, K_2O 0.9, SO_3 3.4, CO_2 4.4 mass-%) and 50 % reaction of the fly ash (CaO 4.8, SiO_2 59, Al_2O_3 34, MgO 0.9, Na_2O 0.7, K_2O 0.9 mass-%) is assumed. Al/Si in C-S-H = 0.05.

of strätlingite has been observed in blends where portlandite was still present, e.g. by Escalante-Garcia and Sharp for a one year old 70–30 mass-% PC–FA blend [64], again indicating an inhomogeneous microstructure, which cannot be captured with thermodynamic models of the bulk composition.

Another important effect of the presence of fly ash is the higher availability of aluminium; class F fly ashes contain between 15–35 % Al_2O_3. This increase of the availability of aluminium results in an increase of amount of AFm phases as shown in Fig. 4.7, which has also been observed experimentally [48, 49, 64, 65]. In the absence of limestone (Fig. 4.7 (a)), ettringite is destabilized to monosulfate and the formation of a small amount of C_3AH_6 is predicted if a part of the PC is replaced by fly ash. Above approximately 40 % of fly ash in the cement, monosulfate (and monocarbonate in the presence of limestone, Fig. 4.7 (b)) are destabilized while strätlingite is calculated to be the dominant AFm phases. In the presence of limestone, monocarbonate and ettringite are calculated to be present in up to 40 % of fly ash.

Again, the same trends as in the binary plots in Fig. 4.7 can also be observed in the 2 ternary phase sub-diagrams given in Fig. 4.4, although some minor differences in the amount of AFm phases are obtained due to the different assumption of aluminium uptake in C-S-H (between Al/Si = 0.05 and 0.1 in the ternary phase diagrams and fixed at Al/Si = 0.05 in the thermodynamic modelling). The squares in Fig. 4.4 illustrate how the replacement of Portland cement with fly ash lowers the quantity of CaO in the blend and affects the relative amount of monosulfate and ettringite.

4.3.4 Ternary plots

More trends resulting from the replacement of Portland cement by limestone and/or fly ash can be seen if the quantities of the hydrates obtained from thermodynamic modelling are plotted in ternary graphs displaying the amount of Portland cement, limestone, and fly ash (Fig. 4.8). These ternary graphs illustrate the amount of hydrates formed if different amounts of Portland cement, limestone, and fly ash are present. For Portland cement, complete hydration was assumed and for fly ash 50 % of reaction, while calcite was allowed to react freely.

Fig. 4.8 (a) shows the total volume of the solid phases and indicates a volume maximum at 2 mass-% of limestone and 0 to 10 mass-% of fly ash. In the presence of up to 8 mass-% of limestone or 22 mass-% of fly ash, the calculated volume equals or surpasses the volume of plain Portland cement without calcite or fly ash additions. Up to approximately 40 mass-% of fly ash, the calculated volume decreases by less than 5 vol.% compared to plain Portland cements. This agrees with strength measurements in PC-fly ash blends, where the presence of up to 35 mass-% of fly ash decreases the early compressive strength, but has no significant effect on the strength after 3 months or longer [48, 56]. A significantly lower volume of solids is calculated only if higher fractions of calcite and/or fly ash are present.

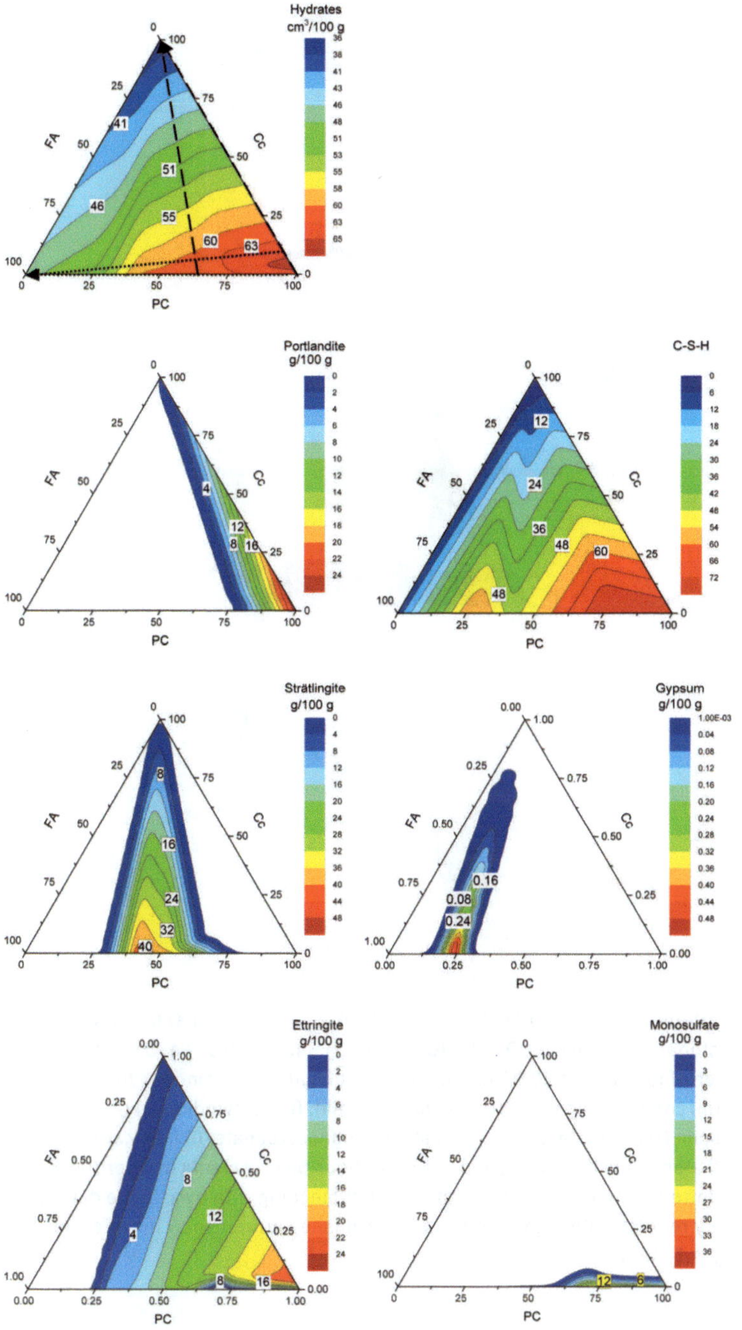
Hydrates
cm³/100 g
Portlandite
g/100 g
C-S-H
Strätlingite
g/100 g
Gypsum
g/100 g
Ettringite
g/100 g
Monosulfate
g/100 g
FA
Cc
PC

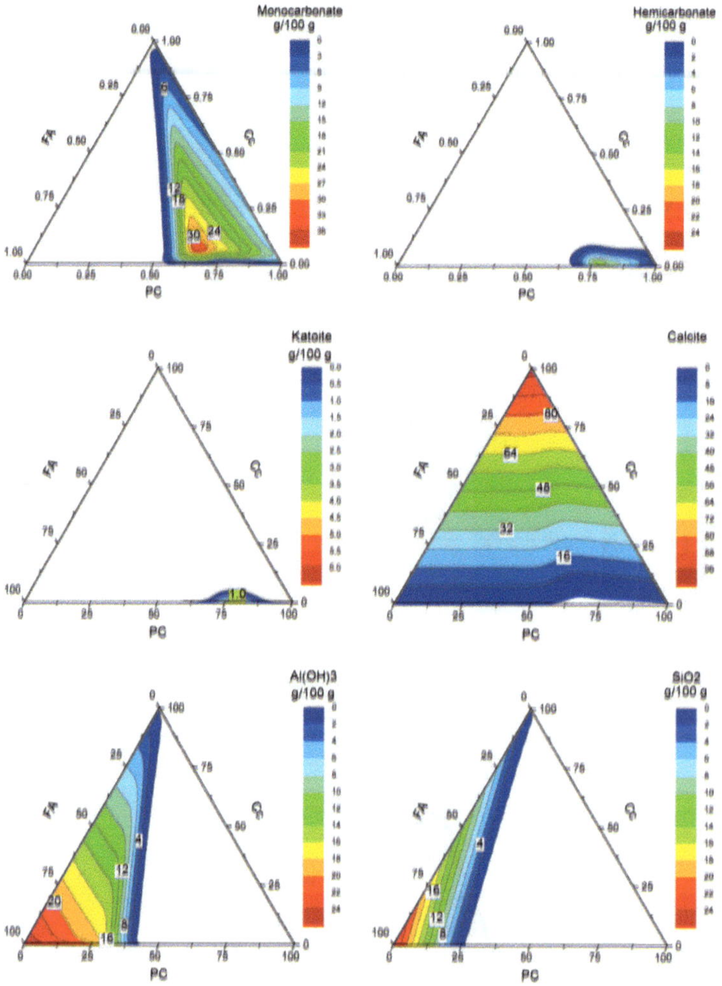

Fig. 4.8: Ternary plot Portland cement (CaO 64, SiO_2 23, Al_2O_3 6.2, MgO 2.0, Na_2O 0.3, K_2O 1.0, SO_3 3.8 mass-%), limestone, and fly ash (CaO 4.8, SiO_2 59, Al_2O_3 34, MgO 0.9, Na_2O 0.7, K_2O 0.9 mass-%, 50 % degree of reaction) at 20 °C: (a) total volume of hydrates in cm^3 per 100 g unhydrated solid, (b)–(m) contents in g per 100 g unhydrated solid; (b) portlandite, (c) C-S-H, (d) strätlingite, (e) gypsum, (f) ettringite, (g) monosulfate, (h) monocarbonate, (i) hemicarbonate, (j) katoite, (k) calcite, (l) aluminium hydroxide, and (m) amorphous SiO_2 contents in g per 100 g unhydrated solid. The dotted arrow in (a) indicates the effect of blending with fly ash; the data is the same as in the binary representation given in Fig. 4.7, the dashed arrows show the effect of limestone addition (Fig. 4.5, 4.6).

Fig. 4.8 (b)–(m) gives the calculated weights of the different hydrates and calcite in g per 100 g of unhydrated solid. The maximum amount of ettringite is found in the range of 4 to 10 mass-% of calcite and 0 to 6 mass-% of fly ash. Monosulfate, katoite, and hemicarbonate are calculated to be stable only in the presence of less than 10 % of limestone and at more than 50 mass-% of PC. In the presence of more calcite, monocarbonate is stable, in the presence of more fly ash, strätlingite forms instead in mixes where portlandite is absent.

In the absence of limestone, portlandite is calculated to be present only if less than 25 % of the PC is replaced by fly ash. When the calcite content is higher, this fraction increases somewhat. The stability field of strätlingite is observed in mixes with more fly ash where portlandite is destabilized. In the presence of very high quantities of fly ash (> 60 mass-%), the formation of aluminium hydroxide, gypsum, and eventually SiO_2 is predicted. C-S-H is stable over almost the entire range, although its composition varies considerably from Ca/Si = 1.6 in the presence of portlandite to Ca/Si = 0.67 in the presence of silica fume.

4.4 Calcium sulfoaluminate cements

4.4.1 Overview

Calcium sulfoaluminate (CSA) cements contain ye'elimite ($C_4A_3\bar{S}$) as a major cementing phase (30–70 %) [67, 68]. Ye'elimite was patented by Alexander Klein as an expansive or shrinkage compensating addition to cementitious binders ("Klein's compound"). CSA cements have been produced, used, and standardized in China for about 30 years, where they are known as the "third cement series" [69]. In recent years, several European cement producers started research and development activities in this field and entered the market as well [68].

CSA cements are currently receiving increasing interest because they promise to provide a low-CO_2 alternative to Portland cement [70]. Compared to alite, ye'elimite releases only approximately one third of the CO_2 per ml of cementing phase when calcined from the anhydrous raw materials. The firing temperature used to produce CSA clinker is typically about 200 °C lower than for Portland cement clinker. In addition, CSA clinker is more friable and thus easier to grind than Portland cement clinker. The clinker is generally blended or ground together with approximately 10–25 mass-% calcium sulfate, usually anhydrite. Its main hydration products are ettringite, monosulfate, aluminium hydroxide, and other hydrate phases such as strätlingite or C-S-H, which originate from the reaction of belite, which is usually present in the CSA clinker as well.

Several types of CSA cements are on the market [68]. Calcium sulfoaluminate cements with high ye'elimite contents (> 50 mass-%) are mostly used in combination with Portland cement to manufacture fast-setting, rapid hardening, and/or shrinkage compensated construction materials. Calcium sulfoaluminate cements with less ye'elimite (in the order of 25–50 mass-%) contain significant amounts of belite (30–50 mass-%) and ferrite (5–20 mass-%). They are suggested to be a sustainable replacement material for Portland cement. In addition, expansive agents based on CSA are used e.g. for self-stressing or expansive concrete. Another recent type of CSA is alite calcium sulfoaluminate cement.

4.4.2 Hydration of ye'elimite in the presence of calcium sulfate

The optimization of the calcium sulfate content is one of the key issues concerning the properties of CSA cements (such as rapid-hardening or expansive) with respect to specific applications. The amount of calcium sulfate added to ye'elimite determines the ratio between ettringite and monosulfate in the hydration products, see equations (4.2) and (4.3).

$$C_4A_3\bar{S} + 18H \rightarrow C_3A{\cdot}C\bar{S}{\cdot}12H + 2AH_3 \tag{4.2}$$

$$C_4A_3\bar{S} + 2C\bar{S}H_x + (38 - x)H \rightarrow C_3A{\cdot}3C\bar{S}{\cdot}32H + 2AH_3 \quad (x = 0; 0.5; 2) \tag{4.3}$$

Using thermodynamic modelling, the stable hydrate assemblages of various mixtures between ye'elimite and anhydrite can be calculated (Fig. 4.9). From a thermodynamic point of view, there is no difference between gypsum and anhydrite addition concerning the types and quantities of the hydrates formed except regarding the amount of water needed. However, it should be mentioned that not only the amount, but also the reactivity of the added calcium sulfate strongly influences hydration kinetics, mainly in the early stages of the first 24 hours [71–78]. Increasing the amount and the reactivity of the added calcium sulfate enhances the formation kinetics of ettringite and thus the expansive properties of CSA cement [79].

The calculations shown in Fig. 4.9 confirm that pure ye'elimite hydrates according to equation (4.2), forming monosulfate and microcrystalline $Al(OH)_3$. In the presence of calcium sulfate, ettringite is formed according to equation (4.3). In the absence of calcium sulfate, low amounts of ettringite are also predicted. The reason for this is that "monosulfate" forms a solid solution with OH–AFm (C_4AH_{13}) [80, 81], such that a part of the sulfate ions is replaced by hydroxide. Thus some sulfate is available to form ettringite according to equation (4.3). With an increased addition of calcium sulfate, the amount of monosulfate decreases and the amount of ettringite increases. Ettringite content reaches its maximum at a molar ratio of calcium sulfate to ye'elimite of 2. At this molar ratio, the total volume of hydrates and the chemical shrinkage reach a maximum. Beyond this, monosulfate is no longer present but instead excess calcium sulfate is present as gypsum.

Corresponding experimental results by means of X-ray diffraction are shown in Fig. 4.10, confirming the results obtained from thermodynamic modelling. Plain ye'elimite (molar ratio gypsum/ye'elimite 0 : 1) mainly forms monosulfate (both the

12- and the 14-hydrate, which is related to the procedure used for hydration stoppage) and microcrystalline $Al(OH)_3$. Traces of ettringite can also be observed as predicted. When calcium sulfate is added at a molar ratio of 1 : 1, ettringite occurs as the main hydration product, whereas monosulfate is hardly visible in the XRD pattern, probably due to its poor crystallinity and overlapping reflections. In the 2 : 1 mixture, no monosulfate is observed, and ettringite and $Al(OH)_3$ are the only hydration products identified. Furthermore, the dissolution of calcium sulfate is not complete, as traces of unreacted gypsum are present in the XRD pattern after 18 h of hydration. Beyond the 2 : 1 gypsum to ye'elimite molar ratio, gypsum is always present, as those mixtures contain a surplus of calcium sulfate as predicted (Fig. 4.9).

The amount of calcium sulfate added not only influences the hydrate assemblage, but also hydration kinetics, as shown in Fig. 4.11. Without gypsum, pure ye'elimite shows two maxima in the heat flow curve. The first occurs directly when the water is added and can be attributed to the heat of wetting and very early hydration reactions. After that, a long induction period of about 10 hours occurs, which is explained in literature by a surface coverage of the clinker grains by early hydration products [75]. The following second heat flow maximum after about 15 hours covers the main part of the hydration reactions and is responsible for the setting and hardening of the paste.

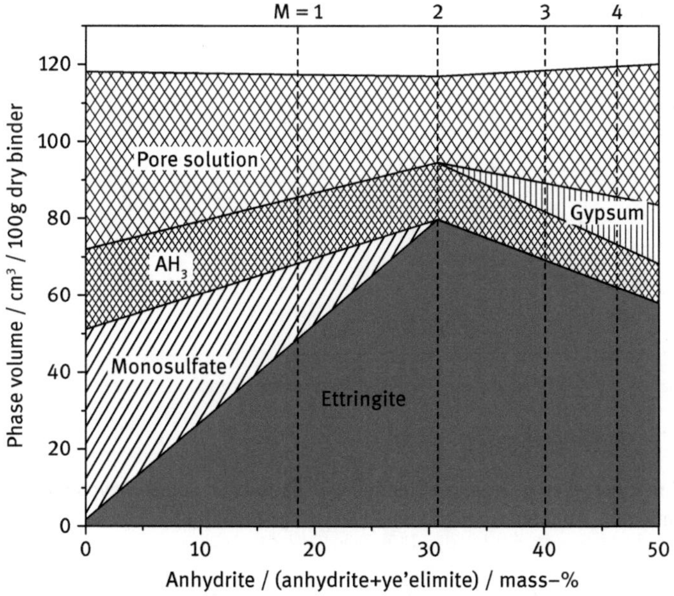

Fig. 4.9: Calculated phase diagram of the thermodynamic stable hydrate assemblages in cm^3 per 100 g of dry solid of the system ye'elimite–anhydrite–water at 20 °C. M is the molar ratio of anhydrite to ye'elimite. Complete hydration at a water/solid ratio of 1 was assumed. Adapted from [76].

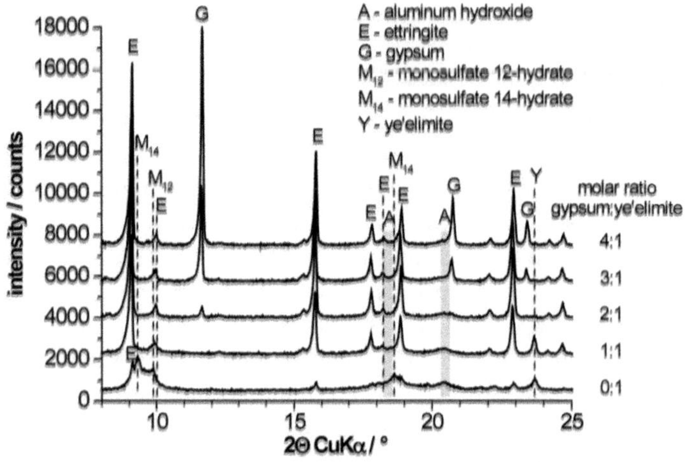

Fig. 4.10: X-ray diffraction analyses of ye'elimite pastes with varying contents of gypsum at a water/solid ratio of 2 after 18 h of hydration at 20 °C. Adapted from Winnefeld & Barlag (2010) [76].

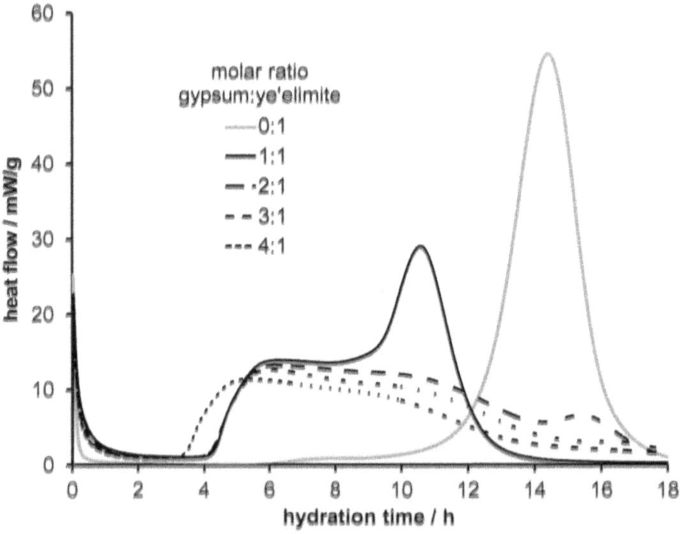

Fig. 4.11: Conduction calorimetry of ye'elimite pastes blended with various amounts of gypsum at 20 °C using a water/solid ratio of 2. Adapted from Winnefeld & Barlag (2010) [76].

All mixtures containing gypsum show a similar very early peak to pure ye'elimite. With gypsum, the induction period is shortened to about 2–3 hours. This acceleration of ye'elimite hydration by the addition of gypsum also occurs in technical CSA cements [75]. The two samples with the lowest gypsum additions exhibit two heat flow maxima after the induction period. The first is in both cases after about 6 hours, the second after 10.5 hours (molar ratio 1 : 1) or 15.5 hours (molar ratio 2 : 1). On the contrary, the samples with higher gypsum contents display only one maximum beyond the induction period. It can be assumed that the first maximum is related to the reaction of ye'elimite with gypsum to ettringite, and that the additional maximum after the induction period indicates the depletion of the gypsum, which results in the formation of monosulfate (according to equation (4.2)), instead of ettringite (equation (4.3)). This second maximum does not occur in the samples with molar ratios 1 : 3 and 1 : 4, as they contain a surplus of gypsum.

4.4.3 Impact of belite on the phase assemblage of hydrated CSA cements

CSA cements generally contain significant amounts of belite as secondary cementing phase. The hydration of belite in CSA leads to the formation of strätlingite and/or C-S-H phases as additional hydration products (equations (4.4)–(4.6)). If belite contents are high, portlandite could also be present. Thus the composition of the hydrate assemblage of CSA is highly variable not only due to varying amounts of calcium sulfate, but also depending on the level of belite present in the CSA clinker and its degree of reaction [67, 82].

$$C_2S + AH_3 + 5H \rightarrow C_2ASH_8 \tag{4.4}$$

$$C_2S + C_2ASH_8 \rightarrow CSH_2 + C_3AH_6 \tag{4.5}$$

$$6C_2S + 2AH_3 + 10H + C_6A\bar{S}_3H_{32} \rightarrow 6CSH_2 + 3C_4A\bar{S}H_{12} \tag{4.6}$$

The stable hydrate assemblages of the subsystem ye'elimite–belite–water in the absence of calcium sulfate can be calculated by thermodynamic modelling and are shown in Fig. 4.12 (a). With increasing amounts of C_2S, first strätlingite is formed under consumption of $Al(OH)_3$, see equation (4.4), and a maximum in volume is reached at approximately 35 mass-% belite. Only in blends with at least 35 mass-% belite, sufficient calcium oxide is available to bind all $Al(OH)_3$ in strätlingite and to enable the formation of C-S-H according to equation (4.5). The uptake of aluminium in C-S-H [44, 83], which is not considered in the present cement database, would extend the stability area of C-S-H to slightly lower belite contents. If more than 35 % belite is present, strätlingite content decreases due to dilution and because strätlingite is destabilized in the presence of sufficient calcium oxide to C-S-H. At > 60 mass-% belite, katoite is additionally calculated to form, according to equation (4.5). At very high C_2S amounts, when strätlingite is no longer occurring, portlandite appears as stable phase. Monosulfate is stable in the entire compositional range (with the exception of hydrated pure C_2S). Ettringite occurs only in very low amounts and is stable only at ye'elimite contents between 100 and 40 mass-%. It is destabilized to monosulfate if sufficient calcium is available, according to equation (4.6). Thus the hydrates occurring in this system are determined to a large extent by the availability of calcium oxide.

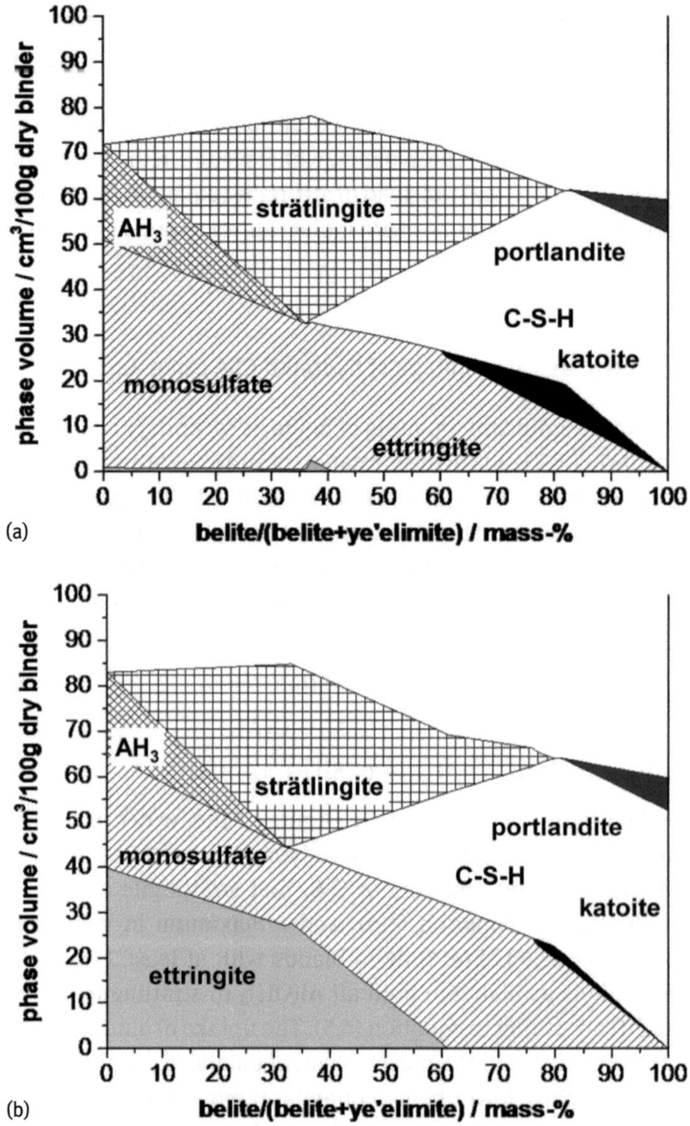

Fig. 4.12: Calculated phase diagram of the thermodynamic stable hydrate assemblages in cm^3 per 100 g of dry solid at 20 °C of (a) the system ye'elimite–belite–water, and (b) the system ye'elimite–calcium sulfate–belite–water at a fixed molar ratio anhydrite/ye'elimite of 0.8 (anhydrite/ye'elimite = 15 : 85 by mass). Complete hydration at a water/solid ratio of 1 was assumed. Adapted from Winnefeld & Lothenbach (2016) [82].

Fig. 4.12 (b) presents the hydrate assemblages of the ternary blend ye'elimite–belite–calcium sulfate at a fixed molar ratio of anhydrite/ye'elimite = 0.8 (15/85 by mass), which is a typical ratio used in non-expansive, high strength CSA cements with high ye'elimite content [67]. As Fig. 4.12 (a) shows, it can be seen that at approximately 30–40 mass-% belite content a maximum volume of hydrates occurs. The changes in the phase assemblage depending on the mixing ratio of ye'elimite/belite are similar to the system without calcium sulfate (Fig. 4.12 (a)), but with much higher quantities of ettringite and less monosulfate occurring due to the additional presence of calcium sulfate. With the increasing replacement of ye'elimite by belite, more strätlingite forms and $Al(OH)_3$ is consumed. Ettringite and strätlingite are no longer stable at belite contents beyond 60 % and 80 %, respectively, while C-S-H and monosulfate are stabilized due to the availability of sufficient calcium oxide from belite according to equations (4.5) and (4.6). Katoite and portlandite occur beyond approximately 75–80 mass-% belite.

Further trends can be seen if the quantities of the hydrate phases are plotted in ternary plots $C_4A_3\bar{S}$–C_2S–$C\bar{S}$. Fig. 4.13 (a)–(h) shows the amounts of ettringite, monosulfate, C-S-H, strätlingite, katoite, gypsum, aluminium hydroxide, and portlandite formed upon complete hydration of the ternary blend. Similar plots can be obtained for ternary blends of CSA cement with Portland cement and calcium sulfate as discussed in detail in Winnefeld & Lothenbach (2014) [84].

Ettringite (Fig. 4.13 (a)) is stable in almost the entire range of the ternary system with the exception of a small region where belite contents above 50 % and very low calcium sulfate contents below 5 mass-% are present. Maximum ettringite contents are located around 20–40 mass-% ye'elimite, 20–40 mass-% belite, and 40–50 mass-% anhydrite. The presence of belite shifts the area of maximum ettringite formation above the calcium sulfate/ye'elimite ratio of 2 (indicated by the 2 : 1 line in Fig. 4.13 (a)), as calcium oxide from belite promotes the formation of monosulfate from ettringite and AH_3 (equation (4.6)). Monosulfate is only present at low calcium sulfate contents (Fig. 4.13 (b)).

C-S-H contents decrease towards lower belite contents (Fig. 4.13 (c)), and C-S-H is absent in the region with both low belite (< 30 mass-%) and low anhydrite (< 30 mass-%) contents. Under these conditions, the silica provided by the belite is incorporated into strätlingite (Fig. 4.13 (d)), which is present only when there are low anhydrite contents and not too high belite contents. Katoite (Fig. 4.13 (e)) only occurs at high belite and very low ye'elimite and anhydrite contents. Surplus gypsum is present at high anhydrite contents (Fig. 4.13 (f)), aluminium hydroxide decreases strongly with increasing belite and anhydrite contents (Fig. 4.13 (g)), and portlandite is only present with belite contents approximately > 80 mass-% (Fig. 4.13 (h)).

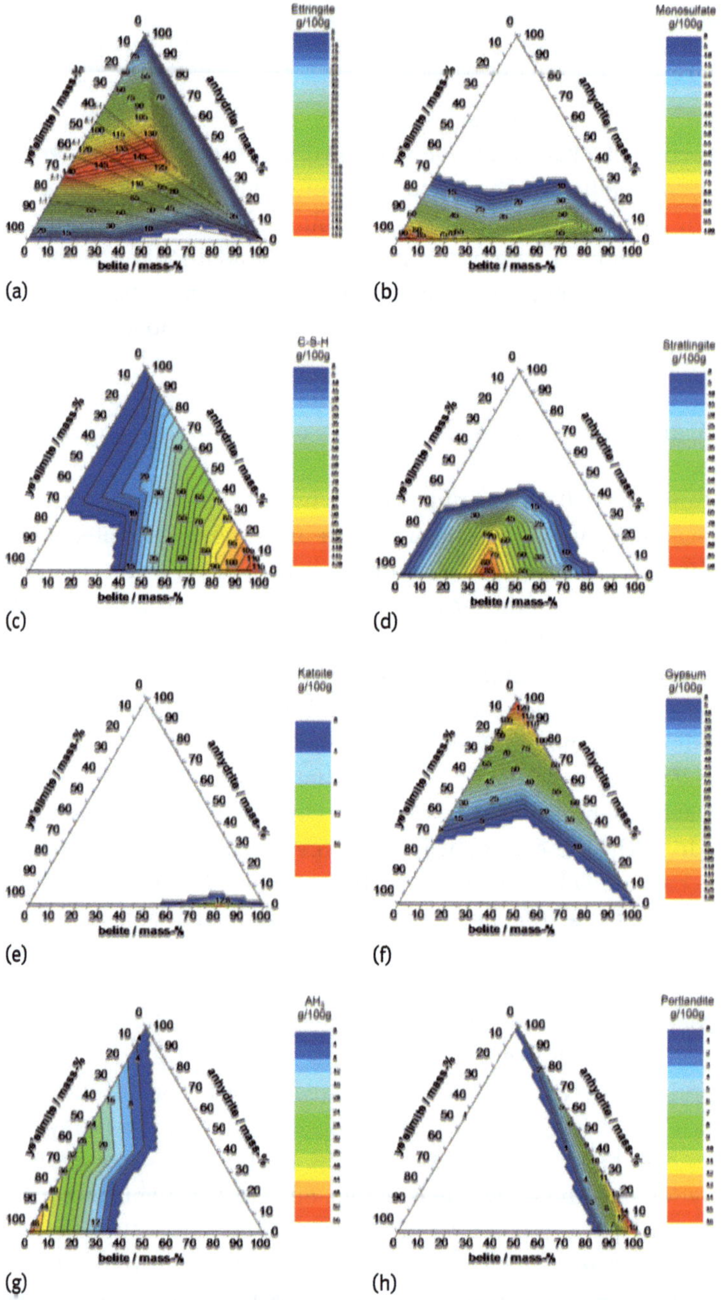
Ettringite g/100g
Monosulfate g/100g
C-S-H g/100g
Stratlingite g/100g
Katoite g/100g
Gypsum g/100g
AH_3 g/100g
Portlandite g/100g
ye'elimite / mass-%
anhydrite / mass-%
belite / mass-%
(a)
(b)
(c)
(d)
(e)
(f)
(g)
(h)

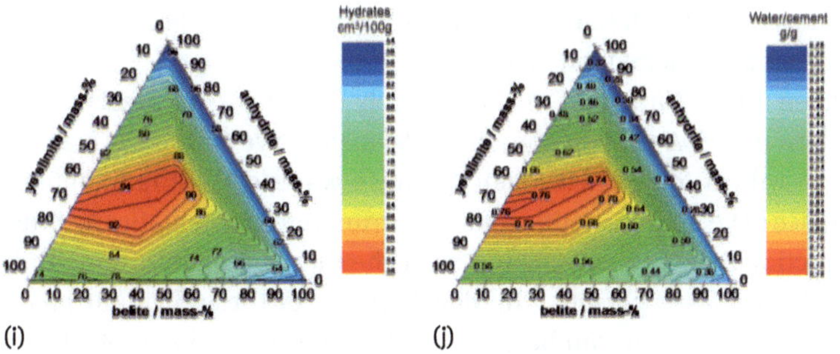

Fig. 4.13: Ternary plots of the hydrates present in the system $C_4A_3\bar{S}$–C_2S–$C\bar{S}$–H_2O at 20 °C. (a) ettringite, (b) monosulfate, (c) C-S-H, (d) strätlingite, (e) katoite, (f) gypsum, (g) aluminium hydroxide, and (h) portlandite contents in g per 100 g unhydrated solid, (i) total volume of hydrates and (j) *w/c* needed for complete hydration at 20 °C. Complete hydration at a water/solid ratio of 1 was assumed. The dotted lines in a) represent different molar ratios between anhydrite and ye'elimite. Adapted from Winnefeld & Lothenbach (2016) [82].

The total volume of hydrates (Fig. 4.13 (i)) and the water/cement ratio needed for complete hydration (Fig. 4.13 (j)) are strongly related to the amount of ettringite in the hydrate assemblage, as ettringite is the phase incorporating the highest amount of water among the hydrates present in the system [82, 84]. The highest water demand occurs for the binary system ye'elimite-anhydrite at a molar ratio of calcium sulfate/ye'elimite of 2 (see also Fig. 4.9).

From Fig. 4.13 (a)–(h), it is evident that some phases can coexist in the stable hydrate assemblage and some phases do not occur together. For example, C-S-H is the only phase compatible with all the other phases. Ettringite does not occur together with katoite, and monosulfate is not stable in the presence of gypsum (or anhydrite) as this combination would react to ettringite. The other main hydrate phases of CSA cements, strätlingite and $Al(OH)_3$ both do not coexist with gypsum or portlandite. The full compatibility table and a ternary plot of the stable phase assemblages are given in Winnefeld & Lothenbach (2016) [82]. The modelled phase assemblages agree well with experimental data from various authors [74, 76, 82, 85–88].

The modelled data only considers thermodynamically stable phase assemblages. In CSA cements, metastable CAH_{10} is reported to occur, especially in systems with a low ratio of calcium sulfate to ye'elimite such as plain CSA clinker [75, 82, 89]. It forms mainly during the first days of hydration according to equation (4.7), and starts to decompose at later ages.

$$3C_4A_3\bar{S} + 98H \rightarrow C_3A{\cdot}3C\bar{S}{\cdot}32H + 6CAH_{10} + 2AH_3 \tag{4.7}$$

In the previous examples of thermodynamic modelling of CSA cements, complete hydration was assumed. In practise, it is very often also of interest to study the development of hydrate phases with hydration time. In this case, the dissolution kinetics of the anhydrous phases need to be determined, which is generally done by fitting experimental data using a mathematical model.

An example taken from Winnefeld & Lothenbach (2010) [77] is presented in Fig. 4.14. The hydration of a CSA cement containing 54 mass-% ye'elimite, 19 mass-% belite, and 21 mass-% anhydrite was investigated using a water/cement ratio of 0.80. The dissolution kinetics of ye'elimite, belite, and anhydrite were mathematically fitted based on X-ray diffraction data and used as input for the thermodynamic modelling. With the densities of all phases present, the development of the individual phase volumes as well as the total volume can be calculated. The dissolution of the above-mentioned anhydrous phases leads to the formation of ettringite, microcrystalline $Al(OH)_3$, and strätlingite, while the quantity of pore solution decreases. Monosulfate forms only after about two days of hydration, when the main part of ye'elimite and anhydrite has dissolved. At this time, the formation of $Al(OH)_3$ also stops, and its amount decreases slightly while more strätlingite is formed. At later stages, ettringite and strätlingite are the main hydration products, and $Al(OH)_3$ and monosulfate are present as minor hydrate phases. The hydration leads to an increase of the volume of solids and to a decrease of the volume of the liquid phase. The total volume decreases with time, reflecting a calculated chemical shrinkage in the order of 11 cm^3/100 g dry cement after 28 days, which is more than twice as high as for Portland cement (Fig. 4.1) [52]. These modelling results agree very well with the experimental data from X-ray diffraction, thermogravimetric analysis [77], SEM-EDX on polished sections [90], and measurements of chemical shrinkage [91].

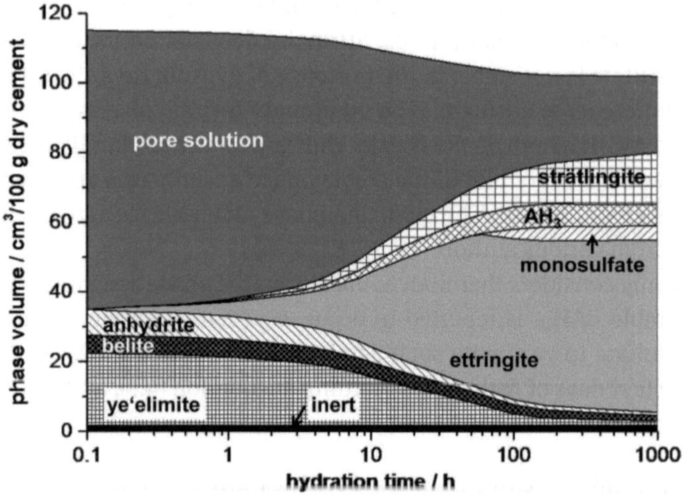

Fig. 4.14: Modelled changes of phase volumes during hydration of a calcium sulfoaluminate cement hydrated at a water/cement ratio of 0.80 at 20 °C. Adapted from Winnefeld & Lothenbach (2010) [77].

The hydration of CSA cements may also take place under different thermal conditions, such as the application at cold or warm temperatures. Furthermore, elevated temperatures may occur, for example during early hydration (self-heating) due to the rapid development of heat of hydration [77], and can lead to changes in the hydrate assemblage depending on the phase composition of the CSA cement. The expected changes of the hydrate assemblages of CSA cements depending on temperature have been calculated by thermodynamic modelling under water-saturated conditions [92].

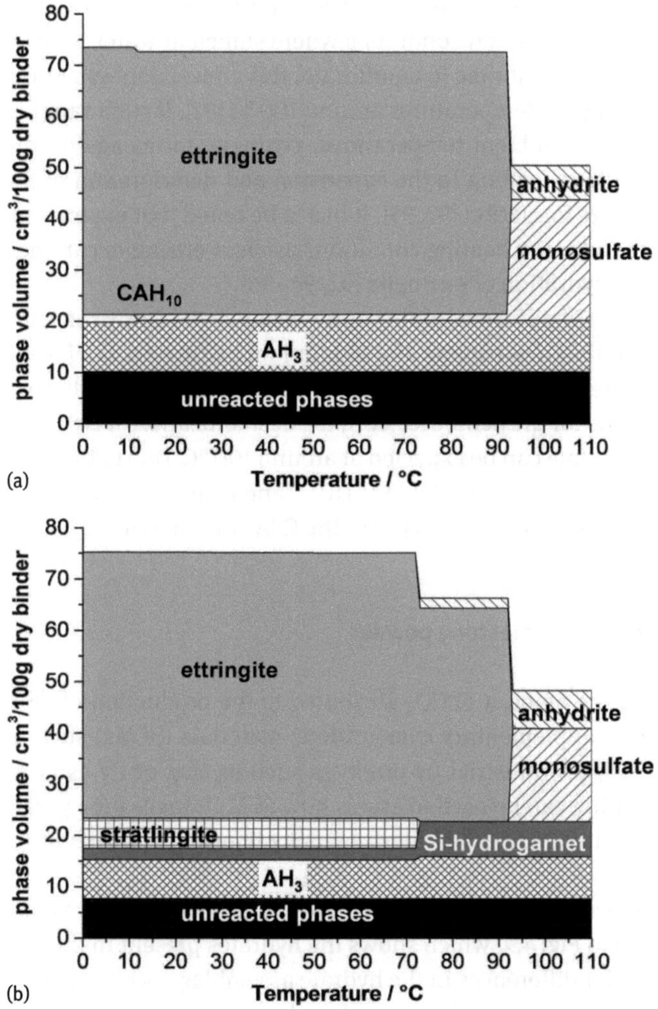

Fig. 4.15: Influence of temperature on the phase assemblage predicted for (a) hydrated CSA cement containing no significant amounts of belite and (b) hydrated CSA cement containing 18 % belite under water-saturated conditions. Adapted from Kaufmann et al. (2016) [92].

In a CSA sample without significant amounts of belite (Fig. 4.15 (a)) using a molar ratio of calcium sulfate to ye'elimite of 2.0, mainly the presence of ettringite and aluminium hydroxide is expected, while a small quantity of monosulfate (or of CAH_{10} below 10 °C) is predicted. C_2AH_8 is less stable than monosulfate or CAH_{10} [24] and is thus not predicted to be a stable phase. Under water saturated conditions, ettringite is calculated to be stable up to approximately > 90 °C, which is significantly higher than for Portland cements containing limestone, where the decomposition temperature of ettringite and monocarbonate to monosulfate and calcite is calculated to be below 50 °C [19]. In the absence of limestone, ettringite is expected to destabilize to monosulfate and calcium sulfate only above 90 °C. Under saturated conditions where sufficient water is available and if the samples have sufficient time to equilibrate, this conversion was found experimentally at somewhat higher temperatures around 100 °C [92]. If such samples come into contact with water at ambient temperatures, ettringite forms again from monosulfate and calcium sulfate, leading to the expansion and deterioration of the sample due to delayed ettringite formation [92, 93]. It has to be noted that experimental and results are different under dry heating conditions as meta-ettringite can then be formed from the partial dehydration of ettringite [92, 94–96].

Thermodynamic modelling of a CSA containing approximately 18 mass-% of belite (Fig. 4.15 (b)) predicts, in addition to ettringite and aluminium hydroxide, the formation of strätlingite. Strätlingite is calculated to be unstable with respect to siliceous hydrogarnet at around 70 °C. As for the belite-free CSA, the destabilization of ettringite to monosulfate and calcium sulfate can be expected at around 100 °C. Due to the presence of siliceous hydrogarnet, which is also stable at 110 °C and above, the changes in volume are somewhat less distinct than in the case of the CSA without belite.

4.4.4 Blending CSA cements with limestone powder

In order to reduce costs and the amount of CO_2 attributed to the production of CSA, they are often blended with supplementary cementitious materials (SCMs) such as limestone powder [88, 97, 98] or industrial by-products such as slag or fly ash [99, 100]. Thermodynamic modelling can be used to assess how SCM changes the hydrate assemblage of CSA. This is illustrated in the following using limestone powder as an example.

Fig. 4.16 (a) shows the calculated stable hydrate assemblages of different ye'elimite/calcite blends. Compared to Fig. 4.9, which shows the hydrates present in the system ye'elimite/anhydrite, major differences in the hydrate assemblage occur. With an increased addition of calcite, the amount of monosulfate decreases, while increasing amounts of ettringite and monocarbonate form. Ye'elimite can react with water in the presence of calcite to ettringite, monocarbonate, and $Al(OH)_3$ (equation (4.8)):

$$3C_4A_3\bar{S} + 2C\bar{C} + 72H \rightarrow C_3A{\cdot}3C\bar{S}{\cdot}32H + 2C_3A{\cdot}C\bar{C}{\cdot}11H + 6AH_3 \qquad (4.8)$$

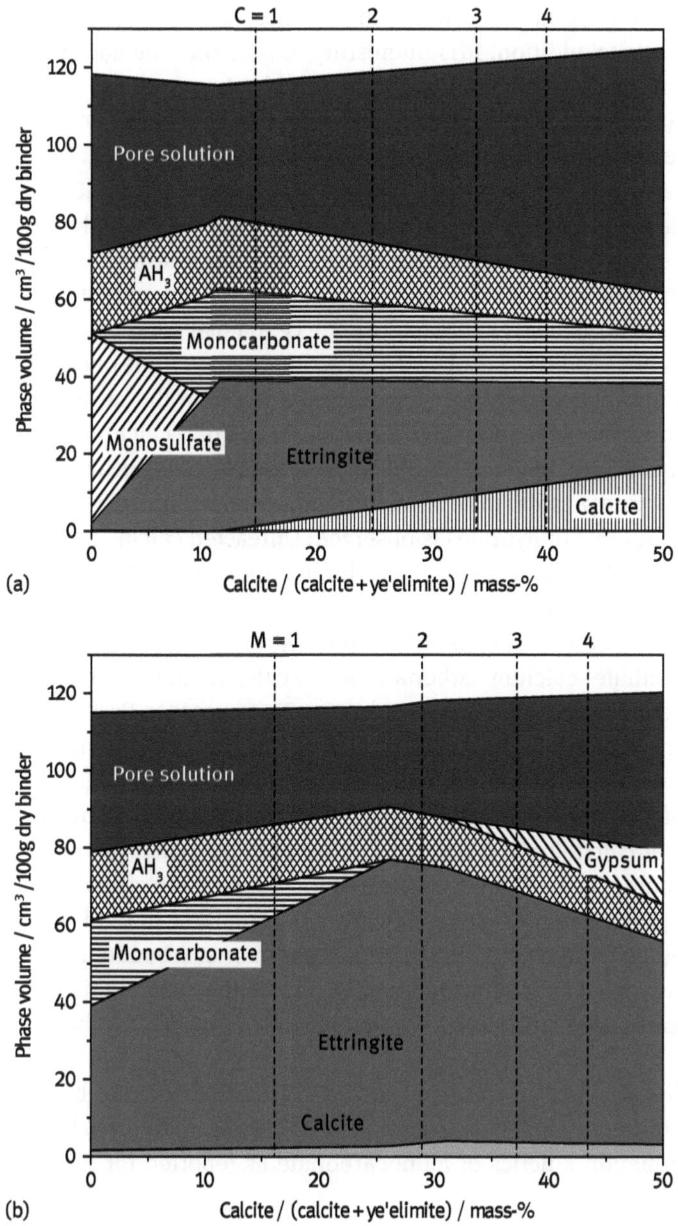

Fig. 4.16: Calculated phase diagram of the thermodynamic stable hydrate assemblages in cm^3 per 100 g of dry solid at 20 °C of (a) the system ye'elimite–calcite–water, and (b) the system ye'elimite–calcite–calcium sulfate–water at a fixed molar ratio calcium sulfate/ye'elimite of 1 (anhydrite/ye'elimite = 14 : 86 by mass). C = molar ratio of calcite to ye'elimite. M is the molar ratio of anhydrite to ye'elimite. Complete hydration at a water/solid ratio of 1 was assumed. Adapted from Winnefeld & Lothenbach (2013) [84].

Beyond a calcite addition of 10 %, monosulfate is no longer stable, and excess calcite is present beyond 12 % calcite addition. It is interesting to note that the addition of approximately 10 % of limestone leads to a maximum of solid volume, similar to what is observed in Portland (Fig. 4.5) and blended (Fig. 4.6) cements. In Portland cements, the volume maximum corresponds to a maximum in compressive strength [12]. In CSA systems, strength increases due to limestone additions are reported as well [88, 97], however the situation is more complicated as hydration kinetics are strongly influenced by the admixtures used which has a strong impact on compressive strength.

The stable hydrate phases in the ternary system ye'elimite–calcite–anhydrite are shown in Fig. 4.16 (b). A mixture of ye'elimite and calcite (1 : 1 molar ratio = 86 : 14 by mass, surplus calcite is always present) is blended with different amounts of calcium sulfate. Without the addition of calcium sulfate, ye'elimite reacts in the presence of calcite to ettringite, monocarbonate, and $Al(OH)_3$ (equation (4.8)). If calcium sulfate is added, the amount of ettringite increases. Monocarbonate decreases until it disappears completely, in this case at a molar ratio of calcium sulfate to ye'elimite M of 1.6. At this ratio, the maximum volume of hydrates is observed. Unreacted calcite increases with increasing molar ratio of calcium sulfate to ye'elimite, until, at a molar ratio of 2 and beyond, no calcite takes part in the reactions.

Based on the previously described subsystems, the hydration of the ternary system ye'elimite–calcium sulfate–calcium carbonate can be calculated. Fig. 4.17 shows a binary plot of the different stable hydrate assemblages. In contrast to Portland cement (Fig. 4.2) [81], where portlandite and C-S-H are always present, in CSA dominated cements, portlandite and hemicarbonate are never calculated to be stable. Monocarbonate occurs instead of hemicarbonate. Ettringite and AH_3 are always present. From Fig. 4.17, it is evident that calcite only takes part in the reactions when the ratio of calcium sulfate to ye'elimite is below 2. In this case, monocarbonate is formed instead of monosulfate, which provides sulfate ions allowing the formation of (additional) ettringite. With a low addition of calcite, monosulfate and monocarbonate coexist, depending on the amount of calcium sulfate added. The modelling also predicts that significant amounts of carbonate can be incorporated into ettringite [97], as shown in Fig. 4.18.

However, hemicarbonate is often observed instead of monocarbonate by XRD [88, 97]. According to equation (4.9), The formation of hemicarbonate probably occurs due to the slow formation kinetics of monocarbonate as reported for Portland cements [101].

$$6C_4A_3\bar{S} + C\bar{C} + 135H \rightarrow 2C_3A{\cdot}3C\bar{S}32H + 2C_3A{\cdot}0.5C\bar{C}{\cdot}0.5CH{\cdot}11.5H + 14AH_3 + 5CH \qquad (4.9)$$

The modelled phase stabilities fit well to the experimental X-ray diffraction data obtained on a commercial CSA clinker blended with calcite and/or anhydrite [88] after 90 days of hydration (Fig. 4.19), with the exception that hemicarbonate is observed in-

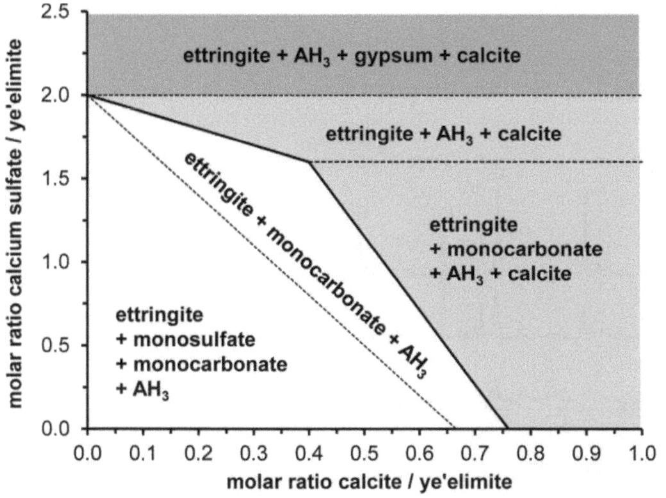

Fig. 4.17: Calculated stable phase assemblages in the system ye'elimite–calcium sulfate–calcium carbonate–water at 20 °C and a water/binder ratio of 1. In the light grey area excess calcite is present, in the dark grey area excess calcite and gypsum are present. Adapted from Winnefeld & Lothenbach (2013) [84].

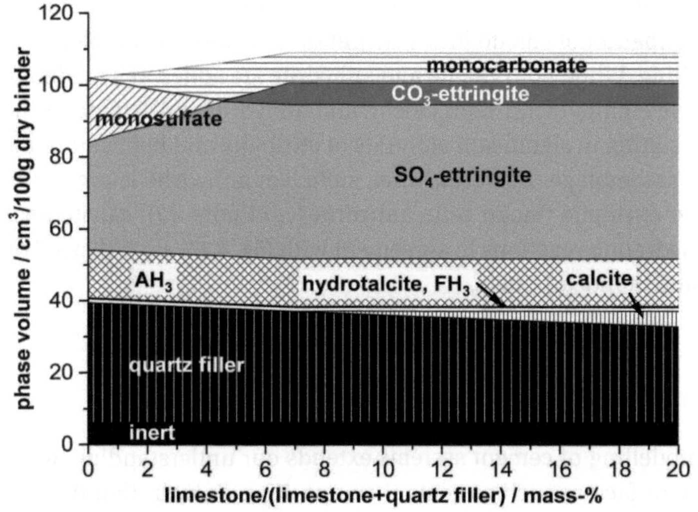

Fig. 4.18: Calculated stable hydrate assemblages at 20 °C in a CSA paste depending on the replacement of quartz filler by limestone filler on a mass basis. Total filler content is 47 mass-%. The CSA cement contains gypsum with a ratio calcium sulfate/ye'elimite of 1.45. Full hydration at a water/cement ratio of 0.80 was assumed. Ettringite is formed as a single phase (solid solution) and is subdivided into SO_4-ettringite and CO_3-ettringite in the graph to emphasize that, in the presence of calcite, some carbonate can be incorporated in the solid solution. Adapted from Pelletier-Chaignat et al. (2013) [97].

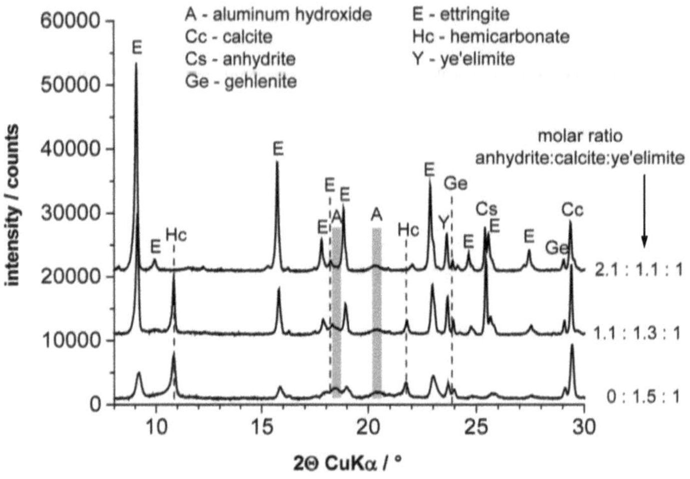

Fig. 4.19: Influence of calcium sulfate and calcium carbonate on the hydrate assemblage determined by X-ray diffraction of a CSA cement hydrated 90 d at 20 °C using a water/binder ratio of 0.74. Adapted from Martin et al. (2015) [88].

stead of monocarbonate. When only calcite is added at a molar ratio calcite/ye'elimite of 1.5 : 1, ettringite, AH_3, and hemicarbonate are identified in the hydrate assemblage. This confirms the participation of calcite in the hydration reactions, as predicted by thermodynamic modelling. When both calcite and anhydrite are added at a moderate molar ratio anhydrite/ye'elimite of 1.1, both calcite and anhydrite participate in the hydration reactions, resulting in significant amounts of ettringite and hemicarbonate present in the hydrate assemblage. However, when sufficient anhydrite is present to convert all ye'elimite to ettringite (molar ratio anhydrite/ye'elimite ≥ 2), calcite does not participate in the hydration reactions in agreement with Fig. 4.17, as evidenced by the absence of hemi- and monocarbonate.

4.5 Conclusions

The thermodynamic modelling of cement systems extends our understanding of the consequences of different factors such as cement composition or hydration time on the properties of hydrated cementitious systems. The stable hydrate assemblage in the system Portland cement, fly ash, and limestone and in the system ye'elimite, calcium sulfate, and belite has been predicted by thermodynamic modelling and by using ternary phase diagrams. Changes in the overall chemical composition affect the amount as well as the kind of solid phases formed.

The effect of blending PC with limestone and/or fly ash can be summarized as follows:

- In the absence of limestone, the stable phase assemblage in hydrated Portland cement includes C-S-H, portlandite, ettringite, hydrotalcite, and monosulfate. Calcium carbonate stabilizes monocarbonate plus ettringite instead of monosulfate, which leads to a higher volume of hydrated cement [6–8, 54, 55]. It has been shown, both experimentally as well as by thermodynamic modelling, that the presence of up to 5 % calcite reduces the porosity of hydrated cements and can increase the compressive strength compared to a PC without limestone. Replacement ratios above 10 mass-% dilute the cement, increase the porosity in the hydrated pastes, and reduce compressive strength.
- The replacement of Portland cements by fly ash results in the destabilization of portlandite and the formation of more C(-A)-S-H. In the presence of more than approximately 30–40 mass-% of fly ash, strätlingite replaces monocarbonate or monosulfate after long hydration times resulting in less C(-A)-S-H, which has little effect on the measured compressive strength after 3 months and longer. Although modelling predicts strätlingite in fly ash blends, it is rarely observed in experimental studies as the reaction degree of fly ash is often too low.
- A volume maximum is calculated for Portland cement containing ≈ 2 mass-% of limestone and up to 10 mass-% of fly ash. In the presence of 0–8 mass-% limestone or 0–22 mass-% fly ash, the calculated volume equals or surpasses the volume of the plain Portland cement, which agrees with the compressive strength measured after 3 months or longer in such blends [48, 56]. A reduction of volume is calculated for higher fractions of calcite and/or fly ash.

The effect of calcium sulfate, calcium carbonate, and belite on the hydrates formed from calcium sulfoaluminate (CSA) cements can be summarized as follows:

- In the absence of calcium sulfate, monosulfate and $Al(OH)_3$ are the main hydrates in CSA cements. The addition of calcium sulfate results in the stabilization of ettringite, less monosulfate, and a higher volume of hydrate phases up to a molar ratio calcium sulfate/ye'elimite (M) of 2. Above $M = 2$, gypsum is stable.
- The reaction of belite adds calcium oxide to the hydrates which results at low additions in the stabilization of strätlingite. High contents of belite are expected to lead to the formation of C-S-H.
- In the absence of gypsum, the presence of calcite leads to the formation of monocarbonate and to the stabilization of ettringite. At $M \geq 2$, calcite acts as filler only.

Many studies have applied thermodynamic modelling in its different forms to Portland cement systems and the modelling results generally agree very well with the extensive experimental evidence available for Portland cement systems. Much less experimental data and experience of long-term behavior are available for the ternary system Portland cement, fly ash, and limestone and for the system ye'elimite–belite–

calcite–calcium sulfate. Thermodynamic calculations, which allow for easy and fast parameter variation, are thus the ideal instrument to plan and to complete systematic experimental studies on the effects of blending different clinkers and SCMs. The use of thermodynamic models together with experimental investigations provides a sound scientific basis for the efficient development of new cementitious materials and design-blended cements with optimal properties.

The results of such calculations depend on the quality and completeness of the thermodynamic data. While reliable data are available for the main hydrates, the development of dedicated models for aluminium uptake in C-S-H or an adequate description of the formation kinetics of siliceous hydrogarnet may somewhat modify the diagrams presented in this paper.

The results presented here for the phase assemblages in the systems Portland cement, fly ash, and limestone and the system $C_4A_3\bar{S}$–$C\bar{S}$–C_2S or $C\bar{C}$ show the high potential of thermodynamic equilibrium calculations to explain the formation and the coexistence of different hydrates depending on the relative amount and the reaction degree of the anhydrous material. The ternary plots calculated and shown in this paper can also be used directly to assess the long-term composition of such blended cements.

Although thermodynamic modelling helps to deepen our understanding of cementitious systems and to interpret experimental observations, thermodynamic modelling cannot replace the experiments, as in many cases kinetic limitations (slow dissolution, prevention of precipitation) occur. Thus, thermodynamic modelling should, as far as possible, always be verified against experimental data.

References

[1] Taylor HFW. Cement Chemistry. London: Thomas Telford Publishing; 1997.

[2] Lothenbach B, Winnefeld F, Alder C, Wieland E, Lunk P. Effect of temperature on the pore solution, microstructure and hydration products of Portland cement pastes. Cement and Concrete Research. 2007; 37(4): 483–491.

[3] Lothenbach B, Winnefeld F. Thermodynamic modelling of the hydration of Portland cement. Cement and Concrete Research. 2006; 36(2): 209–226.

[4] Reardon EJ. Problems and approaches to the prediction of the chemical composition in cement/water systems. Waste Management. 1992; 12: 221–239.

[5] Lee JH, Roy DM, Mann B, Stahl D. Integrated approach to modeling long-term durability of concrete engineered barriers in LLRW disposal facility. Material Research Society Symposium Proceedings. 1995; 353: 881–889.

[6] Lothenbach B, Le Saout G, Gallucci E, Scrivener K. Influence of limestone on the hydration of Portland cements. Cement and Concrete Research. 2008; 38(6): 848–860.

[7] Juel I, Herfort D, Gollop R, Konnerup-Madsen J, Jakobsen HJ, Skibsted J. A thermodynamic model for predicting the stability of thaumasite. Cement & Concrete Composites. 2003; 25: 867–872.

[8] Matschei T, Lothenbach B, Glasser FP. The role of calcium carbonate in cement hydration. Cement and Concrete Research. 2007; 37(4): 551–558.
[9] Nielsen EP, Herfort D, Geiker MR. Phase equilibria of hydrated Portland cement. Cement and Concrete Research. 2005; 35: 109–115.
[10] Herfort D, Lothenbach B. Ternary phase diagrams. In: Scrivener KL, Snellings R, Lothenbach B, editors. A Practical Guide to Microstructural Analysis of Cementitious Materials. Oxford, UK: CRC Press; 2016. 487–504.
[11] Lothenbach B. Thermodynamic equilibrium calculations in cementitious systems. Materials and Structures. 2010; 43: 1413–33.
[12] Damidot D, Lothenbach B, Herfort D, Glasser FP. Thermodynamics and cement science. Cement and Concrete Research. 2011; 41(7): 679–695.
[13] Parkhurst DJ, Appelo CAJ. User's Guide to PHREEQC (version 2) – A computer program for speciation, batch reaction, one dimensional transport, and inverse geochemical calculations. Water-Resources Investigation Report. Denver, Colorado; 1999.
[14] van der Lee J, De Windt L. CHESS tutorial and cookbook. Updated for version 3.0. Ecole Nationale Supérieure des Mines de Paris; 2002.
[15] Kulik D, Wagner T, Dmytrieva S, Kosakowski G, Hingerl F, Chudnenko K, et al. GEM-Selektor geochemical modeling package: revised algorithm and GEMS3K numerical kernel for coupled simulation codes. Comput Geosci. 2013; 17(1): 1–24.
[16] Wagner T, Kulik DA, Hingerl FF, Dmytrieva SV. GEM-Selektor geochemical modeling package: TSolMod library and data interface for multicomponent phase models. Canadian Mineralogist. 2012; 50: 1173–1195.
[17] Blanc P, Bourbon X, Lassin A, Gaucher EC. Chemical model for cement-based materials: Thermodynamic data assessment for phases other than C-S-H. Cement and Concrete Research. 2010; 40(9): 1360–1374.
[18] Blanc P, Bourbon X, Lassin A, Gaucher EC. Chemical model for cement-based materials: Temperature dependence of thermodynamic functions for nanocrystalline and crystalline C-S-H phases. Cement and Concrete Research. 2010; 40(6): 851–866.
[19] Lothenbach B, Matschei T, Möschner G, Glasser FP. Thermodynamic modelling of the effect of temperature on the hydration and porosity of Portland cement. Cement and Concrete Research. 2008; 38(1): 1–18.
[20] Matschei T, Lothenbach B, Glasser FP. Thermodynamic properties of Portland cement hydrates in the system CaO-Al_2O_3-SiO_2-$CaSO_4$-$CaCO_3$-H_2O. Cement and Concrete Research. 2007; 37(10): 1379–1410.
[21] Schmidt T, Lothenbach B, Romer M, Scrivener KL, Rentsch D, Figi R. A thermodynamic and experimental study of the conditions of thaumasite formation. Cement and Concrete Research. 2008; 38(3): 337–349.
[22] Möschner G, Lothenbach B, Rose J, Ulrich A, Figi R, Kretzschmar R. Solubility of Fe-ettringite ($Ca_6[Fe(OH)_6]_2(SO_4)_3 \cdot 26H_2O$). Geochimica et Cosmochimica Acta. 2008; 72(1): 1–18.
[23] Dilnesa BZ, Lothenbach B, Renaudin G, Wichser A, Kulik D. Synthesis and characterization of hydrogarnet $Ca_3(Al_xFe_{1-x})_2(SiO_4)_y(OH)_{4(3-y)}$. Cement and Concrete Research. 2014; 59: 96–111.
[24] Lothenbach B, Pelletier-Chaignat L, Winnefeld F. Stability in the system CaO–Al2O3–H2O. Cement and Concrete Research. 2012; 42(12): 1621–1634.
[25] Dilnesa BZ, Lothenbach B, Renaudin G, Wichser A, Wieland E. Stability of monosulfate in the presence of iron. Journal of the American Ceramic Society. 2012; 95(10): 3305–3316.
[26] Dilnesa BZ, Lothenbach B, Le Saout G, Renaudin G, Mesbah A, Filinchuk Y, et al. Iron in carbonate containing AFm phases. Cement and Concrete Research. 2011; 41(3): 311–323.
[27] Kulik DA. Improving the structural consistency of C-S-H solid solution thermodynamic models. Cement and Concrete Research. 2011; 41(5): 477–495.

[28] Rozov KB, Berner U, Kulik DA, Diamond LW. Solubility and thermodynamic properties of carbonate-bearing hydrotalcite-pyroaurite solid solutions with a 3 : 1 Mg/(Al+Fe) mole ratio. Clays and Clay Minerals. 2011; 59(3): 215–232.

[29] Balonis M, Lothenbach B, Le Saout G, Glasser FP. Impact of chloride on the mineralogy of hydrated Portland cement systems. Cement and Concrete Research. 2010; 40(7): 1009–1022.

[30] Matschei T, Glasser FP. Thermal stability of thaumasite. Materials and Structures. 2015; 48: 2277–2289.

[31] Balonis M. The influence of inorganic chemical accelerators and corrosion inhibitors on the mineralogy of hydrated Portland cement systems. Aberdenn, GB: University of Aberdeen; 2010.

[32] Hummel W, Berner U, Curti E, Pearson FJ, Thoenen T. Nagra/PSI Chemical Thermodynamic Data Base 01/01. Universal Parkland, Florida: Publishers/uPublishcom; 2002.

[33] Thoenen T, Kulik DA. Nagra/PSI Chemical Thermodynamic Data Base 01/01 for the GEM-Selektor (V.2- PSI) Geochemical Modeling Code: Release 28–02–03. Internal Report TM-44–03–04. Available from: http://leswebpsich/software/GEMS-PSI/thermodata/indexhtml; 2003.

[34] Garvin D, Parker VB, White HJ. CODATA thermodynamic tables. Selections for some compounds of calcium and related mixtures: a prototype set of tables. Berlin: Springer Verlag; 1987.

[35] Babushkin VJ, Matveyev GM, Mchedlov-Petrossyan OP. Thermodynamics of Silicates. Berlin: Springer-Verlag; 1985.

[36] Robie RA, Hemingway BS. Thermodynamic properties of minerals and related substances at 298.15 K and 1 bar (105 Pascals) pressures and at higher temperatures. US Geol Surv Bull. 1995; 2131: 1–461.

[37] Helgeson HC, Delany JM, Nesbitt HW, Bird DK. Summary and critique of the thermodynamic properties of rock-forming minerals. American Journal of Science. 1978; 278-A: 1–229.

[38] Möschner G, Lothenbach B, Ulrich A, Figi R, Kretschmar R. Solid solution between Al-ettringite and Fe-ettringite ($Ca_6[Al_{1-x}Fe_x(OH)_6]_2(SO_4)_3 \cdot 26H_2O$). Cement and Concrete Research. 2009; 39: 482–489.

[39] Corazza E, Sabelli C. The crystal structure of syngenite, $K_2Ca(SO_4)_2 \cdot H_2O$. Zeitschrift für Kristallographie. 1967; 124: 398–408.

[40] Boysen H, Lerch M, Stys A, Senyshyn A. Structure and oxygen mobility in mayenite ($Ca_{12}Al_{14}O_{33}$): a high-temperature neutron powder diffraction study. Acta Crystallographica. 2007; B63: 675–682.

[41] Hörkner W, Müller-Buschbaum H. Zur Kristallstruktur von $CaAl_2O_4$. Journal of Inorganic and Nuclear Chemistry. 1976; 38(5): 983–984.

[42] Goodwin DW, Lindop AJ. The crystal structure of $CaO \cdot 2Al_2O_3$. Acta Crystallographica Section B. 1970; 26(9): 1230–1235.

[43] Myers R, Bernal SA, Provis JL. A thermodynamic model for C-(N-)A-S-H gel: CNASH_ss. Derivation and validation. Cem Concr Res. 2014; 66: 27–47.

[44] Haas J, Nonat A. From C-S-H to C-A-S-H: Experimental study and thermodynamic modelling. Cement and Concrete Research. 2015; 68: 124–138.

[45] Hong S-Y, Glasser FP. Alkali binding in cement pastes. Part I. The C-S-H phase. Cement and Concrete Research. 1999; 29: 1893–1903.

[46] Lothenbach B, Nonat A. Calcium silicate hydrates: solid and liquid phase composition. Cement and Concrete Research. 2015; 78: 57–70.

[47] Antoni M, Rossen J, Martirena F, Scrivener K. Cement substitution by a combination of metakaolin and limestone. Cement and Concrete Research. 2012; 42(12): 1579–1589.

[48] De Weerdt K, Ben Haha M, Le Saout G, Kjellsen KO, Justness H, Lothenbach B. Hydration mechanisms of ternary Portland cements containing limestone powder and fly ash. Cement and Concrete Research. 2011; 41(3): 279–291.
[49] Deschner F, Winnefeld F, Lothenbach B, Seufert S, Schwesig P, Dittrich S, et al. Hydration of a Portland cement with high replacement by siliceous fly ash Cement and Concrete Research. 2012; 42: 1389–1400.
[50] Parrot LJ, Killoh DC. Prediction of cement hydration. British Ceramic Proceedings. 1984; 35: 41–53.
[51] Lothenbach B, Wieland E. A thermodynamic approach to the hydration of sulphate-resisting Portland cement. Waste Management. 2006; 26(7): 706–719.
[52] Geiker M, Knudsen T. Chemical shrinkage of portland cement pastes. Cement and Concrete Research. 1982; 12(5): 603–610.
[53] Jansen D, Goetz-Neunhoeffer F, Lothenbach B, Neubauer J. The early hydration of Ordinary Portland Cement (OPC): An approach comparing measured heat flow with calculated heat flow from QXRD. Cement and Concrete Research. 2012; 42(1): 134–138.
[54] Kuzel H-J, Pöllmann H. Hydration of C_3A in the presence of $Ca(OH)_2$, $CaSO_4 \cdot 2H_2O$ and $CaCO_3$. Cement and Concrete Research. 1991; 21: 885–895.
[55] Bonavetti VL, Rahhal VF, Irassar EF. Studies on the carboaluminate formation in limestone filler-blended cements. Cement and Concrete Research. 2001; 31: 853–859.
[56] De Weerdt K, Kjellsen KO, Sellevold E, Justness H. Synergy between fly ash and limestone powder in ternary cements. Cement and Concrete Composites. 2011; 33(1): 30–38.
[57] Steenberg M, Skibsted J, Poulsen SL, Damtoft JS, Herfort D. Composite cement based on Portland cement clinker, limestone and calcined clay. Proceedings of the 13th International Congress on the Chemistry of Cement. Madrid; 2011. 97–104.
[58] Herfort D, Rasmussen S, Jons E, Osbaeck B. Mineralogy and performance of cement based on high sulfate clinker. Cem Concr Aggreg. 1999; 21(1): 64–72.
[59] De Weerdt K, Justness H, Kjellsen KO, Sellevold E. Fly ash-limestone ternary composite cements: synergetic effect at 28 days. Nordic Concrete Research. 2010; 42(2): 51–70.
[60] Lothenbach B, Scrivener K, Hooton RD. Supplementary cementitious materials. Cement and Concrete Research. 2011; 41(12): 1244–1256.
[61] Hanehara S, Tomosawa F, Kobayakawa M, Hwang KR. Effects of water/powder ratio, mixing ratio of fly ash, and curing temperature on pozzolanic reaction of fly ash in cement paste. Cement and Concrete Research. 2001; 31(1): 31–19.
[62] Glasser FP, Tyrer M, Quillin K, Ross D, Pedersen J, Goldthorpe K, et al. The chemistry of blended cements and backfills intended for use in radioactive waste disposal. UK Environment Agency; 1998.
[63] Girao AV, Richardson IG, Taylor R, Brydson RMD. Composition, morphology and nanostructure of C-S-H in 70 % white Portland cement – 30 % fly ash blends hydrated at 55 °C. Cem Concr Res. 2010; 40: 1350–1359.
[64] Escalante-Garcia JI, Sharp JH. The chemical composition and microstructure of hydration products in blended cements. Cement & Concrete Composites. 2004; 26: 967–976.
[65] Luke K, Lachowski E. Internal composition of 20-year-old fly ash and slag-blended ordinary Portland cement pastes. Journal of the American Ceramic Society. 2008; 91(12): 4084–4092.
[66] Tishmack JK, Olek J, Diamond S, Sahu S. Characterization of pore solutions expressed from high-calcium fly-ash-water pastes. Fuel. 2001; 80: 815–819.
[67] Juenger MCG, Winnefeld F, Provis JL, Ideker JH. Advances in alternative cementitious binders. Cement and Concrete Research. 2011; 41(12): 1232–1243.
[68] Ludwig H-M, Zhang W. Research review of cement clinker chemistry. Cement and Concrete Research. 2015; 78, Part A: 24–37.

[69] Zhang L, Su MZ, Wang YM. Development of the use of sulfo- and ferroaluminate cements in China. Advances in Cement Research. 1999; 11(1): 15–21.

[70] Gartner E. Industrially interesting approaches to "low-CO_2" cements. Cement and Concrete Research. 2004; 34(9): 1489–1498.

[71] Majling J, Znasik R, Gabrisova A, Svetik S. The influence of anhydrite activity upon the hydration of calcium sulfoaluminate cement clinker. Thermochim Acta. 1985; 92(SEP): 349–352.

[72] Hanic F, Kapralik I, Gabrisova A. Mechanism of hydration reactions in the system C4A3S-CS-CaO-H2O referred to hydration of sulfoaluminate cements. Cement and Concrete Research. 1989; 19(5): 671–682.

[73] Sahu S, Havlica J, Tomkova V, Majling J. Hydration behavior of sulfoaluminate belite cement in the presence of various calcium sulfates. Thermochim Acta. 1991; 175(1): 45–52.

[74] Sahu S, Tomkova V, Majling J, Havlica J. Influence of particle sizes of individual minerals on the hydration processes in the system C_2S-$C_4A_3S \cdot CS$. Cement and Concrete Research. 1993; 23(3): 693–699.

[75] Winnefeld F, Barlag S. Influence of calcium sulfate and calcium hydroxide on the hydration of calcium sulfoaluminate clinker. Zkg Int. 2009; 62(12): 42–53.

[76] Winnefeld F, Barlag S. Calorimetric and thermogravimetric study on the influence of calcium sulfate on the hydration of ye'elimite. J Therm Anal Calorim. 2010; 101(3): 949–957.

[77] Winnefeld F, Lothenbach B. Hydration of calcium sulfoaluminate cements – Experimental findings and thermodynamic modelling. Cement and Concrete Research. 2010; 40(8): 1239–1247.

[78] Allevi S, Marchi M, Scotti F, Bertini S, Cosentino C. Hydration of calcium sulphoaluminate clinker with additions of different calcium sulphate sources. Materials and Structures. 2016; 49(1/2): 453–466.

[79] Chen IA, Hargis CW, Juenger MCG. Understanding expansion in calcium sulfoaluminate-belite cements. Cement and Concrete Research. 2012; 42(1): 51–60.

[80] Pöllmann H. Syntheses, properties and solid solution of ternary lamellar calcium aluminate hydroxi salts (AFm-phases) containing SO_4^{2-}, CO_3^{2-} and OH^-. Neues Jahrbuch für Mineralogie - Abhandlungen. 2006; 182: 173–181.

[81] Matschei T, Lothenbach B, Glasser FP. The AFm phase in Portland cement. Cement and Concrete Research. 2007; 37(2): 118–130.

[82] Winnefeld F, Lothenbach B. Phase equilibria in the system $Ca_4Al_6O_{12}SO_4$–Ca_2SiO_4–$CaSO_4$–H_2O referring to the hydration of calcium sulfoaluminate cements. RILEM Technical Letters. 2016; 1: 10–16.

[83] Myers RJ, L'Hôpital E, Provis JL, Lothenbach B. Effect of temperature and aluminium on calcium (alumino)silicate hydrate chemistry under equilibrium conditions. Cement and Concrete Research. 2015; 68: 83–93.

[84] Winnefeld F, Lothenbach B. Thermodynamic modeling of the hydration of calcium sulfoaluminate cements blended with mineral additions. 1st International Conference on Sulphoaluminate Cement: Materials and Engineering Technology, Wuhan, China. 2013: 212–228.

[85] Collepardi M, Turriziani R, Marcialis A. The paste hydration of $4CaO \cdot 3Al_2O_3 \cdot SO_3$ in presence of calcium sulphate, tricalcium silicate and dicalcium silicate. Cement and Concrete Research. 1972; 2(2): 213–223.

[86] Smrckova E, Palou MT, Tomkova V. Application of conduction calorimetry for study of the reactivity of C_2S in the system C_2S-C_4A_3S-CS-H. J Therm Anal Calorim. 1996; 46(2): 597–605.

[87] Bullerjahn F, Zajac M, Ben Haha M. CSA raw mix design: effect on clinker formation and reactivity. Materials and Structures. 2015; 48(12): 3895–3911.

[88] Martin LHJ, Winnefeld F, Müller CJ, Lothenbach B. Contribution of limestone to the hydration of calcium sulfoaluminate cement. Cement and Concrete Composites. 2015; 62: 204–211.

[89] Berger S, Cau Dit Coumes C, Le Bescop P, Damidot D. Hydration of calcium sulfoaluminate cement by a $ZnCl_2$ solution: Investigation at early age. Cement and Concrete Research. 2009; 39(12): 1180–1187.

[90] Winnefeld F, Ben Haha M, Lothenbach B. Hydration mechanisms of calcium sulfoaluminate cements assessed by scanning electron microscopy and thermodynamic modeling. 13th International Congress on the Chemistry of Cement, Madrid, Spain, July, 3–8, 2011, 7 pp.

[91] Lura P, Winnefeld F, Klemm S. Simultaneous measurements of heat of hydration and chemical shrinkage on hardening cement pastes. J Therm Anal Calorim. 2010; 101(3): 925–932.

[92] Kaufmann J, Winnefeld F, Lothenbach B. Stability of ettringite in CSA cement at elevated temperatures. Advances in Cement Research. 2016; 28(4): 251–261.

[93] Taylor HFW, Famy C, Scrivener KL. Delayed ettringite formation. Cement and Concrete Research. 2001; 31(5): 683–693.

[94] Skoblinskaya NN, Krasilnikov KG, Nikitina LV, Varlamov VP. Changes in crystal structure of ettringite on dehydration. Cement and Concrete Research. 1975; 5(5): 419–431.

[95] Zhou Q, Lachowski EE, Glasser FP. Metaettringite, a decomposition product of ettringite. Cement and Concrete Research. 2004; 34(4): 703–710.

[96] Baquerizo LG, Matschei T, Scrivener KL. Impact of water activity on the stability of ettringite. Cement and Concrete Research. 2016; 79: 31–44.

[97] Pelletier-Chaignat L, Winnefeld F, Lothenbach B, Mueller CJ. Beneficial use of limestone filler with calcium sulphoaluminate cement. Construction and Building Materials. 2012; 26(1): 619–627.

[98] Hargis CW, Telesca A, Monteiro PJM. Calcium sulfoaluminate (Ye'elimite) hydration in the presence of gypsum, calcite, and vaterite. Cement and Concrete Research. 2014; 65: 15–20.

[99] Zivica V. Properties of blended sulfoaluminate belite cement. Construction and Building Materials. 2000; 14(8): 433–437.

[100] Garcia-Mate M, De la Torre AG, Leon-Reina L, Aranda MAG, Santacruz I. Hydration studies of calcium sulfoaluminate cements blended with fly ash. Cement and Concrete Research. 2013; 54: 12–20.

[101] Zajac M, Rossberg A, Le Saout G, Lothenbach B. Influence of limestone and anhydrite on the hydration of Portland cements. Cement & Concrete Composites. 2014; 46: 99–108.

Part II: **Special cement and binder mineral phases**

G. Artioli*, M. Secco, A. Addis, and M. Bellotto

5 Role of hydrotalcite-type layered double hydroxides in delayed pozzolanic reactions and their bearing on mortar dating

Abstract: Double-layer hydroxide minerals are part of a very interesting group of natural and synthetic compounds with trigonal or hexagonal symmetry and a flexible layered crystal structure. They are formed extremely frequently in geologic, industrial, and synthetic processes. The ease of formation is related to the possibility of accommodating divalent and trivalent cations in the structure, together with a range of anionic species. Some compounds of the group, namely those based on hydrotalcite chemistry, are invariably found as products of the pozzolanic reaction between lime and clays in ancient mortars and modern binders that serve as alternatives to Portland clinker. The present review wishes to relate the structural properties of hydrotalcite-type compounds to the crystal-chemical mechanisms taking place during long-term pozzolanic processes. The kinetics of CO_3 exchange between the hydroxide and the atmosphere has important negative consequences for the radiocarbon dating of ancient mortars.

Keywords: layered double hydroxides, LDH, hydrotalcite, ancient mortars, pozzolanic reaction

5.1 Introduction

A large number of layered double hydroxides (LDH) share common crystal-chemical and structural features, based on simple di- and tri-octahedral layers, commonly referred to in the mineralogical literature as gibbsite- and brucite-type octahedral layers. Their importance is related to their frequency of formation and widespread occurrence, derived from the possibility of incorporating a rich combination of cation and anion chemistry into the structure. Double layered hydroxides are therefore important in a number of diverse technical applications, such as cements and binders, paper formulation, metal corrosion, catalysis, and environmental applications, among others

***Corresponding author: G. Artioli,** CIRCe Centre and Dipartimento di Geoscienze, Università di Padova, Padova, Italy
M. Secco, CIRCe Centre and Dipartimento di Ingegneria Civile, Edile ed Ambientale, Università di Padova, Padova, Italy
A. Addis, CIRCe Centre and Dipartimento di Geoscienze, Università di Padova, Padova, Italy
M. Bellotto, Chemiplastica Specialties S.p.A., Italy

DOI 10.1515/9783110473728-006

[1, 2]. In the field of catalysis and intercalation compounds, LDHs are defined as anionic clays.

Of all the known double hydroxide compounds, we are here interested in those occurring in binder systems, both ancient and modern. As a matter of fact, (Mg,Al) LDHs of the hydrotalcite group or (Ca,Al) LDHs of the hydrocalumite group are almost ubiquitously formed during long-term pozzolanic reactions in ancient hydraulic mortars [3], in modern pozzolanic cements [4], in slag cement systems [5–7], and in alternative binder materials based on calcined clays [8]. The Al-rich LDH phases forming in the portlandite and sulphate saturated system of cement are commonly called AFm (i.e. Al_2O_3-Fe_2O_3-mono) phases, including monosulphoaluminate, monocarbonate, and their solid solutions [9–12]. The AFt phases, including ettringite [13–15], are first crystallized during cement hydration, and then they convert with time into AFm phases.

The interest in LDH-type as products of pozzolanic reactions is therefore related on the one hand to the need of understanding the long-term behavior of the pozzolanic materials employed in ancient times, especially concerning their contribution to the methodologies proposed for the radiocarbon dating of ancient mortars [16]. On the other hand, the present drive towards clinker-free cements [17] faces the challenging use of alternative pozzolanic materials, mostly based on calcined clays, which invariably yield LDH-type AFm phases as reaction products. Furthermore, the ability of the LDH structure to intercalate a variety of inorganic and organic molecules makes them reference compounds in intercalation chemistry [1, 2, 18, 19], and potentially important players in the control of the superplasticizing effect during cement hydration [20].

5.2 Crystal structural features

The crystal structures of trigonal LDH have been mainly investigated in mineral members of the pyroaurite [21, 22] and hydrotalcite [23] groups (Tab. 5.1 and Mills et al. (2011) [24]). The crystal structure is based on the ·ABC·ABC· sequence of the basic brucite layer (Fig. 5.1) [13, 25]. The stacking yields a threefold periodicity with respect to the basal layer thickness (cell parameters: $a \approx 3.1$ Å, $c \approx 23.4$ Å) and a rhombohedral space group symmetry (R-3m or R-3).

Many of the trigonal mineral phases have the corresponding hexagonal polytype [26], based on the ·AB·AB· sequence of layers. They all share the doubling of the interlayer basal spacing (cell parameters: $a \approx 3.1$ Å, $c \approx 15.6$ Å) and a hexagonal space group ($P6_3/mmc$). The hexagonal phases (such as sjögrenite, manasseite, barbertonite) are not discussed here.

Synthetic LDHs prepared by co-precipitation frequently show extensive stacking faults as the result of the intimate random intergrowth of the rhombohedral and hexagonal polytypes. The layer stacking arrangement is a function of the extent of Coulombic layer-interlayer interactions, as opposed to hydrogen bonding and the nature of the interlayer anion [27]. High temperature hydrothermal treatment does not affect the stacking fault density, for entropic reasons, while low temperature treatments favor the rhombohedral polytype [28].

The general formula of the double hydroxides with 2(OH) and carbonate as the interlayer anion can be defined as:

$$[M^{+2}_{1-x}M^{+3}_{x}(OH)_2](CO_3)_{x/2}(H_2O)_m$$

The flexibility of the LDH structure derives from the ample substitution of divalent and trivalent cations in the brucite-type layer. The cations are homogeneously distributed in the structure without segregation, both in cases of disordered distributions or ordering patterns, which are frequent for the $M^{+2}:M^{+3}$ ratio of 2. The recurring ratio of $M^{+2}:M^{+3} = 0.75:0.25$ (i.e. $x = 0.25$, Tab. 5.1) is the one ensuring an optimal distribution of Al^{+3} cations in the brucite sheet, so that there are no neighboring Al atoms and repulsive forces are minimized [29]. In the (Mg,Al) system (hydrotalcite), higher concentrations of Mg produce areas with higher density of Mg and the nucleation of brucite, whereas higher concentrations of Al should cause the precipitation of gibbsite or bayerite. Similarly, in the Fe-rich system, higher concentrations of Fe or variations in the oxidation state of iron cause the precipitation of Fe hydroxides. In the latter case, the stability of so-called green rust is also related to the stabilizing effect of the

Tab. 5.1: Mineral members of the group of trigonal LDHs (space group R-3m or R-3). Cation and anion contents are referred to the general formula based on $(OH)_2$ (see text). For a more complete list of mineral species see Mills et al. (2012) [24].

Mineral name	M^{+2}	M^{+3}	A^{-n}	H_2O
Hydrotalcite	$Mg_{0.75}$	$Al_{0.25}$	$(CO_3)_{0.125}$	$(H_2O)_{0.5}$
Meixnerite	$Mg_{0.75}$	$Al_{0.25}$	$(OH)_{0.25}$	$(H_2O)_{0.5}$
Motukoreaite	$Mg_{0.67}$	$Al_{0.33}$	$Na_{0.11},(SO_4)_{0.22}$	$(H_2O)_{1.33}$
Pyroaurite	$Mg_{0.75}$	$Fe_{0.25}$	$(CO_3)_{0.125}$	$(H_2O)_{0.5}$
Coalingite	$Mg_{0.83}$	$Fe_{0.17}$	$(CO_3)_{0.08}$	$(H_2O)_{0.17}$
Iowaite	Mg	$Fe_{0.25}$	(O,Cl)	$(H_2O)_{0.5}$
Desaultesite	$Mg_{0.75}$	$Mn_{0.25}$	$(CO_3)_{0.125}$	$(H_2O)_{0.5}$
Stichtite	$Mg_{0.75}$	$Cr_{0.25}$	$(CO_3)_{0.125}$	$(H_2O)_{0.5}$
Woodalllite	$Mg_{0.75}$	$Cr_{0.25}$	$(Cl)_{0.25}$	$(H_2O)_{0.5}$
Mössbauerite		$Fe_{0.25}$	$O,(CO_3)_{0.25}$	$(H_2O)_{0.75}$
Trébeurdenite	$Fe_{0.4}$	$Fe_{0.8}$	$O_{0.4},(CO_3)_{0.2}$	$(H_2O)_{0.6}$
Reevesite	$Ni_{0.75}$	$Fe_{0.25}$	$(CO_3)_{0.125}$	$(H_2O)_{0.5}$
Honessite	$Ni_{0.75}$	$Fe_{0.25}$	$(SO_4)_{0.125}$	$(H_2O)_{0.5}$
Takovite	$Ni_{0.75}$	$Al_{0.25}$	$(CO_3)_{0.125}$	$(H_2O)_{0.5}$
Comblainite	$Ni_{0.75}$	$Co_{0.25}$	$(CO_3)_{0.125}$	$(H_2O)_{0.5}$
Woodwardite	$Cu_{0.67}$	$Al_{0.33}$	$(SO_4)_{0.17}$	$(H_2O)_{0.67}$
Caresite	$Fe_{0.67}$	$Al_{0.33}$	$(CO_3)_{0.17}$	$(H_2O)_{0.5}$
Shigaite	$Mn_{0.64}$	$Al_{0.36}$	$(SO_4)_{0.18}$	$(H_2O)_{0.72}$
Nikischerite	$Na_{0.11}Fe_{0.67}$	$Al_{0.33}$	$(SO_4)_{0.22}$	$(H_2O)_{1.23}$

inter-layer cations [30–32]. The amount and distribution of anions and water in the interlayer are affected by the ordered distribution of the cations in the brucite layer [27, 33].

The phase hydrocalumite $Ca_{0.67}Al_{0.33}(OH)_2(OH,CO_3,Cl)(H_2O)$ is the corresponding structure in the (Ca,Al) system. It can be described as a stacking of distorted portlandite layers, with additional water and anions in the interlayer [13, 34]. Several members of the hydrocalumite series are formed during the reaction of Portland cement with water, having a complex relationship with monosulfate in the SO_4 loaded system [9, 11, 36–38]. The (Ca,Al) LDH structure can reach high hydration states, up to a full additional layer of water, and also incorporate a substantial amount of organic molecules [18, 19, 39, 40], so that the structure has been proposed for metal immobilization [41].

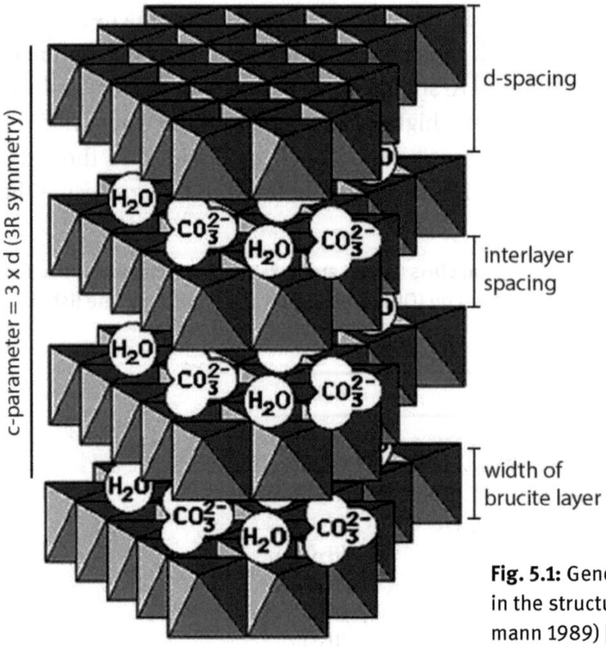

Fig. 5.1: General stacking of brucite layers in the structure of LDH (modified after Pöllmann 1989) [19].

5.3 Formation of hydrotalcite during pozzolanic reaction

The complex crystal chemistry and the extreme flexibility of the LDH structure makes their nucleation very frequent during the water solid-reactions taking place in mortars, cements, and binding materials. LDH phases are often observed as products of the pozzolanic reactions occurring in ancient hydraulic mortars [3, 42, 43]. The pozzolanic reaction is based on the success of Roman *opus caementitium*, which indeed revolutionized the architecture of the ancient world in the beginning of the second century BC. The pozzolanic reaction in ancient times was consciously triggered by mixing lime with natural volcanic materials containing reactive glass [3, 4], although, when volcanic material was not locally available the mortar formulation included fragmented or ground pottery (*cocciopesto*) as ubiquitously available reactive silica source. Sometimes the hydraulic reaction was stimulated by the addition of reactive silica of plant origin, and even unusual clays with high reactivity were employed, especially in late Roman times [44].

Concerning the properties of the LDH phases formed by pozzolanic reaction, their crystal-chemistry is generally controlled by the chemical composition of the employed pozzolanic additive, as showed by the XRD patterns of several representative mortars reported in Fig. 5.2. When the Vitruvian tradition of employing acid (i.e. high silica) pyroclastic products is respected, the chemistry of the pozzolanic products is controlled by the dominant $CaO–SiO_2–Al_2O_3$ composition, and thus the precipitation of hydrocalumite-type LDH's like Ca-hemicarbonate is favored (d-spacing of the basal peak ≈ 8.0 Å). On the other hand, the presence of Mg,Al-containing materials such as natural clays, fired pottery, or more basic pyroclastic products shifts the reaction equilibria from a pure portlandite/limestone system to a more complex chemical en-

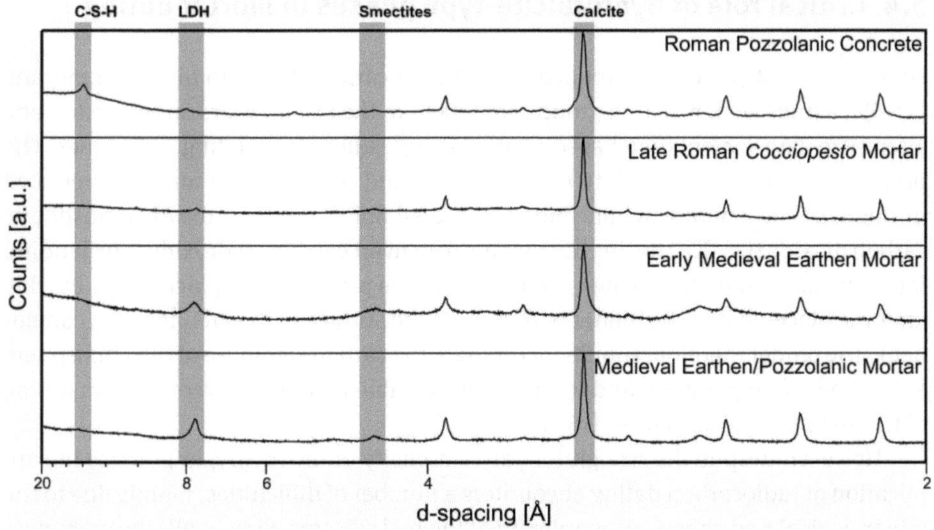

Fig. 5.2: XRPD patterns of ancient mortars formulated with different pozzolanic materials, showing the ubiquitous formation of AFm phases. From top to bottom: Roman pozzolanic concrete from the Republican city walls of Aquileia (North-Eastern Italy); late Roman cocciopesto mortar from the baptismal font of the San Giovanni Baptistery in Lomello (Pavia, North-Western Italy); early Medieval earthen mortar based on pure saponite clays from the Torba monastery (Castelseprio, Varese, North-Western Italy); Medieval mortar based on smectite clays and basic pyroclastic hyaloclastites from the ruins of the Sachuidic castle (Forni di Sopra, North-Eastern Italy).

vironment, where hydrotalcite-type LDH phases are stabilized (d-spacing of the basal peak ≈ 7.7 Å). Hydrotalcite is the most common LDH phase found in ancient mortars (Fig. 5.2), as it is the phase likely to be formed from fully hydrolized cations [45, 46].

It is interesting to note that, following the ancient tradition of pozzolanic reactions involving cocciopesto, there is at present a stream of active research towards using calcined clays as pozzolanic materials [8]. The drive for this is mainly sustainability [17, 47, 48], in order to move industrial cement production away from the present clinker technology, which is responsible for large CO_2 emissions [49]. The focus here is on understanding and optimizing clay reactivity [50, 51], in the attempt to employ natural clays in the formulation of alternative cements. At this stage, reasonable success has been obtained by the substitution of calcined clays in place of industrial supplementary cementitious materials (i.e. blastfurnace slags, fly-ash, etc.). In the future, thermally, chemically, or mechanically activated clays may be the core of alternative binder technology. Again, LDH phases are common products of alkali activated or geopolymeric formulations. Understanding the formation of hydrotalcite phases during the hydration of alternative binders is of course mandatory for the control of their rheological, chemical, and mechanical properties.

5.4 Critical role of hydrotalcite-type phases in mortar dating

Following a few pioneering applications of radiocarbon dating to historical mortars [52–55], a number of research groups embarked in the last 40 years on the systematic attempt to date ancient lime-based mortars using radiocarbon dating techniques. The principle is rather simple: lime-based mortars and plasters were and are produced through the carbonation of lime putty (i.e. slaked lime), and thus absorb atmospheric carbon dioxide (i.e. CO_2) during the process to convert calcium hydroxide (portlandite) into calcium carbonate (calcite). Conceptually, the process takes place rather quickly after the placement of the binder within or on the surface of the architectural component. The binder therefore should incorporate the carbon signature of the atmosphere at the time of preparation, and it represents a viable material for radiocarbon dating of the architectural structures [16, 56].

However, despite the straightforward simplicity of the theory, in practice the application of radiocarbon dating encounters a number of difficulties, mainly due to the mineralogical and chemical complexity of the real systems. First of all, the carbonate present in the sample may be contaminated by geologic carbonate (i.e. old carbonate or fossil carbonate) [57]. Even minute quantities of geologically old calcite can produce large errors in the resulting dates. Furthermore, the carbonation process after the emplacement may not be so rapid, so that the obtained dates are from calcite produced quite some time after the production (i.e. younger carbonate). Or else there might be subsequent "younger" generations of calcite contaminating the original carbonate, due to percolating water, organic impurities, or subsequent recrystallization in contact with younger fluids. For this reason, when available some laboratories rely on the dates obtained on the "lime lumps" embedded in the mortar matrix, which are supposed to be relicts from the original slaking process [58]. It is however known that the carbonation of the lump material may take place quite some time after the original emplacement of the mortar, depending on the density of the portlandite nodule.

The real problem therefore lies in the careful extraction, separation, characterization, and processing of the carbonate fractions present in the matrix that correspond to the carbonation of the "original" lime putty of the binder. A number of research laboratories have developed different protocols to solve this problem [59–62]. The issue is far from being resolved, and at present each laboratory has different protocols for the extraction of the calcite fractions (thermal step treatments, acidification, cryofracturing, etc.). This may result in rather different dates from the same samples if using different protocols, and therefore a complex interpretation of the results. The major conceptual problem at present is that many laboratories tend to separately date the different extracted calcite fractions (thereby obtaining a wide range of dates), and then select the resulting dates that best fit the expectations based on rather subjective and arbitrary considerations.

In our laboratories, we focus on the careful preparation and full mineralogical characterization of the calcite fractions extracted from the samples. The aim is to define the criteria and treatments able to separate (physically and chemically) the different generations of calcite in the mortar samples. Then each fraction is carefully checked by mineralogical and spectroscopic characterization in order to decide before hand (i.e. before the dating protocol) whether the sample is suitable for dating or contaminated by older or younger carbon. At this stage, contamination by old carbon can be safely detected by cathodoluminescence-induced spectroscopy (Fig. 5.3). The protocol is very efficient and it has been tested on samples from different environments and chronological periods. We may safely state that fossil carbonate can be efficiently detected, although sometimes it is too fine-grained to allow a complete removal. However, the protocol may at least justify the exclusion of the sample on physical basis.

The contamination of younger carbon is more difficult to detect. Sometimes, delayed carbonation or slow pozzolanic reactions occur at much later times, so that younger calcite is formed, and carbonate anions are exchanged into the interlayer of hydrotalcite-type phases. Interaction with percolating fluids may also dissolve some of the originally formed calcite and re-precipitate the carbonate in younger secondary calcite. This later calcite can sometimes be removed by physical sieving or light acidation of the binder fraction, although there are not yet appropriate tests to evaluate the process.

A common problem when extracting the finer fraction of the original calcite binder of ancient hydraulic mortars for dating purposes is that it is frequently contaminated by hydrotalcite or hydrocalumite phases, depending on the type of pozzolanic material and the extent of the pozzolanic reaction. The binder fraction to be dated is therefore intimately mixed with LDH compounds with the same grain size. The ther-

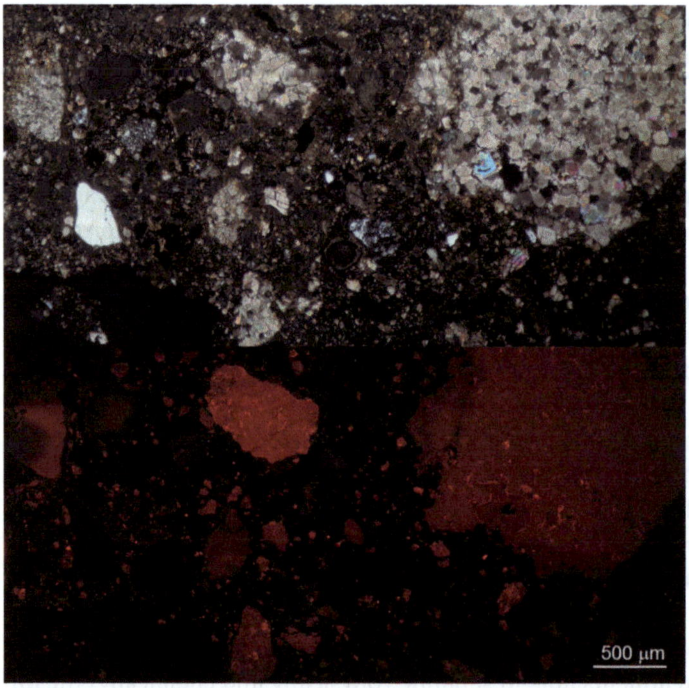

Fig. 5.3: Above: polarized optical microphotograph of a thin section of an historical mortar (crossed Nicols), clearly showing polycrystalline geologic carbonate embedded in the mortar matrix. Below: the same area measured in catodoluminescence spectroscopy. The orange luminescence clearly shows that geologic carbonate is present both in large fragments as aggregate and as fine particles intermixed with the mortar binder.

mal decomposition of hydrotalcite [63–66] is tentatively being exploited to separate the two contributions: the younger CO_2 released during the decomposition of hydrotalcite (at about 350–400 °C) is eliminated by heating in vacuum conditions; then the sample is brought to a higher temperature and the CO_2 released by the calcite decarbonation is reduced to graphite and radiocarbon measured. This improved protocol seems to be a very promising method for the reliable dating of ancient mortars that have undergone pozzolanic reactions to some degree.

Acknowledgment: The authors wish to acknowledge their stimulating and fruitful collaboration with the research group at the CIRCE Centre accelerator, Caserta on the project related to the radiocarbon dating of ancient mortars.

References

[1] Cavani F, Trifirò F, Vaccari A. Hydrotalcite-type anionic clays: Preparation, properties and applications. Catalysis Today. 1991; 11: 173–301.

[2] Khan AI, O'Hare D. Intercalation chemistry of layered double hydroxides: recent developments and applications. Journal of Materials Chemistry. 2002; 12: 3191–3198.

[3] Brandon CJ, Hohlfelder RL, Jackson MD, Oleson JP. Building for eternity: the history and technology of Roman concrete engineering in the sea. Oxford and Philadelphia: Oxbow Books; 2014.

[4] Massazza F. Pozzolana and pozzolanic cements. In: Hewlett P, ed. Lea's chemistry of cement and concrete. 4th Edition. Elsevier-Butterworth-Heinemann; 1998: 471–635.

[5] Wang SD, Scrivener KL. 29Si and 27Al NMR study of alkali-activated slag. Cement and Concrete Research. 2003; 33: 769–774.

[6] Richardson IG, Groves GW. Microstructure and microanalysis of hardened cement pastes involving ground granulated blast-furnace slag. Journal of Materials Science. 1992; 27: 6204–6212.

[7] Whittaker M, Zajac M, Ben Haha M, Bullerjahn F, Black L. The role of the alumina content of slag, plus the presence of additional sulfate on the hydration and microstructure of Portland cement-slag blends. Cement and Concrete Research. 2014; 66: 91–101.

[8] Scrivener K, Favier A, editors. Calcined Clays for Sustainable Concrete. RILEM, Springer Netherlands; 2015.

[9] Pöllmann H. Syntheses, properties and solid solution of ternary lamellar calcium aluminate hydroxi salts (AFm-phases) containing SO_4^{2-}, CO_3^{2-} and OH^-. Neues Jahrbuch für Mineralogie-Abhandlungen: Journal of Mineralogy and Geochemistry. 2006; 182: 173–181.

[10] Matschei T, Lothenbach B, Glasser, FP. The AFm phase in Portland cement. Cement and Concrete Research. 2007; 37: 118–130.

[11] Matschei T, Lothenbach B, Glasser, FP. Thermodynamic properties of Portland cement hydrates in the system $CaO–Al_2O_3–SiO_2–CaSO_4–CaCO_3–H_2O$. Cement and Concrete Research. 2007; 37: 1379–1410.

[12] Lothenbach B, Le Saout G, Gallucci E, Scrivener K. Influence of limestone on the hydration of Portland cements. Cement and Concrete Research. 2008; 38: 848–860.

[13] Taylor HFW. Crystal structures of some double hydroxide minerals. Mineralogical Magazine. 1973; 39: 377–389.

[14] Taylor HFW. Cement chemistry. 2nd edition. London: Thomas Telford; 1997.

[15] Pöllmann H, Kuzel HJ, Wenda R. Compounds with ettringite structure. Neues Jahrbuch für Mineralogie-Abhandlungen. 1989; 160: 133–158.

[16] Hale J, Heinemeier J, Lancaster L, Lindroos A, Ringbom Å. Dating ancient mortar. American Scientist. 2003; 91: 130–137.

[17] Kurtis KE. Innovations in cement-based materials: Addressing sustainability in structural and infrastructure applications. MRS Bulletin. 2015; 40: 1102–1109.

[18] Pöllmann H. Organische Derivate des TCAH mit Schichtstrukturen vom Typus $Ca_2[Al(OH)_6]^+$ $[RCOO·nH_2O]^-$. Zeitschr. Krist. 1989; 186: 236–237.

[19] Pöllmann H. Study of the hydration mechanisms and formation of new hydrates applying organic additives to the aluminate phase of cement. In: Bayles J, Gouda GR, Nisperos A, editors. Proc. 11th Int. Conf. on Cem. Micr., 1989 New Orleans. 287–305.

[20] Raki L, Beaudoin JJ, Mitchell L. Layered double hydroxide-like materials: nanocomposites for use in concrete. Cement and Concrete Research. 2004; 34: 1717–1724.

[21] Ingram L, Taylor HFW. The crystal structures of sjögrenite and pyroaurite. Mineral Mag. 1967; 36: 465–479.

[22] Allmann R. The crystal structure of pyroaurite. Acta Crystallographica Section B: Structural Crystallography and Crystal Chemistry. 1968; 24: 972–977.
[23] Allmann R, Jepsen HP. Structure of hydrotalcite. Neues Jahrb Mineral Monatsh. 1969; 12: 544–551.
[24] Mills SJ, Christy AG, Génin JMR, Kameda T, Colombo F. Nomenclature of the hydrotalcite supergroup: natural layered double hydroxides. Mineralogical Magazine. 2012; 76: 1289–1336.
[25] Rives V, Ulibarri MA. Layered double hydroxides (LDH) intercalated with metal coordination compounds and oxometalates. Coordination Chemistry Reviews. 1999; 181: 61–120.
[26] Bookin AS, Drits VA. Polytype diversity of the hydrotalcite-like minerals. I. Possible polytypes and their diffraction features. Clays and Clay Minerals. 1993; 41: 551–557.
[27] Bellotto M, Rebours B, Clause O, Lynch J, Bazin D, Elkaïm E. A reexamination of hydrotalcite crystal chemistry. The Journal of Physical Chemistry. 1996; 100: 8527–8534.
[28] Pausch I, Lohse HH, Schuermann K, Allmann R. Syntheses of disordered and Al-rich hydrotalcite-like compounds. Clays and Clay Minerals. 1986; 34: 507–510.
[29] Brindley GW, Kikkawa S. A crystal-chemical study of Mg, Al and Ni, Al hydroxy-perchlorates and hydroxy-carbonates. American Mineralogist. 1979; 64: 836–843.
[30] Bernal JD, Dasgupta DR, Mackay AL. The oxides and hydroxides of iron and their structural inter-relationships. Clay Miner Bull. 1959; 4: 15–30.
[31] Refait P, Drissi SH, Pytkiewicz J, Génin JM. The anionic species competition in iron aqueous corrosion: role of various green rust compounds. Corrosion science. 1997; 39: 1699–1710.
[32] Génin JMR, Aïssa R, Géhin A, Abdelmoula M, Benali O, Ernstsen V, Ona-Nguema G, Upadhyay C, Ruby C. Fougerite and Fe II–III hydroxycarbonate green rust; ordering, deprotonation and/or cation substitution; structure of hydrotalcite-like compounds and mythic ferrosic hydroxide Fe(OH)(2+x). Solid State Sciences. 2005; 7: 545–572.
[33] Hofmeister W, Platen HV. Crystal chemistry and atomic order in brucite-related double-layer structures. Crystallography Reviews. 1992; 3: 3–26.
[34] Rousselot I, Taviot-Guého C, Leroux F, Léone P, Palvadeau P, Besse JP. Insights on the structural chemistry of hydrocalumite and hydrotalcite-like materials: Investigation of the series $Ca_2M^{3+}(OH)_6Cl \cdot 2H_2O$ (M^{3+}: Al^{3+}, Ga^{3+}, Fe^{3+}, and Sc^{3+}) by X-ray powder diffraction. Journal of Solid State Chemistry. 2002; 167: 137–144.
[35] Christensen AN, Jensen TR, Hanson JC. Formation of ettringite, $Ca_6Al_2(SO_4)_3(OH)_{12} \cdot 26H_2O$, AFt, and monosulfate, $Ca_4Al_2O_6(SO_4) \cdot 14H_2O$, AFm-14, in hydrothermal hydration of Portland cement and of calcium aluminum oxide—calcium sulfate dihydrate mixtures studied by in situ synchrotron X-ray powder diffraction. Journal of Solid State Chemistry. 2004; 177: 1944–1951.
[36] Allmann R. Die Doppelschichtstruktur der plättchenförmigen Calcium-Aluminium-H_2O-Salze am Beispiel des $(CaO)_3 \cdot Al_2O_3 \cdot CaSO_4 \cdot (H_2O)_{12}$. Neues Jahrb Mineral Monatsh. 1968: 140–144.
[37] Allmann R. Doppelschichtstrukturen mit brucitähnlichen Schichtionen. Chimica. 1970; 24(3): 99–108.
[38] Matschei T. Thermodynamics of cement hydration [dissertation]. University of Aberdeen; 2007.
[39] Vieille L, Moujahid EM, Taviot-Guého C, Cellier J, Besse JP, Leroux F. In situ polymerization of interleaved monomers: a comparative study between hydrotalcite and hydrocalumite host structures. Journal of Physics and Chemistry of Solids. 2004; 65: 385–393.
[40] Plank J, Yu B. Preparation of hydrocalumite-based nanocomposites using polycarboxylate comb polymers possessing high grafting density as interlayer spacers. Applied Clay Science. 2010; 47: 378–383.
[41] Chrysochoou M, Dermatas D. Evaluation of ettringite and hydrocalumite formation for heavy metal immobilization: literature review and experimental study. Journal of Hazardous Materials. 2006; 136: 20–33.

[42] Vola G, Gotti E, Brandon C, Oleson JP, Hohlfelder RL. Chemical, mineralogical and petrographic characterization of Roman ancient hydraulic concretes cores from Santa Liberata, Italy, and Caesarea Palestinae, Israel. Periodico di Mineralogia. 2011; 80: 317–338.
[43] Jackson MD, Landis EN, Brune PF, Vitti M, Chen H, Li Q, Kunz M, Wenk H-R, Monteiro PJM, Ingraffea AR. Mechanical resilience and cementitious processes in Imperial Roman architectural mortar. Proceedings of the National Academy of Sciences. 2014; 111: 18484–18489.
[44] Secco M, Addis A, Artioli G. Characterization of lime-stabilized earthen mortars from historic masonry structures. In: Modena C, da Porto F, Valluzzi MR, eds. Proceedings of IB2MAC 2016 – 16th International Brick and Block Masonry Conference, Padova, 26–30 June 2016. CRC Press/Balkema. 2016: 1889–1896.
[45] Taylor RM. The rapid formation of crystalline double hydroxy salts and other compounds by controlled hydrolysis. Clay Miner. 1984; 19: 591–603.
[46] Sato T, Fujita H, Endo T, Shimada M, Tsunashima A. Synthesis of hydrotalcite-like compounds and their physico-chemical properties. Reactivity of Solids. 1988; 5: 219–228.
[47] Flatt RJ, Roussel N, Cheeseman CR. Concrete: An eco material that needs to be improved. Journal of the European Ceramic Society. 2012; 32: 2787–2798.
[48] Scrivener KL. Options for the future of cement. Indian Concr J. 2014; 88: 11–21.
[49] Worrell E, Price L, Martin N, Hendriks C, Meida LO. Carbon dioxide emissions from the global cement industry 1. Annual Review of Energy and the Environment. 2001; 26: 303–329.
[50] Fernandez R, Martirena F, Scrivener KL. The origin of the pozzolanic activity of calcined clay minerals: a comparison between kaolinite, illite and montmorillonite. Cement and Concrete Research. 2011; 41: 113–122.
[51] Tironi A, Trezza MA, Scian AN, Irassar EF. Assessment of pozzolanic activity of different calcined clays. Cement and Concrete Composites. 2013; 37: 319–327.
[52] Labeyrie J, Delibrias G. Dating of old mortars by the carbon-14 method. Nature. 1964; 201: 742.
[53] Delibrias G, Labeyrie J. The dating of mortars by the carbon-14 method. In Proceedings from the 6th International Conference on 14C and Tritium Dating. Washington, DC: Clearinghouse for Federal Scientific & Technological Information, National Bureau of Standards. US Dept. Commerce. 1965: 344–347.
[54] Baxter MS, Walton A. Radiocarbon dating of mortars. Nature. 1970; 225: 937–938.
[55] Folk RL, Valastro S. Successful technique for dating of lime mortar by carbon-14. Journal of Field Archaeology. 2013; 3: 195–201.
[56] Heinemeier J, Jungner H, Lindroos A, Ringbom Å, von Konow T, Rud N. AMS 14 C dating of lime mortar. Nuclear Instruments and Methods in Physics Research Section B: Beam Interactions with Materials and Atoms. 1997; 123: 487–495.
[57] Van Strydonck M, Dupas M, Dauchot-Dehon M, Pachiaudi C, Marechal J. The influence of contaminating (fossil) carbonate and the variations of deltacarbon-13 in mortar dating. Radiocarbon. 1986; 28: 702–706.
[58] Pesce GL, Ball RJ. Dating of Old Lime Based Mixtures with the "Pure Lime Lumps" Technique. INTECH Open Access Publisher; 2012.
[59] Sonninen E, Jungner H. An improvement in preparation of mortar for radiocarbon dating. Radiocarbon. 2001; 43: 271–274.
[60] Marzaioli F, Lubritto C, Nonni S, Passariello I, Capano M, Terrasi F. Mortar radiocarbon dating: preliminary accuracy evaluation of a novel methodology. Analytical chemistry. 2011; 83: 2038–2045.
[61] Ortega LA, Alonso-Olazabal A, Ibañez A, Insausti M, Zuluaga MC, Murelaga X. Improved Sample Preparation Methodology on Lime Mortar for Reliable 14C Dating. INTECH Open Access Publisher; 2012.

[62] Ringbom A, Lindros A, Heinemeier J, Sonck-Koota P. 19 Years of Mortar Dating: Learning from Experience. Radiocarbon. 2014; 56: 619–635.
[63] Brindley GW, Kikkawa S. Thermal behavior of hydrotalcite and of anion-exchanged forms of hydrotalcite. Clays and Clay Minerals. 1980; 28: 87–91.
[64] Bellotto M, Rebours B, Clause O, Lynch J, Bazin D, Elkaïm E. Hydrotalcite decomposition mechanism: a clue to the structure and reactivity of spinel-like mixed oxides. The Journal of Physical Chemistry. 1996; 100: 8535–8542.
[65] Rives V. Characterisation of layered double hydroxides and their decomposition products. Materials Chemistry and Physics. 2002; 75: 19–25.
[66] Vágvölgyi V, Palmer SJ, Kristóf J, Frost RL, Horváth E. Mechanism for hydrotalcite decomposition: A controlled rate thermal analysis study. Journal of Colloid and Interface Science. 2008; 318: 302–308.

R. Kaden* and H. Poellmann

6 Setting control of CAC by substituted acetic acids and crystal structures of their calcium salts

Abstract: Derivatives of acetic acid with different substitutes (halogens, methyl and hydroxyl groups) and formic acid are tested as admixtures to control the setting of Secar 51 as a typical calcium aluminate cement (CAC). All tested admixtures were found to act as retarders. Owing to the different effects of the substitutes on the acid strength and their additional interaction potential with the cement paste, they significantly affect the retarding effect. The delayed setting times of the onsets t_{onset} and of the heat flow maxima t_{max} in calorimetry fit well with polynomial functions, allowing a pre-adjustment of an intended setting time. Moreover, by the exponential function of the logarithm $\ln t_{onset}$ and $\ln t_{max} = A \cdot e^{mc}$, the retarding effect can be measured by m as a retardation power factor which only depends on the admixture type. This value m allows for a comparison of the retarding effects of the different admixtures, e.g. in relation to their acid strength. Among these, the hydroxyl acids glycolic, glyoxylic, lactic, and pyruvic acid were found to be the strongest retarders. For applied admixture concentrations $c_{admix} \leq 0.05$ M, the impact on the hydration products is minor. For higher concentrations, the formation of the main hydration product CAH_{10} becomes reduced the stronger the setting is delayed and the Ca-richer phase C_2AH_{8-x} increases and is accompanied by a small straetlingite formation. Deviating from this pattern, glycolic, glyoxylic, and pyruvic acid do not lead to a significant increase of C_2AH_{8-x} or a straetlingite formation.

Crystallographic data of the calcium salts of the different admixtures tested are given to serve for the phase identification of possible reaction products for higher admixture concentrations or for a possible use of the calcium salts as admixtures. A short overview of crystal structural features is presented, giving hints on the possible interactions of the substitutes with the cement paste. The strongest interactions by direct covalent bonds with calcium were found for the hydroxyl acetic acids, which also acted as the strongest retarders. In addition to that, the crystal structural data of calcium glyoxylate is provided.

Keywords: substituted acetic acids, calcium salts, crystal structures, setting control, heat flow calorimetry

***Corresponding author: R. Kaden,** Department of Mineralogy and Geochemistry, Martin Luther University Halle-Wittenberg, Halle (Saale), Germany, rkaden@calucem.com
H. Poellmann, Department of Mineralogy and Geochemistry, Martin Luther University Halle-Wittenberg, Halle (Saale), Germany

DOI 10.1515/9783110473728-007

6.1 Introduction

In the formulation and processing of cement pastes, mortars and concrete admixtures play an essential role in optimizing properties such as rheology, workability, setting, water demand, and, moreover, the performance of hardened concrete in strength development, shrinkage behavior, air-entraining, and freezing protection [1–4]. The influence of different admixtures on the setting behavior of CAC can strongly differ from that of OPC, especially in the magnitude and direction of acceleration and retardation, also depending on the applied concentration [1, 5]. Typical retarders and accelerators for OPC have been recently summarized [6–8]: among numerous inorganic and organic chemicals such as inorganic salts of halides, carbonates, and nitrates, as well as organic salts like calcium formate, calcium acetate, calcium propionate, and calcium butyrate. Typical retarders used are sugar, sugar derivatives, and carboxylic acids.

For CAC, the setting can typically be controlled by different alkaline and alkaline earth cations and NH_4^+ as well as anions OH^-, halogenides, nitrate, formate and acetate [1, 5, 9]. The most effective accelerators are lithium compounds such as Li_2CO_3, LiOH, LiCl, $LiNO_3$, and Li_2SO_4 [1, 10–13]. Carboxylic acids such as acetic acid are effectively applicable and well investigated retarders which can act either by obstructing the clinker dissolution or the nucleation of hydration products [12, 14–16]. Particularly for acetic acid, the retarding effect was reported to be caused by the formation of positively charged calcium acetate complexes adsorbed to the negatively charged clinker grains, which obstructs their dissolution [15]. A very similar mechanism was described for tartaric acid by the formation of surface layers on the clinker grains by chelate-like calcium complexes precipitating after initial dissolution of the reactive clinker phases [17]. High admixture concentrations of acetic acid can accelerate the setting by the formation of calcium acetate monohydrate [15]. In addition to that, the formation of an acetate-Afm phase or gel formations have also been reported [14]. Other, different carboxylic acids and their calcium salts were reported to influence the setting of CAC as retarders such as formic, propionic, benzoic, tartaric, oxalic, citric, gluconic, lactic, glycolic acid and calcium mono-, di-, and trichloroacetate [14, 18–23]. Moreover, for alkaline earth benzoates, naphthoates, glycolates, and chloroacetates, an evident influence of the cation type on the strength of the retarding effect was reported [24–27]. A special focus for applicable admixtures for CAC, OPC, and also OPC-CAC mixtures is on the hydroxyl carboxylic acids citric, tartaric, lactic, and glycolic acid owing to their capability to chelate Ca^{2+} or also Al^{3+} cations, also involving the hydroxyl groups similar to sugars [28–34]. Moreover, such α-hydroxyl acids and their salts are used to modify the setting, rheology, and mechanical performance of calcium phosphate cements [35, 36] and are potential retarders for calcium-sulpho-aluminate cements as recently summarized by Zajac et al. (2016) [37]. The numerous reports and applications of carboxylic acids and their calcium salts demonstrate their potentials for the setting control of different types of cement. The present study aims to give a comparison of the effect strength on the setting of a CAC (Secar 51) for 20 different carboxylic acids, which are systematically selected as substituted derivatives of acetic acid. As reported for acetic acid, high concentrations of such admixtures can lead to a crystallization of the corresponding calcium salt [14] or the calcium salt can be alternatively applied as admixture. Therefore crystallographic data of the calcium salts of some of the investigated substituted acetic acids are provided in addition.

6.2 Experimental methods

6.2.1 Analytical methods

Powder X-ray diffraction (PXRD) was carried out on an X'Pert Pro X-ray diffractometer with X'Celerator RTMS detector in Bragg-Brentano-geometry (PANalytical) using Cu-Kα radiation, $\lambda = 1.5418$ Å, within a range of $3° \leq 2\theta \leq 70°$, with a step size of 0.016° 2θ and an irradiation time of 19.96 s/step. Silicon or corundum served as an internal reference for the correction of the specimen displacement. The PXRD data evaluation and refinements were done using the software HighScore Plus (PANalytical). Single crystal X-ray diffraction (sc-XRD) data were collected at T = 298 K in a SMART 1K 3-circle goniometer with CCD area detector (Bruker) using Mo-Kα radiation ($\lambda = 0.71073$ Å) in a range of $6.0° \leq 2\theta \leq 56.4°$, in steps of 0.3° with an irradiation time of 20–60 s depending on the crystal size. The crystal structure was solved and refined using the program JANA2006 with implemented structure solution routine SUPERFLIP [38]. Crystal structure images were drawn using the software Diamond 3.2k [39].

The time curves of the heat development during the hydration reaction of a CAC (Secar51) with solutions of the selected substituted acetic acids in different molarities M were recorded by an isoperibolic heat flow calorimeter after Kuzel [40, 41]. For all mixtures, a liquid/cement ratio $w/c = 0.5$ was applied at an ambient temperature of 21(1) °C. The mixtures were prepared by injection method, i.e. the crucibles filled with 1 g of the CAC were placed in the calorimeter and separately the syringes were filled with 0.5 ml admixture solution or deionized water beside them. Both were allowed to acclimatize for at least 10 hours before injecting the admixtures into the crucibles to start the reaction.

Chemical analyses of the cement paste solutions were performed with an ICP-OES Ultima2 (Horiba Jobin Yvon). Therefore, $w/c = 2$ with deionized water (H_2O_{deion}) and selected admixture concentrations were applied. The pastes were filtrated with cellulose filters (Rotilabo® type 115A) and 1 ml of the filtrate was acidified with 0.5 ml HNO_3 supra pure in 8.5 ml H_2O_{deion}. Typical spectral lines to determine the concentrations were used for Al (396.152 nm), Ca (317.933 nm), and I (182.976 nm).

6.2.2 Materials

CAC

A Secar 51 was chosen as typical calcium aluminate cement. The chemical and mineralogical composition and quantities of the clinker phases determined by PXRD and Rietveld refinement as well as the calorimetric data of hydration are given in Tab. 6.1.

Tab. 6.1: Composition and calorimetric data of hydration of Secar 51 used in this work.

Oxide	m%	Oxide	m%	Phase	m%
SiO_2	4.80	P_2O_5	0.18	CA	62.4(5)
Al_2O_3	48.26	TiO_2	2.10	gehlenite	18.5(4)
Fe_2O_3	2.14	MnO	0.06	perovskite	4.5(3)
CaO	40.43	Cr_2O_3	0.05	C_3FT	3.9(3)
MgO	0.37	ZrO_2	0.10	pleochroite	3.8(3)
SO_3	0.06	SrO	0.05	amorphous	6.9(3)
K_2O	0.27	LOI	0.99		
Na_2O	0.11	Σ	99.95		
Heat flow calorimetry characteristics, $w/c = 0.5$, $T = 21(1)$ °C, ave. of 13 measurements					
t_{onset} (h)	t_{max} (h)	H_{max} (mW/g)	Q_{max} (J/g)	Hydration products	
4.3(3)	5.4(3)	27.8(5)	395(15)	CAH_{10}, traces of C_2AH_{8-x}	

Admixtures – substituted acetic acids

Depending on the formula weight of the acids, the admixture concentrations applied, 0.001 M, 0.002 M, 0.005 M, 0.01 M, 0.02 M, 0.05 M, and 0.1 M for the chosen ratio $w/c = 0.5$, correspond to 0.0023–1.484 m% relative to the cement fraction or 0.0015–0.989 m% relative to the cement paste. 20 derivatives of acetic acid, including formic and acetic acid, were selected with different type and number of substitutes X according to $CH_{3-n}X_nCOOH$, $n = 0–3$, X = F, Cl, Br, I, –OH, =O, $-CH_3$, $=CH_2$, Tab. 6.2. The different substitutes affect a change of the acid strength measured by the pK_a value [42, 43]. The substitution of hydrogen by methyl groups with their electron-donating character leads to a positive inductive effect +I towards the carboxylic group. As a result, the acidic proton becomes more difficult to remove and the acid strength becomes reduced from formic to acetic towards pivalinic acid, $+I^1$ in Fig. 6.1. By substituting $-CH_3$ with the electron-withdrawing double bonded $=CH_2$, the negative mesomeric effect increases the acid strength, i.e. reduces the pK_a value from propionic to acrylic and from iso-butyric to methacrylic acid, $-M^1$ and $-M^2$ in Fig. 6.1. Electron negative, i.e. electron-withdrawing substitutes such as halogens and hydroxyl lead to a negative inductive effect –I, increasing the acid strength, $-I^1$ and $-I^2$ in Fig. 6.1. An additional substitution by $-CH_3$, as from glycolic to lactic acid or from acrylic to methacrylic acid has an electron-pushing effect and reduces the acid strength, $+I^2$ and $+I^3$ in Fig. 6.1. The exchange of the hydroxyl group of lactic acid by a ketone group =O leads to a negative mesomeric effect from lactic to the stronger acidic pyruvic acid, $-M^3$ in Fig. 6.1. Based on this systematic, the influence of the substituted acetic acids on the setting of Secar 51 is evaluated in this study.

Tab. 6.2: Substituted acetic acids with formula, 2D-structures and acronym used in this work. 2D-structures of fluoro-, chloro-, bromo-, and iodoacetic acids are similar to the ones presented for methylated acetic acids.

Non-substituted acids	Acetic acid – Ac CH_3COOH	Formic acid – Fo HCOOH	
Substituted acids	**mono-**	**di-**	**tri-**
Fluoroacetic acids	*FCH_2COOH, not used*	$F_2CHCOOH$ – DFA	F_3CCOOH – TFA
Chloroacetic acids	$ClCH_2COOH$ – MCA	$Cl_2CHCOOH$ – DCA	Cl_3CCOOH – TCA
Bromoacetic acids	$BrCH_2COOH$ – MBA	$Br_2CHCOOH$ – DBA	Br_3CCOOH – TBA
Iodoacetic acid	ICH_2COOH – MIA	*not available*	*not available*
Methylated acetic acids	Propionic – MmA CH_3CH_2COOH 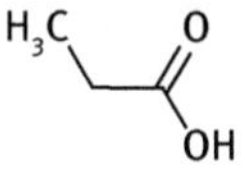	Iso-butyric – DmA $(CH_3)_2CHCOOH$ 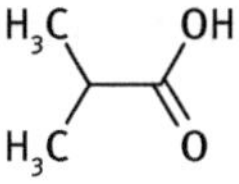	Pivalinic – TmA $(CH_3)_3COOH$ 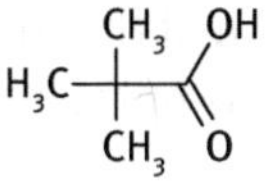
Acrylic acids	Acrylic – Acr $CH_2CHCOOH$ 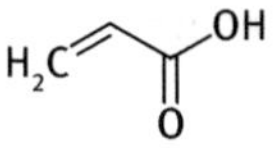	Methacrylic – MAcr $(CH_3,CH_2)CCOOH$ 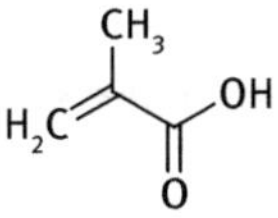	
Hydroxylated acetic acids	Glycolic – Glyc $HOCH_2COOH$ 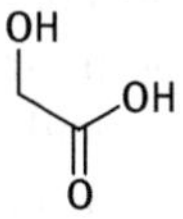	Glyoxylic – Glyox $*(HO)_2CHCOOH$ 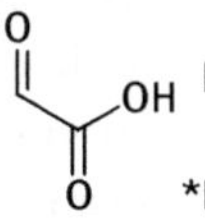 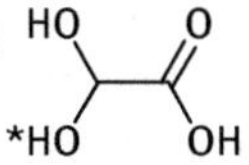	
Methylated and hydroxylated/oxylated acetic acids	Lactic – Lact (D,L) $(HO,CH_3)CHCOOH$ 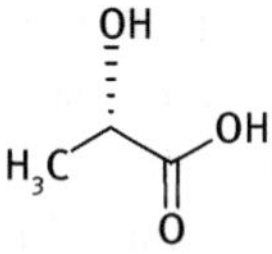 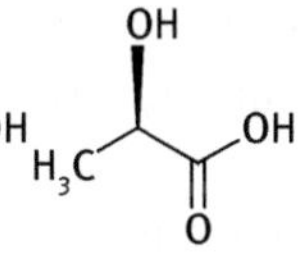	Pyruvic – Pyruv $(O,CH_3)CCOOH$ 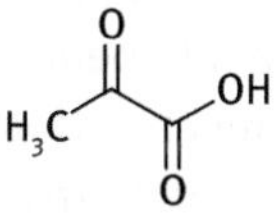	

* glyoxylic acid, which is always formed in aqueous solutions as it is monohydrate

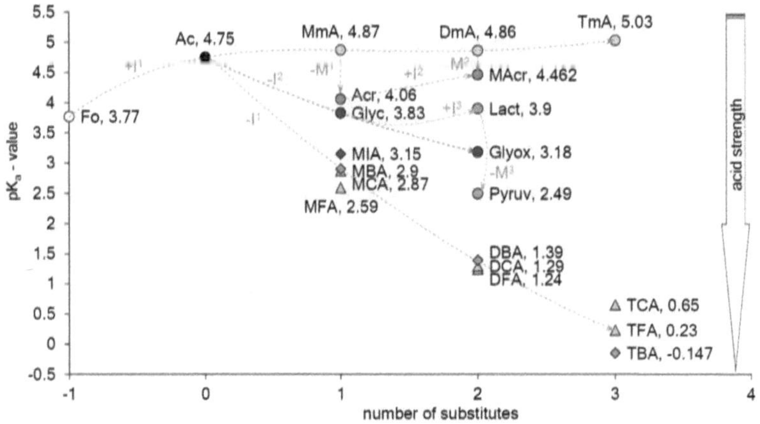

Fig. 6.1: Acid strength, pK_a values of the selected substituted acetic acids (Tab. 6.2) and formic acid with tendencies of inductive (I) and mesomeric (M) effects on pK_a.

The calcium salts of the substituted acetic acids were synthesized by the reaction of calcium carbonate and the acid in aqueous solution. Subsequently, the slow evaporation of the solvent water at room temperature led to the crystallization of the calcium salts or their hydrates.

6.3 Results

6.3.1 Setting control of CAC using substituted acetic acids

Within the applied admixture concentration range 0.001–0.1 M, the substituted acetic acids and formic acid retard the setting of Secar 51. This retarding effect always increases with increasing admixture concentration. For the most of the selected admixtures larger concentrations do not lead to setting within 200 hours. The examples of the calorimetric heat flow time curves for chlorinated acetic acids (MCA, DCA, TCA), monobromoacetic acid (MBA), formic (Fo), acetic (Ac), lactic (Lact), and pyruvic acid (Pyruv) in Fig. 6.2 demonstrate the significant influence resulting from the type and number of the substitutes and, thus, of the acid strength on the setting. The retarding effect for chlorinated acetic acids is reduced from MCA to TCA with increasing acid strength. The fluorinated acids behave similarly. MBA and also MIA show a strong retarding effect, but for concentrations $c > 0.02$ M no setting was observed within 200 hours. Formic, acetic, and methylated acetic acids (MmA, DmA, TmA), together

▸ **Fig. 6.2:** Calorimetric heat flow time curves of Secar 51 with H_2O_{deion} and substituted acetic admixtures (Tab. 6.2) in different molarities M, $w/c = 0.5$, $T = 21(1)$ °C.

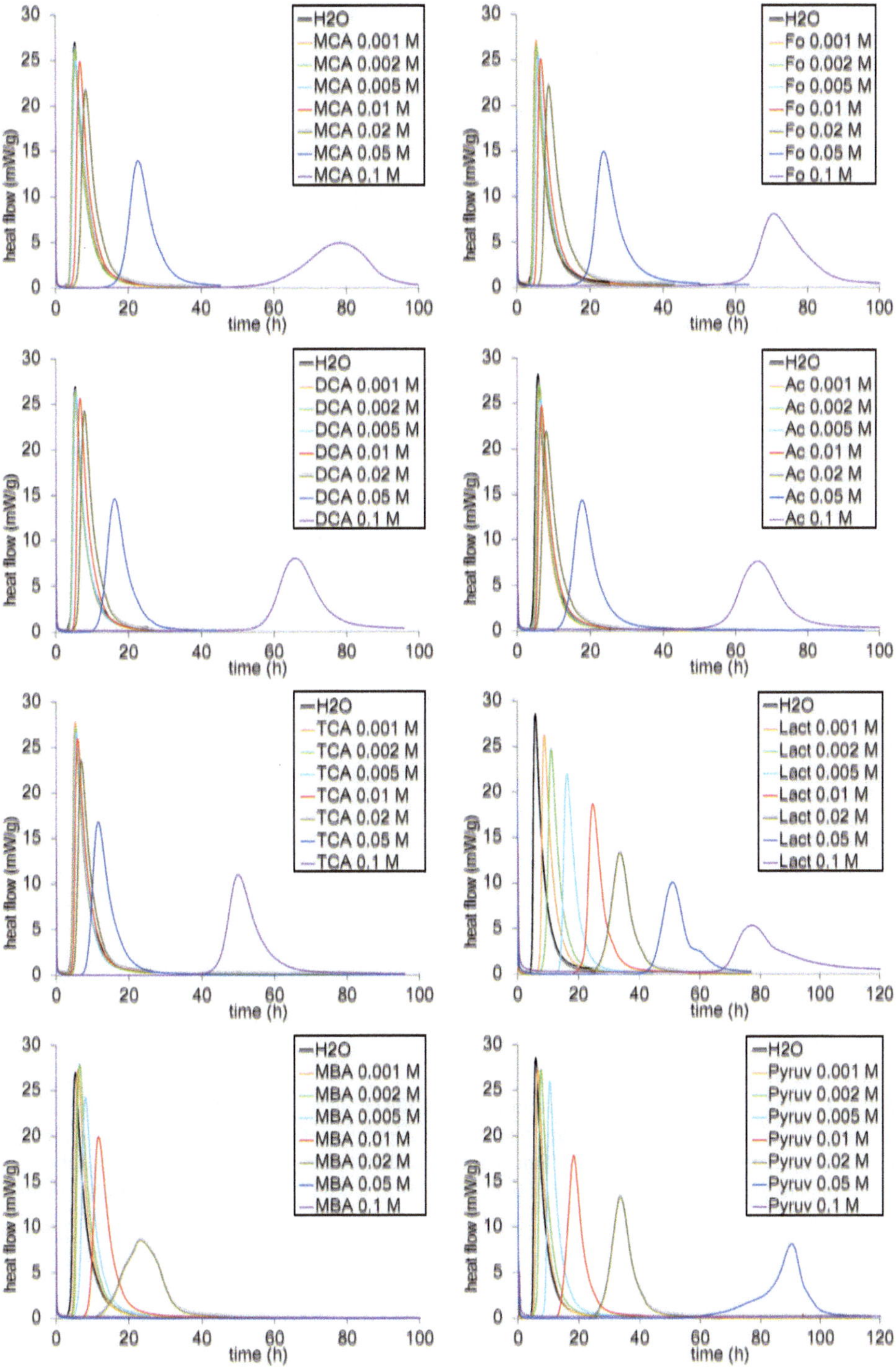
heat flow (mW/g)
time (h)
H2O
MCA 0.001 M
MCA 0.002 M
MCA 0.005 M
MCA 0.01 M
MCA 0.02 M
MCA 0.05 M
MCA 0.1 M
Fo 0.001 M
Fo 0.002 M
Fo 0.005 M
Fo 0.01 M
Fo 0.02 M
Fo 0.05 M
Fo 0.1 M
DCA 0.001 M
DCA 0.002 M
DCA 0.005 M
DCA 0.01 M
DCA 0.02 M
DCA 0.05 M
DCA 0.1 M
Ac 0.001 M
Ac 0.002 M
Ac 0.005 M
Ac 0.01 M
Ac 0.02 M
Ac 0.05 M
Ac 0.1 M
TCA 0.001 M
TCA 0.002 M
TCA 0.005 M
TCA 0.01 M
TCA 0.02 M
TCA 0.05 M
TCA 0.1 M
Lact 0.001 M
Lact 0.002 M
Lact 0.005 M
Lact 0.01 M
Lact 0.02 M
Lact 0.05 M
Lact 0.1 M
MBA 0.001 M
MBA 0.002 M
MBA 0.005 M
MBA 0.01 M
MBA 0.02 M
MBA 0.05 M
MBA 0.1 M
Pyruv 0.001 M
Pyruv 0.002 M
Pyruv 0.005 M
Pyruv 0.01 M
Pyruv 0.02 M
Pyruv 0.05 M
Pyruv 0.1 M

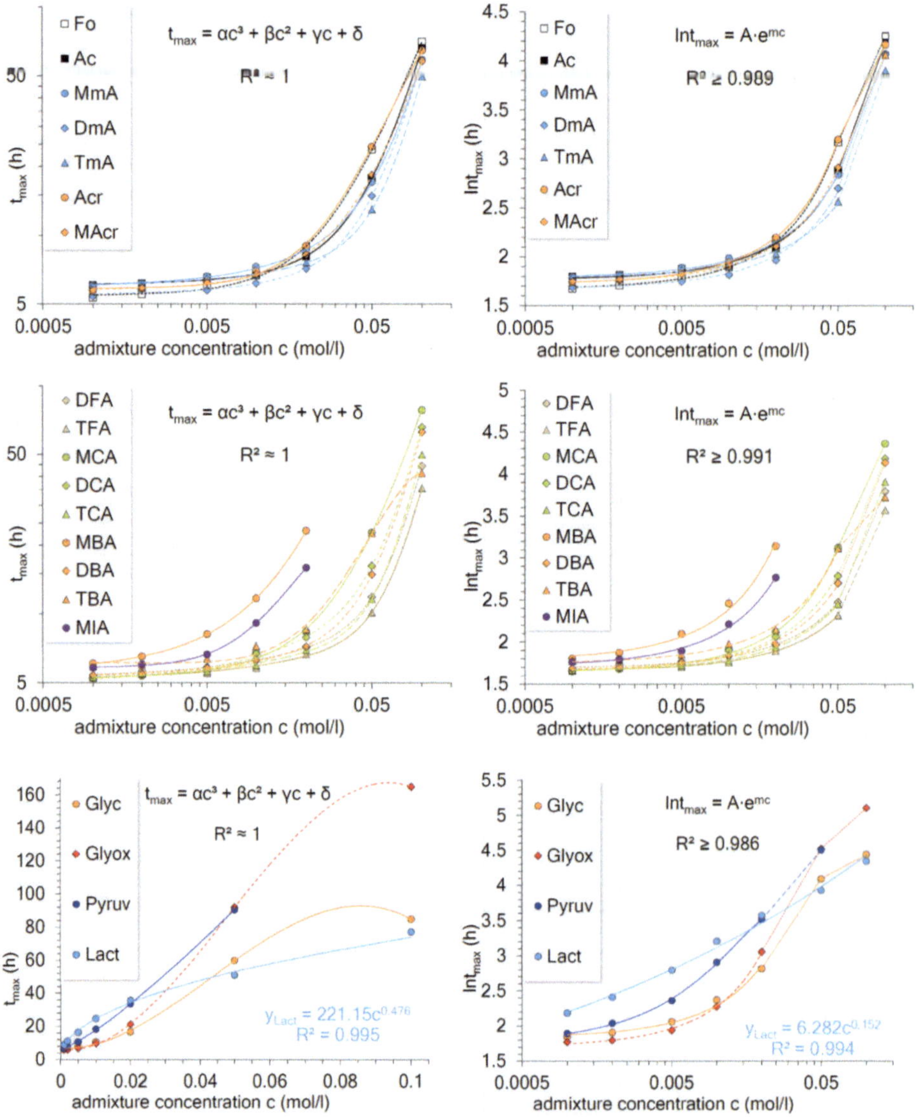

Fig. 6.3: Times of heat flow maxima tmax of the hydration of Secar 51 and substituted acetic acids used as additives from isoperoblic heat flow calorimetry presented in Fig. 6.2, $w/c = 0.5$, $T = 21(1)$ °C. Left column: polynomial functional dependence of t_{max} on admixture concentration, right column: exponential dependence of ln t_{max} on c for $c \leq 0.05$ M and $c \leq 0.02$ M for Glyc, Glyox, Lact, Pyruv.

with acrylic and methacrylic acid, affect a retarding effect which grows with the acid strength. Fig. 6.2 shows the setting delays for formic acid are larger than for acetic acid. The strongest retarding effects, especially for $c \leq 0.01$ M, were found for the hydroxylated/oxylated acids (Glyc, Glyox, Lact, Pyruv), whereas for all the other tested admixtures the effect significantly rises for $c \geq 0.01$ M. The heat flow time curves for those 4 admixtures (Lact and Pyruv in Fig. 6.2) become slightly split for $c \geq 0.05$ M and thus for strong setting delays, similar to what has been reported for alkaline earth glycolates and glycolic acid [26]. Pyruvic acid applied as admixture with $c = 0.1$ M retards the setting by more than 200 hours.

As previously reported for alkaline earth chloroacetates, glycolates, and glycolic acid used as admixtures for CAC [26, 27], the dependence of the retarding effect on the concentration can be fitted by a polynomial function. According to this, for all admixtures investigated in this study within the applied concentration range 0.001–0.1 M, the dependence of the setting delay on the concentration, onsets, as well as heat flow maxima follows a polynomial function, demonstrated for t_{max} in Fig. 6.3, left column, Tab. 6.3 and Tab. 6.4:

$$t = \alpha c^3 + \beta c^2 + \gamma c + \delta. \tag{6.1}$$

The polynomic coefficients α, β, and γ fit the increase of the setting delay with the concentration, whereas δ corresponds to the setting time of a Secar 51 cement paste for $c =$ 0 M, i.e. without admixture. For MBA and MIA, no setting was observed within 200 h for $c > 0.02$ M, thus the scope of the function ends at this concentration. For glycolic and glyoxylic acid, the determined function is found to be valid only for $c \leq 0.05$ M. Exceeding this concentration, the increase of the setting delay becomes strongly reduced. The significant exception is lactic acid with a power function of the onset and heat flow maximum times of the form $t = n \cdot c^x$, Fig. 6.3, bottom. The exponent x of nearly $\frac{1}{2}$ indicates a square root functional development of the onset and maximum times with the concentration. This dependence of the retarding effect, which differs strongly from all other tested admixtures, may allow a more convenient dosage of lactic acid, as can also be estimated from Fig. 6.2.

Nevertheless, for all the tested admixtures the functions of onset and heat flow maximum times allow an appropriate adjustment of the setting of Secar 51 cement pastes. The determined functions with a correlation coefficient R^2 better than 0.999 are given in Tab. 6.3 for the onsets and in Tab. 6.4 for the heat flow maximum times.

However, to compare the retarding power of the different tested admixtures, the polynomial coefficients fitting the best possible match deviate too much. Therefore, the data were abstracted to find another function sufficiently fitting the dependence of t_{onset} and t_{max} on the admixture concentration in ranges for the setting times reasonable for applications. The best appropriate fit results from an exponential function for the logarithms of the setting times:

$$\ln t = A \cdot e^{mc}. \tag{6.2}$$

Tab. 6.3: Functions of onset time t_{onset} depending on admixture concentration c.

admixture	pK_a	$t_{onset} = \alpha c^3 + \beta c^2 + \gamma c + \delta$, $c \leq 0.1$ M, $R^2 \approx 1$				$\ln(t_{onset}) = A \cdot e^{m \cdot c}$, $c \leq 0.05$ M		
		α	β	γ	δ	A	m	R^2
Fo	3.77	1313.8	5429	34.84	4.20	1.42(1)	14.9(2)	0.999
Ac	4.75	42351	675	32.83	4.77	1.51(3)	10.6(6)	0.982
MmA	4.87	41316	−163	66.82	4.94	1.58(1)	9.6(3)	0.994
DmA	4.86	52005	−904	30.63	4.33	1.41(4)	9.6(9)	0.953
TmA	5.03	53179	−1937	48.64	4.97	1.57(3)	6.6(6)	0.954
Acr	4.06	−23776	7823	−28.03	4.81	1.50(1)	13.8(3)	0.997
MAcr	4.462	27921	1555	29.48	4.64	1.49(2)	11.0(5)	0.989
DFA	1.24	52315	−3141	106.92	3.99	1.42(1)	7.6(2)	0.994
TFA	0.23	33983	−1787	67.84	4.05	1.41(1)	6.7(3)	0.988
MCA	2.87	9552.5	4318	43.35	3.93	1.36(2)	15.2(4)	0.995
DCA	1.29	69267	−3019	146.07	3.84	1.41(2)	11.6(5)	0.988
TCA	0.65	64099	−3317	97.43	4.07	1.42(1)	8.3(2)	0.997
MBA	2.9	−1162051	−35872	225.84	4.57	1.61(6)	26.0(9)	0.946
DBA	1.39	60348	−1508	73.27	4.02	1.38(2)	11.3(4)	0.992
TBA	−0.147	−57048	9432	−60.29	4.97	1.51(2)	13.0(4)	0.994
MIA	3.15	−666265	28827	13.94	4.72	1.52(2)	23.4(8)	0.995
Glyc	3.83	−241271	32438	−94.85	5.07	1.64(4)	24.5(9)	0.976
Glyox	3.18	−455102	66529	−601.72	4.59	1.42(3)	32.8(9)	0.988
Pyruv	2.49	−430866	30873	709.70	4.74	1.53(3)	34.5(9)	0.934
Lact	3.9	$t_{onset} = 204.02 \cdot c^{0.489}$, $R^2 = 0.996$				$\ln t_{onset} = 6.336 \cdot c^{0.165}$		0.990

Tab. 6.4: Functions of heat flow maximum time t_{max} on admixture concentration c.

admixture	pK_a	$t_{max} = \alpha c^3 + \beta c^2 + \gamma c + \delta$, $c \leq 0.1$ M, $R^2 \approx 1$				$\ln(t_{max}) = A \cdot e^{m \cdot c}$, $c \leq 0.05$ M		
		α	β	γ	δ	A	m	R^2
Fo	3.77	−7497.4	6800	44.82	5.42	1.67(1)	12.9(1)	0.999
Ac	4.75	36134	1868	50.91	6.06	1.76(2)	9.7(3)	0.992
MmA	4.87	39789	123	116.70	5.95	1.79(1)	9.2(1)	0.999
DmA	4.86	41576	544	56.80	5.47	1.67(2)	9.5(4)	0.990
TmA	5.03	49064	−1441	89.74	5.98	1.78(1)	7.2(2)	0.995
Acr	4.06	−30050	8824	4.84	5.85	1.72(1)	12.4(1)	1.000
MAcr	4.462	16016	3024	61.20	5.74	1.72(1)	10.4(1)	0.999
DFA	1.24	47761	−1941	108.04	5.34	1.69(1)	7.6(1)	0.998
TFA	0.23	37793	−1567	82.88	5.21	1.66(1)	6.7(2)	0.994
MCA	2.87	19787	4673	66.64	5.24	1.64(1)	12.9(2)	0.998
DCA	1.29	63454	−1824	155.11	5.07	1.67(2)	10.4(3)	0.994
TCA	0.65	60601	−2737	114.60	5.15	1.66(1)	7.8(1)	0.999
MBA	2.9	388686	15473	412.23	5.64	1.80(3)	28.4(9)	0.990
DBA	1.39	56115	794	90.00	5.33	1.66(1)	9.6(2)	0.996
TBA	−0.147	−71679	11257	−56.53	6.18	1.74(1)	11.6(3)	0.997
MIA	3.15	−996507	47516	−50.91	5.87	1.71(2)	24.2(8)	0.995
Glyc	3.83	−285383	37277	−94.31	6.33	1.77(3)	21.7(9)	0.981
Glyox	3.18	−426173	61438	−296.97	5.89	1.70(1)	29.4(5)	0.999
Pyruv	2.49	−163453	21054	1056.60	5.40	1.76(4)	32.0(4)	0.930
Lact	3.9	$t_{onset} = 221.15 \cdot c^{0.476}$, $R^2 = 0.995$				$\ln t_{onset} = 6.282 \cdot c^{0.152}$		0.994

As demonstrated for t_{max} in Fig. 6.3, right column, this function fits for nearly all of the tested admixtures for $c \leq 0.05$ M. Towards $c = 0.1$ M, the increase of the retarding effect becomes reduced in most cases for t_{max} as well as for t_{onset}. For the hydroxylated acid admixtures Glyc, Glyox, and Pyruv, the validity range of the exponential function is reduced to $c \leq 0.02$ M. The only exception is lactic acid, which leads to a square root-like correlation of the setting times with the admixture concentration. These correlations are best described by power functions. The determined functions are given in Tab. 6.3 for the onsets and in Tab. 6.4 for the heat flow maximum times.

The parameter A of the exponential function (equation (6.2)) corresponds to the natural logarithm of t_{onset} or t_{max} for a Secar 51 cement paste without an admixture. Thus it is given as a constant by the cement within experimental deviations. As a result, the only free parameter affected by the admixture is m, which can be mentioned as the retardation power factor. This single parameter m describes the increase of the setting delay with the concentration depending only on the admixture type. Thus, m allows a comparison between all of the tested admixtures except Lact, Fig. 6.4.

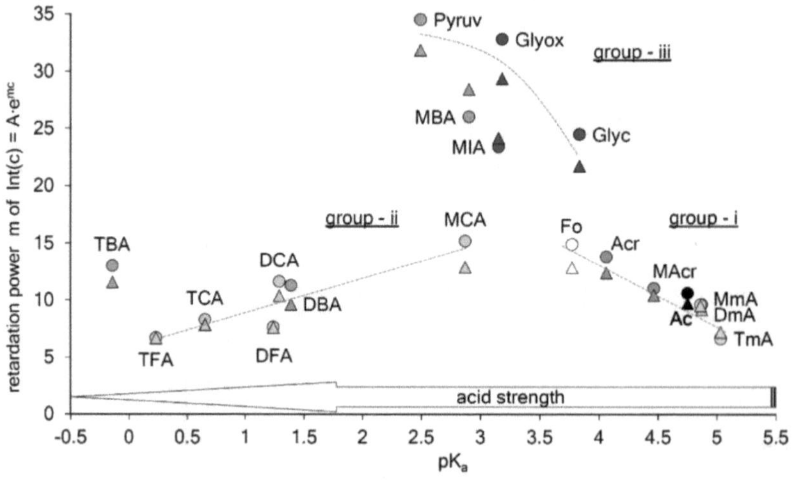

Fig. 6.4: Retardation power factor m of different substituted acetic acids related to their acid strength, determined for the setting of Secar 51, $w/c = 0.5$, $T = 21(1)\,°C$, $c_{admix} \leq 0.05$ M. Circles correspond to t_{onset}, triangles t_{max}. Dashed lines highlight general tendencies.

As mentioned in Section 6.2.2, the different types and numbers of substitutes for acetic acid derivatives lead to different acid strengths. The retardation power m as a specific parameter on the setting can therefore be correlated to the pK_a value as a specific parameter of the admixture, Fig. 6.4. This plot allows distinguishing between three different groups of tested admixtures:

(i) Fo, Ac, MmA, DmA, TmA, Acr and MAcr,
(ii) halogenated acetic acids and
(iii) hydroxylated/oxylated acetic acids Glyc, Glyox, Pyruv.

(i) The admixtures of group (i) with non-polar substitutes together with acetic and formic acid can only act with the carboxyl group with the clinker grains and the cations, especially Ca^{2+}. With decreasing acid strength, these interactions become weaker, as does the dissociation of the acid in aqueous solution. Consequently, the retarding power decreases with decreasing acid strength. Propionic (MmA) and iso-butyric acid (DmA) with similar pK_a values also have a similar retarding power.

(ii) The polar substitutes of the halogenated acetic acids lead to increased acid strengths. In addition to the carboxyl group, electron negative halogen atoms can interact with cations and water molecules, as will be demonstrated by the crystal structures of the calcium salts in Section 6.3.2. However, the retarding power of group (ii) decreases with increasing acid strength of the admixtures from MCA towards TCA and TFA. This trend may be due to the facilitated solubility of the calcium halogenoacetate complexes with increasing acid strength, owing to increased interactions between the electron negative substitutes with the water molecules in aqueous solutions. TBA creates an exceptionally larger retarding effect than corresponding to the tendency. The mono-halogenated admixtures MBA and MIA lead to very high retarding power values. In general, halogenated acetic acids tend to release the halogen as halide ion by hydrolysis in aqueous solutions [44], especially in a basic milieu as given in the cement paste. For sodium salts in aqueous solutions in particular, monochloroacetate was found to be much more stable than monoiodo- and monobromoacetate [44]. The hydrolysis leads to the formation of the halide ion and glycolic acid, e.g.:

$$BrCH_2COOH + OH^- \rightarrow HOCH2COOH + Br^- . \tag{6.3}$$

Chemical analysis with $AgNO_3$ testing of the cement paste solution 1 hour after mixing for MCA, MBA, and MIA, $w/c = 2$, filtrated and acidified with HNO_3 to $pH = 0$, lead to a precipitation of AgCl and a much stronger precipitation of AgBr and AgI. ICP-OES analysis of cement paste solutions with MIA used as admixture in concentrations of 0.01 M and 0.02 M revealed the iodine concentration had slightly reduced from the initial admixture concentration to 0.009 M and 0.017 M, respectively. In contrast to this, the calcium concentration is increased from 0.016 M for a cement paste without admixture to 0.02 M and 0.023 M for the admixture concentrations of 0.01 M and 0.02 M, respectively. The aluminum concentrations are slightly dropped from 0.03 M to 0.027 M for both admixture concentrations compared to water. These findings indicate the strong retarding effect of MIA and MBA is a result of the hydrolysis releasing glycolate and iodide or bromide. Owing to glycolate as well as the halides retard the setting of CAC [1, 9, 26], the concentrations of the retarding agents become nearly doubled and the retarding effects of both interfere.

(iii) The hydroxylated acetic acids of group (iii) and lactic acid achieve the strongest retarding effects of all tested admixtures, increasing with increasing acid strength in the order Glyc, Glyox, Pyruv. Although the retarding effect of lactic acid exceptionally follows another functional dependence for elevated concentrations, it is comparable to that of glycolic acid, Fig. 6.3, bottom left. The strong retarding effect of this admixture group obviously results from the direct involvement of the hydroxyl groups in a complexation of Ca^{2+} similar to tartaric acid [33]. Moreover, in the crystal structures of alkaline earth glycolates [26] and calcium glyoxylate (Section 6.3.2), the hydroxyl oxygen forms direct, covalent bonds to the cation. ICP-OES analysis of Secar 51 cement paste solutions 1 h after mixing with glycolic acid admixture (Glyc), $w/c = 3$, revealed a drastic reduction of the calcium concentration from 0.016 M (without Glyc) to 0.008 M (0.05 M Glyc) and to 0.004 M (0.1 M Glyc). In contrast to this, the aluminum concentration was increased from 0.024 M to 0.03 M and 0.057 M, respectively. This result attests to glycolic acid's strong fixation of Ca^{2+}, despite an obviously sufficient initial dissolution of the clinker grain surfaces to release the high aluminum concentrations compared to the cement paste without admixture. The doubling of the hydroxyl groups from glycolic to glyoxylic acid obviously leads to an increased calcium complexation and thus a stronger retardation. The strongest retarding effect in this group observed for pyruvic acid can be addressed to the chemical behavior of pyruvate in aqueous solutions. For increased concentrations, the ketone group can hydrate to a dihydroxy form as 2,2-dihydroxypropionate: $CH_3{-}CO{-}COO^- + H_2O \rightarrow CH_3{-}C(OH)_2{-}COO^-$, which is similar to glyoxylate. In particular for calcium pyruvate, the formation of complexes $[Ca(pyruv)]^+$ in aqueous solutions was found similar to the calcium acetate and tartrate complexes reported in cementitious systems [15, 17, 33, 45]. Moreover, a cross-linking between $[Ca(pyruv)]^+$-pyruvate and Ca^{2+}-pyruvate and water leads to an aggregation of several units with aggregate lengths between 0.5 µm and 10 µm for small concentrations [45]. With the increase of concentration, a turn-over into a gel was observed. Especially basic pH values as given in cementitious systems lead to a polymerization of the pyruvate, owing to the high activity of the hydroxyl groups. It can be assumed that the strong retarding effect results from the formation of $[Ca(pyruv)]^+$ complexing at the surface on the clinker grains, similar to tartrate, and the aggregation and polymerization [17].

As reported in many previous works, the retardation of the setting of CAC leads to a decrease of the maximum heat flow of the hydration reaction and an elongation of the hydration reaction period [6, 7, 9, 10, 14, 18, 19, 22, 24–28]. In the most cases, i.e. for small to medium admixture concentrations, no significant deviation of the integral reaction heat was reported.

The admixtures tested in this study conform to these findings. Moreover, the maximum heat flow H_{max} was found to depend more on the time of its event t_{max} than on the admixture type. As reported for alkaline earth chloroacetates [27], H_{max} can be expressed as a function of t_{max} with small deviations. Including all the different tested admixtures with their different retarding powers and thus their different setting times depending on the concentration, H_{max} primary depends on t_{max}, Fig. 6.5 right. The correlation can be expressed as a power function $H_{max} = 61.2 \cdot t_{max}^{-0.472}$ or as a linear function $\ln H_{max} = -0.472 \cdot \ln t_{max} + 4.11$, both with a correlation coefficient $R^2 = 0.903$. The maximum heat values of the hydroxylated/oxylated acetic acids deviate most strongly; those of all the others match well even for high setting delays, marked with a white dot in Fig. 6.5. This functional dependence allows for an estimation of the maximum heat development for an intended retarded set of Secar 51, largely regardless of the admixture type.

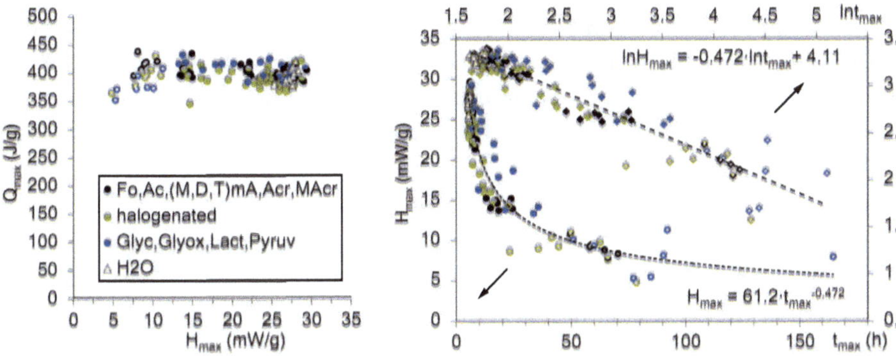

Fig. 6.5: Left: integral heat release of the hydration reaction of Secar 51 with substituted acetic acids used as admixtures in different molarities and H_2O, $w/c = 0.5$, $T = 21(1)$ °C. Right: the corresponding maximum heat flow values H_{max} as function of t_{max}, both with $R^2 = 0.903$. The legend symbols apply for both graphs. The symbols with a white dot correspond to those high admixture concentrations that exceed the validity range of equation (6.2), i.e. $c = 0.1$ M together with 0.05 M for Glyc, Glyox, Lact, and Pyruv.

Despite this, the integral heat of the main reaction Q_{max} including the main dissolution of the clinker phases and formation of the hydration products primarily corresponds to that of the Secar 51 cement pastes with water, $Q_{max} = 395(15)$ J/g, Fig. 6.5 left. The deviations lie within the range of experimental errors. For the strongly reduced heat flow maxima of the high setting delays, Q_{max} becomes slightly reduced. These values, marked with a white dot in Fig. 6.5, correspond to the applied admixture concentrations $c = 0.1$ M and especially to $c \geq 0.05$ M for the hydroxylated/oxylated acids together with $c = 0.02$ M for MBA with the slightly split heat flow profile shown in Fig. 6.2.

In previous works on this system of Secar 51 with alkaline earth chloroacetate and glycolate admixtures, $w/c = 0.5$, $T = 21(1)$ °C, no significant influence was reported on Q_{max} and the formation of the hydration products, which are dominated by CAH_{10} and traces of C_2AH_{8-x} [26, 27]. These findings are confirmed for the tested admixtures in this study for $c \leq 0.05$ M and $Q_{max}(\text{admix}) \approx Q_{max}(H_2O)$, i.e. for those concentrations that lie within the validity range of the retardation power factor m. Strongly retarded cement pastes, especially those with admixtures with a strong retardation power and $c = 0.1$ M, in some cases possess a significantly changed phase formation, as demonstrated in Fig. 6.6. While formic acid in concentrations ≤ 0.05 M generates CAH_{10} as its main hydration product, traces of C_2AH_{8-x} slightly rise in appearance for $c = 0.02$ M to 0.05 M, Fig. 6.6, left. For 0.1 M formic acid, CAH_{10} formation is strongly reduced whereas the content of C_2AH_{8-x} is strongly increased. In addition to that, traces of formic ettringite were detected by PXRD and reflections of a possible lamellar formic Afm-phase at 10.51(2) and 16.89(2)° 2θ [46]. Similar to Fo, MBA and MIA in high admixture concentrations lead to C_2AH_{8-x} as the main hydration product. The higher the retardation effect of an admixture, the stronger the formation of C_2AH_{8-x}, as can be seen in Fig. 6.6, right, for TmA towards Fo or from TCA towards MCA. Moreover, for the strong retarding effects, minor amounts of straetlingite are formed in addition.

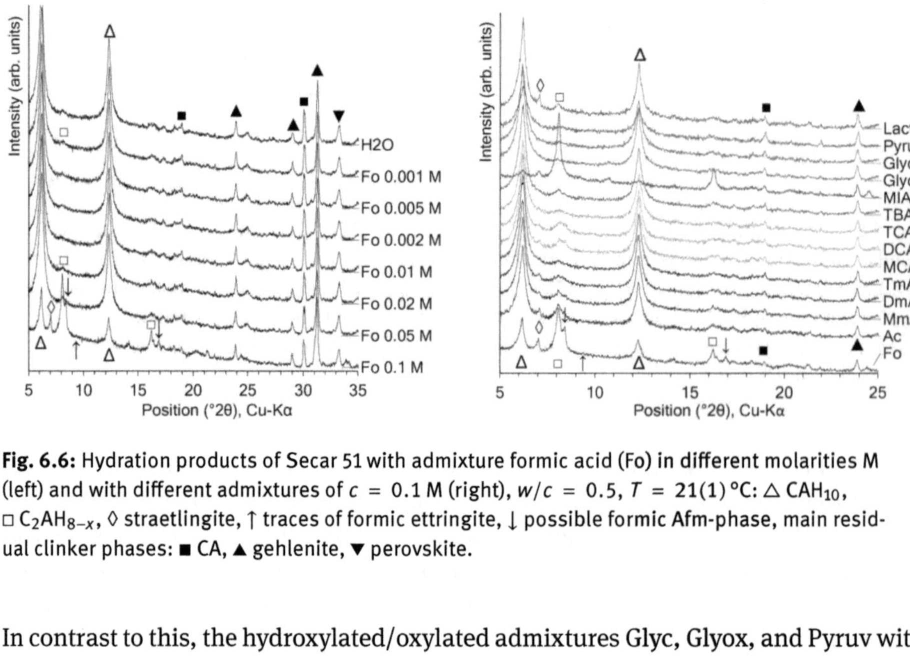

Fig. 6.6: Hydration products of Secar 51 with admixture formic acid (Fo) in different molarities M (left) and with different admixtures of $c = 0.1$ M (right), $w/c = 0.5$, $T = 21(1)$ °C: △ CAH_{10}, □ C_2AH_{8-x}, ◊ straetlingite, ↑ traces of formic ettringite, ↓ possible formic Afm-phase, main residual clinker phases: ■ CA, ▲ gehlenite, ▼ perovskite.

In contrast to this, the hydroxylated/oxylated admixtures Glyc, Glyox, and Pyruv with the highest determined retarding power factors do not generate a significant increase of the formation of C_2AH_{8-x} or straetlingite, except for lactic acid. Among this, 0.1 M pyruvic acids lead to a strongly reduced reaction of CA and excess water in the hydrated product.

A semi-quantitative estimation of the crystalline phase ratios by peak intensity ratio method in the hydration products after heat flow calorimetry allows for a comparison of the impact of the different tested admixtures with $c = 0.1$ M on the phase formation. The peak areas I of selected reflections of hydration products and the main residual clinker phases were therefore normalized to the perovskite (CT) peak area at 33.3° 2θ: $I_{\circ 2\theta}(\text{phase})/I_{\circ 2\theta}(\text{CT})$. The 3 groups are distinguished by the retarding power factor m: (i) Fo, Ac, (M,D,T)mA, Acr, and MAcr, (ii) halogenated acetic acids and (iii) Glyc, Glyox, Pyruv, and also Lact act with different strengths on the phase formation, too. The selected intensity ratios of the hydration products for the different admixtures are plotted in Fig. 6.7 on the corresponding t_{max}, expressing the setting delay, since $c = 0.1$ M exceeds the validity range of m. The group (i) and (ii) admixtures strongly reduced the formation of CAH_{10} with increased t_{max} and, in opposition to this, enhanced the formation of C_2AH_{8-x}. These changes of the hydrate phase formation distinctly follow in intensity the order of m and are stronger for group (i) than for the halogenated group (ii) admixtures. The group (iii) admixtures also drastically reduce the CAH_{10} formation with increased t_{max} from Lact, Glyc to Glyox, and Pyruv. But in contrast to group (i) and (ii), the formation of C_2AH_{8-x} is relatively low except for Lact.

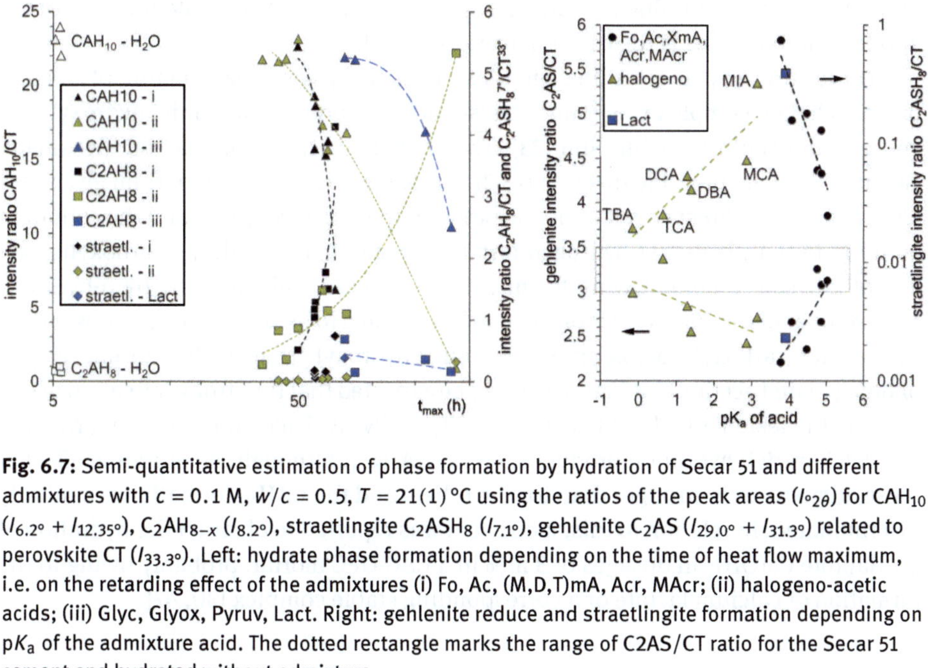

Fig. 6.7: Semi-quantitative estimation of phase formation by hydration of Secar 51 and different admixtures with $c = 0.1$ M, $w/c = 0.5$, $T = 21(1)$ °C using the ratios of the peak areas ($I_{\circ 2\theta}$) for CAH_{10} ($I_{6.2^\circ} + I_{12.35^\circ}$), C_2AH_{8-x} ($I_{8.2^\circ}$), straetlingite C_2ASH_8 ($I_{7.1^\circ}$), gehlenite C_2AS ($I_{29.0^\circ} + I_{31.3^\circ}$) related to perovskite CT ($I_{33.3^\circ}$). Left: hydrate phase formation depending on the time of heat flow maximum, i.e. on the retarding effect of the admixtures (i) Fo, Ac, (M,D,T)mA, Acr, MAcr; (ii) halogeno-acetic acids; (iii) Glyc, Glyox, Pyruv, Lact. Right: gehlenite reduce and straetlingite formation depending on pK_a of the admixture acid. The dotted rectangle marks the range of C2AS/CT ratio for the Secar 51 cement and hydrated without admixture.

The formation of straetlingite observed allows to conclude an attack of gehlenite. Indeed, the intensity ratios for gehlenite are reduced when straetlingite is formed, Fig. 6.7, right. The entire group (i) admixtures affect this reduction of gehlenite and an increased straetlingite formation with increasing acid strength, i.e. decreased pK_a, and thus, increasing retardation power m. Lactic acid is the only member of group (iii) with observed straetlingite formation and matches well this tendency on

pK_a. In opposition to this, the retarding effect of the halogenated group (ii) admixtures in general increases towards higher pK_a, i.e. with decreasing acid strength. Owing to the straetlingite formation for this group, straetlingite increases in the same manner with decreased acid strength (Fig. 6.7). The acid strength is less the driving force for the attack of gehlenite and straetlingite generation than the retardation effect on the hydration reaction.

A possible explanation for the observed effects of the different tested admixtures on the phase formations by hydration of Secar 51 can be approached with the findings that the carboxylic acids complex the calcium ions as a surface layer on the clinker grains, removing Ca^{2+} from the reacting system for the period of retardation [15, 17]. During that time, the released Al^{3+} precipitates as rather amorphous $Al(OH)_3$, visible as whitish muddy precipitation in the residual liquid of the calorimetry samples with elevated w/c for $c_{admix} \geq 0.05$ M. In the following main hydration reaction, the resulting Al-deficiency, especially for high c_{admix}, can then lead to the formation of Ca-richer hydration products C_2AH_{8-x} or possible Afm-phases or calcium salts of the applied admixture, as is the case for acetic acid [15].

For tartaric acid as retarding agent in a gypsum system, the structure of calcium tartrate chelate complex was found to also include bonds between the hydroxyl oxygen and the central calcium ion [33]. The crystal structures of calcium glyoxalate, calcium glycolate bromide (Section 6.3.2) and of the strontium and barium glycolates [26] always have direct covalent bonds between the hydroxyl group and the cation. In addition to that, hydroxyl hydrogen forms hydrogen bonds with the carboxylic oxygen of adjacent anions. Owing to this, the bonding of calcium ions by α-hydroxyl carboxylic acids/anions as complexes can be assumed to be significantly stronger than by the other carboxylic acids/anions without a hydroxyl group such as formic, acetic, or halogenated acetic acids/anions. It may be expected that the stronger fixation of calcium by the tested α-hydroxyl acids Glyc, Glyox, Pyruv, Lact supresses the formation of a calcium-rich hydration product such as C_2AH_{8-x}, although the retarding effect of these admixtures is stronger than for the others and the CAH_{10} formation is similarly discriminated. In relation to this, lactic acid was reported to obstruct the hydration of portlandite $Ca(OH)_2$ in dosages of 3 m% in a CAC-OPC mortar, probably owing to the formation of calcium lactate or a corresponding stable complex [30, 47].

6.3.2 Calcium salts of substituted acetic acids

Among typical and potential admixtures for the setting control of different types of cement, organic acids as well as their salts with lithium, sodium, or calcium are applicable [1, 4–9, 22, 24–27, 35–37]. The dominant impact of organic acids on the setting of CAC affecting a retardation was reported to be mainly addressed to a complex of calcium forming a surface layer on the clinker grains that obstructs their dissolution [15, 17, 33]. Thus, it seems worth inspecting the crystal structures and providing crystallographic data of the calcium salts of the substituted acetic acids for phase analysis and purity control as potential admixture materials as well as to obtain possible hints about their action in a cement paste.

The calcium salt crystal structures of many of the substituted acetic acids tested in this study are not available in literature or databases. Calcium formate $Ca(CHOO)_2$ exists in 3 different modifications α, β, and γ as reported and summarized elsewhere [48]. The triclinic structure of calcium acetate monohydrate $Ca(CH_3COO)_2(H_2O)$ has been published [49]. For Calcium propionate monohydrate, only the unit cell parameters have been reported [50, 51]. Calcium acrylate was found to crystallize from aqueous solution as dihydrate $Ca(CH_2CHCOO)_2 \cdot 2H_2O$ in a monoclinic structure with SG $P2_1/a$ determined at 183 K [52]. The monoclinic calcium monochloroacetate monohydrate $Ca(ClCH_2COO)_2(H_2O)$, SG $P2_1/c$, was reported to possess attractive interactions between the electron negative chlorine substitute and the central calcium ion [53]. In the monoclinic structure, SG $P2_1/c$, of another halogeno-substituted acetic calcium salt $Ca(F_3COO)_2(H_2O)$, such interactions between fluorine and calcium are missing [54]. The structure of calcium trichloroacetate tetrahydrate $Ca(Cl_3CCOO)_2(H_2O)_4$ has previously been described [27, 55].

To gain insight into the possible interactions of the tested substituted acetic acids with the calcium in cement paste, the crystal structures of the calcium salts were determined from sc-XRD data at room temperature. Complementing the set of the available crystallographic data, an overview is given in Tab. 6.5. As a peculiarity within this set, the synthesis of calcium monobromoacetate from aqueous solution did not lead to the desired compound. Owing to the low stability of monobromoacetic acid in aqueous solution [44], a hydrolysis in the presence of calcium resulted in the formation of fine fibrous aggregates of calcium glycolate hydrate, identified by PXRD, and hydro bromic acid. By the further slow evaporation of the solvent water, the glycolate was redissolved again by reaction with the HBr, resulting in the crystallization of the addition compound calcium glycolate bromide hydrate $CaBr(HOCH_2COO)(H_2O)_3$. This conversion demonstrates the strong tendency of the MBA admixture to hydrolyse and to act more like glycolic and hydro bromic acid in a cement paste, as mentioned in Section 6.3.1. Monoiodoacetic acid seems to behave similarly in a basic cement paste. However, the calcium monoiodoacetate hydrate could be crystallized from aqueous solution but is highly sensitive to elevated temperatures and light.

Assessing the atomic configuration of the coordination sphere of the central calcium ion with the surrounding acetate ligands and coordinatively bonded water, another classification than the 3 groups identified for the retarding effect (Section 6.3.1) becomes obvious, Fig. 6.8. Among that systematic, α-calcium formate and calcium acetate hydrate are particularly different. The high number of ligands (6) for calcium formate points to the strong affinity of formate to interact with Ca^{2+}. In group (i) (Section 6.3.1), formic acid has the strongest retarding effect. The calcium acetate structure has 2 different Ca-sites, one with 5 acetate ligands and another with 4 and 2 H_2O molecules, Fig. 6.8 (b). Except for the compounds with hydroxyl acetic

Tab. 6.5: Crystallographic data of calcium salts of formic, acetic, and substituted acetic acids tested as admixtures (A), referred to in the literature or our own data marked by [#].

A	Calcium salt from aqueous solution	SG	*Z*	*a* (Å)	*b* (Å)	*c* (Å)	*α* (°)	*β* (°)	*γ* (°)	Ref.
Fo	α-$Ca(HCOO)_2$	Pbca	12	13.407(3)	10.192(2)	6.282(3)				[48]
Ac	$Ca(CH_3COO)_2(H_2O)$	P-1	4	6.750(5)	11.076(8)	11.782(9)	116.49(5)	92.40(9)	97.31(8)	[49]
MmA	$Ca(CH_3CH_2COO)_2(H_2O)$	$P2_1/a$	4	24.375(1)	6.8124(1)	5.9143(1)		95.320(2)		[51]
		$P2_1/c$	4	5.9215(7)	6.8169(9)	24.435(3)		95.289(2)		[#]
DmA	$Ca((CH_3)_2CHCOO))_2(H_2O)_3\cdot 2H_2O$	$P2_1/c$	4	10.495(2)	24.361(4)	6.662(1)		108.45(1)		[#]
TmA	$Ca((CH_3)_3CCOO))_2(H_2O)_3\cdot 2H_2O$	Aea2	8	24.359(3)	22.130(3)	6.541(1)				[#]
Acr	$Ca(CH_2CHCOO)_2(H_2O)_2$	$P2_1/a$	4	14.44(2)	6.63(1)	11.76(2)		118.4(1)		[52]
MAcr	$Ca((CH_2,CH_3)CCOO)_2(H_2O)_3$	$P2_1/c$	4	7.3824(7)	25.764(3)	6.5658(6)		90.919(2)		[#]
DFA	$Ca(F_2CHCOO)_2(H_2O)$	$P2_1/c$	4	6.8366(8)	20.072(2)	6.0673(7)		94.348(2)		[#]
TFA	$Ca(F_3CCOO)_2(H_2O)$	$P2_1/n$	2	9.465(2)	9.360(3)	16.565(7)		90.79(3)		[54]
	$Ca(F_3CCOO)_2(H_2O)_2$	P-1	2	7.848(3)	8.204(4)	8.318(4)	87.55(1)	84.61(1)	82.45(1)	[#]
MCA	$Ca(ClCH_2CO)_2(H_2O)$	$P2_1/c$	4	6.186(1)	20.992(2)	6.804(1)		92.990(9)		[53]
DCA	$Ca(Cl_2CHCOO)_2(H_2O)$	$P2_1/c$	4	6.9620(7)	6.2499(6)	23.922(3)		97.334(2)		[#]
TCA	$Ca(Cl_3CCOO)_2(H_2O)_4$	$P2_1/c$	4	11.890(1)	21.166(1)	5.9358(4)		97.337(1)		[55]
MBA*	$CaBr(HOCH_2COO)(H_2O)_3$	$P2_1/c$	8	18.565(6)	9.174(3)	10.003(3)		95.706(5)		[#]
DBA	$Ca(Br_2CHCOO)_2(H_2O)_2$	Pccn	8	17.535(4)	19.413(4)	7.832(2)				[#]
TBA	$Ca(Br_3CCOO)_2(H_2O)_4$	P-1	2	6.127(1)	6.336(1)	24.051(4)	85.874(3)	83.681(3)	61.438(3)	[#]
MIA	$Ca(ICH_2COO)_2(H_2O)$	$Pca2_1$	4	26.195(3)	5.9914(1)	6.813(1)				[#]
Glyox	$Ca((HO)_2CHCOO)_2$	I4cm	4	9.0212(8)		10.267(1)				[#]

* the instable monobromoacetic acid hydrolyzed in aqueous solution in the presence of calcium releasing HBr and forms a calcium glycolate bromide hydrate

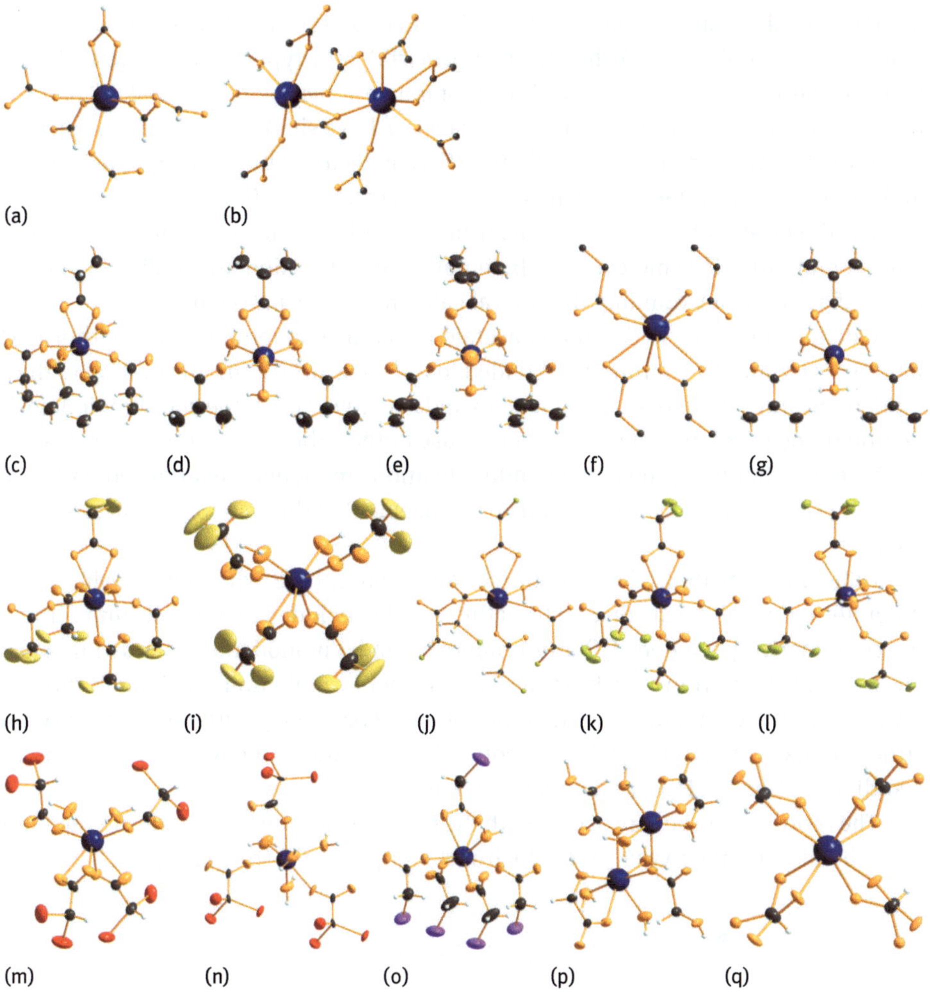

Fig. 6.8: Coordination spheres of Ca^{2+} with the directly bonded ligands and coordinative water molecules of calcium (a) formate [48], (b) acetate hydrate [49], (c) propionate hydrate, (d) iso-butyrate hydrate, (e) pivalate hydrate, (f) acrylate hydrate [52], (g) methacrylate hydrate, (h) difluoroacetate hydrate, (i) trifluoroacetate hydrate, (j) monochloroacetate hydrate [53], (k) dichloroacetate hydrate [27], (l) trichloroacetate hydrate [27, 55], (m) dibromoacetate hydrate, (n) tribromoacetate hydrate, (o) monoiodoacetate hydrate, (p) glycolate bromide hydrate, (q) glyoxylate. Colors: blue, central = Ca, black = C, orange = O, light blue, small = H, yellow = F, green = Cl, red = Br, violet = I.

anions, all other calcium salts can be classified by the number of coordinatively bonded water molecules. Although they have different types and numbers of substitutes, the monohydrates of calcium propionate ("monomethylacetate"/"MmA"), difluoro-, mono-, dichloro-, and monoiodoacetate have similar coordination spheres, including 1 terminal standing coordinative H_2O molecule, 1 bidentate bonded ligand, and 4 monodentate attached ligands, Fig. 6.8 (c), (h), (j), (k), (o).

The dihydrates of calcium acrylate, trifluoro- and dibromoacetate include 2 terminally coordinative H_2O molecules, 2 bidentate and 2 monodentate bonded ligands in the configuration surrounding the central calcium ion Fig. 6.8 (f), (i), (m).

Calcium trichloro- and tribromoacetate crystallize as tetrahydrates with 4 H_2O molecules terminally coordinative bonded to Ca^{2+}, Fig. 6.8 (l), (n). In addition to that, 1 bidentate and 2 monodentate bonded ligands are involved in the coordination polyhedron of the trichloroacetate salt. In the case of the tribromoacetate salt, instead of the bidentate bonding, the corresponding ligand is monodentate attached with free standing second carboxylic oxygen. The arrangement of the water molecules also differs.

Different from that, the calcium coordination sphere with methacrylate, isobutyrate, and pivalate, i.e. di- and trimethylacetate, includes 2 terminally and 2 bridging coordinatively bonded H_2O and 1 bidentate and 3 monodentate bonded ligands, Fig. 6.8 (g), (d), (e). However, from the 2 anions per formula unit, only one is directly involved in this configuration. The other one is separately positioned in the crystal structure, fixed by strong hydrogen bonds between the carboxylic oxygen and the coordinative bonded H_2O, as demonstrated in Fig. 6.9.

The coordination sphere of Ca^{2+} with the hydroxyl acetate ligands always includes direct bonds with the oxygen of the hydroxyl group, Fig. 6.8 (p) and (q).

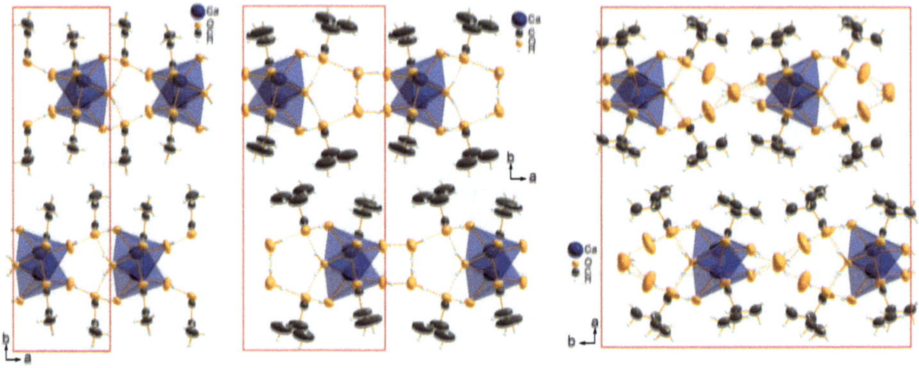

Fig. 6.9: Projections of the crystal structures of calcium methacrylate (left), iso-butyrate (middle), and pivalate (right) hydrate. The red rectangles mark the unit cell; dashed bonds are hydrogen bonds H···O (orange).

The configuration of the calcium coordination spheres of methacrylate, iso-butyrate, and pivalate spread as infinite chains along the crystallographic c-direction, in Fig. 6.9 ordered with increasing pK_a from left to right. The separately positioned second anion per formula is fixed by hydrogen bonds with the coordinative bonded water molecules in general. Moreover, the calcium salts of the weaker acids additionally contain free constitutional water also fixed by hydrogen bonds with the coordinative bonded H_2O molecules, with the separate anion and hydrogen bonds between the constitutional water molecules. Via these hydrogen bonds, the coordination polymeric chains are interlinked to layers stacked along the crystallographic b-direction or a-direction, in the case of the pivalate salt.

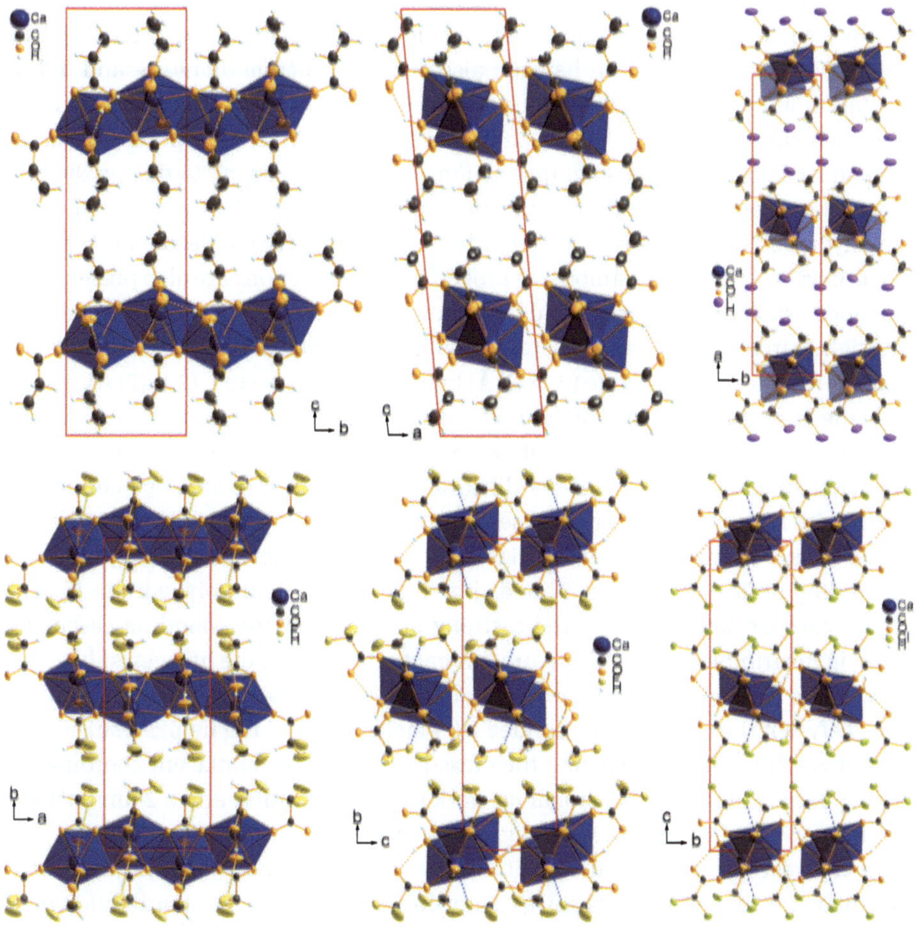

Fig. 6.10: Projections of the crystal structures of the monohydrates of calcium propionate (top, left, middle), monoiodoacetate (top, right), difluoroacetate (bottom, left, middle), and dichloroacetate (bottom, right). The red rectangles mark the unit cell; dashed bonds are hydrogen bonds H···O (orange), H···F (yellow), H···Cl (green), or F/Cl···Ca^{2+} interactions (blue).

The occurrence of the separate anion is not present in all inspected structures of the calcium salts of the relatively weak acids. For calcium acetate hydrate and propionate hydrate, the later with a pK_a of propionic that is similar to that of iso-butyric acid, both anions per formula are directly bonded to Ca^{2+}. However, it may indicate a less pronounced interaction with calcium especially for iso-butyric and pivalinic acid.

The crystal structures of the monohydrate salts of calcium propionate, difluoro-, mono-, dichloro-, and monoiodoacetate have very similar construction schemes. The coordination spheres of calcium also spread as coordination polymeric chains interlinked by bridging carboxyl groups of an acetic ligand, Fig. 6.10. Owing to the different types, sizes, and numbers of substitutes, the layers are differently shifted azimuthally to the stacking, leading to deviating unit cell parameters and positions of the monoclinic inclination for the different compounds.

However, as reported for calcium monochloroacetate hydrate [53], coulombic interactions between Ca^{2+} and a halogen also occur in calcium dichloro- and difluoroacetate monohydrate, marked as blue dashed bonds in Fig. 6.10. The bond lengths of these interactions are 3.206(1) Å [53], 3.0664(4) Å, and 2.7037(2) Å, respectively. For the monoiodoacetate structure, such interactions are missing, owing to the overly large distance between the central Ca^{2+} and the iodine substitute. Moreover, in addition to the hydrogen bonds between the coordinative bonded H_2O molecule and the carboxylic oxygen of the acetic anions, hydrogen bond interactions are also present with the halogen substitutes, marked as dashed bonds in Fig. 6.10. These bond lengths become tighter from relatively undirected for monochloroacetate (3.102 Å) to more direct interactions for dichloroacetate (2.38(2) Å) towards difluoroacetate (2.10(2) Å).

In the di- and also the tetrahydrate structures of calcium dibromo- and trifluoroacetate and trichloro- and tribromoacetate, the coordination polyhedra form separated polymeric chains. In the case of the dihydrates, the polyhedra are edge-connected. Those of the tetrahydrates consist of separate polyhedra interlinked by bridging carboxylic groups of one acetic ligand, Fig. 6.11. Owing to the deviating configuration of the building units, two hydrogen bonds interlink the adjacent polyhedra in the calcium trichloroacetate structure and a strong one with the free standing carboxylic oxygen in the tribromoacetate structure. Interactions between the halogen substitutes and Ca^{2+}, as observed for the monohydrates, are also present. However, hydrogen bonds between the coordinative bonded H_2O molecules and the halogen substitutes are strongly pronounced, realizing the cross-linking between adjacent coordination polymeric chains to a 3-dimensional network for the dihydrates and a 2-dimensional layer-network for the tetrahydrates, Fig. 6.11.

In contrast to the halogen substitutes, the hydroxyl groups always form direct covalent bonds to the central cation and are always involved in the coordination polyhedra as also reported for strontium and barium glycolate [26]. The bond lengths are comparable to those of the carboxylic oxygen to the central cation. Thus the glycolate anion $HOCH_2COO^-$ in calcium glycolate bromide hydrate has 3 binding arms for directly interlinking neighboring calcium coordination polyhedra to a 2-dimensional

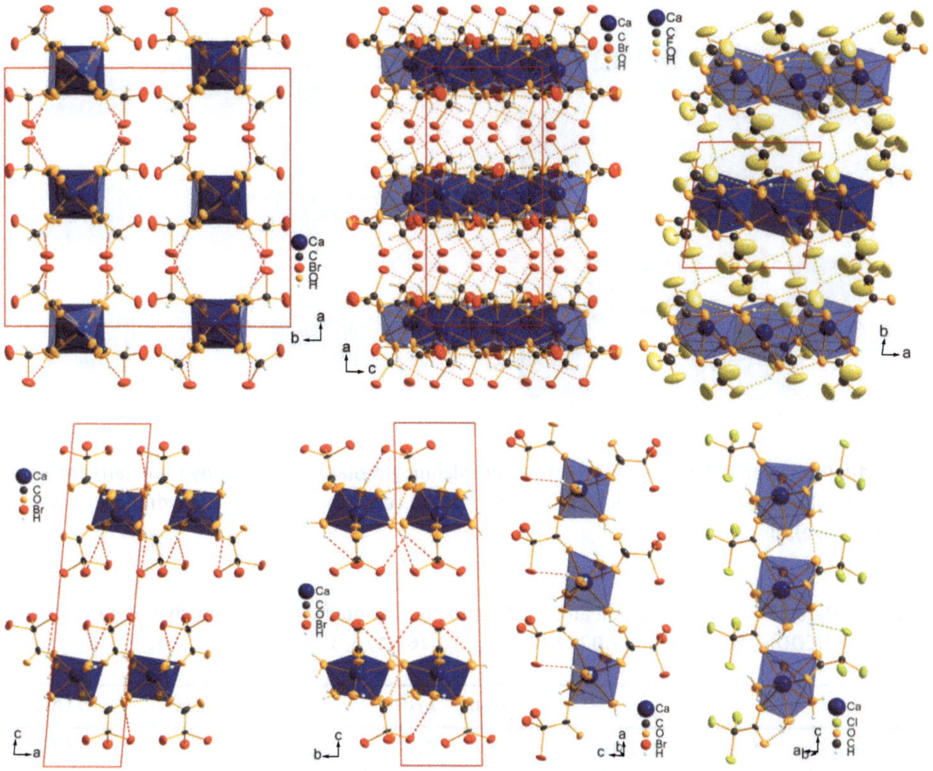

Fig. 6.11: Projections of the crystal structures of calcium dibromoacetate (top, left and middle) and trifluoroacetate dihydrate (top, right), tribromoacetate tetrahydrate (bottom, left, middle), and of a single chain fragment compared to that of calcium trichloroacetate tetrahydrate (bottom, right). The red rectangles mark the unit cell; dashed bonds are hydrogen bonds H···O (orange), H···F (yellow), H···Br (red), H···Cl (green).

network, Fig. 6.12. The free bromide ions in the structure are fixed in position by hydrogen bonds with the coordinatively bonded water molecules and the hydrogen of the hydroxyl group.

Calcium glyoxylate crystallizes anhydrously from aqueous solution, or more precise as the monohydrate of glyoxylic acid. The double bonded oxygen of glyoxylic acid hydrates with water to two hydroxyl groups. Both hydroxyl oxygen directly bond to Ca^{2+} in addition to the carboxylic ones, connecting the two neighboring calcium positions and resulting in a 2-dimensional network. The hydroxyl groups of all glyoxylate molecules are directed in one direction $-c$ or $+c$ depending on the enantiomeric form of the polar structure (SG I4cm) forming strong hydrogen bonds (1.75(1) Å) with the carboxyl groups of the next layer, Fig. 6.12. The crystal structural data for calcium glyoxylate are given in Tab. 6.6.

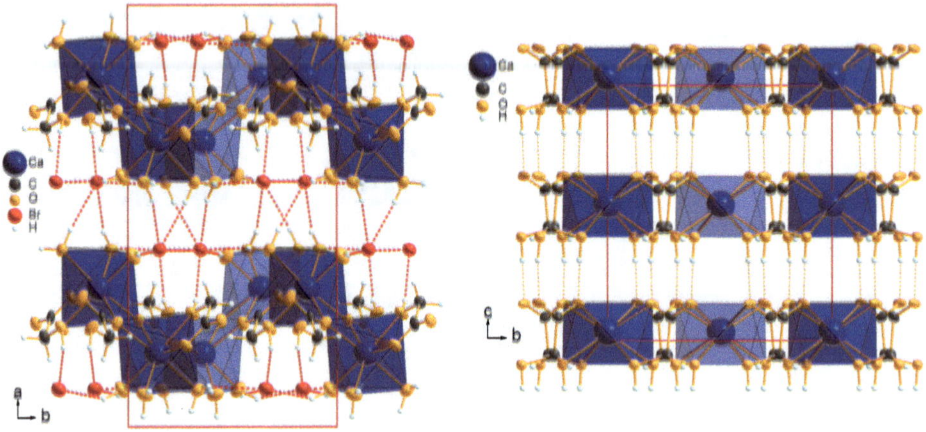

Fig. 6.12: Projections of the crystal structures of calcium glycolate bromide trihydrate (left) and calcium glyoxylate (right). The red rectangles mark the unit cell; dashed bonds are hydrogen bonds H···O (orange), H···Br (red).

Tab. 6.6: Data of sc-XRD acquisition and crystal structure refinement of calcium glyoxylate $Ca((HO)_2CHCOO)_2$, SG I4cm, a = 9.0212(8) Å, c = 10.267(1) Å, V = 835.6(1), Z = 4.

		Atom	x/a	y/b	z/c	U (Å³)
M (g/mol)	222.16	Ca1	0.5	0.5	0.5304(1)	0.0142(2)
d_{calc} (g/cm³)	1.7661	C1	0.2763(4)	0.2237(4)	0.5884(4)	0.0224(9)
$\mu_{Mo\text{-}K\alpha}$ (mm⁻¹)	0.766	C2	0.2625(4)	0.2375(4)	0.4396(5)	0.0228(5)
Crystal size (μm)	60 × 60 × 170	O1	0.3704(3)	0.3044(3)	0.6454(2)	0.0259(7)
Transmission min | max	0.903 | 0.958	O2	0.2999(3)	0.3828(3)	0.4060(2)	0.0204(5)
Reflections N_{total} | N_{unique}	2644 | 2378	H1C2	0.333(4)	0.167(4)	0.400(4)	0.019(1)
R_{sym} | R_{int} (%)	2.9 | 3.2	H1O2	0.309(3)	0.377(3)	0.316(1)	0.017(1)
Miller Index limits within 6.38° ≤ 2θ ≤ 56.35°, 98.66 % completeness	$-9 \le h \le 11$, $-11 \le k \le 8$, $-13 \le l \le 13$					
R_{all} | wR_{all} (%)	3.4 | 4.3					

As shown by the short overview of the crystal structural features of the calcium salts of the different substituted acetic acids tested as admixtures, different interactions occur, especially between the halogen and hydroxyl substitutes with calcium and the coordinatively bonded water molecules. The interactions with Ca^{2+} in the structure can give a hint of the possible contribution of the substitutes in addition to the reported calcium complexation, i.e. for acetate [15], of the carboxyl group in a cement paste. On the other hand, the hydrogen bonds of the halogen substitutes with the coordinatively bonded H_2O molecules give a hint on possible interactions of the electron negative part of the halogenoacetic molecules with the cement paste solution. Owing to the different bond lengths, the bond valence is a more valid expression for the strength of the observed structural interactions of the different atom types [56, 57]. Therefore, the bond valences (*bv*, in valence units vu) for the observed interactions were calculated [56]:

$$bv = \exp[(R_0 - R)/B], \tag{6.4}$$

with $B = 0.37$; R is the bond length; R_0 is the length of a bond of unit valence taken from Breese and O'Keefe (1991) [57]. The resulting bond valences regarded for interactions with $bv \geq 0.005$ vu are graphically presented in Fig. 6.13.

The halogen-hydrogen bonds (H···halogen) in general increase in number with increasing acid strength, i.e. with decreasing pK_a. Maximum values for single strong hydrogen bonding are found for calcium difluoro- and dichloroacetate hydrates with $bv =$ 0.044(2) vu and 0.050(2) vu, respectively. For the other compounds, the values are between 0.005(2) vu and 0.024(2) vu. Accumulating the single values to a total interaction between the halogen substitutes and the coordinatively bonded water molecules, a tendency arises showing increasing total interaction for decreasing pK_a values of the corresponding acid, Fig. 6.13. The only exception is the sum of H···halogen bond valences of calcium dibromoacetate dehydrate, which is significantly larger than the corresponding tendency.

However, the trend of halogen-water interactions increasing with acid strength in general correlates with the decrease of the retardation power for the halogenated acetic acid admixtures shown in Fig. 6.4.

Coulombic interactions between calcium and halogen substitutes appear for the monohydrates of mono- and dichloroacetate and difluoroacetate, Fig. 6.13. The bond valences of 0.105(1) vu, 0.152(1) vu, and 0.097(1) vu, respectively, are somewhat weaker than the Ca–O bonds of the bidentate attached ligand with 0.151(1) and 0.263(1) vu,

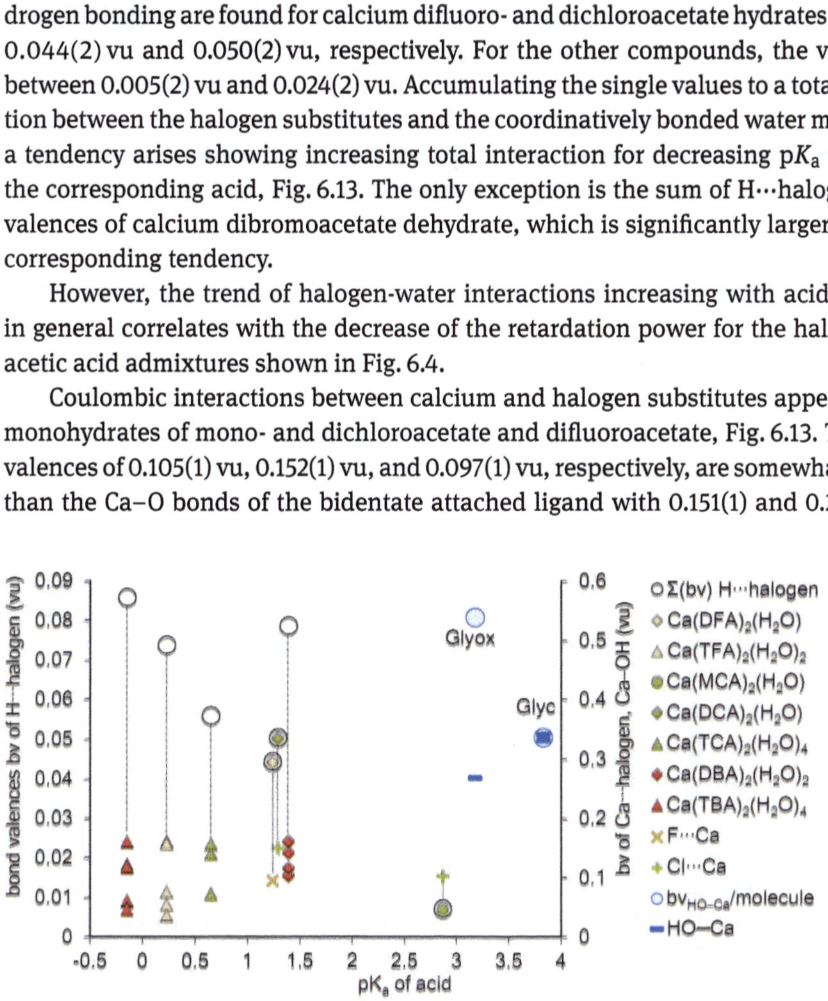

Fig. 6.13: Bond valences bv of halogen-hydrogen (H···halogen) and halogen-calcium interactions (F,Cl···Ca) of calcium salt hydrates of the different halogen-substituted acetic acids. Blue: hydroxyl-calcium bonds (HO–Ca) of calcium glyoxylate (Glyox) and the glycolate molecule (Glyc) of calcium glycolate bromide hydrate.

0.146(1) and 0.241(1) vu, 0.157(1) and 0.189(1) vu, respectively. The other carboxylic oxygen and the water molecule of the coordination polyhedra are bonded with valences in a range of 0.279(1)–0.373(1) vu.

The hydroxyl groups of glycolate and glyoxylate ligands form direct covalent bonds to the central Ca^{2+}, as described above and shown in Fig. 6.8 and 6.12. The bond valences of these bonds, 0.334(1) and 0.343(1) vu for calcium glycolate bromide hydrate and 0.270(1) vu for calcium glyoxylate, lie within the range of the Ca–O bonds to the carboxylic oxygen and coordinative bonded water of these structures with 0.255(1)–0.347(1) vu and 0.291(1) vu, respectively. Because glyoxylate has 2 hydroxyl groups, the sum of the bond valences for one molecule with 0.541(1) vu addresses a stronger potential and strength for bonding interactions with calcium than for a glycolate molecule. Compared to the coulombic interactions of the halogen substitutes with the central calcium ion, these bonds are significantly stronger and potentially able to realize stable calcium complexes, obviously leading to the observed strong retarding effects on the setting of the Secar 51 cement pastes, as mentioned in Section 6.3.1.

6.4 Conclusions

The different substituted acetic acids tested as admixtures for Secar 51 cement pastes all retard the setting. The setting delay is described well by polynomial functions on the applied admixture concentration within the tested range $0.001\,\mathrm{M} \leq c_{\mathrm{admix}} \leq 0.1\,\mathrm{M}$, allowing an appropriate adjustment of the setting point. Moreover, the increase of the retarding effect can be measured by an exponential function of the logarithm of the onset and maximum heat flow times on c_{admix}: $\ln t = A \cdot \mathrm{e}^{mc}$ for $c_{\mathrm{admix}} \leq 0.05\,\mathrm{M}$. Whereas A corresponds to the setting time of the cement paste without admixture, m depends only on the admixture type. Thus m has the character of a retardation power factor of the different admixtures. By this factor, the tested admixtures can be classified in 3 different groups:

(i) formic, acetic, methylated acetic acids, acrylic, and methacrylic acid,
(ii) halogenated acetic acids,
(iii) hydroxylated/oxylated acetic acids (glycolic, glyoxylic, lactic, pyruvic acid).

The retarding power of group (i) rigorously decreases with decreasing acid strength whereas for group (ii) the setting delays generally decrease with increasing acid strength. The strongest retarding effects were found for the group (iii) admixtures. Among that, the setting delays of lactic acid follow a square root function.

The crystal structures of the calcium salts of the different substituted acetic acids crystallized from aqueous solution can give possible hints as to the interactions of the acetic anions and the substitutes with the calcium ions and the water molecules. The non-polar substitutes of group (i) together with formate and acetate show neither interactions with Ca^{2+} nor water. Thus the retarding effect is only affected by the acid strength and the dissociation ability of the acid.

The polar halogen substitutes in some cases show coulombic interactions with Ca^{2+}, enabling a possible contribution to the calcium complexation, which leads to the reported surface layer on the clinker grains obstructing their dissolution. The retarding effect of these admixtures, e.g. dichloro- and difluoroacetic acid, is comparable to that of the unsubstituted acetic acid or stronger for monochloroacetic acid. However, all halogen substitutes form hydrogen bonds with the water molecules in the crystal structures increasing in strength with increasing acid strength. This tendency is directly opposed to the increase of the retardation power. As a result, a similar interaction of the halogenoacetic anions with the water molecules of the cement pastes is conceivable, probably counteracting the complexation of Ca^{2+} and/or the formation of a surface layer on the clinker grains. In contrast to this, the hydroxyl groups of the group (iii) admixtures form direct covalent bonds with Ca^{2+} in equivalent bond valences as the carboxyl groups. This leads to a much stronger fixation of calcium by the complexation in a surface layer and thus to strong retarding effects.

For all tested admixtures, the heat flow maxima H_{max} of the hydration reaction decreases with increasing setting delay. It can be shown that H_{max} correlates with the time of the heat flow maximum t_{max} by a functional dependence, allowing a prediction of the maximum heat development for a pre-adjusted setting delay. The total reaction heat Q_{max} deviates within the bounds of experimental error for admixture concentrations within the validity range of the found retarding power factor m and is similar to that of the cement paste without admixture. For these cement paste mixtures, the resulting hydration products are all similar, dominated by CAH_{10} with traces of C_2AH_{8-x}. Higher concentrations can slightly reduce Q_{max} and affect the phase formation by reducing CAH_{10} and increasing C_2AH_{8-x}. This effect on the hydration products increases with increasing retarding effect and increased setting delay and may be related to the complexation of Ca^{2+} beside the release of Al^{3+}. When the aluminum precipitates as $Al(OH)_3$ before a strongly delayed main hydration, the resulting calcium excess leads to the preferred formation of a calcium richer phase. In contrast, the strong Ca^{2+} complexation by the hydroxyl carboxylic acids seems to be less reversible under the conditions in the Secar 51 cement paste. Although these admixtures have the strongest retarding effects affecting the strongest setting delays and a visible precipitation of $Al(OH)_3$ for elevated w/c, the CAH_{10} formation is reduced, similar to the other admixtures but without a significant increase of the Ca-richer C_2AH_{8-x}. Only for lactic acid was a C_2AH_{8-x} formation observed, which was less pronounced compared to the other admixtures. In addition to that, for strong setting delays a slight dissolution of gehlenite and a formation of straetlingite was found in the hydration products. The degree of straetlinge formation is more correlated to the setting delay than to the acid strength of the applied admixture, except for glycolic, glyoxylic, and pyruvic acid, not leading to a straetlingite formation.

Acknowledgment: The authors wish to thank A. Boretski from the Institute for Agricultural and Nutrition Science, Department of Soil Science and Conservation for performing the ICP-OES measurements. We are indebted to J. Sieksmeier and R. Oberste-Padtberg from ARDEX GmbH for providing cement samples.

References

[1] Taylor HFW. Cement Chemistry, 2nd edition. London: Academic Press; 1997.
[2] Bensted J. Special Cements. In: Hewlett P C. Lea's Chemistry of Cement and Concrete. 4th edition. London: Arnold; 1998. 837–901.
[3] Kurdowski W. Cement and Concrete Chemistry. Springer. 2014; DOI 10.1007/978-94-007-7945-7.
[4] Gelardi G, Mantellato S, Marchon D, Palacios M. Chemistry of chemical Admixtures. In: Aïtcin P-C, Flatt RJ, editors. Science and Technology of Concrete Admixtures. Elsevier; 2016. 149–218.
[5] Myrdal R. SINTEF REPORT SBF BK A07025. Advanced cementing materials. Controlling hydration development. SINTEF Building and Infrastructure; 2007.
[6] Marchon D, Flatt RJ. Impact of chemical admixtures on cement hydration. In: Aïtcin P-C, Flatt RJ, editors. Science and Technology of Concrete Admixtures. Elsevier; 2016. 279–299.
[7] Aïtcin P-C. Retarders. In: Aïtcin P-C, Flatt RJ, editors. Science and Technology of Concrete Admixtures. Elsevier; 2016. 395–404.
[8] Aïtcin P-C. Accelerators. In: Aïtcin P-C, Flatt RJ, editors. Science and Technology of Concrete Admixtures. Elsevier; 2016. 405–412.
[9] Currell BR, Grzeskowlak R, Midgley HG, Parsonage JR. The acceleration and retardation of set high alumina cement by additives. Cem Con Res. 1987; 17: 420–432.
[10] Rodger SA, Double DD. The chemistry of hydration of high alumina cement in the presence of accelerating and retarding admixtures. Cement and Concrete Research. 1984; 14: 73–82.
[11] Scian A N, Porto López J M, Pereira E. Mechanochemical activation of high alumina cements – Hydration behaviour. I, Cement and Concrete Research. 1991; 21: 51–60.
[12] Scrivener KL, Capmas A. Calcium Aluminate Cements. In: Hewlett PC, editor. Lea's Chemistry of Cement and Concret. 4th edition. London: Arnold; 1998. 709–778.
[13] Estival J, Mahiaoui J, Watt V, Taquet P, Colombet JF. A new setting accelerator upon lithium sulfate technology for calcium aluminates based materials. In: Fentiman C, Mangabhai R, Scrivener K, editors. Calcium Aluminates: Proceedings of the International Conference 2014. IHS BRE Press; 2014. EP104, ISBN 978-1-84806-316-7: 321–332.
[14] Poellmann H. Study of the hydration mechanisms and formation of new hydrates applying organic additives to the aluminate phase of cement. Proc 11th Int Conf on Cem Micr 1989. 286–305.
[15] Smith A, El Hafiane Y, Bonnet JP. Role of a Small Addition of Acetic Acid on the Setting Behavior and on the Microstructure of a Calcium Aluminate Cement. J Am Ceram Soc. 2005; 88: 2079–2084.
[16] El Hafiane Y, Smith A, Bonnet JP, Tanouti B. Effect of a carboxylic acid on the rheological behaviour of an aluminous cement paste and consequences on the properties of the hardened material. J Eur Ceram Soc. 2005; 25: 1143–1147.
[17] Bishop M, Barron AR. Cement hydration inhibition with Sucrose, Tartaric Acid, and Ligninsulfonate: Analytical and Spectroscopic Study. Ind Eng Chem Res. 2006; 45: 7042–7049.
[18] Poellmann H. Effect of organic additives on the hydration behaviour of the aluminate phase of cement. Proc 10th Int. Conf on Cem Micr 1988. 324–343.

[19] Poellmann H, Micheaux M, Nelson EB. Study on the influence of citric acid on the retardation and formation of new hydrates in cement. Proc 12th Int Conf on Cem Micr. 1990: 303–322.

[20] Young JF. Effect of Organic Compounds on the Interconversions of Calcium Aluminate Hydrates: Hydration of Tricalcium Aluminate. J Am Ceram Soc. 1970; 53: 65–69.

[21] Sugama T, Carciello NR. Alkali carbonation of calcium aluminate cements: influence of set-retarding admixtures under hydrothermal conditions. J Mater Science. 1992; 27: 4909–4916.

[22] Schmidt M, Pöllmann H. Hydratation behaviour of CAC at various temperatures by isoperibolic heat flow calorimetry using acetates. In: Fentimen CH, Mangabahai RJ, Scrivener KL, editors. Calcium Aluminate Cements. Proc Int Symp, HIS BRE Press, Bracknell; 2008. 93–108.

[23] Wang Y, Zhu B, Li X, Chen P. Effect of dispersants on the hydrate morphologies of spinel-containing calcium aluminate cement and on the properties of refractory castables. Ceram Int. 2016; 42: 711–720.

[24] Pöllmann H, Sturm P, Kaden R. Hydration control of Portland cement and Calcium Aluminate cement using different earth alkali benzoate hydrates. Proc 35th Int Conf on Cem Micr 2013. 302–356.

[25] Pöllmann H, Fischer K, Kaden R. Hydration control of Portland cement and Calcium Aluminate cement using different alkaline earth naphtoate hydrates. Proc 36th Int Conf on Cem Micr 2014. 123–160.

[26] Kaden R, Pöllmann H. Alkaline Earth Glycolates and their Application as CAC Admixture. In: Fentiman C, Mangabhai R, Scrivener K, editors. Calcium Aluminates: Proceedings of the International Conference; 2014. IHS BRE Press, 2014, EP104, ISBN 978-1-84806-316-7: 396–404.

[27] Kaden R, Poellmann H. Application of Alkaline Earth Chloroacetates for the Setting Control of Calcium Aluminate Cements. Proc 37th Int Conf on Cem Micr. 2015: 27–69.

[28] Moeschner G, Lothenbach B, Figi R, Kretzschmar R. Influence of citric acid on the hydration of Portland cement. Cem Con Res. 2009; 39: 275–282.

[29] Plank J, Winter Ch. Competitive adsorption between superplasticizer and retarder molecules on mineral binder surface. Cem Con Res. 2008; 38: 599–605.

[30] Kastiukas G, Zhou X, Castro-Gomes J, Huang S, Saafi M. Effects of lactic and citric acid on early-age engineering properties of Portland/calcium aluminate blended cements. Con Build Mat. 2015; 101: 389–395.

[31] Schulz RJ. Admixtures and method for accelerating the setting of portland cement compositions. US Patent US4264368; 1977.

[32] Schumacher G, Patel R. Lactate activated cement activator compositions. US Patent US8349072 B2; 2013.

[33] Hill JR, Plank J. Retardation of setting of plaster of Paris by organic acids: Understanding the mechanism through molecular modelling. J Comput Chem. 2004; 25: 1438–1448.

[34] Cody AM, Lee H, Cody RD, Spry PG. The effects of chemical environment on the nucleation, growth, and stability of ettringite $[Ca_3Al(OH)_6]_2(SO_4)_3 \cdot 26H_2O$. Cem Conc Res. 2004; 34: 869–881.

[35] Barralet JE, Tremayne M, Lilley KJ, Gbureck U. Modification of Calcium Phosphate Cement with α-Hydroxy Acids and Their Salts. Chem Mater. 2005; 17: 1313–1319.

[36] Mariño FT, Torres J, Hamdan M, Rodríguez CR, Cabarcos EL. Advantages of Using Glycolic Acid as a Retardant in a Brushite Forming Cement. Biomed Mater Res Part B: Appl Biomater. 2007; 83B: 571–579.

[37] Zajac M, Skocek J, Bullerjahn F, Ben Haha M. Effect of retarders on the early hydration of calcium-sulpho-aluminate (CSA) type cements. Cem Con Res. 2016; 84: 62–75.

[38] Petricek V, Dusek M, Palatinus L. JANA2006, version 25/10/2015, Institute of Physics, ASCR, Praque, CZ; 2015.

[39] Brandenburg, K. Diamond version 3.2k, Crystal Impact CbR, Bonn, Germany; 2014.

[40] Kuzel H-J. Ein leistungsfähiges Wärmeleitungskalorimeter. TIZ-Fachberichte. 1984; 108: 46–50.

[41] Pöllmann H, Kuzel H-J, Meyer HW. Heat-flow calorimetry in cement chemistry Construction and Application of a low cost high sensitive calorimeter. 13th International Conference on Cement Microscopy, Florida, USA; 1991. 303–313.

[42] Agreda VH, Zoeller JR. Acetic Acids and its Derivatives. CRC Press; 1992.

[43] Worsham PR. Halogenated Derivatives. In: Agreda VH, Zoeller JR, editors. Acetic Acids and its Derivatives. CRC Press; 1992. 285–304.

[44] Drushel WA, Simpson GS. The Relative Stability of Halogen Substituted Aliphatic Acids in Water Solution. J Am Chem Soc. 1917; 39: 2453–2460.

[45] Paradies HH, Quitschau P, Pischel I. Structure and Properties of Calcium Pyruvate in Aqueous solutions. Z Phys Chem. 2000; 214: 201–311.

[46] Poellmann H. Calcium Aluminum Oxide Formate Hydrate, ICDD reference code 00-041-0726. JCPDS 1990.

[47] Singh NB, Prabha Smt S, Singh AK. Effect of lactic acid on the hydration of Portland cement. Cem Conc Res. 1986; 16: 545–553.

[48] Watanabé T, Matsui M. A Redetermination of the Crystal Structures of α-Calcium Formate, α-Strontium formate and Barium formate by X-ray Analysis. Acta Cryst. 1978; B34: 2731–2736.

[49] Klop EA, Schouten A, van der Sluis P, Spek AL. Structure of Calcium Acetate Monohydrate $Ca(C_2H_3O_2)_2 \cdot H_2O$. Acta Cryst. 1984; C40: 51–53.

[50] Charbonier F, Gobert-Ranchoux F. Crystal data for two salts of propionic acid: $Ca(C_2H_5CO_2)_2 \cdot H_2O$ and $Sr(C_2H_5CO_2)_2 \cdot H_2O$. J Appl Cryst. 1977; 10: 357–358.

[51] Valor A, Reguera E, Sánchez-Sinencio F. Synthesis and X-ray diffraction study of calcium salts of some carboxylic acids. Powder Diff. 2002; 17. DOI 10.1154/1.1414011.

[52] Le Page Y, Fortier S, Donnay G. Redetermination of the Crystal Structure of Calcium Acrylate Dihydrate at 183 K. J Polymer Sci: Polymer Chemistry Edition. 1978; 16: 2265–2273.

[53] Karipides A, Peiffer K. Direct C-Cl···Ca^{2+} Binding: Crystal Structure of Calcium Chloroacetate Hydrate. Inorg Chem. 1988; 27: 3255–3256.

[54] Khristov M, Peshev P, Angelova O, Petrova R, Macicek J. Preparation, Thermal Behaviour, and Structure of Calcium Trifluoroacetate Monohydrate. Monatshefte für Chemie. 1998; 129: 1093–1102.

[55] Singh S, Saini D, Kaur G, Mehta SK, Kaur R, Ferretti V. Synthesis and characterization of alkaline earth trichloroacetates. Inorg Chim Acta. 2014; 419: 13–18.

[56] Brown I D. The Bond-Valence Method: An empirical Approach to Chemical Structure and Bonding. In: O'Keefe M, Navrotsky A., ed. Structure and bonding in crystals, Vol. II. Academic Press, Inc.; 1981. ISBN 0-12-525102-5.

[57] Breese NE, O'Keefe M. Bond-Valence Parameters for Solids. Acta Cryst. 1991; B47: 192–197.

S. Stöber* and H. Pöllmann

7 Crystallography and crystal chemistry of AFm phases related to cement chemistry

Abstract: Ettringite and AFm phases (aluminum-iron-mono phases) are the main hydration products of aluminate and ferrate phases during the reaction of cementitious materials. The different AFm phases can incorporate many anions, structurally unnecessary water, and uncharged molecules. The paper summarizes the crystal chemistry of the entire family of potentially occurring AFm phases in hydrated cements and newly formed phases during cement deterioration reactions.

Keywords: AFm phases, crystal structure, cement hydration

7.1 Introduction

AFm phases can be formed in cementitious systems under various different conditions. Some of the main instances in which they occur and their formation conditions are mentioned below:
- AFm phases as hydration products of anhydrous cement minerals
- AFm phases caused by the influence of admixtures [1]
- AFm phases formed during the carbonation of cementitious materials
- AFm phases caused by defrosting salts
- AFm phases formed as minor compounds
- AFm phases formed by the hydration of special cements
- AFm phase formation in landfills

Previous literature has mainly discussed sulfate containing phases with respect to cements, and has sometimes also considered their interaction with carbonate [2]. This chapter gives some additional insight into other possible formations of AFm phases in cementitious materials and also describes the different structure types and arrangements of these layered structures.

Minerals of the layered double hydroxide family are composed of positively charged main layers which are compensated by negatively charged interlayers. The ratios of positively charged metals can vary and, as a result of these variations, the amount of anions can also vary. The interlayer can additionally contain structurally

***Corresponding author: S. Stöber,** Institute for Geological sciences, mineralogy/geochemistry, Martin-Luther-University Halle, Halle (Saale), Germany, stefan.stoeber@geo.uni-halle.de
H. Pöllmann, Institute for Geological sciences, mineralogy/geochemistry, Martin-Luther-University Halle, Halle (Saale), Germany

DOI 10.1515/9783110473728-008

unnecessary water molecules which can be removed and introduced without destroying the main structural units. The water content is variable at different temperatures and also depends on the type of anion. The different possibilities of solid solutions within these structural units can be described schematically as follows. Depending on the formation conditions, additional alkali sulphates can also be stored in the interlayer.

The overall formula can be given by:

$$[\mathrm{Me}^{2+}_{(1-x)}\mathrm{Me}^{3+}_{(x)}(\mathrm{OH})_2]^{x+}[(\mathrm{An})^{y}_{(x/y)} \cdot m \cdot \mathrm{Me}X \cdot z\,\mathrm{H_2O}]^{x-}.$$

- Incorporation in the main layer:

$$[\mathrm{Me}^{2+}_{(1-x)}\mathrm{Me}^{3+}_{(x)}(\mathrm{OH})_2]^{x+}$$

- Incorporation in the interlayer:

$$[(\mathrm{An})^{y}_{(x/y)} \cdot m \cdot \mathrm{Me}X \cdot z\,\mathrm{H_2O}]^{x-}$$

- Additional incorporation in the interlayer: m
- Incorporation of additional alkali salts: MeX
- Incorporation instead of interlayer water: $z\,H_2O$
- Surface adsorption

Several structure determinations of layered structures have been performed on sulfate, carbonate and chloride containing AFm phases [5, 8, 9, 43, 53, 92, 102, 192, 197].

Further important structure descriptions listed below:
- Hemichloridechromate $C_3A\cdot\frac{1}{2}CaCl_2\cdot\frac{1}{2}CrO_4\cdot12H_2O$ [3] with C = CaO, A = Al_2O_3 [4]
- Monoiodide $C_3A\cdot CaI_2\cdot10H_2O$ [5–7]
- Monobromide $C_3A\cdot CaBr_2\cdot10H_2O$ [8]
- Hemibromatechloride $C_3A\cdot\frac{1}{2}CaBr_2\cdot\frac{1}{2}CaCl_2\cdot10H_2O$ [9]
- Monogalliumchloride: $[Ca_2Ga(OH)_6]^+$ $[Cl\cdot2H_2O]^-$ [10]
- Monochromate $[Ca_2Al(OH)_6]^+$ $[\frac{1}{2}CrO_4\cdot3H_2O]^-$ [5]
- Monoperchlorate $[Ca_2Al(OH)_6]^+$ $[2ClO_4\cdot2H_2O]^-$ [5]

7.2 Layered double hydroxides, $[M^{II}_{(1-x)}M^{III}_{x}(OH)_2]^{x+}[(An)^{y}_{(x/y)}\cdot m\cdot MX\cdot z\,H_2O]^{x-}$ with variable $x = M^{II}/(M^{III}+M^{II})$ ratios

7.2.1 Origin of LDHs in cement pastes and hardened cement pastes

LDH solid solutions were first formed as secondary phases in contact zones between clay and cementitious materials in the barrier systems of nuclear waste disposals [12].

The phase analysis of different 20 year old ordinary Portland cement (OPC) ground granulated blast furnace slag blends (0 to 100 % slag) yielded different concentrations of Mg–Al layered double hydroxide phases [13]. The addition of certain $CaMg(CO_3)_2$ contents increased the amount of hydrotalcite and the mechanical properties (comprehensive strength) of hydrated cement pastes at 40 and 60 °C [14]. Hydrotalcite was also observed in mortars exposed to sulfate solutions [15]. Such LDHs are promising phases for adding to concrete in order to improve its durability in aggressive environments [16]. Currently, many papers and research projects are concerned with the possibility of substituting different types of pozzolanic materials in cements against clinker material in order to decrease CO_2 production during cement manufacturing [17–24].

7.2.2 Crystal chemistry of layered double hydroxides with variable ion ratios

The general structural properties of LDHs can be summarized as follows. Metal ions are six-fold coordinated by hydroxide ions in the so called main layer $[M^{II}_{(1-x)}M^{III}_x(OH)_2]^{x+}$, forming a brucite like lattice. In the layered crystal structure of brucite [25], octahedra are compressed along the stacking axis, so that the local geometry is lowered, which means that O–O and Mg–Mg distances parallel to (001) increase to 3.142 Å, the layer distance c' shrinks to 2.112 Å, and the O–Mg–O angles in the $[Mg(OH)_2]$ layer become 96.7° and 83.3°, rather than a rectangular angle of 90° (Fig. 7.1).

The distortion does not influence its hexagonal symmetry ($a = b = 3.142$ Å, $c = 4.766$ Å, and SG P$\bar{3}$m1) [26–28]. The lattice parameter a can be calculated according to $a = 2\sqrt{d}(M–O)$ [29, 30] (bond length O and M^{II} in the brucite layer) or $a =$

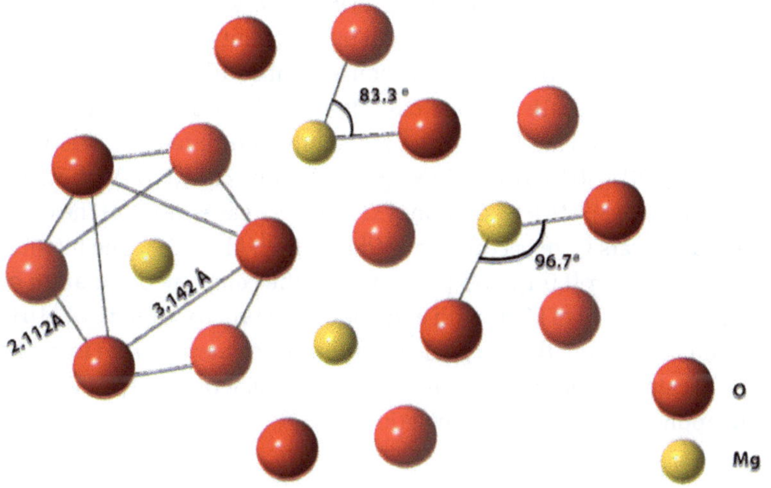

Fig. 7.1: Characteristic bonds and angles between Mg and O in a single brucite layer.

2(M–O) sin($\alpha/2$) with α = O–M–O [29], provided that an ideal atomic arrangement exists in the brucite lattice [$Mg(OH)_2$].

Due to the presence of certain divalent and trivalent ion concentrations in octahedral coordination, the layers are positively charged $x = M^{II}/(M^{III} + M^{II})$. The metal-oxygen bond length d(M–O) is related to the ionic radii by the equation $d(M\text{–}O) = (1 - x)r(M^{II}) + xr(M^{III})$ or $\delta_a/\delta_x = -\sqrt{2}[r(M^{II}) - r(M^{III})]$. Due to the distributions of trivalent cations in the main layer, super cells with special M^{II}/M^{III} and a/a_0 ratios were detected in LDHs [31, 32]. Typical representatives for such phases are hydrotalcite [33, 34], $a = a_0 = 3.054$; chlormagaluminite, [35] $\sqrt{3}a_0 = 5.29$ Å; quintinite-2H [36], $2\sqrt{3}a_0 = 10.55$ Å; stichtite [37], $2a_0 = 6.169$ Å; and wermlandite [38], $3a_0 = 9.303$ Å.

It has not been described in literature that the metal ratio $x = M^{II}/(M^{III} + M^{II})$ of LDHs can vary between $0 \le x \le 1$, but, according to the limits of their unit cells, the ratio varies in the range $0.2 \le x \le 0.4$ [39, 40]. The upper limit is restricted by electrostatic repulsion between neighboring trivalent metals in the layers, which is unavoidable if $x > \frac{1}{3}$ [41–43]. The lower limit results in distances between inter lamellar anions that are too high, leading to a collapse of the interlayer. In most cases, values with $x = \frac{1}{3}$ and $x = \frac{1}{4}$ are detected for LDHs. But the existence of LDHs with range $0.20 < x < 0.33$ [44, 45], $x = 0.41$–0.48 for $M^{II}/M^{III} - CO_3$ LDHs (M^{II} = Mg, Ni, Co, Cu; M^{III} = Al, V) [46] and 0.5 for Fe^{II}/Fe^{III} LDHs were proposed [47], however impurities [48], amorphous contents, leeching of M^{II} ions at low pHs [49] and many more errors lead to the assumption that LDHs outside $0.20 < x < 0.33$ exist.

The charge density is balanced by the fixation of anions in the interlayer $[X^{m-}_{x/m} \cdot n\,H_2O]^{x-}$. Water molecules, either essential or non-essential for the stability of the LDH structure, are present in the interlayer. Such molecules can be exchanged reversibly as a result of variable temperatures and relative humidities. Anions, water molecules, and hydroxide ions form a complex disordered network in order to bond main and interlayer [40]. For cement chemistry, LDHs with anions like OH^- [50], CO_3^{2-} [51], SO_4^{2-} [51], NO_3^- [51], Cl^- [51] must be taken into account [50].

The theory of the LDH crystal structure derivation as a substitution of M^{+III} against M^{+II} is slightly different for Al-LDHs. In this case, two thirds of octahedral coordinated sites in dioctahedral gibbsite [γ-$Al(OH)_3$] are empty and different LDHs with the composition $[M^{II}Al_4(OH)_{12}](NO_3)_2 \cdot nH_2O$ (M^{II} = Co, Ni, Cu, Zn) were synthesized using activated gibbsite with $M^{II}(NO_3)_2$ solutions [52]. In these cases, 50 % of the octahedral sites in the main layers were occupied by M^{II} ions (M^{II}/Al = 0.25). The fixation of Li^+ ions in the gibbsite layer $[LiAl_2(OH)_6]^+$ resulted in the complete occupation of the octahedral sites, as is typical of LDHs with a trioctahedral brucite layer. Li-Al LDHs have been studied intensively and their crystal structures were characterized in detail [53–58]. Li-Al LDHs became important as calcium aluminate cement (CAC) pastes were homogenized with applicable high Li concentrations.

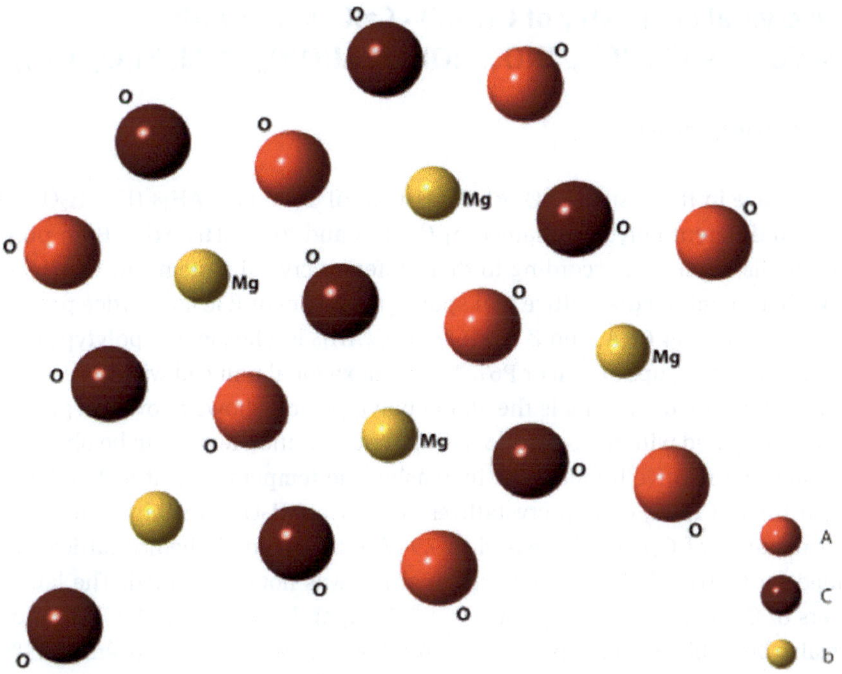

Fig. 7.2: Description of the main layer as a close packing of spheres with three different positions with A, B, and C.

Due to the mutual orientation of octahedra in opposite main layers the formation of different LDH polytypes are possible. For instance, two polytypes of Mg-Al-CO_3 LDHs: hydrotalcite (3R) [59] – manasseite (2H) [60] or of Mg-Fe-CO_3 LDHs: pyroaurite (3R) [61, 62] and sjoegrenite (2H) [62, 63] exist [32]. For the notation of those polytypes, a special nomenclature was established [36, 64, 65]. With the description of the main layer as a close packing of spheres, three different positions A, B, and C exist. Oxide atoms of hydroxide ions occupy the first brucite layer positions A and C, B is occupied by M^{II}, and the occupation by cations is indicated using a small letter, here b (Fig. 7.2). The symbol of the initial layer is "AbC" or shortened "AC". If in the underlying sheet oxygen atoms reside either in A or B, the hydroxide ions form an octahedra (O-Type), if oxygen occupies position C hydroxide ions form a prism (P-Type). According to this principle, all theoretically possible arrangements in different LDH polytypes were derived.

7.3 The crystal chemistry of $C_3(A,F) \cdot CaX \cdot n\,H_2O$ with $X = CO_3^{2-}, SO_4^{2-}, 2Cl^-, 2NO_3^-, 2OH^-, 2Al(OH)_4^-, 2(Al, Si)O_2(OH)_4^-$

7.3.1 $C_3A \cdot Ca(OH)_2 \cdot nH_2O$

The stable phase in the system $CaO–Al_2O_3–H_2O$ until 50 °C is C_4AH_{19} (H = H_2O) [4] (Fig. 7.3). Two different polytypes named α_1-C_4AH_{19} and α_2-C_4AH_{19} exist. Both polytypes can be distinguished according to their different crystal systems. α_1-C_4AH_{19} is trigonal with a rhombohedral lattice and space group $R\bar{3}m$ or R3c and lattice parameters a = 5.77 Å, c = 64.08 Å, and Z = 3 [66]. α_2-C_4AH_{19} is a hexagonal polytype (1H) with possible space groups $P6_3/m$ or $P6_322$ and a hexagonal unit cell with a = 5.77 Å and c = 21.37 Å [66]. α_2-C_4AH_{19} is the stable polytype and shows a lower degree of disordering compared with α_1-C_4AH_{19} [67]. Different hydration levels can be obtained by decreasing the relative humidity or increasing the temperature. At 81 % relative humidity (r.h.) and 25 °C, C_4AH_{13} crystallizes since 6 mol H_2O were dehydrated [68]. The layer distance of C_4AH_{19} decreased to 7.94 Å [68]. A rhombohedral lattice was determined for C_4AH_{13}, but a possible space group was not determined. The lattice parameters of C_4AH_{13} are a = 5.752 Å, c = 95.27 Å, and Z = 6 [69]. This hydration level can also be stabilized if temperatures of 40 °C and 25 % r.h. were chosen. At 25 °C and a storage over anhydrous $CaCl_2$, C_4AH_{11} is the stable hydration level. The layer distance c' decreased to 7.35 Å [68, 70] caused by the release of 2 mol H_2O. Finally, C_4AH_7, is the lowest hydration level, without any interlayer water. Its layer distance is 5.6 Å [68, 70] and can be prepared by storing the compound in an atmosphere of P_2O_5 or at 110–120 °C. All hydration levels can rehydrate to C_4AH_{13}, if the relative humidity is raised higher than 81 %.

Fig. 7.3: SEM image of $C_3A \cdot Ca(OH)_2 \cdot 12H_2O$ (TCAH) crystals.

The Fe analogue of C_4AH_{19} [71] was synthesized and characterized [71, 72]. A layer distance of 7.9 Å was determined for C_4FH_{19} and the complete miscibility with C_4AH_{19} was also investigated [73].

7.3.2 $C_3A \cdot CaSO_4 \cdot nH_2O$

7.3.2.1 Structure

The natural occurrence of monosulfate was named as kuzelite [74]. The crystal structure of monosulfate-12 hydrate [75] is composed of sequences containing $[Ca_2Al(OH)_6]^+$ and $[SO_4 \cdot 3H_2O]^-$ stacks perpendicular [001]. According to space group R$\bar{3}$, three sequences of $[Ca_2Al(OH)_6]^+$ $[SO_4 \cdot 3H_2O]^-$ occur per unit cell. The unit cell has the dimensions $a = 5.7586(3)$ Å ($\approx \sqrt{3} \cdot 3.3$ Å of the brucite layer) and $c = 26.7946(12)$ Å. The layer distance is $c' = \frac{1}{3}c = 8.931$ Å.

$[Ca_2Al(OH)_6]^+$ forms the main layer. Al is six-fold coordinated but the bonds Al–OH are quite different (2 × 2.82 Å, 2 × 3.27 Å, and 2 × 4.01 Å), which results in a buckled octahedron in c-direction. Ca ions are coordinated by 6 oxygens with three distances Ca–OH = 2.357 Å and three distances Ca–OH = 2.455 Å. Because Ca is shifted from its centre along [001] towards the inner layer $[SO_4 \cdot 3H_2O]^-$, Ca is coordinated by an additional oxygen of a water molecule $H_2O(1)$, which results in a tight bonding of 2/3 of the water molecules to Ca. Therefore, the coordination of Ca is seven. The inner layer or interlayer is strongly disordered, the sulphate anion is only present in every second cell, an oxygen ion of a water molecule occupies the position 0, 0, 1/2 instead of the oxygen O(2) of the SO_4^{2-} group. The interlayer contains different types of water molecules. $H_2O(1)$ is responsible for the charge compensation between the main layer and the interlayer, together with O(2) of the SO_4^{2-} group. $H_2O(2)$ are space fillers and do not contribute to the stability of the crystal structure because they do not belong to the hydrogen bond network. Buttler et al. (1959) have demonstrated that these water molecules can easily be removed at temperatures higher than 50 °C [70].

The dehydration process has been investigated extensively by Pöllmann (1984) [76]. Monosulfate has been dried at 35 % r.h. in order to stabilize the 12-hydrate. At non ambient temperatures, monosulfate 12-hydrate releases water in the range of 43–200 °C. The water released from the interlayer is subdivided into different dehydration processes at 43 °C, 57 °C, 68 °C, 87 °C, 145 °C, and 200 °C. A total of 6 mol H_2O were dehydrated. The layer distance c', which is equal to one third of the lattice parameter c of monosulfate 12 hydrate, decreases from 8.94 Å to 7.91 Å at 80 °C. The layer distance could not be investigated at higher temperature levels by XRD because the dehydration of the interlayer overlaps with the incipient dehydration of the main layer with the removal of $3H_2O$ ($6OH^-$) of the main layer. Dosch (1962) [77] and Kuzel (1966) [78] investigated the dehydration process of monosulfate 12 hydrate earlier and found those hydration levels too, but could not detect $C_3A \cdot CaSO_4 \cdot 11H_2O$ and $C_3A \cdot CaSO_4 \cdot 7H_2O$.

Not only the variation of temperature has a huge influence on the properties of the interlayer, the precise setting of the relative humidity is a key factor too. At temperatures above 40 °C and a relative humidity of 90 %, a H_2O-rich monosulfate $C_3A \cdot CaSO_4 \cdot 14H_2O$ was detected by XRD [75]. Kuzel (1969) [79] synthesized large single crystals (max. 4 mm) of $C_3A \cdot CaSO_4 \cdot 12H_2O$ and investigated the dehydration process applying a temperature controlled precession camera. At 90 % r.h., $C_3A \cdot CaSO_4 \cdot 14H_2O$ is the stable hydration level crystallizing in space group R$\bar{3}$. The unit is a cell with dimensions $a = 5.751(1)$ Å and $c = 28.672(2)$ Å, compared with the unit cell of monosulfate 12 hydrate extended about 2 Å in c-direction. Dehydration in the air at 37 % r.h. forces the dehydration to $C_3A \cdot CaSO_4 \cdot 12H_2O$ without the stabilization of an intermediate hydrate level. The dehydration-rehydration process $12H_2O$–$14H_2O$ is reversible and can be repeated as often as desired. At relative humidities higher than 90 % and at a temperature level of 0 °C, a hydrate level with the chemical composition $C_3A \cdot CaSO_4 \cdot 16H_2O$ was stabilized by Turriziani et al. (1955) [80]. Due to the significant broadening of hkil reflections, only the pseudo cell with $a = 5.753(3)$ Å and $c' = 10.30(1)$ Å was determined. Pöllmann et al. (1993) synthesized and investigated $C_3A \cdot CaSO_4 \cdot 16H_2O$ [75, 81]. They refined lattice parameters $a = 5.730$ Å and $c = 30.671$ Å from XRD powder data at 35 % r.h. Unlike the dehydration-rehydration process between 12 and $14H_2O$, hydrothermally synthesized $C_3A \cdot CaSO_4 \cdot 12H_2O$ cannot be converted to $C_3A \cdot CaSO_4 \cdot 16H_2O$ in air at 100 % r.h. caused by cooling [75, 82, 83].

7.3.3 $C_3F \cdot CaSO_4 \cdot nH_2O$ ($F = Fe_2O_3$) [4]

Malquori & Caruso (1938) let Fe^{3+}-salt solutions react with $Ca(OH)_2$ solutions and obtained the phase $C_3F \cdot CaSO_4 \cdot 12H_2O$ [82]. Schippa (1958) reported that $C_3F \cdot CaSO_4 \cdot 12H_2O$ synthesized in a saturated solution or above 95 % relative humidity contains 13–$14H_2O$ [85]. Its layer distance c' would be greater than that of $C_3F \cdot CaSO_4 \cdot 16H_2O$ [86]. The Fe analogue to monosulfate was synthesized and investigated by Kuzel (1969) [79, 87]. $C_3F \cdot CaSO_4 \cdot 12H_2O$ crystallizes in a R lattice with extinction conditions $-h + k + l = 3n$ ($n = 1, 2, 3, \ldots$) for all hkls. Possible space groups were R3, R$\bar{3}$, R32, and R$\bar{3}$m. Therefore, three layer packages $Ca_2Fe(OH)_6 \cdot \frac{1}{2}SO_4 \cdot 3H_2O$ exist per unit cell perpendicular [001]. The lattice parameters were refined as $a = 5.888(1)$ Å and $c = 26.625(9)$ Å ($c' = 8.875$ Å). Further investigations of $C_3F \cdot CaSO_4 \cdot 12H_2O$ showed that its space group is R$\bar{3}$ with $a = 5.889$ Å and $c = 26.669$ Å [88]. First attempts to solve the structure revealed slightly different lattice parameters with $a = 5.891$ Å and $c = 26.682$ Å ($c' = 8.894$ Å) [89]. One-dimensional Fourier synthesis and projections demonstrated that the structure of $C_3F \cdot CaSO_4 \cdot 12H_2O$ is very similar to the crystal structure of $C_3A \cdot CaSO_4 \cdot 12H_2O$ [89], only the electron density changed at the position of the cation caused by the increased scattering power of Fe [89]. The crystal structure of iron monosulfate was investigated at 20 °C and 50 °C using samples with certain concentration of impurities equilibrated for 680 and 360 days [90]. The lattice parameters were refined using LeBail fits. At 20 °C, iron monosulfate crystallized trigonal with a rhombohedral bravais lattice and space group R$\bar{3}$. The lattice parameters were refined as $a = 5.8864(2)$ Å and $c = 26.614(2)$ Å at 20 °C and at 50 °C as $a = 5.8862(3)$ Å and $c = 26.635(1)$ Å. The structure refinement was done using the Rietveld technique [91], taking the structure of monosulfate [92] as a starting model and refining the iron substitution in the main layer and applying soft constraints on S–O distances. Iron monosulfate is typically composed of a $[Ca_2Fe(OH)_6]^+$ main layer with standard coordinations of Ca and Fe by oxygen atoms O(1) of the hydroxide ions and Ca with an additional O O(2) belonging to a water molecule in the interlayer. In the inter layer region $[\frac{1}{2}SO_4 \cdot 3H_2O]^-$, the sulfate position is occupied by 50 % sulfate anions (up and down orientation) and 50 % weakly bonded water molecules.

7.3.4 $C_3A \cdot CaSO_3 \cdot nH_2O$

Motzet investigated sulfite containing AFm phases in the system $CaO–Al_2O_3–SO_2–H_2O$ [93]. The synthesis of CA + 3CaO + Na_2SO_3 + H_2O in excess yielded pure $C_3A \cdot CaSO_3 \cdot nH_2O$ after a reaction time of 6 months (Fig. 7.4) [94]. The chemical analysis showed that only 0.3 % Na_2O were determined by atomic absorption spectrometry. There are also many descriptions in literature of AFm phases with the compositions $C_3A \cdot CaSO_3 \cdot 7H_2O$ [95], $C_3A \cdot CaSO_3 \cdot 11H_2O$ [93, 96, 97], and $C_3A \cdot CaSO_3 \cdot 12H_2O$ [98]. Unlike $C_3A \cdot CaSO_4 \cdot 12H_2O$ [92], the lattice of monosulfite is trigonal primitive. Its lattice parameters $a = 5.7709(4)$ Å and $c = 51.284(5)$ Å were refined using 2θ values from powder data [99]. The determination of water concentrations by TG analysis in combination with high temperature X-ray measurements showed that $C_3A \cdot CaSO_3 \cdot 11H_2O$

Fig. 7.4: Platy, shaped hexagonal crystals of $C_3A \cdot CaSO_3 \cdot 12H_2O$.

at 35 % r.h. dehydrated stepwise at 20 °C, 40 °C, 85 °C, 110 °C, losing 1, 1, 1, and 2 mol of structurally necessary water. At 260 °C, the dehydration process of $3H_2O$ from the main layer $[Ca_2Al(OH)_6]^+$ was initiated.

7.3.5 $C_3F \cdot CaSO_3 \cdot nH_2O$

Iron monosulfite was synthesized purely during a reaction time of 2 weeks at a permanent temperature of 60 °C [100]. In contrast to the Al analogue [93, 94], $C_3F \cdot CaSO_3 \cdot nH_2O$ contains 12 mol of water at 35 and 100 % r.h. and the primitive lattice of $C_3A \cdot CaSO_3 \cdot 11H_2O$ has changed to a rhombohedral lattice ($R\bar{3}c$) in $C_3F \cdot CaSO_3 \cdot 12H_2O$. Lattice parameter a = 5.903 Å of $C_3F \cdot CaSO_3 \cdot 12H_2O$ is larger than that of $C_3A \cdot CaSO_3 \cdot 11H_2O$ (a = 5.771 Å) as a result of the iron contents in the main layer. For completeness, the layer distance of $C_3F \cdot CaSO_3 \cdot 12H_2O$ is c = 51.284 Å [100, 101]. DSC experiments of $C_3F \cdot CaSO_3 \cdot 12H_2O$ yielded 2 endothermic and 2 exothermic reactions, whose DSC signals were produced by the oxidation of sulfite to sulfate and the crystallization of C_2F. The endothermic reactions symbolize 2 different dehydration effects at 110 and 135 °C. At 110 °C, the loss of 1 mol H_2O in the interlayer is responsible for the decrease from c' = 8.42 Å to 7.70 Å. At 135 °C, the final dehydration of 5 mol H_2O is initiated, whereby the dehydration of the main layer proceeds simultaneously.

7.3.6 $C_3A \cdot CaCO_3 \cdot nH_2O$

7.3.6.1 Ordered structure of $C_3A \cdot CaCO_3 \cdot 11H_2O$ "O-$C_4A\bar{C}H_{11}$" ($\bar{C}$ = CO_2)

In order to solve the crystal structure of monocarbonate (Fig. 7.5), single crystals were synthesized by hydrothermal methods, mixing stoichiometric concentrations of $Ca(OH)_2$, $AI(OH)_3$, and $CaCO_3$ in a 3.5 : 2 : 0.5 ratio at 393 K for one month [102].

The space group of monocarbonate $C_3A \cdot CaCO_3 \cdot 11H_2O$ is P1 with lattice parameters a = 5.7747(14) Å, b = 8.4689(11) Å, c = 9.923(3) Å, α = 64.77(2)°, β = 82.75(2)°, γ = 81.43(2)° [102]. The lattice parameters were determined earlier by Fischer & Kuzel (1982) [69]. Sequences of layers with the compositions $[Ca_2Al(OH)_6]^+$ and $[CO_3 \cdot 5H_2O]^-$ are arranged perpendicularly (011). The layer distance between 2 adjacent $[Ca_2Al(OH)_6]^+$ stacks is 7.55 Å. In the main layer, the coordination of Al is 6 and of Ca is 7, a common feature of all AFm phases. Four different Ca-environments exist in the structure. Ca(1) is coordinated by O(1), O(3), O(4), O(5), O(10), O(12), and O(13) with bonds in the range of 2.351–2.546 Å. Ca(2) has O(2), O(3), O(4), O(6), O(9), O(11), O(14) as its next neighbors with bonds in the range of 2.349–2.518 Å. Ca(3) is coordinated by O(2), O(6), O(7), O(8), O(9), O(11), and O(19) with bonds in the range of 2.346–2.515 Å and Ca(4) is coordinated by O(1), O(5), O(7), O(8), O(10), O(12), and O(15) with bonds in the range of 2.348–2.553 Å. O(13), O(14), O(15) belong to water molecules of the interlayer and the coordination polyhedron of Ca(3) contains O(19) as part of the

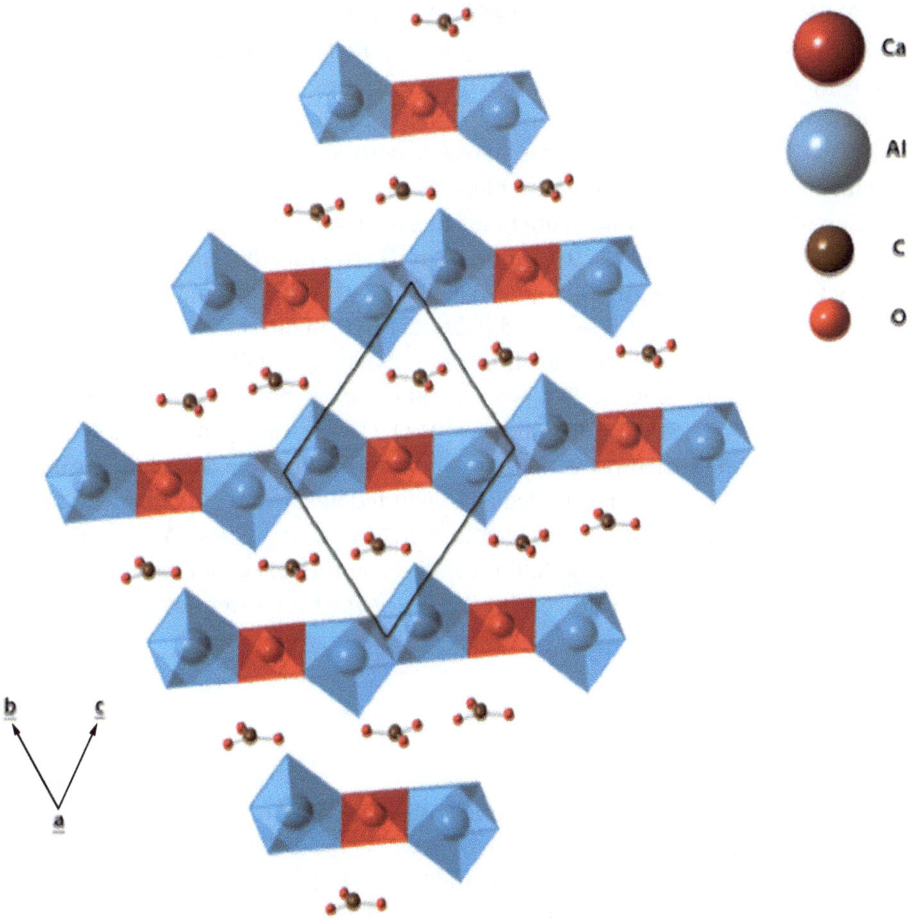

Fig. 7.5: Crystal structure of "ordered" $C_3A \cdot CaCO_3 \cdot 11H_2O$ "O-$C_4A\check{C}H_{11}$".

carbonate group. All Ca-atoms are shifted between 0.52–0.61 Å perpendicularly (011) from their centre in the coordination polyhedron. Al(1) and Al(2) occupy the centre of two distorted octahedra with the oxygen atoms of hydrogen groups, respectively. The bonds are 1.916–2.511 Å and 1.343–2.020 Å.

In the interlayer $[CO_3 \cdot 5H_2O]^{2-}$, the hydrogen network is quite complicated. Hydrogen-oxygen bonds in the range of 1.7–1.9 Å play the dominant role in the stability of this layer. Two water molecules $H_2O(16)$ and $H_2O(14)$ each connect 2 carbonate groups. $H_2O(15)$ is linked to one carbonate group and another interaction between one H and O of two H_2O molecules $H_2O(17)$ and $H_2O(14)$ exists. The whole framework is connected with the main layer bonding Ca(3)–CO_3(19), Ca(2)–$H_2O(14)$, and Ca(4)–$H_2O(15)$. Because of the missing adhesion of $H_2O(17)$ in the interlayer O–H network, this will be the first water molecule which will be released at non-ambient temperatures.

7.3.6.2 Disordered structure of $C_3A{\cdot}CaCO_3{\cdot}11H_2O$ "D-$C_4A\bar{C}H_{11}$"

D-$C_4A\bar{C}H_{11}$ was synthesized and investigated by Renaudin et al. (1999) [103]. The authors found that D-$C_4A\bar{C}H_{11}$ has many similarities with an ordered monocarbonate structure, for instance the composition and coordination of the main layer $[Ca_2Al(OH)_6]^+$ [103]. The differences between both structures are the different arrangements of water and carbonate molecules in the interlayer, which is responsible for different types of stacking sequences. In the O-structure, the layer sequence is $CO_3^{2-}{\cdot}3H_2O$–$[Ca_4Al_2(OH)_{12}]^{2+}$–$2H_2O$ and in the D-structure the arrangement of the layers is $CO_3^{2-}{\cdot}3H_2O$–$[Ca_4Al_2(OH)_{12}]^{2+}$–$\frac{1}{2}(2CO_3^{2-}{\cdot}6H_2O)$. The ordered structure is traditionally built of $[Ca_4Al_2(OH)_{12}]^{2+}$ and $[CO_3^{2-}{\cdot}2H_2O]^{2-}$ in order to balance the charge of the main layer and the D-structure is composed of $[Ca_4Al_2(OH)_{12}{\cdot}2CO_3{\cdot}6H_2O]^{2-}$ and $[Ca_4Al_2(OH)_{12}{\cdot}4H_2O]^{2+}$ sheets. The layer distances c' in both structures are almost identical with $d = 7.554$ Å (perpendicular (011), O-structure) and $d = 7.536$ Å (perpendicular (002), D-structure). The authors pointed out that the disordered structure with space group $P\bar{1}$ can be interpreted as a distorted hexagonal structure with lattice parameter dimensions of $a = 5.7422$ Å, $b = 5.7444$ Å, $c = 5.091$ Å, $\alpha = 92.290°$, $\beta = 87.450°$, $\gamma = 119.54°$, close to the conditions of a hexagonal base ($a = b \neq c$, $\alpha = \beta = 90°$, $\gamma = 120°$). The lattice of O-$C_4A\bar{C}H_{11}$ shows a clearer deviation from the hexagonal base [103].

7.3.7 $C_3F{\cdot}CaCO_3{\cdot}nH_2O$

For the synthesis of iron, monocarbonate stoichiometric concentrations of C_2F, CaO, and $CaCO_3$ were applied. After a reaction of 14 weeks, pure $C_3F{\cdot}CaCO_3{\cdot}12H_2O$ was obtained. The lattice parameters $a = 5.92$ Å and $c = 47.852$ Å were refined using powder data of samples kept at 35 and at 100 % r.h. [104]. It could be shown that under both relative humidities, equal layer distances were present. The stacking of iron monocarbonate is that of a 3R polytype with space group $R\bar{3}c$ [104].

Another option to synthesize iron monocarbonate is to use $Fe(OH)_3$ instead of C_2F [100]. The same space group and comparable lattice parameters of $a = 5.917$ Å and $c = 47.679$ Å were published in 1998 [100, 105]. The dehydration process is characterized by 2 endothermic events at 110 and 135 °C, plus an exothermic reaction at approximately 400 °C, indicating the crystallization of C_2F. Both endothermic reactions indicate the dehydration of the interlayer with 1.5 and 4.5 mol H_2O, respectively. The dehydration of the main layer overlaps with the final dehydration reaction of the interlayer and is finished at about 500 °C. With the dehydration of the interlayer area, two different layer distances of different hydration levels with 7.24 Å and 6.01 Å were identified by high temperature XRD measurements. Another hydration level of iron monocarbonate between 55 °C and 90 °C could not be investigated separately [100].

Iron monocarbonate was also synthesized in order to provide solubility products of Fe-containing monocarbonate [106]. The structure of a sample containing

89 wt%. iron monocarbonate was investigated by synchrotron powder diffractometry. The refined chemical composition yielded the formula $3CaO \cdot Fe_2O_3 \cdot CaCO_3 \cdot 12H_2O$. Its trigonal structure with space group $R\bar{3}c$ consisted of main layer $[Ca_2Fe(OH)_6]^+$ and interlayer $[\frac{1}{2}CO_3 \cdot 3.10(2)H_2O]^-$ with lattice parameters $a = 5.9196(1)$ Å and $c = 47.8796(10)$ Å. The coordination of Ca^{2+} is seven and Fe^{3+} is in accordance with Al AFm phases six. The application of EXAFS showed that Fe has 6 neighboring O-atoms at a distance of 2.02 Å and six neighboring Ca atoms at 3.47 Å [106]. The interlayer of iron monocarbonate is a network of carbonate ions and water molecules which are distributed statistically in this area [106]. The authors mentioned the presence of an unrealistic Ow2–Ow2 distance of 2.208 Å, which reflects the statistical distribution of one carbonate anion with two water molecules at the centre of an interlayer. A comparable situation was also detected for the disordered D-C_4ACH_{11} structure [103]. Monocarbonate and iron monocarbonate exhibit certain differences. Due to the complete substitution of Al with Fe, the layer distance a is larger than that of the monocarbonate unit cell. The authors doubt the existence of a complete solid solution series $Ca_4[(Al_xFe_{1-x})_2(OH)_{12}] \cdot CO_3 \cdot (6-x)H_2O$ because of significant symmetry differences and different locations and coordination conditions of carbonate ions in the interlayer of both structures.

The thermal analysis of iron monocarbonate was applied in order to determine the water content of the interlayer. 5.8 mol water were specified, which is in accordance with previous findings [100]. The dehydration process was done at about 240 °C. Further weight losses were caused by the dehydroxilation of the main layer, dehydration of CH impurities and decarbonation of $C\bar{C}$ at 700 °C.

7.3.8 $C_3A \cdot CaCl_2 \cdot nH_2O$

Two different structures of monochloride α-$C_3A \cdot CaCl_2 \cdot 10H_2O$ (r.h. 35 %), more commonly known as "Friedel's salt", were solved [43, 107, 108]. The low temperature 2M polytype (LT-structure) [43] is monoclinic with space group C2/c and lattice parameters $a = 9.979(3)$ Å, $b = 5.751(2)$ Å, $c = 16.320(6)$ Å, $\beta = 104.53(3)°$ (Fig. 7.6). The structure is composed of typical sequences $[Ca_2Al(OH)_6]^+$ and $[Cl \cdot 4H_2O]^-$. The main layer $[Ca_2Al(OH)_6]^+$ with Ca ions is in seven-fold (H_2O and 6OH) and Al in six-fold coordination with six hydroxide ions. The interlayer $[Cl \cdot 4H_2O]^-$ forms a primitive hexagonal net, where chloride ions are coordinated by four hydrogen of water molecules. Parallel to (010), the Cl–H_2O arrangement is bonded by hydrogen ions. According to Terzis et al. (1987) [43], two different bonding principles are important for the stability of the LT-structure. One oxygen atom of a water molecule is part of the coordination sphere of Ca. A further six hydrogen bonds from two adjacent $[Ca_2Al(OH)_6]^+$ layers increase the coordination of chlorine to ten so that chlorine is coordinated as a distorted antiprism. Terzis et al. (1987) have indicated that a phase transition takes place at 32 °C [43]. The LT-structure is strongly related to a rhombohe-

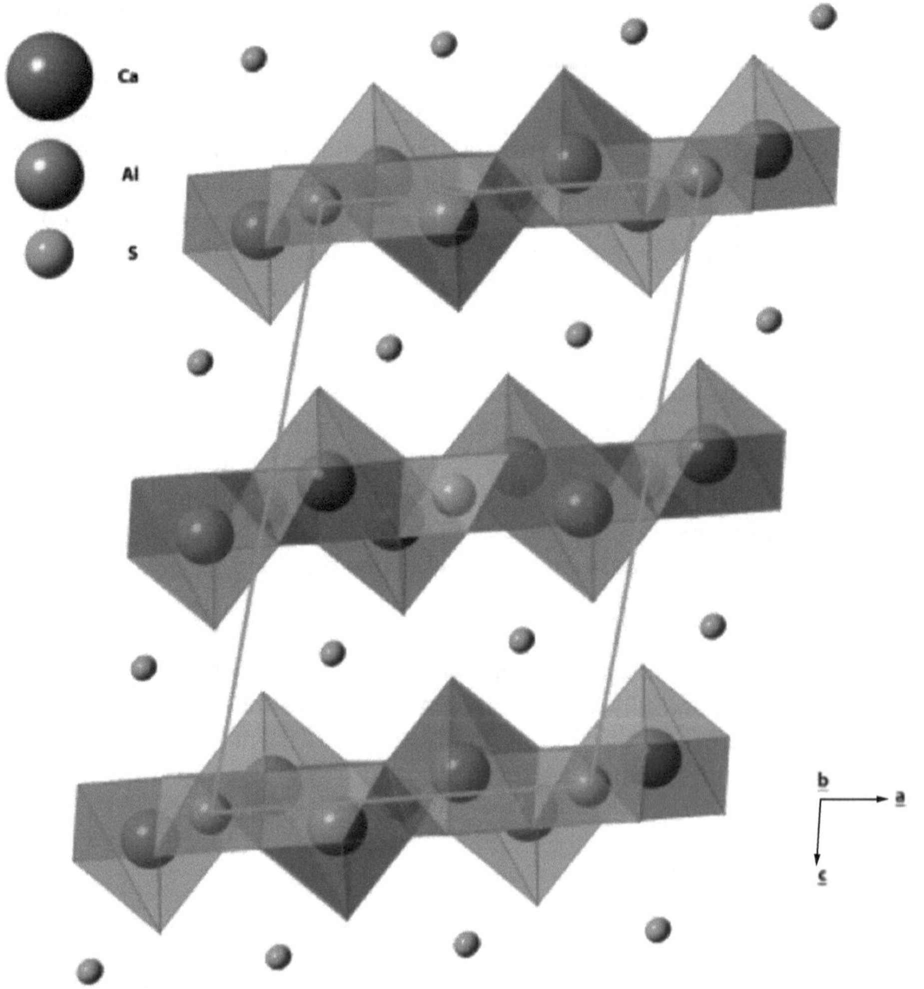

Fig. 7.6: Crystal structure of Friedel's salt α-monochloride $C_3A \cdot CaCl_2 \cdot 10H_2O$.

dral β-phase (HT-structure) transforming the basis like $a_{1h} = \frac{1}{2}(b - a)$, $a_{2h} = \frac{1}{2}(b + a)$, and $c_h = -(a + 3c)$ (a_{1h}, a_{2h}, and c_h: lattice parameters of hexagonal base).

The HT-structure of monochloride $C_3A \cdot CaCl_2 \cdot 10H_2O$ was solved in 1999 (Fig. 7.7) [109]. The first order phase transition monoclinic–trigonal is reversible and the transition temperature on heating the temperature stages of the polarization microscope was 34.2(3) °C and on cooling 32.0(5) °C. With the aid of a polarization microscope, a temperature level was stabilized to keep low and high temperature domains in one crystal. By means of combined synchrotron and XRD measurements at temperatures in the range of 0–40 °C, the first order character of the phase transition was confirmed, the powder pattern contained two species of reflections at 30 °C which were indexed

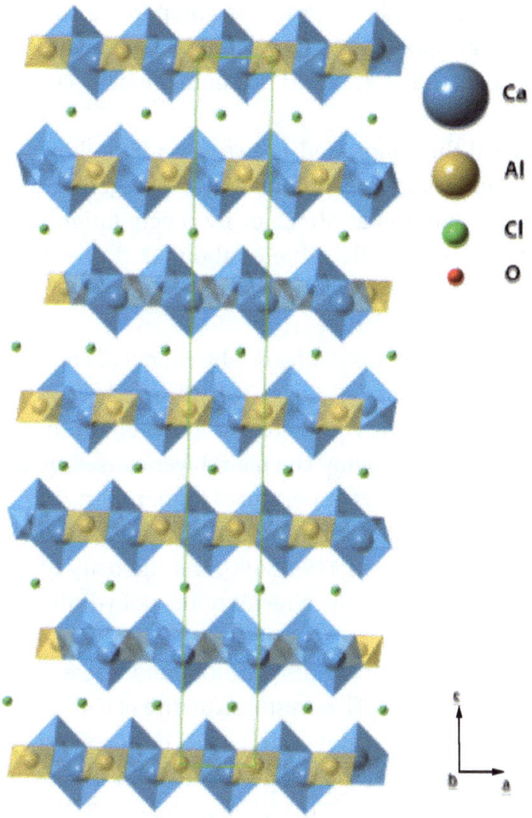

Fig. 7.7: Crystal structure of β-monochloride $C_3A{\cdot}CaCl_2{\cdot}10H_2O$.

monoclinically or hexagonally [110]. At 40 °C, the powder pattern contained only peaks of the rhombohedral polytype. The lattice parameters of the HT-monochloride $C_3A{\cdot}CaCl_2{\cdot}10H_2O$ are a = 5.7358(6) Å and c = 46.849(9) Å with space group R$\bar{3}$c [111]. Further experiments on the α–β phase transition over the temperature range from −121 °C to 109 °C confirmed the structural R-phase transition for Friedel's salt at about 34 °C [112].

7.3.9 $C_3F{\cdot}CaCl_2{\cdot}nH_2O$

Kuzel (1969) synthesized iron-monochloride $C_3F{\cdot}CaCl_2{\cdot}nH_2O$ at RT to 100 °C [79]. He indicated that the compound became unstable between 50–150 °C. $C_3F{\cdot}CaCl_2{\cdot}10H_2O$ (35 % r.h.) is trigonal with rhombohedral lattice a = 5.858(1) Å, c = 23.267(5) Å, and c' = 7.756 Å.

The dehydration process $C_3F \cdot CaCl_2 \cdot 10H_2O$ at 35 % r.h. is comparable to $C_3A \cdot CaCl_2 \cdot 10H_2O$ in the temperature range 25–150 °C. The four interlayer water of $C_3F \cdot CaCl_2 \cdot 10H_2O$ may not be equivalent, because even at 20 °C about 2 mol H_2O were dehydrated.

Götz-Neunhoeffer (1996) synthesized the compound in two ways [89]. Stoichiometric concentrations of CaO, $CaCl_2$ and freshly precipitated $Fe(OH)_3$ together with saturated $Ca(OH)_2$ solution reacted for 3 months at 40 °C, or by precipitation from $FeCl_3 \cdot nH_2O$ and $CaCl_2 \cdot 2H_2O$ with $1n$Na(OH) at a pH value of about 12. The evaluation of the XRD data showed comparable lattice parameters of $a = 5.861(1)$ Å, $c = 23.279(2)$ Å, $c' = 7.760$ Å, and a unit cell with R-lattice [113].

The chemical composition $C_3F \cdot CaCl_2 \cdot 10H_2O$ was confirmed by chemical and thermal analysis [79, 114]. The dehydration process of $C_3F \cdot CaCl_2 \cdot 10H_2O$ includes two dehydration reactions of 4 mol water, equal with the complete water content of the interlayer and a second endothermic reaction destroying the main layer caused by the removal of 6 mol water. The chlorine is released at temperatures higher than 700 °C. The dehydration process of the interlayer is coupled to the variation in layer thickness from 7.760 Å to 6.842 Å. In the temperature range of 115–234 °C, the intensity of the reflections decreases until the structure is destroyed as a result of the dehydration of the main layer.

In opposition to the Al-analogue, DTA-measurements did not yield a reversible monoclinic-trigonal phase transformation at 35 °C. The crystal structure of $C_3F \cdot CaCl_2 \cdot 10H_2O$ was determined by Rapin (2002) and Rousselot et al. (2002) [5, 10]. Both authors determined space group R$\bar{3}$ for the compound but refined different lattice parameters: $a = 5.9000(3)$ Å, $c = 23.740(6)$ Å, $V = 715.7(3)$ Å^3 [5] and $a = 5.873(1)$ Å, $c = 23.362(2)$ Å, $V = 697.85$ Å^3 [10]. The structure is composed of a sequence of $[Ca_2Fe(OH)_6]^+$ layers with interstitial $[Cl \cdot 2H_2O]^-$ layers. Further details and a general representation of the structure can be seen at [5, 10]. Iron-monochloride crystallizes as the HT-modification [40] with trigonal symmetry and space group R$\bar{3}$ at room temperature, whereas the HT-modification of monochloride has space group R$\bar{3}$c. The mutual arrangement of main layers in the AFm-structure is responsible for this condition [36, 64, 65]. The sites of hydroxide ions are described by the capital letters A, B, C in the main layer and the anion site (in this case of chloride) with lower case letter b. According to their nomenclature, 3R polytypes with space group R$\bar{3}$ possess the sequence AC = BC = BA = AC, where OH-groups form prisms and in the 6R polytype with space group R$\bar{3}$c the arrangement of hydroxide groups results in the formation of octahedra. Drits et al. (1993) proposed different OH-site arrangements with ...AC-AC-BA-BA-CB-CB-AC... or ...AC-AB-CB-CA-BA-BC-AC.... if ions with larger ionic radii like I^- ($r_{I^-} = 2.20$ Å) [115] or SO_4^{2-} are fixed in the interlayer where the 3R polytype is present. The fixation of intermediate ions like Br^- ($r_{Br^-} = 1.96$ Å) [115] and Cl^- ($r_{Cl^-} = 1.81$ Å) [115] exhibit both polytypes [10]. The presence of 3R-iron-monochloride at room temperatures in contrast to 2M-monochloride has been discussed by Renaudin et al. (2015) [116]. Their iron-monochloride sample contained certain concentrations of calcite. Further investigations by Raman spectroscopy yielded a fixation of CO_3^{2-} ions in the interlayer at site 3b as a substitution for Cl^- ions. With the fixation of carbonate ions in the interlayer, the transition temperature 2M ⇔ 3R (6R) decreases [117], which is responsible for the observation of the HT-modification. In order to explain the different phase transition behaviors low temperature XRD experiments in the range of −130 °C to 25 °C were executed [116].

Peak shifts and broadening effects indicated a variation of lattice parameters a and c, indicating a phase transition at approximately −15 °C, which has also been proofed by DSC-experiments (differential scanning calorimetry). In order to receive high quality data, high resolution synchrotron experiments must be initiated. It can be concluded that, in contrast to the monochloride HT polytype (6R, $R\bar{3}c$), an LT polytype (2M, C2/c) is linked and the LT polytype structure of iron-monochloride (possibly P) has not yet been solved [116]. The synthesis conditions, choice of halide ions, and the Al/Fe ratio of the main layer are responsible for the presence of those polytypes.

7.3.10 $C_3A \cdot Ca(NO_3)_2 \cdot nH_2O$

Mononitrate can result from exposure to nitrate-containing solutions on concrete [118]. According to XRD analysis, mononitrate single crystals with the chemical composition $C_3A \cdot Ca(NO_3)_2 \cdot 9.5H_2O$ crystallize in space group $P\bar{3}$or P3 as a one layer polytype (1P) with $a = 5.743(1)$ Å and $c = 8.623(3)$ Å [79]. Mononitrate is the only known AFm phase which is optical uniaxial positive [76]. At room temperature, 9–10 mol of water were determined for $C_3A \cdot Ca(NO_3)_2 \cdot nH_2O$ at 35 % r.h. Previous investigations concerning the water content yielded 9, 9.3, 9.5, and 10.4 mol of water [119–122]. The crystal structure of $C_3A \cdot Ca(NO_3)_2 \cdot 10H_2O$ was determined by Renaudin et al. (1999) [123]. The lattice of $C_3A \cdot Ca(NO_3)_2 \cdot 10H_2O$ is primitive with SG $P\bar{3}c1$ and lattice parameters $a = 5.7445(8)$ Å, $c = 17.235(5)$ Å. Typically, $[Ca_2Al(OH)_6]^+$ sheets are stacked subsequently, perpendicular [001] with the $[NO_3 \cdot 2H_2O]^-$ layer, in order to compensate positive charges of the main layer. The interlayer $[NO_3 \cdot 2H_2O]^-$ consists of different types of H_2O molecules (OW1 and OW2) and different nitrate groups (ON1, ON2, ON3). A dynamic disorder of NO_3^- anions and free H_2O molecules (OW2) are present in the interlayer because nitrate ions and OW2 can rotate freely around [001] [119]. In order to model the dynamic disorder in the interlayer, nitrogen atoms are shifted from a special position 12c to a general position. Nitrate ions with a generally flat constitution (O–N–O angle of 120°) are arranged perpendicular to the main layer. 50 of them are linked to Ca atoms, whereas the other 50 % Ca atoms are bonded with OW1 molecules.

The dehydration process of mononitrate is similar to the one of monochloride or AFm phases containing halogen ions as interlayer anion [123]. Exceeding room temperature, mononitrate starts to dehydrate. At 87 °C, c' decreases by about 0.6 Å to 8.02 Å, where lattice parameter a is constant. At approximately 100 °C, $C_3A \cdot Ca(NO_3)_2 \cdot$

$9.5H_2O$ has lost the entirety of the water in the interlayer. Its chemical formula must be written as $C_3A{\cdot}Ca(NO_3)_2{\cdot}6H_2O$, with a layer distance $c' = 7.91$ Å. At 200 °C, the dehydration of the main layer is initiated, where the remaining 6 mol H_2O of the main layer are released. Together with the decomposition of the nitrate anion, the intensities of 00l-reflections decrease and the phase becomes XRD amorphous. If the dehydration process does not decompose the main layer but just reduces the water content of the interlayer, $C_3A{\cdot}Ca(NO_3)_2{\cdot}6H_2O$ can rehydrate over a saturated $CaCl_2$ solution. The treatment of $C_3A{\cdot}Ca(NO_3)_2{\cdot}9.5H_2O$ in moist CO_2-free atmospheres generated by saturated salt solutions or certain chemicals has different effects than the temperature treatment by thermal analysis. If H_2SO_4 or P_2O_5 (<1% r.h.) are used as drying agents, it causes a new dehydration product with unit cell dimensions $a = 5.71(1)$ Å and $c' = 7.30$ Å to stabilize. The water content of $C_3A{\cdot}Ca(NO_3)_2{\cdot}6H_2O$ was determined by thermogravimetry (TG). XRD experiments at 90 °C and 1% r.h. yielded hkls of the 8.02 Å hydrate unit cell. The increase of the ambient temperature to 120 °C stabilized pure 8.02 Å-hydrate. Both phases, the "8.02 Å-hydrate" and the "7.91 Å-hydrate" are linked by a reversible phase transition and the two phases are described as α-$C_3A{\cdot}Ca(NO_3)_2{\cdot}6H_2O$ ($c' = 8.02$ Å) and β-$C_3A{\cdot}Ca(NO_3)_2{\cdot}6H_2O$ ($c' = 7.91$ Å).

Recent investigations on the thermal decomposition and studies of the dehydrations were conducted by Renaudin et al. (2000) [123]. They investigated the dehydration of $C_3A{\cdot}Ca(NO_3)_2{\cdot}10H_2O$ by micro Raman experiments and thermogravimetry and confirmed the previous results [79, 124, 125]. Furthermore, they extended the investigated temperature range until 1300 °C so that further weight losses of nitrate reduction, loss of nitrate, and dehydroxylation processes were detected by thermogravimetry.

For the structural study of $C_3A{\cdot}Ca(NO_3)_2{\cdot}8H_2O$, data sets of single crystals heated at 70 °C were applied. In opposition to $C_3A{\cdot}Ca(NO_3)_2{\cdot}10H_2O$, $C_3A{\cdot}Ca(NO_3)_2{\cdot}8H_2O$, a 6R polytype, crystallizes with trigonal symmetry, space group $R\bar{3}c$. The layer parameters are $a = 5.731(2)$ Å and $c = 48.32(1)$ Å.

The structural arrangements are similar to different AFm phase structures with the chemical composition $[Ca_2Al(OH)_6]^+$ comprising 6-fold coordinated AlO_6 octahedra and CaO_7 polyhedra. Due to the increase of temperature to 70 °C, "the main layers undergo a sideways translatory motion" [123], resulting in a shrinkage of the layer distance from 8.62 Å to 8.05 Å, which has also been recognized and interpreted by Kuzel (1969, 1970). Finally, the seventh coordination of Ca are oxygen atoms of nitrate groups. The remaining water molecules are just bonded by H–O interactions in the interlayer and can be regarded as unnecessary for the stability of the structure. Nitrate ions are also dynamically disordered, as pointed out for the room temperature structure [125].

7.3.11 $C_3F \cdot Ca(NO_3)_2 \cdot nH_2O$

Kuzel (1969) synthesized and investigated $C_3F \cdot Ca(NO_3)_2 \cdot nH_2O$ [79]. He distinguished between two different hydration levels, $C_3F \cdot Ca(NO_3)_2 \cdot 9.5H_2O$ with $a = 5.88(1)$ Å and $c = 8.70(2)$ Å, as well as $C_3F \cdot Ca(NO_3)_2 \cdot nH_2O$ ($n > 10$) with $a = 5.89(1)$ Å and $c = 10.33(2)$ Å. The phase was also synthesized, applying freshly precipitated ferrihydrate together with stoichiometric concentrations of CaO, $Ca(NO_3)_2 \cdot 4H_2O$, and CO_2-free lime-saturated H_2O [79, 89, 124]. The XRD experiments of the samples gave comparable results with the ones investigated at 100 % r.h.: $a = 5.887(2)$ Å and $c = 10.321(2)$ Å – at 35 % r.h.: 5.888(1) Å and $c = 8.636(1)$ Å [89, 126]. The dehydration process of $C_3F \cdot Ca(NO_3)_2 \cdot 10H_2O$ [89] starts with three partly overlapping weight losses at 60, 85, and 100 °C. The layer distance c' decreases from 8.636 Å to 8.00 Å. In analogy to mononitrate, β-$C_3F \cdot Ca(NO_3)_2 \cdot 6H_2O$ becomes stable at 85 °C and a further increase to 100 °C α-$C_3F \cdot Ca(NO_3)_2 \cdot 6H_2O$ is present with $c' = 7.33$ Å until 170 °C. At this temperature, α-$C_3F \cdot Ca(NO_3)_2 \cdot 6H_2O$ transforms into β-$C_3F \cdot Ca(NO_3)_2 \cdot 6H_2O$, indicated by the increase of c'. At 190 °C, the main layer dehydrates, the structure decomposes at 420 °C due to the decomposition of the nitrate group.

7.3.12 $C_3A \cdot CaHBO_3 \cdot 12H_2O$

Borate ions are important set retarders and cements are used to stabilize borate containing wastes [127–130]. $C_3A \cdot CaHBO_3 \cdot 12H_2O$ was synthesized by mixing stoichiometric concentrations (1 : 1 : 1) of CaO, C_3A, and pure boric acid in decarbonated water [131]. After a reaction time of 7 days at room temperature, the precipitate (platy hexagonal crystals) contained small impurities of katoite and minor concentrations of carbonate (Raman spectroscopy). The unit cell is trigonal with lattice parameters $a = 5.7764(1)$ Å, $c = 49.5499(9)$ Å, $c' = 8.26$ Å, and space group $R\bar{3}c$, indicating a layer sequence of a 6R polytype [131]. Beside the characteristic structural constitution of the main layer [131], trigonal borate anions and water molecules distributed statistically on site 18e with a combined occupancy of 0.88. Such a static disorder was proofed by micro Raman and agreed well with the findings of the Rietveld refinement.

7.3.13 $C_3A \cdot CaHBO_3 \cdot 11.5H_2O$

The mixture of CaO, Bayerite, B_2O_3, and decarbonated and deionized H_2O ($w/s = 10$) yielded single crystals with 0.2–0.3 mm diameters [132]. The single crystals investigated by polarization microscopy showed that six zones under crossed polarizers were present and two adjacent fields were in extinction at the same time. The single observation of such a homogeneous zone in the conoscopic beam path revealed that $C_3A \cdot CaHBO_3 \cdot 11.5H_2O$ was two axial positive (bisectrix cut).

If the two axial crystals were heated up to 100 °C, the crystals showed a homogeneous extinction under crossed polarizers and the optical character changed to an optic sign of uniaxial minerals at 60 °C. The investigations by XRD in this temperature range didn't indicate the change of the layer distance of $C_3A \cdot CaHBO_3 \cdot 11.5H_2O$. $C_3A \cdot CaHBO_3 \cdot 11.5H_2O$ crystallizes with monoclinic symmetry SG C2/c or Cc, as a 2M-polytype with a = 10.005(2) Å, b = 576.5(1) Å, c = 16.840(3) Å, and β = 101.30(1)° and is transformed to a 6R polytype with rhombohedral lattice, space group R3c or R$\bar{3}$c, and lattice parameters a = 5.77.6(1) Å and c = 49.511(7) Å, quite close to the results for $C_3A \cdot CaHBO_3 \cdot 12H_2O$ [131].

The dehydration process of $C_3A \cdot CaHBO_3 \cdot 11.5H_2O$ starts with the removal of $5.5H_2O$ in the temperature range between 20 °C and 160 °C [132]. DSC and TG experiments indicated 4 different dehydration reactions at 20, 60, 65, and 90 °C. Due to the loss of water molecules in the interlayer, its distance c' decreased from the initial value of 8.25 Å to 6.50 Å. Above 200 °C, the decomposition of the main layer started, which is equal to the destruction of the AFm phase.

7.3.14 $C_3F \cdot CaHBO_3 \cdot 12H_2O$

After a reaction time of 15 months, samples with the chemical composition of $4CaO \cdot Fe_2O_3 \cdot xH3BO_3 \cdot nH_2O$ with $0.5 \le x \le 2.3$, hexagonally platy shaped crystals of 3–8 µm were detected by SEM. XRD experiments and indexing of the reflections exhibited a trigonal cell of a 6R polytype. The refinement of their lattice parameters yielded values for the compound $C_3F \cdot CaHBO_3 \cdot 12\text{–}14H_2O$: a = 5.902(1) Å and c = 61.122(18) Å (100 % r.h.) and, as a result of drying in an atmosphere of 35 % r.h., the unit cell of $C_3F \cdot CaHBO_3 \cdot 10H_2O$ has values of a = 5.959(1) Å and c = 48.299(5) Å [89, 133]. The dehydration process of $C_3F \cdot CaHBO_3 \cdot 10H_2O$ starts at 100 °C, where 12.4 weight equal to 4 mol water were dehydrated. The new hydration level $C_3F \cdot CaHBO_3 \cdot 6H_2O$ has a layer distance of 6.9–6.7 Å. Further increases of the temperature let the interlayer shrink subsequently. The decomposition of the main layer is initiated at approximately 190 °C.

7.3.15 C_2AH_8

C_2AH_8 is an important hydration phase at early reaction times of CACs. In dependence on the temperature, calcium aluminate (CA) and H_2O reacts to CAH_{10} at temperatures below 27 °C. Below 20 °C, only CAH_{10} is formed [75]. The increase of temperature encourages the crystallization of C_2AH_8 and AH_3 in favor of CAH_{10}. CAH_{10} transforms to C_2AH_8 with the release of water. The stability of CAH_{10} and C_2AH_8 decreases with rising temperatures. Above 50–70 °C, the dominant hydrates are C_3AH_6 and AH_3. Different analytical techniques like thermal analysis, NMR IR, and XRD were performed in order to identify different C_2AH_x phases according to their different water contents and to refine lattice parameters and ultimately solve the crystal structures of different C_2AH_x phases.

C_2AH_x has a complex chemical composition

$$[Ca_2Al(OH)_6]^+[\tfrac{m}{n}X^{n-} \cdot (1-m)Al(OH)_4 \cdot mAl(OH)_3 \cdot yH_2O)],$$

indicating solid solutions in the system TCAH–C_2AH_x [134–136].

According to Taylor (1997) [137], the AFm phase C_2AH_8 or α-C_2AH_8 with containing aluminium species in the interlayer is metastable in contact with aqueous solution at 18 °C. C_2AH_8 or α-C_2AH_8 becomes unstable after reducing the relative humidity to 45 % with $T > 26$ °C [138]. $C_2AH_{7.5}$ was identified as the stable AFm phase under those conditions. $C_2AH_{7.5}$, described as β-C_2AH_8 [67, 138], can be stabilized even at 24 % r.h. [139]. Either further drying at 12 % r.h., applying P_2O_5 or anhydrous $CaCl_2$, or raising the temperature to 120 °C yields C_2AH_5. $C_2AH_{7.5}$ and C_2AH_5 can be rehydrated without problems to C_2AH_8 at 81 % r.h. At temperatures higher than 120 °C, the presence of C_2AH_4 was postulated [68], which cannot be rehydrated to C_2AH_8 at a relative humidity of 81 %.

The dehydration process of C_2AH_8 was investigated intensively, proposing different dehydration reactions and temperatures for that phase [140–147]. Recent investigations concerning the dehydration of C_2AH_8 were conducted by Ukrainczyk et al. (2007) [148]. DSC and TG-experiments have demonstrated that the dehydration process of C_2AH_8 is composed of 4 main reactions. At 110 °C, physisorbed water and 3 mol of interlayer water were separated. The layer distance c' was reduced from 10.7 to 8.7 Å. C_2AH_5 is the stable phase. At 175 °C, the so-called "crafting process" of $Al(OH)_4^-$ tetrahedra on the main layer $[Ca_2Al(OH)_6]^+$ with a simultaneous removal of hydroxide from $Al(OH)_4^-$ is initiated. The chemical composition of the AFm phase was determined as C_2AH_4. The layer distance c' decreased from 8.7 Å to 7.4 Å. Further dehydroxylation reactions occurred between 200 °C and 240 °C, which overlapped with the dehydroxylation of the main layer, resulting in the destruction of its crystal structure.

A complete description for the crystal structure of C_2AH_8 has not been published. Ukrainczyk et al. (2007) [148] combined the results of different papers [137, 139, 149, 150]. They summarized that C_2AH_8 are composed of a typical layered structure containing different main layers $[Ca_2Al(OH)_6]^+$ and inter layers $[Al(OH)_4 \cdot 3H_2O]^-$. C_2AH_8 lattice parameters of $a = 9.946$ Å, $b = 5.733$ Å, $c = 43.138$ Å, and $\beta = 97.96°$ were published by Richard et al. (1995) [139]. In opposition to the monoclinic cell C_2AH_8 with an hexagonal cell, $a = 5.75$ Å and $c' = 10.7$ Å was postulated [138]. Precise hexagonal lattice parameters of $a = 5.79$ Å and $c = 64.696$ Å were published [151]. Recent investigations on C_2AH_8 yielded lattice parameters for the unit cell of a higher hydration level $C_2AH_{8.2}$ [152, 153] of $a = 5.775(2)$Å, $c = 65.274(4)$ Å and $c' = 10.88$ Å – and for C_2AH_8 ($a = 5.784(2)$ Å, $c = 64.486(2)$ Å, $c' = 10.75$ Å) [152, 154].

Typical coordination conditions of Ca and Al and a standard atomic ratio of Ca : Al = 2 : 1 for AFm phases were determined, too. In the interlayer area, a further

50 % of the Al content are fixed as $Al(OH)_4^-$ ions. However, just every second Al site is occupied. Al voids in the interlayer are "coordinated" by 3 hydroxide ions and a H_2O molecule. For the coordination of Al in the interlayer, different results were determined and published by different NMR techniques, but the results suggesting an exclusive fourfold coordination of Al in C_2AH_8 were the strongest [139, 148, 150, 155].

$C_2AH_{7.5}$ was indexed originally with an hexagonal cell ($a = 5.7$ Å and $c = 10.4$ Å) [137], but a monoclinic cell with space group C2/c or Cc and with lattice parameters $a = 9.93$ Å, $b = 5.74$ Å, $c = 42.2$ Å, $\beta = 97°$ were refined [137, 139], which were confirmed by Raab (2010) [152, 156]. The stacking of the layers in the $C_2AH_{7.5}$ is different from that in the 8 H_2O hydrate. Scheller et al. (1974) investigated $C_2AH_{7.5}$ and demonstrated the principle structure properties, but did not publish crystal data (for instance the *xyz* coordinates of Ca, Al, or OH ions) [137].

The structure of C_2AH_5, the hydration state with a water free interlayer, was partially solved [137]. C_2AH_5 is trigonal with space group R$\bar{3}$m and lattice parameters $a = 5.75$ Å and $c = 52.2$ Å. The stacking arrangement of main and interlayer is equal to a 6R polytype of a hexagonal layered brucite structure. Six formula units $[Ca_2Al(OH)_6]^+$ $[Al(OH)_4]^-$ exist per unit cell. Typical bond lengths Al–O and Ca–O (O = oxygen atoms of hydroxide ions) and coordination spheres with Al and Ca were identified in the octahedral layer. In the interlayer, $Al(OH)_4$-tetrahedra are positioned parallel to the ternary axis and the four hydroxide ions of the tetrahedra are distributed statistically on six different positions [137, 157].

7.3.16 C_2ASH_8 (S = SiO_2) [4]

Strätlingite occurs in systems where portlandite and calcium sulfate are absent as e.g. in alkali activated slags or in Portland cements blended with high fractions of metakaolin, or gehlenite hydrate, nominally C_2ASH_8 [158]. Strätlingite is a mineral named after W. Strätling. Different occurrences for instance from the Bellerberg near Mayen (Germany) [32], in a marly inclusion of a phonolitic lava at the Campomorto quarry near Montalto di Castro (Italy) [159], Strätlingite associated with Vertumnite, C-S-H phases, and zeolites at Colle Fabbri (Central Italy) [160] and Emmelberg (Germany) were described earlier [161].

Strätlingite from the Bellerberg with the chemical composition $2CaO \cdot Al_2O_3 \cdot SiO_2 \cdot 8H_2O$ is a mineral with uniaxial negative optical character. The indexing of powder and single crystal data showed that strätlingite is trigonal with space group R3 or R$\bar{3}$. The unit cell parameters are $a = 5.737(5)$ Å and $c = 37.59(5)$ Å [32]. Further investigations were conducted on synthetic strätlingite single crystals (up to 200 µm), which were prepared by crystallization from hydrated gehlenite (C_2AS) glasses [162]. XRD experiments showed that the phase is trigonal with space group R3 or R$\bar{3}$. The unit cell parameters are $a = 5.747(1)$ Å and $c = 37.64(1)$ Å. Because of $Z = 3$, three double layers with the composition $[Ca_2Al(OH)_6]^+$ $[AlSi_3(OH)_{12} \cdot 4H_2O]^-$ exist per unit cell.

Different authors have mentioned a monoclinic cell [163, 164]. Dehydration reactions were determined by applying TG and DTA techniques (differential thermal analysis). Three different endothermic reactions at 131 °C, 181 °C, and 226 °C were detected in the temperature range 25–300 °C. The weight loss at 131 °C represents 2 mol water which is equal to the chemical composition of C_2ASH_4. XRD patters in the temperature range of 24–131 °C were quite similar to those of the initial compound at room temperature but the refinement of the lattice parameters a and c showed a certain decrease of the unit cell volume. At 135 °C the lattice parameters dropped significantly and the powder pattern became diffuse. In opposition to the reversible dehydration-rehydration process in the temperature range of 24–131 °C concerning the water content in the interlayer, the pronounced change of lattice parameters at 135 °C is not reversible. In the range of 140–210 °C, the dehydration process proceeded to stabilize the phase C_2ASH_2. At this temperature, the powder patterns indicated an almost amorphous phase at 990 °C. A sharp exothermic signal indicated the crystallization of gehlenite and the remaining hydroxide ions in the amorphous sample were dehydrated until 1000 °C. Early crystal structure concepts of strätlingite assumed that the tetrahedral coordination polyhedra are arranged in six rings with the apical OH-group in the direction of a Ca atom in order to appear as the seventh coordination atom. Further water molecules fill the cavities of the sheet cantered at the threefold axis. These theses are in accordance with the small decrease of c' during heating [162].

The crystal structure of the mineral strätlingite from Montalto di Castro (VT), Italy was solved by Rinaldi et al. (1990) (Fig. 7.8) [166]. The positively charged brucite-type layers consisting of Al octahedra and Ca are in a seven fold coordination, the interlayer is composed of double tetrahedral sheets and described as "Zweier-Doppelschicht" [165]. Al atoms can also be substituted by Fe [167]. Electron micro probe analysis of specimens of different occurrences at Mayen and Montaldo di Castro yielded comparable formulas with $(Ca_{1.94}Sr_{0.03}Ba_{0.02}Na_{0.01})_2Al(Al_{1.02}Si_{1.03})O_{1.85}(OH)_{10.85} \cdot 2.25H_2O$ and $(Ca_{1.90}Sr_{0.04}Ba_{0.03}Na_{0.02}K_{0.01})_2Al(Al_{0.93}Si_{1.24})O_{2.13}(OH)_{10.44} \cdot 2.25H_2O$. The reduced chemical formula is $Ca_2Al(Al,Si)O_2(OH)_{10} \cdot 2.25H_2O$, according to crystal structure refinements. The structure analysis of a single crystal from Montaldo di Castro supported the trigonal symmetry with space group R$\bar{3}$m [162] and lattice parameters a = 5.753(6) Å; c = 37.82(5) Å. Three formula units with $[Ca_2Al(OH)_6]^+$ $[(Al,Si)O_2(OH)_4 \cdot 2.25H_2O]^-$ were detected per unit cell, indicating a 3R polytype [64]. The interlayer of strätlingite is a complex arrangement of two different corner shared tetrahedra, where their central positions T1 and T2 are occupied by Si and Al atoms. The coordination spheres of T1 and T2 contains three O atoms (Oh1) of different hydroxide ions perpendicular (001) with bond distances T–Oh1 = 1.64(4) Å. The fourth bond T1–O1 or T2–O1 (1.81(2) and 1.73(2) Å, respectively) connects two adjacent tetrahedra T1 and T2 generating the tetrahedra double sheet $[(T,\square)_4(OH,O)_8 \cdot 2.25H_2O]^-$. In the structure of strätlingite, a statistical distribution of Si and Al on T1 and T2 positions in the double layer exists, resulting in a double layer symmetry of $\bar{3}$m. The occupancy of both T1 and T2 side with Si and Al together is just 55 %, site O1 is completely occupied

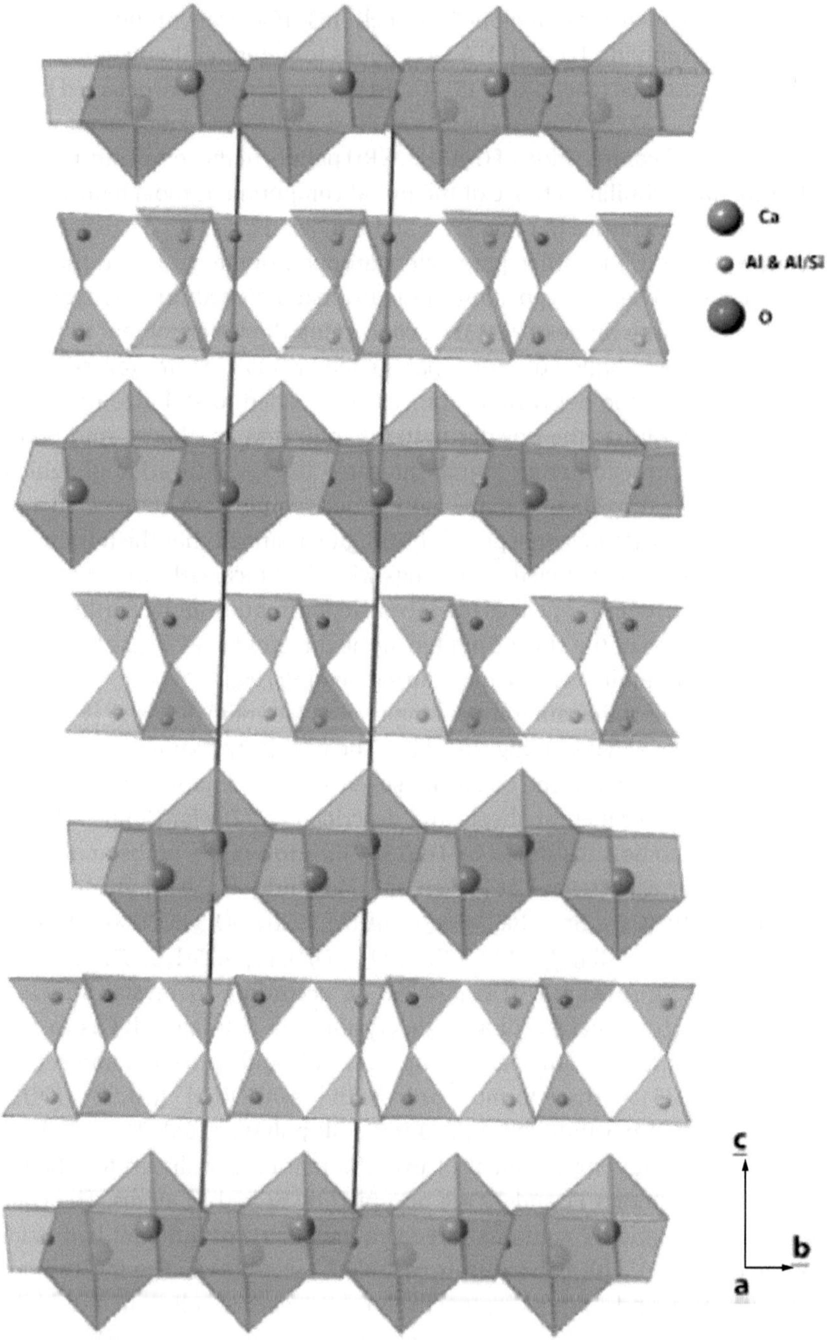

Fig. 7.8: Crystal structure of stratlingite $Ca_2Al(Al,Si)O_2(OH)_{10} \cdot 2.25H_2O$.

and site Oh1 just of 50 %. Another water molecule (Ow1) is fixed at the centre of the six ring in the double layer. Furthermore, an unusual T–O–T angle of 180° was detected, which might be supported by the high concentration of 45 % vacancies in the double tetrahedra layer. NMR experiments confirm most of the findings, but reveal that another arrangement of Al with hydroxide and water molecules is present in the structure [168], which is responsible for an alteration of the six ring network to form an arrangement of different silicate ring structures which occasionally bond to each other [168]. Recent investigations concerning the crystal structure of strätlingite were performed by NMR techniques, thermal analysis, and synchrotron X-ray powder diffraction [169]. The sample material was synthesized strätlingite. The refined structure model exhibits the key features reported by Rinaldi et al. (1990) [166] and supports the results of NMR experiments by Kwan et al. (1995) [168].

The crystal structure of vertumnite is closely related to that of strätlingite [166, 173]. Vertumnite was discovered at the Campomorto quarry, near Montalto di Castro [166]. This AFm phase is not a typical phase present in cement pastes, but should be mentioned because its chemical composition and crystal structure is closely related to strätlingite. Vertumnite, with the idealized chemical formula $Ca_4AI_4Si_4O_6(OH)_{24} \cdot 3H_2O$, was also discovered in a phonolitic rock at Campomorto Montaldo di Castro [171]. In opposition to the crystal structure of strätlingite, vertumnite is monoclinic with $a = 5.744(5)$ Å, $b = 5.766(5)$ Å, $c = 25.12(1)$ Å, $\beta = 119.72(5)°$, and space group $P2_1/m$. Instead of a 3R polytype stacking, there are just two formula units $[Ca_2AI(OH)_6]^+$ $[AlSi_2O_3(OH)_6 \cdot 1.5H_2O]^-$ per unit cell, which is equal to a 2M polytype [171]. The interlayer is composed similarly to that of strätlingite, and also consists of a double tetrahedra layer with different occupation densities of T1 and T2 and hydroxide contents than strätlingite.

7.4 The fixation of alkali ions in AFm phases

Different LDHs with a complex interlayer chemistry containing alkali ions were identified and investigated. Important minerals of that species are shigaite $[Mn_6Al_3(OH)_{18}]$ $[(SO_4)_2Na \cdot 12H_2O]$[172] (Fig. 7.9), motukoreite [173], mountkeithite $[(Mg,Ni)_9(Fe^{3+},Cr,AI)_3(OH)_2]^{3+}$ $[(CO_3,SO_4)_{1.5}(Mg,Ni)_2(SO_4)_2(H_2O)_{11}]^{3-}$ [174], or wermlandite $[Mg_7(Al_{0.57}Fe^{3+}{}_{0.43})(OH)_{18}]^{2+}$ $[(Ca_{0.6},Mg_{0.4})(SO_4)_2(H_2O)_{12}]^{2-}$ [175]. Different alkali substituted shigaite type phases (Li^+, Na^+, K^+, Rb^+, $NH_4{}^+$) have also been synthesized and investigated recently [176, 177]. The ability to incorporate alkali ions into the crystal structure of AFm phases was not reported frequently. Structure determination and refinements were just performed for alkali free phases. One of the earlies descriptions of alkali containing AFm phase was published in 1963 [178]. This phase is of technical interest for cement-based systems containing high amounts of Na_2SO_4 [179–181], for steam-cured concretes containing densified silica fume, for various alkali levels [182], and for the thermal stability of ettringite in alkaline solutions levels [183].

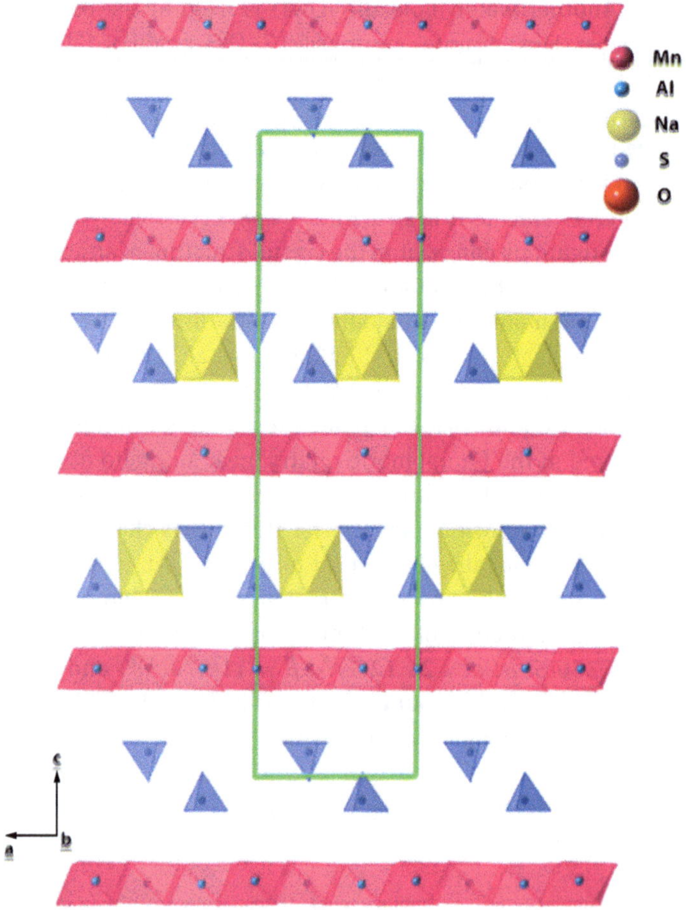

Fig. 7.9: Crystal structure of shigaite $[Mn_6Al_3(OH)_{18}][(SO_4)_2Na \cdot 12H_2O]$.

The so-called "U phase" must be considered as a unique AFm phase which contains sodium ions fixed in the crystal lattice [178] and differs from monosulfate not only in the alkali concentration but also in its higher SO_3/Al_2O_3 ratio of 1.25 (monosulfate SO_3/Al_2O_3 ratio of 1). A potential composition of

$$\{Ca_4Al_2(OH)_{12}\}^{2+}\{(SO_4)yNa_2SO_4 \cdot aq\}^{2-}$$

or, in oxidic notation [178],

$$4CaO \cdot (1 - x)Al_2O_3 \cdot (1 + y)SO_3 \cdot (3 + y)Na_2O$$

was formulated. For the final chemical composition of the U phase, $4CaO \cdot 0.9Al_2O_3 \cdot 1.1SO_3 \cdot 0.5Na_2O \cdot aq$ was postulated [178]. Just like most of the AFm phases, different hydration levels exist, depending on the relative humidity. In the case of the U phase, three different hydration levels of $16\,H_2O$ (81 % r.h.), $12\,H_2O$ (22 % r.h.), and $8\,H_2O$ (0.01 % r.h.) were detected. Furthermore, lattice parameters of $a = 5.74–5.76$ Å and $c(81\,\%\text{ r.h.}) = 10.0$ Å, $c(22\,\%\text{ r.h.}) = 9.3$ Å, and $c(0.01\,\%\text{ r.h.}) = 8.12$ Å were determined. Precise lattice parameters were published for the phases $C_3A{\cdot}CaSO_4{\cdot}0.5Na_2SO_4{\cdot}15H_2O$ ($a = 5.745$ Å, $c = 30.070$ Å, $P^*\bar{3}$ (147)) [184] and $C_3A{\cdot}CaSO_4{\cdot}0.5K_2SO_4{\cdot}15H_2O$ ($a = 5.755$ Å, $c = 30.330$ Å, $P^*\bar{3}$ (147)) [185]. Due to the technical interest and a lack of data, the different methods of the pure U phase synthesis were investigated, but the authors did not publish any lattice parameters [186]. Recently, the U phase was investigated by ^{27}Al and ^{23}Na NMR spectroscopy in order to demonstrate the ettringite conversion to the U phase in NaOH solution and to identify the area where sodium ions are chemically fixed in the crystal structure of the U phase [183].

7.5 Binary systems and intermediate AFm phases

7.5.1 The system $C_3A{\cdot}CaSO_4{\cdot}nH_2O–C_3A{\cdot}Ca(OH)_2{\cdot}nH_2O$

First investigations of the system $CaO–Al_2O_3–CaSO_4–H_2O$ were performed by Jones (1944) [187], D'Ans & Eick (1953) [188], and Dosch (1962) [77]. They postulated that monosulfate is stable in the presence of C_4AH_X (TCAH), however they doubted whether solid solutions exist between the end members monosulfate and TCAH. Later, hemisulfate was described by Keller (1971) [96]. The formation of solid solutions between monosulfate and TCAH was shown by XRD [40, 43, 68, 172, 188]. In 1968, Seligmann & Greening mentioned the existence of $C_4A{\cdot}\frac{3}{5}SO_3{\cdot}H_2O$ [189]. Pöllmann (1984) investigated the system $C_3A{\cdot}CaSO_4{\cdot}nH_2O–C_3A{\cdot}Ca(OH)_2{\cdot}nH_2O$ [76, 190]. He synthesized different solid solutions with the formula

$$C_3A \cdot (1-x) \cdot CaSO_4 \cdot xCa(OH)_2 \cdot nH_2O, \quad 0 \le x \le 1.$$

After a reaction time of 16 weeks, solid solutions with

$$C_3A \cdot (1-x) \cdot CaSO_4 \cdot xCa(OH)_2 \cdot nH_2O, \quad 0 \le x \le 0.5,$$

were identified by XRD at 100 % r.h. Lattice parameters are almost constant (5.759–5.757 Å). Lattice parameters c decreased in the system from 26.763 Å to 26.383 Å and the layer distances c' became smaller from 8.921 Å to 8.794 Å as a result of a steady substitution of SO_4^{2-} for $2OH^-$. At 0.48 mole-% $SO_4^{2-}/Al(OH)_3$, a miscibility gap was detected in the system $C_3A{\cdot}CaSO_4{\cdot}nH_2O–C_3A{\cdot}CaSO_4{\cdot}nH_2O$. A further decrease in the sulfate concentration stabilized TCAH in the system. TCAH and $C_3A{\cdot}\frac{1}{2}CaSO_4{\cdot}\frac{1}{2}Ca(OH)_2{\cdot}nH_2O$ coexist in samples with $SO_4^{2-}/Al(OH)_3 = 0.5–0.9$ mole-%. At 35 % r.h., C_4AH_{19} dehydrated to C_4AH_{13} with a constant layer distance of 7.94 Å and $C_3A{\cdot}\frac{1}{2}CaSO_4{\cdot}\frac{1}{2}Ca(OH)_2{\cdot}nH_2O$ dehydrated to $C_3A{\cdot}\frac{1}{2}CaSO_4{\cdot}\frac{1}{2}Ca(OH)_2{\cdot}12H_2O$ with a final $c' = 8.76$ Å. At 35 % r.h., the phase equilibria in the system do not change. Due to the reduction of the relative humidity to 35 % r.h., the AFm phases lose water in the interlayers. The layer distances and lattice parameters c of the solid solutions

$$C_3A \cdot (1-x)CaSO_4 \cdot xCa(OH)_2 \cdot nH_2O, \quad 0 \le x \le 0.5,$$

decrease to 8.93–8.75 Å and 26.812–26.250 Å, respectively. Lattice parameters a stay in the system in the range of 5.760–5.755 Å.

The increase of the synthesis temperature to 45 °C yielded again pure monosulfate at $SO_4^{2-}/Al(OH)_3 = 1\,mol$ – and solid solutions inside the range $SO_4^{2-}/Al(OH)_3 = 0.9–0.66\,mol$ – at 100 % r.h. [76, 190]. Neither hemihydrate nor TCAH are stable at 45 °C. TCAH decomposed to CH and C_3AH_6, which are the two phases which coexist with $C_3A \cdot 0.66CaSO_4 \cdot 0.34Ca(OH)_2 \cdot nH_2O$ in the miscibility gap. The same phase relations exist at the lower r.h. of 35 %.

At 60 °C, monosulfate and a reduced solid solution stability field were detected by XRD in the range of

$$C_3A \cdot (1-x)CaSO_4 \cdot xCa(OH)_2 \cdot nH_2O, \quad 0.1 \le x < 0.17,$$

at 100 % r.h. At $x = 0.17$, $C_3A \cdot 0.83 \cdot CaSO_4 \cdot 0.17Ca(OH)_2 \cdot nH_2O$ coexists again with hydrogrossular and CH. The composition of the solid solution series has the composition

$$C_3A \cdot (1-x)CaSO_4 \cdot xCa(OH)_2 \cdot 12H_2O, \quad 0.1 \le x < 0.17.$$

At 80 °C, no solid solutions $C_3A \cdot (1-x)CaSO_4 \cdot xCa(OH)_2 \cdot nH_2O$ were stable. Just monosulfate and C_3AH_6 coexisted in the system inside the range $0 \le x \le 1$. Monosulfate was stable up to 280 °C, thereafter it decomposed into $C\bar{S}$, $C_4A_3H_3$, and a sulfate containing C_3AH_6 [190].

7.5.1.1 $C_3A \cdot \frac{1}{2}CaSO_4 \cdot \frac{1}{2}Ca(OH)_2 \cdot nH_2O$

$C_3A \cdot \frac{1}{2}CaSO_4 \cdot \frac{1}{2}Ca(OH)_2 \cdot nH_2O$ (hemisulfate) is the solid solution with the highest OH concentrations in the system $C_3A \cdot CaSO_4 \cdot nH_2O–C_3A \cdot Ca(OH)_2 \cdot nH_2O$. Hemisulfate can be mixed in terms of solid solutions with monosulfate without a miscibility gap, however hemisulfate is immiscible with TCAH. The crystal structure has not been solved yet. It must be considered that hydroxide and sulfate ions are statistically distributed in the interlayer area, therefore hemisulfate must be considered as a solid solution and not as a distinct phase [190]. Superstructure reflections were not detected by XRD [76, 190].

For the trigonal phase, hemisulfate with a R-lattice and space group R*, and lattice parameters $a = 5.755$ Å and $c = 26.249$ Å were refined at 35 % r.h. [76, 190, 191]. The dehydration of hemisulfate is quite complex. In the temperature range 50–1000 °C, four different hydration levels at 50, 80, 110, and 200 °C were detected by TG/DTG and DTA. Between 25 °C and 220 °C, the water content in $C_3A \cdot \frac{1}{2}CaSO_4 \cdot \frac{1}{2}Ca(OH)_2 \cdot nH_2O$ is reduced from 12 to 6 mol H_2O. With the complete dehydration of the interlayer at 200 °C, the dehydration process of the main layer was initiated and proceeded with two endothermic reactions at 260 °C and 480 °C. The refinement of distinct layer distances was only possible for $C_3A\cdot\frac{1}{2}CaSO_4\cdot\frac{1}{2}Ca(OH)_2\cdot 10.5H_2O$ and $C_3A\cdot\frac{1}{2}CaSO_4\cdot\frac{1}{2}Ca(OH)_2\cdot 10H_2O$ with 8.07 Å and 7.92 Å. Because the powder diffraction patterns of hydration levels at lower H_2O concentrations became X-ray amorphous, no further data could be provided for their layer distances.

7.5.2 The system $C_3F\cdot CaSO_4\cdot nH_2O$–$C_3F\cdot Ca(OH)_2\cdot nH_2O$

According to Ecker (1998) [100], the end member $C_3F\cdot Ca(OH)_2\cdot nH_2O$ (TCFH) is not stable under the chosen conditions. The synthesis of solid solutions with the chemical composition

$$4CaO + 2Fe(OH)_3 + (x)Na_2SO_4, \quad 0 \le x \le 1,$$

failed. After a reaction time of 8 weeks, the samples were composed of iron monosulfate, CH, and Fe-hydroxide. The synthesis of the iron analogue $C_3F\cdot\frac{1}{2}CaSO_4\cdot\frac{1}{2}Ca(OH)_2\cdot nH_2O$ did not succeed. With the use of XRD, Ecker (1989) showed that the sample with the targeted composition $C_3F\cdot\frac{1}{2}CaSO_4\cdot\frac{1}{2}Ca(OH)_2\cdot nH_2O$ contained iron monosulfate, CH, and Fe-hydroxide [100].

7.5.3 The system $C_3A\cdot CaSO_4\cdot nH_2O$–$C_3F\cdot CaSO_4\cdot nH_2O$

Early investigations regarding the phase assemblages in the system yielded a broad miscibility gap and an AFm phase with the composition $C_3A_{0.7}F_{0.3}\cdot CaSO_4\cdot nH_2O$ at 25 °C. Only the synthesis at 100 °C resulted in the formation of homogeneous precipitates throughout the solid solution series of $C_3A\cdot CaSO_4\cdot nH_2O$–$C_3F\cdot CaSO_4\cdot nH_2O$ [79, 114]. Recent investigations applying different raw materials for the synthesis

$$4CaO + (2x)Al(OH)_3 + (2-2x)Fe(OH)_3 + Na_2SO_4, \quad 0 \le x \le 1,$$

resulted in single phased precipitates even at 25 °C [90, 100].

7.5.4 The system $C_3A\cdot CaSO_4\cdot nH_2O$–$C_3A\cdot CaCO_3\cdot nH_2O$

Samples with the chemical compositions

$$3CaO + Al(OH)_3 + (1-x)CaSO_4\cdot 2H_2O + xCaCO_3 + nH_2O, \quad 0 \le x \le 1,$$

reacted for three months in the system $C_3A\cdot CaSO_4\cdot nH_2O$–$C_3A\cdot CaCO_3\cdot nH_2O$. XRD analysis of the precipitates yielded the end members $C_3A\cdot CaSO_4\cdot nH_2O$, $C_3A\cdot CaCO_3\cdot nH_2O$, and the intermediate phase hemicarbonate $C_3A\cdot\frac{1}{2}CaCO_3\cdot\frac{1}{2}Ca(OH)_2\cdot nH_2O$ at

100 % r.h. After drying the samples to 35 % r.h., the phase assemblages didn't change, however the layer distances c' of the AFm phases changed. The formation of solid solutions was not detected at 25 °C [76].

7.5.5 The system $C_3F\cdot CaSO_4\cdot nH_2O$–$C_3F\cdot CaCO_3\cdot nH_2O$

Different samples with the composition

$$4CaO + 2Fe(OH)_3 + (1-x)Na_2SO_4 + (1-x)Na_2CO_3 + nH_2O, \quad 0 \le x \le 1,$$

were prepared in the system $C_3F\cdot CaSO_4\cdot nH_2O$–$C_3F\cdot CaCO_3\cdot nH_2O$ [100]. After a reaction time of eight weeks, XRD analysis of the samples yielded phase assemblages of iron monosulfate and iron monocarbonate in particular concentrations according to the weighed in $CaSO_4/CaSO_3$ ratio. At 25 °C, $C_3F\cdot CaSO_4\cdot nH_2O$ and $C_3F\cdot CaCO_3\cdot nH_2O$ were immiscible.

7.5.6 The system $C_3A\cdot CaSO_4\cdot nH_2O$–$C_3A\cdot CaCl_2\cdot nH_2O$

The investigations of precipitates in the system showed that no solid solutions crystallized. In addition to the end members of the system, an intermediate phase with the composition $3CaO\cdot Al_2O_3\cdot \frac{1}{2}CaSO_4\cdot \frac{1}{2}CaCl_2\cdot nH_2O$ was detected in the range $0.2 \le x \le 0.8$ of $C_3A\cdot xCaSO_4\cdot nH_2O(1-x)CaCl_2\cdot nH_2O$. Kuzel (1969) determined 12 mol water for $3CaO\cdot Al_2O_3\cdot \frac{1}{2}CaSO_4\cdot \frac{1}{2}CaCl_2\cdot 12H_2O$ dried over a saturated $CaCl_2$ solution [79]. He managed to synthesize single crystals at 150 °C. The crystal lattice of the intermediate phase is rhombohedral with space group R3c or R$\bar{3}$c with $Z = 6$. The lattice parameters are $a = 5.750(1)$ Å and $c = 100.62(2)$ Å. Furthermore, he pointed out that sulfate and chloride ions are ordered in mono-ionic interlayers perpendicularly [0001].

The system $C_3A\cdot CaSO_4\cdot nH_2O$–$C_3A\cdot CaCl_2\cdot nH_2O$ was reinvestigated partly in order to determine the crystal structure of $3CaO\cdot Al_2O_3\cdot \frac{1}{2}CaSO_4\cdot \frac{1}{2}CaCl_2\cdot 12H_2O$ and to determine the exchange properties of sulfate by chloride [192]. Different samples with interlayer compositions $[Cl^-_{3/4}(SO_4^{2-})_{1/8}]$, $[Cl^-_{1/2}(SO_4^{2-})_{1/4}]$, and $[Cl^-_{1/4}(SO_4^{2-})]$ were each synthesized at different temperatures of 25 °C and 85 °C. At 25 °C, the samples contained higher concentrations of impurities, i.e. monosulfate, Kuzel's salt, alpha and beta polymorphs of monochloride, and ettringite.

7.5.6.1 $C_3A\cdot \frac{1}{2}CaSO_4\cdot \frac{1}{2}CaCl_2\cdot nH_2O$

The structure of Kuzel's salt was determined by applying synchrotron radiation on a pure sample with the targeted chemical formula $[Ca_2Al(OH)_6]^+ [Cl^-_{1/2}(SO_4)_{1/4}\cdot nH_2O]^-$ [192] (Fig. 7.10). Kuzel's salt with the chemical composition $C_3A\cdot \frac{1}{2}CaSO_4\cdot \frac{1}{2}CaCl_2\cdot$

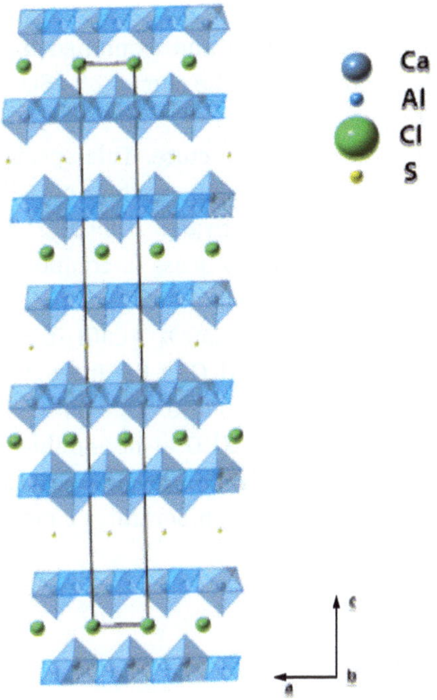

Fig. 7.10: Crystal structure of Kuzel's salt $C_3A \cdot \frac{1}{2}CaSO_4 \cdot \frac{1}{2}CaCl_2 \cdot 11H_2O$.

$11H_2O$ crystallized with trigonal symmetry. The space group is $R\bar{3}$ with lattice parameters $a = 5.7508(2)$ Å and $c = 50.4185(29)$ Å. Three formula units $[Ca_2Al(OH)_6]$ $[\frac{1}{4}SO_4 \cdot \frac{1}{2}Cl \cdot 2.5H_2O]^-$ contain the unit cell. Ca^{2+} and Al^{3+} ions are typically seven and six fold coordinated by oxygens belonging to hydroxide ions and a water molecule (just Ca^{2+}) of the main layer and the interlayer. In the interlayer, chloride and sulfate ions are distributed in different layers, which leads to the unique formation of a super structure of alternating layers $[Cl \cdot 2H_2O]^-$ and $[(SO_4)_{0.5} \cdot 3H_2O]^-$ in the family of AFm phases. The lattice parameter c is approximately twice as large as the stacking of an interlayer of the monosulfate plus the interlayer of monochloride. Low temperature XRD experiments until 100 K provided no evidence of a possible phase transition.

7.5.7 The system $C_3F \cdot CaSO_4 \cdot nH_2O$–$C_3F \cdot CaCl_2 \cdot nH_2O$

After a reaction time of 15 months, the precipitates of different samples with variable $CaCl_2/CaSO_4$ ratios were composed of $C_3F \cdot CaSO_4 \cdot 16H_2O$ [79, 114] (samples with 60–100 mol $CaCl_2$ weight), $C_3F \cdot CaCl_2 \cdot 10H_2O$ (samples with 0–40 mol $CaCl_2$ weight) and the intermediate phase $C_3F \cdot \frac{1}{2}CaSO_4 \cdot \frac{1}{2}CaCl_2 \cdot 10H_2O$ at 100 % r.h. (samples with 20–80 mol $CaCl_2$ weight) [89]. The reduction of the relative humidity to 35 % resulted in the reduction of the interlayer of $C_3F \cdot CaSO_4 \cdot 16H_2O$ by dehydration of 4 mol H_2O. The water content of $C_3F \cdot \frac{1}{2}CaSO_4 \cdot \frac{1}{2}CaCl_2 \cdot 10H_2O$ and $C_3F \cdot CaCl_2 \cdot 10H_2O$ remained stable.

7.5.7.1 $C_3F \cdot \frac{1}{2}CaSO_4 \cdot \frac{1}{2}CaCl_2 \cdot nH_2O$

Kuzel (1969) [79] and Götz-Neunhöffer (1996) [89] synthesized and examined this phase. According to the Al analogue, the lattice parameter c is the sum of $c'[\frac{1}{2}SO_4 \cdot nH_2O]$ and $c'[Cl \cdot nH_2O]$ and therefore it can be concluded that Cl^- and SO_4^{2-} ions are distributed orderly in distinct interlayers in the crystal structure. This special feature has been confirmed for the Al analogue [192]. The crystal structure of $C_3F \cdot \frac{1}{2}CaSO_4 \cdot \frac{1}{2}CaCl_2 \cdot 10H_2O$ is based on a rhombohedral lattice with $a = 5.8823(2)$ Å, $c = 50.143(4)$ Å, and $c' = 16.714$ Å [193]. The dehydration process is composed of two endothermic reactions. At 70 °C two moles H_2O were dehydrated, stabilizing the phase $C_3F \cdot \frac{1}{2}CaSO_4 \cdot \frac{1}{2}CaCl_2 \cdot 8H_2O$. The detection of $C_3F \cdot \frac{1}{2}CaSO_4 \cdot \frac{1}{2}CaCl_2 \cdot 6H_2O$ at 120 °C was not possible, because the remaining two moles H_2O of the interlayer were removed together with the 6 mol H_2O belonging to the main layer. One different hydration stage, $C_3F \cdot \frac{1}{2}CaSO_4 \cdot \frac{1}{2}CaCl_2 \cdot 8H_2O$, was detected by XRD at non-ambient temperatures with a layer distance of 7.860 Å. Additional hydration levels were not detected because the phase became XRD amorphous at higher temperatures caused by the start of the dehydration of the main layer.

7.5.8 The system $C_3A \cdot CaSO_4 \cdot nH_2O$–$C_3A \cdot Ca(NO_3)_2 \cdot nH_2O$

Göske (1998) [194] investigated the system preparing samples with the chemical composition

$$2CaO + CaO \cdot Al_2O_3 + xCaSO_4 \cdot 2H_2O + (1-x)Ca(NO_3)_2 \cdot 4H_2O + nH_2O, \quad 0 \le x \le 1.$$

After a reaction time of 3 months, XRD analysis of those samples indicated that the reaction process was not completed and the results of XRD analyses which were conducted on samples with identical chemical compositions and stored at 40 °C showed comparable results. Pure $C_3A \cdot 0.83 \cdot Ca(NO_3)_2 \cdot 0.17Ca(OH)_2 \cdot 16H_2O$ and monosulfate $C_3A \cdot CaSO_4 \cdot 12H_2O$ exist in the range $0.1 \le x \le 0.9$. Solid solutions with layer distances $c' = 0.828$–0.8401 Å and with an unknown total chemistry were detected in the range $0.1 \le x \le 0.9$ by XRD. Furthermore, ettringite with the chemical composition

$$C_3A \cdot (1-x)3CaSO_4 \cdot (x)3Ca(NO_3)_2 \cdot 30\text{–}36H_2O, \quad 0.33 \le x < 0,$$

was present in the samples $0.2 \le x \le 0.9$. According to Wenda (1984), even small amounts of SO_4^{2-} ions in the samples support the crystallization of sulfate-rich ettringite phases in the system $C_3A \cdot 3CaSO_4 \cdot 30H_2O$–$C_3A \cdot 3Ca(OH)_2 \cdot 33H_2O$ [132]. Pöllmann (1984) examined the system $C_3A \cdot 3Ca(NO_3)_2 \cdot nH_2O$–$C_3A \cdot 3CaSO_4 \cdot nH_2O$ and described ettringite solid solutions with the following phase chemistry [76]

$$[Ca_6[Al_2(OH)_{12} \cdot 24H_2O]^{6+} \cdot [(SO_4)_{(3-x)} \cdot (NO_3)_{(2x)} \cdot nH_2O]^{6-}.$$

7.5.9 The system $C_3A \cdot CaSO_4 \cdot nH_2O$–$C_3A \cdot CaHBO_3 \cdot nH_2O$

For the investigation of the system $C_3A \cdot CaHBO_3 \cdot nH_2O$–$C_3A \cdot CaSO_4 \cdot nH_2O$ [132], stoichiometric concentrations of CaO, bayerite, gypsum, and boric acid were mixed and homogenized with H_2O at a w/s ratio of 10 according to the general chemical formula

$$4CaO \cdot Al_2O_3 \cdot \frac{(1-x)}{2} H_3BO_3 \cdot xSO_3 \cdot nH_2O, \quad 0 \le x \le 1.$$

At 20 °C, the phase composition of the investigated precipitates in the range $0 < x < 0.83$ were composed of $C_3A \cdot CaHBO_3 \cdot 11.5H_2O$ and different hydration levels of monosulfate (16-, 14-, 12-hydrate). Samples with $x \ge 0.83$ contained $C_3A \cdot (1-x) CaHBO_3 \cdot xCaSO_4 \cdot nH_2O$ and finally monosulfate. The established phase assemblages at 35 % r.h. confirmed the insolubility of monosulfate and monoborate 11.5-hydrate in the range $0 < x < 0.83$. In the range $0.83 \le x \le 0.95$, only $C_3A \cdot (1-x)CaHBO_3 \cdot xCaSO_4 \cdot 12H_2O$ was present and at $x = 1$ monosulfate 12-hydrate. Increasing the temperature to 40 °C during the synthesis gave comparable results.

7.5.10 The system $C_3F \cdot CaSO_4 \cdot nH_2O$–$C_3F \cdot CaSO_3 \cdot nH_2O$

The variable fixation of SO_3^{2-} and SO_4^{2-} ions at different ratios were investigated preparing samples with

$$4CaO + 2Fe(OH)_3 + (1-x)Na_2SO_3 + xNa_2SO_4, \quad 0 \le x \le 1.$$

In contrast to the findings for the complete solid solution series in the system $C_3A \cdot CaSO_3 \cdot nH_2O$–$C_3A \cdot CaSO_4 \cdot nH_2O$ [93], the Fe analogue of the system is divided in an area of solid solutions

$$C_3A \cdot (1-x)CaSO_3 \cdot xCaSO_4 \cdot nH_2O, \quad 0 < x < 0.33,$$

$C_3A \cdot \frac{2}{3}CaSO_3 \cdot \frac{1}{3}CaSO_4 \cdot nH_2O$ and the coexistence of the phases $C_3A \cdot \frac{2}{3}CaSO_3 \cdot \frac{1}{3}CaSO_4 \cdot nH_2O$ and $C_3A \cdot CaSO_4 \cdot nH_2O$.

7.5.11 The system $C_3A \cdot Ca(OH)_2 \cdot nH_2O$–$C_3A \cdot CaCO_3 \cdot nH_2O$

Comprehensive studies of this system have already been undertaken [69, 97]. They didn't observe solid solutions between TCAH and monocarbonate but postulated the intermediate phase hemicarbonate. At 45 °C, monocarbonate and hemicarbonate are the stable phases in the binary system. Hemicarbonate is only present in samples with less than 0.35 mol CO_3/Al_2O_3. In the range of $0.35 < x \le 0.75$, CO_3/Al_2O_3 monocarbonate and hemicarbonate are present and at higher CO_3/Al_2O_3 ratios just monocarbonate was detected as a single AFm phase. All samples also contained impurities of C_3AH_6.

The temperature rise to 60 °C destabilized hemicarbonate, which means that the stable AFm phase in the investigated concentration range $0 \leq x \leq 1$ is hemicarbonate with additional contents of CH and C_3AH_6.

7.5.11.1 $C_3A \cdot \frac{1}{2}CaCO_3 \cdot \frac{1}{2}Ca(OH)_2 \cdot nH_2O$

Different investigations on hemicarbonate $C_3A \cdot \frac{1}{2}CaCO_3 \cdot \frac{1}{2}Ca(OH)_2 \cdot 11.5H_2O$ have previously been conducted in order to characterize the AFm phase [68, 69, 119, 195]. Hemicarbonate is stable up to 60 °C and transforms at this temperature to monocarbonate C_3AH_6 and CH. The crystal system of hemicarbonate is trigonal with a space group R3c or R$\bar{3}$c and has a unit cell with $a = 5.77$ Å and $c = 49.16$ Å [69, 196]. The dehydration process was investigated by Pöllmann [76]. DSC and TG experiments showed that $C_3A \cdot \frac{1}{2}CaCO_3 \cdot \frac{1}{2}Ca(OH)_2 \cdot 11.5H_2O$ dehydrates with different endothermic reactions, whereby weight losses of 3.1 %, 6.2 %, and 17.9 % at 40 °C, 105 °C, and 125 °C were determined. In total, 6 mol H_2O were removed from the interlayer with a consecutive destabilization of the hydration stages and shrinkage of the layer distance c'.

- $[Ca_4Al_2(OH)_{12}]^{2+}$ $[\frac{1}{2}CO_3 \cdot OH \cdot 5.5H_2O]^{2-}$ with $c' = 8.20$ Å,
- $[Ca_4Al_2(OH)_{12}]^{2+}$ $[\frac{1}{2}CO_3 \cdot OH \cdot 4.75H_2O]^{2-}$ with $c' = 7.70$ Å,
- $[Ca_4Al_2(OH)_{12}]^{2+}$ $[\frac{1}{2}CO_3 \cdot OH \cdot 4.0H_2O]^{2-}$ with $c' = 7.20$ Å, and
- $[Ca_4Al_2(OH)_{12}]^{2+}$ $[\frac{1}{2}CO_3 \cdot OH \cdot 0H_2O]^{2-}$ with $c' = 6.60$ Å.

At higher temperatures, the structure decomposed as a result of the dehydration of the main layer and further decarbonation processes.

Runcevski et al. (2012) [197] investigated a powder sample of $C_3A \cdot \frac{1}{2}CaCO_3 \cdot \frac{1}{2}Ca(OH)_2 \cdot 11.5H_2O$ and found two different AFm phases $C_3A \cdot \frac{1}{2}CaCO_3 \cdot \frac{1}{2}Ca(OH)_2 \cdot 11.5H_2O$ (hc) and a "carbonated" hc-phase (Fig. 6.9). For both phases, the crystal structures were solved [197]. Hemicarbonate (hc) is trigonal with space group R$\bar{3}$c and lattice parameters $a = 5.7534(1)$ Å and $c = 46.389(1)$ Å. The lattice distance is equal to 1/6 of unit cell parameter c, which means that the 6R polytype is composed of positively charged main layers, $[Ca_4Al_2(OH)_{12}]^{2+}$, and negatively charged interlayers, $[OH_{2n}(CO_3)_{1-n} \cdot 4H_2O]^2$, stacked perpendicularly [001]. In the main layer, characteristic bond lengths and coordination states for Al [6] and Ca [7] in AFm phases were detected. Carbonate and hydroxide anions and water molecules are bonded to calcium atoms of the main layers. 0.25 carbonate and 0.5 hydroxide anions are essential for the charge compensation of the positive main layer. The distribution of carbonate ions in the interlayer is ordered perpendicularly [001] but the hydroxide ions, sited on general positions, are disordered around the position of the carbonate anion. As the authors mentioned, the remaining electron densities in difference Fourier maps after the refinements of hydroxide ions indicate the presence of strongly disordered water molecules in the interlayer.

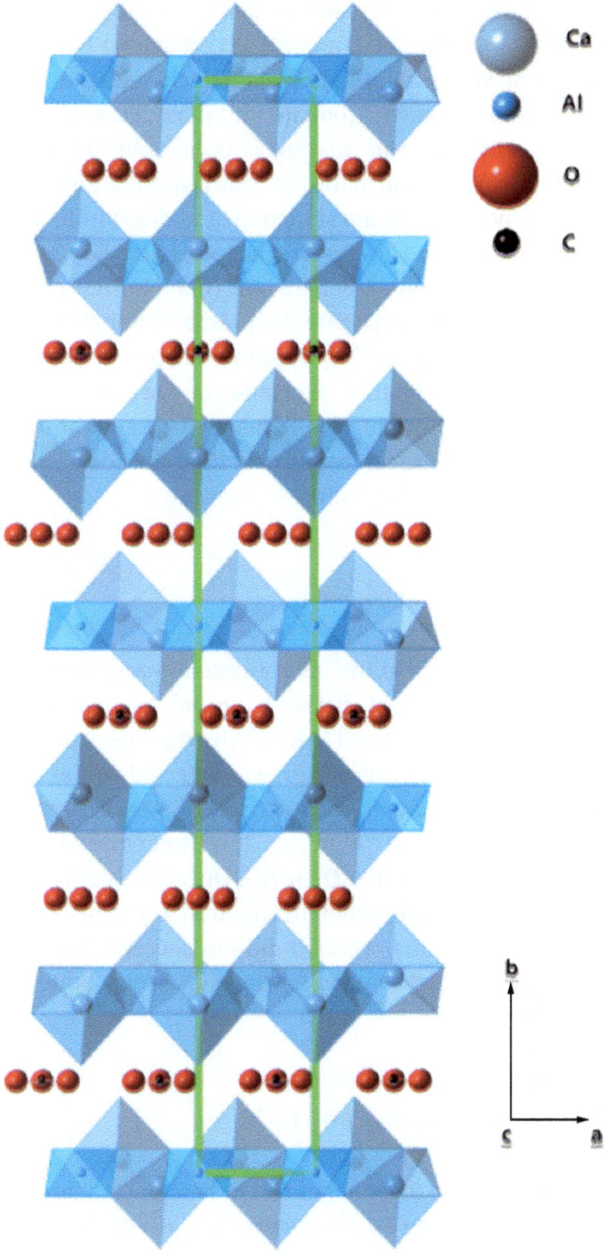

Fig. 7.11: Crystal structure of "carbonated" hemicarbonate $C_3A \cdot \frac{1}{2}CaCO_3 \cdot \frac{1}{2}Ca(OH)_2 \cdot 11.5H_2O$

Those loosely bonded space filling water molecules are removed during the dehydration process at low temperatures [69, 76, 97]. The structure of carbonated hemicarbonate (chc) is quite similar to the hc structure but show a distinctive feature. The refinement of occupancies of the carbon atoms (carbonate ion) provided a value of 0.4 and therefore the hydroxide concentration was adapted in order to establish the charge balance in the crystal structure. In addition to the increased carbonate ion concentration in the interlayer of chc, carbonate ions are shifted up and down from their initial position in the interlayer compared to the hc crystal structure. Micro Raman experiments indicated the presence of non-bound carbonate ions (band at 1086 cm^{-1}), in opposition to the crystal structure of monocarbonate [116]. An additional Raman band (1068 cm^{-1}) indicated the presence of bonded CO_3^{2-} ions to a Ca atom in the main layer, but the reason for this result is the carbonation of crystals on the surface. This assumption was proofed by the authors performing the Raman experiment at open atmosphere under the influence of CO_2. While the intensity of at 1086 cm^{-1} decreased, the intensity of the band at 1068 cm^{-1} increased at higher temperatures.

7.5.12 The system $C_3F \cdot Ca(OH)_2 \cdot nH_2O–C_3F \cdot CaCO_3 \cdot nH_2O$

Early investigations on the phase assemblages in the binary system were performed at 5 and 45 °C [104]. The formation of solid solutions between the end members of the system were not detected, but $C_3F \cdot Ca(OH)_2 \cdot nH_2O$ and $C_3F \cdot CaCO_3 \cdot nH_2O$ were present in the system according to the weight in OH/CO_3 ratio. TCFH was detected only at 5 °C. The system was reinvestigated in 1998 [100]. After a reaction time of 8 weeks, samples with the compositions $4CaO + 2Fe(OH)_3 + (x)Na_2CO_3$, $0 \le x \le 1$, were composed of CH and $Fe(OH)_3$ and the only stable AFm phase $C_3F \cdot CaCO_3 \cdot nH_2O$. Only the sample with a Fe_2O_3/Na_2CO_3 ratio of 1 reacted completely to iron monocarbonate. The synthesis of pure iron hemicarbonate $C_3F \cdot \frac{1}{2}CaCO_3 \cdot \frac{1}{2}Ca(OH)_2 \cdot 11.5H_2O$ did not succeed, however water contents and lattice parameters were determined for the phase equilibrated at 35 % r.h. [106, 198, 199].

7.5.13 The system $C_3A \cdot Ca(OH)_2 \cdot nH_2O–C_3A \cdot CaCl_2 \cdot nH_2O$

The corrosion of reinforcing steel by chloride ions makes this system of technical interest. The formation of solid solutions between Friedel's salt and TCAH occurs when chloride-containing solutions react with the calcium aluminate hydrates in concrete [200]. First investigations into the system were conducted in 1955 [201]. Further authors who analyzed the phase assemblages in the system observed a miscibility gap [68, 96, 111, 202]. For the synthesis of OH–Cl–AFm phases with the targeted composition $C_3A \cdot xCaCl_2 \cdot (1 - x)Ca(OH)_2 \cdot nH_2O$ with $0 \le x \le 1$, different reaction mixtures consisting of C_3A, CaO, and $CaCl_2 \cdot 2H_2O$ and deionized CO_2-free water reacted for six months at 25 °C on a shaker. In the range $0.85 \leq x \leq 1$, monoclinic solid solutions with an α-monochloride structure crystallized. Close to $x = 0.83$, a discontinuity and a crystalline phase of composition $C_3A \cdot \frac{5}{6}CaCl_2 \cdot \frac{1}{6}Ca(OH)_2 \cdot 10.3H_2O$ with trigonal symmetry and a layer thickness of $c' = 784$ pm was identified. The symmetry can be transformed from trigonal to monoclinic by reducing the temperature to 10 ± 5 °C. Successive substitution of chloride by hydroxide yielded again monoclinic phases. The miscibility gap is present at about $x = 0.3$, where TCAH and $C_3A \cdot \frac{2}{6}CaCl_2 \cdot \frac{4}{6}Ca(OH)_2 \cdot 11.4H_2O$ are in equilibrium. The layer distances c' vary at 100 % r.h. between 7.89 and 10.66 Å. After drying the precipitates to 35 % r.h., C_4AH_{19} dehydrates to C_4AH_{13}, the layer distances of the solid solutions exhibiting values in the range of 7.89–7.94 Å. Detailed values were published [202]. A special feature is the monotropic phase transition applicable for all $C_3A \cdot xCaCl_2 \cdot (1-x)Ca(OH)_2 \cdot nH_2O$ with $0 \leq x \leq 1$ from trigonal to monoclinic symmetry caused by grinding. A complete solid solution series between monochloride and hemicarbonate has been proposed. A complete replacement of $[2Cl \cdot 4H_2O]^{2-}$ by $[\frac{1}{2}CO_3 \cdot OH\ 5.5\ H_2O]^{2-}$ can be observed in the interlayer of those solid solutions [203].

7.5.14 The system $C_3A \cdot Ca(OH)_2 \cdot nH_2O$–$C_3A \cdot Ca(NO)_3 \cdot nH_2O$

The influence of nitrate ions on the stability of concrete was investigated and first experiments in the system $C_3A \cdot Ca(OH)_2 \cdot nH_2O$–$C_3A \cdot Ca(NO)_3 \cdot nH_2O$ were conducted in 1971, and solid solutions $[Ca_2Al(OH)_6]^+\ [0.65NO_3 \cdot 0.35OH \cdot 5H_2O]^-$ with a maximal nitrate concentration were reported [96, 118]. In order to obtain more information about the phase stabilities in the system, samples with the chemical compositions

$$4CaO + 2Al(OH)_3 + (x)Ca(NO_3)_2 + (n)H_2O, \quad 0 \leq x \leq 1,$$

were prepared [76]. After a reaction time of 6 months at 25 °C, $[Ca_2Al(OH)_6]^+\ [0.66NO_3 \cdot 0.34OH \cdot nH_2O]^-$ crystallized in the range $0 < x \leq 0.5$. At approximately $x = 0.55$, TCAH occurred as a second phase. The layer distance of $[Ca_2Al(OH)_6]^+\ [0.66NO_3 \cdot 0.34OH \cdot nH_2O]^-$ is stable in the complete concentration range $0 \leq x \leq 1$. After drying the samples to 35 % r.h., $[Ca_2Al(OH)_6]^+\ [0.83NO_3 \cdot 0.17OH \cdot 1.85H_2O]^-$ and $[Ca_2Al(OH)_6]^+\ [0.65NO_3 \cdot 0.35OH \cdot 5H_2O]^-$ were detected as stable phases by XRD. In the range $0.6 \leq x < 1$, a miscibility gap with the phase assemblage TCAH, $[Ca_2Al(OH)_6]^+\ [0.83NO_3 \cdot 0.17OH \cdot 1.85H_2O]^-$, and $[Ca_2Al(OH)_6]^+\ [0.65NO_3 \cdot 0.35OH \cdot 5H_2O]^-$ exists. The pure synthesis of $[Ca_2Al(OH)_6]^+\ [0.65NO_3 \cdot 0.35OH \cdot 5H_2O]^-$ is possible at $x \approx 0.4$ in the binary system.

At 45 °C, $[Ca_2Al(OH)_6]^+\ [0.66NO_3 \cdot 0.34OH \cdot nH_2O]^-$ is the stable phase at 100 % r.h. The crystallization of TCAH did not occur because it was transformed into hydrogrossular and CH. At 35 % r.h., both AFm phases $[Ca_2Al(OH)_6]^+\ [0.83NO_3 \cdot 0.17OH \cdot nH_2O]^-$ and $[Ca_2Al(OH)_6]^+\ [0.65NO_3 \cdot 0.35OH \cdot nH_2O]^-$ were detected again by XRD analysis, however TCAH transformed to C_3AH_6 and CH. A constant temperature during synthesis at 60 °C forced the formation of solid solutions between $[Ca_2Al(OH)_6]^+$

$[0.83NO_3 \cdot 0.17OH \cdot nH_2O]^-$ and $[Ca_2Al(OH)_6]^+$ $[0.65NO_3 \cdot 0.35OH \cdot nH_2O]^-$. Their layer distances decreased with lower nitrate concentrations from 8.86 Å to 8.65 Å. At $x \approx 0.6$, the miscibility gap exited further on, where in addition to the solid solutions, C_3AH_6 and CH crystallized.

7.5.14.1 $C_3A \cdot 0.83Ca(NO_3)_2 \cdot 0.17Ca(OH)_2 \cdot nH_2O$

In order to synthesize mononitrate [194], the precipitate was composed completely of $C_3A \cdot 0.83Ca(NO_3)_2 \cdot 0.17Ca(OH)_2 \cdot 16H_2O$ after a reaction time of 3 months. Chemical analysis performed by ICP/OES ion-chromatography resulted in the conclusion that a maximal hydroxide/nitrate ratio of $\frac{1}{6} : \frac{5}{6}$ can be achieved. Pöllmann (1984) investigated the system $C_3A \cdot Ca(NO_3)_2 \cdot nH_2O$–$C_3A \cdot Ca(OH)_2 \cdot nH_2O$, where $C_3A \cdot 0.83Ca(NO_3)_2 \cdot 0.17Ca(OH)_2 \cdot nH_2O$ crystallized as an intermediate phase [76].

The compound is trigonal with space group P3 or P$\bar{3}$. The lattice parameters of its unit cell were refined to $a = 5.744(3)$ Å and $c = 31.1991(3)$ Å at 100 % r.h. The storage of this material at 35 % r.h. for 3 weeks caused a decrease of interlayer water, leading to the phase $C_3A \cdot 0.83Ca(NO_3)_2 \cdot 0.17Ca(OH)_2 \cdot 16H_2O$ with cell dimensions $a = 5.738(3)$ Å and $c = 31.195(6)$ nm. The space group is again P3 or P$\bar{3}$. $C_3A \cdot 0.83Ca(NO_3)_2 \cdot 0.17Ca(OH)_2 \cdot 16H_2O$ shows a quite complex dehydration process. The storage at different relative humidities, applying saturated salt solutions $Na_2CO_3 \cdot 10H_2O$ (92 % r.h.), KCl (86 % r.h.), NaCl (75 % r.h.), and $CaCl_2 \cdot 6H_2O$ (35 % r.h.) exhibit the loss of 6 mol of interlayer water. Pöllmann (1984) already investigated the dehydration process together with the variation of the layer distance c' [76]. He observed three different endothermic reactions at 50 °C, 80 °C, and 125 °C. Further increases of temperature destroyed the main layer because of the elimination of 6 mol H_2O. Göske (1998) showed that the interlayer of the compound dehydrated completely in the range of 25–110 °C [194]. Due to four subsequent hydration reactions, the phase already lost 1 mol H_2O at room temperature, 2 mol at 35 and 70 °C, and finally 2 mol H_2O at 90 °C. Non-ambient XRD experiments yielded different layer distances c':
9–10 mol H_2O: 8.6(2) Å,
9 mol H_2O: 8.59(2) Å,
8 mol H_2O: 8.56(2) Å,
7 mol H_2O: 8.04(2) Å, and
6 mol H_2O: 7.98(2) Å.

The destruction of the main layer was initiated at 230 °C, which was overlapped by the decomposition of the nitrate ions to NO_x.

7.5.15 The system $C_3A\cdot Ca(OH)_2\cdot nH_2O$–$C_3A\cdot CaHBO_3\cdot nH_2O$

Different samples with the composition $4CaO\cdot Al_2O_3\cdot xB_2O_3\cdot nH_2O$, $0 \le x \le 1$, were investigated by XRD after a reaction time of three months [89, 132]. In order to determine the influence of the temperature during synthesis, different samples were stored at 20 °C, 40 °C, and 60 °C. At 20 °C, the system $C_3A\cdot CaHBO_3\cdot nH_2O$–$C_3A\cdot Ca(OH)_2\cdot nH_2O$ is composed of TCAH (C_4AH_{19}) with $c' = 10.66$ Å [132]. Due to the increase of the B_2O_3 concentration, the quantity of TCAH decreased and the quantity of the coexisting AFm phase $C_3A\cdot CaHBO_3\cdot 11.5H_2O$ increased in the range $0 < x < 1$. Pure $C_3A\cdot CaHBO_3\cdot 11.5H_2O$ was just detected in the sample with a ratio $HBO_3^{2-}/Al_2O_3 = 1$ ($x = 0.5$). Higher B_2O_3 concentrations caused the crystallization of $C_3A\cdot 2Ca[B(OH)_4]_2\cdot Ca(OH)_2\cdot 30H_2O$.

At 40 °C, TCAH was converted to C_3AH_6 + CH with {111} and {100} [132]. Hydrogen borate ions cannot be fixed chemically in the C_3AH_6 structure. Compared to the system at 20 °C, the stability fields of $C_3A\cdot CaHBO_3\cdot 11.5H_2O$ and $C_3A\cdot 2Ca[B(OH)_4]_2\cdot Ca(OH)_2\cdot 30H_2O$ did not change. At 45 °C, $C_3A\cdot 2Ca(B(OH)_4)_2\cdot Ca(OH)_2\cdot 30H_2O$ decomposed and the system at 50 °C just contained $C_3A\cdot CaHBO_3\cdot 11.5H_2O$ as an AFm phase, C_3AH_6, CH, and additional reflections of unidentified calcium borate phases [132].

7.5.16 The system $C_3F\cdot CaHBO_3\cdot nH_2O$–"$C_3F\cdot Ca(OH)_2\cdot nH_2O$"

In order to investigate the phase assemblages in the system, samples with the compositions $4CaO\cdot Fe_2O_3\cdot xH_3BO_3\cdot nH_2O$, $0.5 \le x \le 2.3$, were prepared [89]. As previously reported, $C_3F\cdot Ca(OH)_2\cdot nH_2O$ could not be synthesized, although a reaction time of 15 months was applied. It is possible to synthesize two different borate AFm phases $C_3F\cdot \frac{1}{2}CaHBO_3\cdot \frac{1}{2}Ca(OH)_2\cdot nH_2O$ and $C_3F\cdot CaHBO_3\cdot nH_2O$. The system can be divided in four different ranges.
$0.5 \le x \le 1.1$: $C_3F\cdot \frac{1}{2}CaHBO_3\cdot \frac{1}{2}Ca(OH)_2\cdot nH_2O$ + CH + α-FeOOH.
$x = 1.3$: $C_3F\cdot \frac{1}{2}CaHBO_3\cdot \frac{1}{2}Ca(OH)_2\cdot nH_2O$.
$1.5 \le x \le 1.9$: $C_3F\cdot \frac{1}{2}CaHBO_3\cdot \frac{1}{2}Ca(OH)_2\cdot nH_2O$ and $C_3F\cdot CaHBO_3\cdot nH_2O$.
$x \ge 1.9$: $C_3F\cdot CaHBO_3\cdot nH_2O$.

As a result of the drying process at 35 % r.h., the phase assemblages did not change but the water content of the AFm phases was reduced and therefore $C_3F\cdot CaHBO_3\cdot nH_2O$ was present as two different hydration levels with 10 and 12 H_2O. Wenda (1984) identified the borate ettringite $C_3A\cdot 2Ca[B(OH)_4]_2\cdot Ca(OH)_2\cdot 30H_2O$ in samples with higher hydrogen borate concentrations [132]. In the $C_3F\cdot CaHBO_3\cdot nH_2O$–"$C_3F\cdot Ca(OH)_2\cdot nH_2O$", no iron hydrogen borate ettringite was detected [89].

7.5.16.1 $C_3F\cdot\frac{1}{2}Ca(OH)_2\cdot\frac{1}{2}CaHBO_3\cdot nH_2O$

Iron hemiborate was synthesized byapplying CaO, boric acid, and freshly prepared ferrihydrite [89]. Iron hemiborate crystallizes in the trigonal crystal system with a bravais lattice R and lattice parameters $a = 5.913(1)$ Å and $c = 48.733(5)$ Å [204]. Independent of the relative humidity, only one hydration level with the chemical composition $C_3F\cdot\frac{1}{2}Ca(OH)_2\cdot\frac{1}{2}CaHBO_3\cdot 10H_2O$ exists. Its dehydration process exhibits 3 different processes at 30 °C, 90 °C, and 140 °C, with the removal of 6 mol H_2O in order to receive a water free interlayer, which consists only of OH^- and HBO_3^{2-} ions. Dependant on the dehydration, the layer distance decreases from 8.12 Å to a final value of 6.98–6.65 Å. At 175 °C, the phase becomes XRD amorphous as a result of the dehydration of the main layer.

7.5.17 The system $C_3A\cdot Ca(OH)_2\cdot nH_2O$–$C_3A\cdot CaSO_3\cdot nH_2O$

Samples with the chemical compositions $C_3A\cdot(1-x)CaSO_3\cdot(x)Ca(OH)_2\cdot 11H_2O$, $0 \le x \le 1$, were investigated by XRD and thermal analysis [132]. In the range $0 \le x < 0.66$, the precipitates were composed of solid solutions with $C_3A\cdot(1-x)CaSO_3\cdot(x)Ca(OH)_2\cdot 11H_2O$. With the decrease of the sulphite concentration in the system a solid solution $C_3A\cdot 0.34CaSO_3\cdot 0.66Ca(OH)_2\cdot 11H_2O$ coexisted with C_4AH_{13}.

7.5.18 The system $C_3F\cdot Ca(OH)_2\cdot nH_2O$–$C_3F\cdot CaSO_3\cdot nH_2O$

Different samples with the chemical compositions $4CaO + 2Fe(OH)_3 + xNa_2SO_3$, $0 \le x \le 1$, were prepared. Because of the instability of $C_3F\cdot Ca(OH)_2\cdot nH_2O$, the precipitates contained the AFm phase $C_3F\cdot CaSO_3\cdot 12H_2O$ (35 % r.h.) together with $Fe(OH)_3$ and $Ca(OH)_2$ [100] in the complete system.

7.5.19 The system $C_3A\cdot CaCO_3\cdot nH_2O$–$C_3A\cdot CaCl_2\cdot nH_2O$

The system $C_3A\cdot CaCO_3\cdot nH_2O$–$C_3A\cdot CaCl_2\cdot nH_2O$ was investigated by Pöllmann (1980) [202]. Samples with the chemical compositions $3CaO\cdot 2Al(OH)_3\cdot xCaCO_3\cdot(1-x)CaCl_2\cdot nH_2O$, $0 < x < 1$ and $w/s = 50$, were prepared. After a reaction time of 3 months, the precipitates were composed of solid solutions ($0 \le x < 0.9$) and a solid solution plus small impurities of monocarbonate in sample $x = 0.9$. At 100 %, the unit cell of solid solutions is monoclinic with lattice parameters $a = 9.94$–9.95 Å, $b = 5.74$ Å, $c = 16.2$–16.3 Å, and $\beta = 103$–$104°$. The layer distance c' indicated an increase until $x = 0.8$ and a steep continuous decrease of c until $x = 0.1$. After the drying process, the AFm phases in the range $0.1 \le x \le 0.9$ lost their monoclinic symmetry and became hexagonal.

7.5.19.1 $C_3A\cdot\frac{1}{2}CaCO_3\cdot\frac{1}{2}CaCl_2\cdot nH_2O$

7.5.19.1.1 Hydrocalumite

The type pf hydrocalumite that occurs at Scawt Hill crystallizes with monoclinic symmetry, SG $P2_1$ (a = 9.6 Å, b = 11.4 Å, c = 16.84 Å, β = 111°) together with afwillite, ettringite, and portlandite [48].

Hydrocalumite, $[Ca_8Al_4(OH)_{24}((CO_3)_{0.78}Cl_{1.30}(OH)_{1.14}]\cdot 11.02H_2O$ occurs in phonolitic rocks at Montaldo di Castro, Viterbo (Italy) and is associated with afillite, tobermorite, and katoite [159]. The mineral has a monoclinic symmetry with space group P2/c and lattice parameters a = 10.04(6)7 Å, b = 11.523(6) Å, c = 16.271(9) Å, β = 104.32(5)°. At the Bellerberg near Ettringen (Germany) [111], hydrocalumites with different compositions and optical characters $C_3A\cdot 0.95CaCl_2\cdot 0.05Ca(OH)_2\cdot 10H_2O$ and $C_3A\cdot 0.8CaCl_2\cdot 0.2Ca(OH)_2\cdot 10H_2O$ were detected. Monoclinic $C_3A\cdot 0.95CaCl_2\cdot 0.05Ca(OH)_2\cdot 10H_2O$ is isotypic to Friedel's salt and transforms at 35 °C like synthetic α-$C_3A\cdot CaCl_2\cdot 10H_2O$ to β-hydrocalumite [28]. The trigonal uniaxial hydrocalumite shows strong structural similarities with β-$C_3A\cdot CaCl_2\cdot 10H_2O$ (SG R$\bar{3}$c or R3c).

The hydrocalumite from Boissejour Puy de Dome (France) $C_3A\cdot 0.5CaCl_2\cdot 0.37Ca(OH)_2\cdot 0.13CaCO_3\cdot 11H_2O$ with minor impurities of MgO and alkali is monoclinic SG Pc or P2/c with a = 10.017(4) Å, b = 11.512(4) Å, c = 16.286(3) Å, and β = 104.20(2)° [111].

For the crystal structure analysis of hydrocalumite [205] from Montaldo di Castro, a small crystal with $0.08\times 0.14\times 0.21\,mm^3$ was used to determine space group P2/c and lattice parameters a = 10.04(6)7 Å, b = 11.523(6) Å, c = 16.271(9) Å, and β = 104.32(5)° by applying Weissenberg photographs. Because the reflections with $k \neq 2n$ were weak, data was first collected in the C2/c subcell with $\frac{1}{2}b$. In addition to the well-ordered $[Ca_2Al(OH)_6]^+$ layer, the interlayer consists of two differently ordered sheets containing just Cl^- or CO_3^{2-} ions with a Cl^-/CO_3^{2-} ratio of 2 : 1. The stability of the interlayer is guaranteed by oxygen hydrogen bonds between triangular carbonate ions and water molecules, which are also connected to the upper and lower main layer via Ca atoms. In fact, hydrocalumite is a ternary solid solution containing chloride, hydroxide, and carbonate anions.

7.5.19.1.2 $C_3A\cdot\frac{1}{2}CaCO_3\cdot\frac{1}{2}CaCl_2\cdot 10.5H_2O$

The crystal structure of the intermediate phase $[Ca_2Al(OH)_6]^+$ $[Cl_{\approx 0.5}(CO_3)_{\approx 0.25}\cdot{\approx}2.25\,H_2O]^-$ in the binary system monochloride–monocarbonate has been solved [116]. Its trigonal unit cell (a = 5.7400(4) Å, c = 46.7402(4) Å) contains six formula units and the structure has space group R$\bar{3}$c. In addition to the typical relations in the main layer a network between water molecules and anions in the interlayer exists, which is associated with of Ca atoms of the main layer as the 7th coordination partner. In opposition to hydrocalumite, chloride and carbonate ions are statistically distributed in the interlayer of $C_3A\cdot\frac{1}{2}CaCO_3\cdot\frac{1}{2}CaCl_2\cdot 10.5H_2O$. Site 6a is fully occupied by Cl, C, and O(W2).

Three powder samples with the compositions

$$[Ca_2Al(OH)_6][Cl_{1-x}(CO_3)_{x/2} \cdot 2.25H_2O], \quad x = 0.25, 0.5, \text{ and } 0.75,$$

were synthesized as paste reactions during 4 weeks applying stoichiometric concentrations C_3A, $CaCl_2 \cdot 6H_2O$, and $CaCO_3$.

The phase analysis of those three precipitates revealed that sample AFm-$[Cl^-_{3/4} \cdot (CO_3^{2-})_{1/8}]$ consists of 100 wt% 6R polytype, AFm-$[Cl^-_{1/2} \cdot (CO_3^{2-})_{1/4}]$ 53 wt% 6R-polytype and 47 wt% 2M polytype, and finally sample AFm-$[Cl^-_{1/4} \cdot (CO_3^{2-})_{3/8}]$ is a mixture of 49 wt% 6R polytype and 51 wt% 2M polytype. For those reasons the changes of the different solid solutions structures was investigated between −120 and 50 °C. The transition temperature is dependent on the degree of carbonate concentrations in the solid solutions. The transition temperature of pure α-monochloride $T \approx 35$ °C decreased as a result of the substitution of chloride ions by carbonate ions in the interlayer. In order to expand structural investigations in the solid solution series, new samples with AFm-$[Cl^-_{1-x} \cdot (CO_3^{2-})_x]$, $0.25 \leq x \leq 0.95$, were synthesized at RT and 85 °C. All AFm-$[Cl^-_{1-x} \cdot (CO_3^{2-})_x]$ at −85 °C with $0.25 \leq x \leq 0.95$ had a rhombohedral HT-structure, whereas samples synthesized at 25 °C contained a mixture of LT and HT polytypes [206].

7.5.20 The system $C_3F \cdot CaCO_3 \cdot nH_2O$–$C_3F \cdot CaCl_2 \cdot nH_2O$

Goetz-Neunhoeffer (1996) [89] investigated the phase stabilities in several samples with

$$(3 + x)CaO + Fe(OH)_3 + (1 - x)Na_2CO_3 + xCaCl_{2aq.} + nH_2O, \quad 0 \leq x \leq 1.$$

The results of XRD analysis at 100 and 35 % r.h. yielded solid solutions with different symmetry. In the field with $x = 0$–0.3, the AFm phases were indexed on the basis of a rhombohedral cell with six formula units

$$[Ca_2Fe(OH)_6]^+[(1 - x)(CO_3)_{1/2} \cdot xCl \cdot nH_2O]^-, \quad 11 < n < 12.$$

Solid solutions in the range $0.4 \leq x \leq 0.6$ crystallize with monoclinic symmetry, indicated by the split of the 0012 reflection of a trigonal phase with rhombohedral cell in reflections with hkls 002 and 202 of the monoclinic polytypes. The monoclinic solid solutions contain 11 H_2O per formula unit. The existence of an intermediate phase like Kuzel's salt can be neglected, because the 2θ value of 004 and 202 reflections were shifted with variable chloride concentration. Solid solutions with $x \geq 0.7$ crystallize again as 6R polytypes. Finally, at $x = 1$, the precipitate was composed of 3R-$C_3F \cdot CaCl_2 \cdot 10H_2O$. In order to clarify the discontinuities in the solid solution system, the monoclinic compound $C_3F \cdot \frac{1}{2}CaCO_3 \cdot \frac{1}{2}CaCl_2 \cdot 11H_2O$ was investigated by TG/DTA and nonambient XRD measurements. At 60 °C, the interlayer of $C_3F \cdot \frac{1}{2}CaCO_3 \cdot \frac{1}{2}CaCl_2 \cdot 11H_2O$ loses 1 mol H_2O and the XRD pattern of $C_3F \cdot \frac{1}{2}CaCO_3 \cdot \frac{1}{2}CaCl_2 \cdot 10H_2O$ at 60 °C was indexed hexagonally.

7.5.21 The system $C_3A \cdot CaCO_3 \cdot nH_2O$–$C_3A \cdot Ca(NO_3)_2 \cdot nH_2O$

For the investigations of solid solutions in the system $C_3A \cdot CaCO_3 \cdot nH_2O$–$C_3A \cdot Ca(NO_3)_2 \cdot nH_2O$, samples with the composition

$$3CaO + Al(OH)_3 + (1-x)Ca(NO_3)_2 \cdot 4H_2O + xCaCO_3 + nH_2O, \quad 0 \le x \le 1,$$

reacted for twelve weeks [76]. With the increase of carbonate concentrations, the contents of $C_3A \cdot Ca(NO_3)_2 \cdot nH_2O$ decreased and the base reflections $00l$ of a new compound $C_3A \cdot \frac{2}{3}CaCO_3 \cdot \frac{1}{3}Ca(NO_3)_2 \cdot nH_2O$ were detected. In the precipitate of the sample with a mole ratio 0.33 NO_3^-/CO_3^{2-}, the compound was synthesized purely. Increased CO_3^{2-} concentrations restabilized $C_3A \cdot CaCO_3 \cdot nH_2O$.

7.5.22 The system $C_3A \cdot CaCO_3 \cdot nH_2O$–$C_3A \cdot CaHBO_3 \cdot nH_2O$

Different samples with the chemical compositions

$$4CaO \cdot Al_2O_3 \cdot \left(1 - \frac{x}{2}\right) B_2O_3 \cdot xCO_2 \cdot nH_2O, \quad 0 \le x \le 1,$$

reacted for 3–4 months at 20 °C, 40 °C, and 60 °C [132]. At 20 °C, the precipitates except $x = 1$ are composed of solid solutions with HBO_3^{2-} and CO_3^{2-} in the interlayer. Monocarbonate as a coexisting phase was detected by XRD at about 70 mol monocarbonate. The layer distances decreased with increased carbonate concentrations by about 0.06 Å, lattice parameters a and b remained constant. Due to drying to 35 % r.h., the layer distances decreased by about 0.05–0.07 Å. At 33 mol of monocarbonate, a change in symmetry from monoclinic (2M) to trigonal (6R) was detected by XRD. According to chemical and thermal analysis, the compositions of the investigated solid solutions can be described according to

$$C_3A \cdot (1-x)CaHBO_3 \cdot xCaCO_3 \cdot 11.5 + 2xH_2O, \quad 0 \le x \le 0.67.$$

Their lattice parameters c decreased, although the water contents in the solid solutions increased from $x = 0$ to $x = 0.67$. In the concentration area with two coexisting AFm phases, the carbonate–borate AFm phases had the chemical compositions $[Ca_4Al_2(OH)_{12}]^{2+}[\frac{1}{3}HBO_3 \cdot \frac{2}{3}CO_3 \cdot 6.83H_2O]^{2-}$ and $[Ca_4Al_2(OH)_{12}]^{2+}$ $[\frac{2}{3}HBO_3 \cdot \frac{1}{3}CO_3 \cdot 6.17H_2O]^{2-}$. In comparison with the conditions in the monoborate–hemicarbonate system, the change in symmetry occurred at a comparable composition of the interlayer: system monoborate–hemicarbonate $[\frac{2}{3}HBO_3 \cdot \frac{1}{6}CO_3 \cdot \frac{1}{3}OH\ 5.5H_2O]^{2-}$ – system monoborate–monocarbonate $[\frac{2}{3}HBO_3 \cdot \frac{1}{3}CO_3 \cdot 6.17H_2O]^{2-}$. At 40° and 60 °C, the conditions in the system do not change significantly, just the formation of monocarbonate was detected at $x = 0.2$ (20 mol monocarbonate). Due to the removal of carbonate ions from the solution, the carbonate borate solid solutions in the system incorporate certain hydroxide contents and their compositions became more complicated.

A general composition with

$$[Ca_4Al_2(OH)_{12}]^{2+}[(1-x)HBO_3 \cdot (x-y)CO_3 \cdot (2y)OH \cdot mH_2O]^{2-}, \quad 0 \leq x \leq 0.67, \ y > 0,$$

C_3AH_6 and impurities of CH were detected.

7.5.23 The system $C_3A \cdot CaHBO_3 \cdot nH_2O$–$C_3A \cdot \frac{1}{2}CaCO_3 \cdot \frac{1}{2}Ca(OH)_2 \cdot 11.5H_2O$

One of the important CO_2-containing AFm phases is hemicarbonate. If B_2O_3 or boric acid is applied as cement admixture, it would influence the crystal chemistry of $C_3A \cdot \frac{1}{2}CaCO_3 \cdot \frac{1}{2}Ca(OH)_2 \cdot 11.5H_2O$. Therefore, samples with the chemical composition

$$4CaO \cdot Al_2O_3 \cdot \frac{(1-x)}{2}B_2O_3 \cdot \frac{x}{2}CO_3 \cdot nH_2O, \quad 0 \leq x \leq 1,$$

were prepared [132]. At a reaction temperature of 20 °C, just solid solutions with the composition

$$C_3A \cdot (1-x)CaHBO_3 \cdot \tfrac{1}{2}xCaCO_3 \cdot \tfrac{1}{2}xCa(OH)_2 \cdot 11.5H_2O$$

were investigated in the different precipitates $0 < x < 1$. However, the stacking sequences and symmetry changed between $0.3 < x < 0.4$ from monoclinic (2M) to trigonal (6R).

At 40 °C, different 2M and 6R solid solutions with the general composition

$$C_3A \cdot (1-x)CaHBO_3 \cdot \tfrac{1}{2}xCaCO_3 \cdot \tfrac{1}{2}xCa(OH)_2 \cdot 11.5H_2O$$

also exist. The change of symmetry lies again in the area between $0.3 < x < 0.4$. But with the increase of the reaction temperature, the stability of hemicarbonate decreased, so that small concentrations of C_3AH_6, CH, and monocarbonate were identified by XRD analysis. Carbonate ions must be substituted by 2 OH^- ions, so that the general formula is

$$C_3A \cdot (1-x)CaHBO_3 \cdot (0.5x-y)CaCO_3 \cdot (0.5x+y)Ca(OH)_2 \cdot nH_2O, \quad 0 \leq x \leq 1, \ y > 0.$$

At 60 °C, hemicarbonate transforms to monocarbonate + C_3AH_6 + CH. The carbonate-rich area of the system contains only monocarbonate (1T) as an AFm phase. In the range of $0 < x < 0.7$, solid solutions with a 2M and a 6R stacking arrangement are stable. The chemical composition of the solid solutions can only be approximated as

$$C_3A \cdot (1-x)CaHBO_3 \cdot (0.5x-y)CaCO_3 \cdot (0.5x+y)Ca(OH)_2 \cdot nH_2O, \quad 0 \leq x \leq 1, \ y > 0.$$

7.5.24 The system $C_3F \cdot CaCO_3 \cdot nH_2O$–$C_3F \cdot CaSO_3 \cdot nH_2O$

The phase analysis of samples belonging to the system $C_3A \cdot CaSO_3 \cdot nH_2O$–$C_3A \cdot CaCO_3 \cdot nH_2O$ demonstrated that an intermediate phase with the chemical composition $C_3A \cdot$

$\frac{1}{3}CaSO_3 \cdot \frac{2}{3}CaCO_3 \cdot 11H_2O$ was present [93]. Ecker (1998) [100] investigated samples in the system $C_3F \cdot CaSO_3 \cdot nH_2O$–$C_3F \cdot CaCO_3 \cdot nH_2O$ which were synthesized from different mixtures with

$$4CaO + 2Fe(OH)_3 + (1 - x)Na_2SO_3 + xNa_2CO_3, \quad 0 \le x \le 1.$$

The evaluation of XRD analysis indicated that the end members of the system are completely miscible. The precipitates were single phased with the chemical composition

$$C_3F \cdot xCaSO_3(1 - x)CaCO_3 \cdot nH_2O$$

at 100 and 35 % r.h.

7.5.25 The system $C_3A \cdot CaCl_2 \cdot nH_2O$–$C_3A \cdot Ca(NO_3)_2 \cdot nH_2O$

Kuzel (1969) investigated $C_3A \cdot \frac{1}{2}Ca(NO_3)_2 \cdot \frac{1}{2}CaCl_2 \cdot nH_2O$ crystals, which were synthesized hydrothermally [79]. He mentioned that the unit cell was composed of six double layers $[Ca_4Al_2(OH)_{12}]^{2+}$ $[Cl \cdot NO_3 \cdot 4H_2O]^{2-}$ with $a = 5.744(1)$ Å, $c = 98.28(3)$ Å and $c' = 16.380$ Å. Further data on phase assemblages or the formation of solid solutions in the system were not published. Göske (1998) investigated the system at 100 and 35 % r.h. [194]. Different samples with the general formula

$$2CaO + CaO \cdot Al_2O_3 + (1 - x)CaCl_2 \cdot 6H_2O + (x)Ca(NO_3)_2 \cdot 4H_2O + nH_2O, \quad 0 \le x \le 1,$$

were prepared. The analysis of the liquid phase showed that in samples with high nitrate concentration up to 15 mg/l nitrate ions and in chloride-rich samples, 7–8 mg/l chloride ions were not fixed in the interlayer of the AFm phase and remained in solution. At 100 % r.h., ternary solid solutions in the range of

$$2CaO + CaO \cdot Al_2O_3 + (1 - x)CaCl_2 \cdot 6H_2O + (x)Ca(NO_3)_2 \cdot 4H_2O + nH_2O, \quad 1 < x < 0.7,$$

containing Cl^-, NO_3^-, and OH^- ions in the interlayer are present. In the area

$$2CaO + CaO \cdot Al_2O_3 + (1 - x)CaCl_2 \cdot 6H_2O + (x)Ca(NO_3)_2 \cdot 4H_2O + nH_2O, \quad 0.7 \le x < 0.5,$$

two different AFm phases with $C_3A \cdot 0.83Ca(NO_3)_2 \cdot 0.17Ca(OH)_2 \cdot 16H_2O$ and $C_3A \cdot 0.66CaCl_2 \cdot 0.34Ca(OH)_2 \cdot 10H_2O$ coexisted. In between $0.5 \le x < 0.3$, $C_3A \cdot 0.83Ca(NO_3)_2 \cdot 0.17Ca(OH)_2 \cdot 16H_2O$ and $C_3A \cdot 0.83CaCl_2 \cdot 0.17Ca(OH)_2 \cdot 10H_2O$ were the two stable phases. Samples with $x = 0.2$ and 0.1 were composed of a single ternary AFm phase containing Cl^-, NO_3^-, and OH^- ions. At $x = 1$, $C_3A \cdot 0.83CaCl_2 \cdot 0.17Ca(OH)_2 \cdot 10H_2O$ was determined by XRD and chemical analysis.

The results of the XRD analysis at 35 % r.h. in the system show comparable relations. However, the drying procedure to 35 % r.h. induced lower water concentration in the interlayer area, so that the layer distance dimensions c' were reduced.

7.5.26 The system $C_3A \cdot CaCl_2 \cdot nH_2O$–$C_3A \cdot CaHBO_3 \cdot nH_2O$

For the investigation of the system $C_3A \cdot CaHBO_3 \cdot nH_2O$–$C_3A \cdot CaCl_2 \cdot nH_2O$, stoichiometric concentrations of CaO, bayerite, $CaCl_2 \cdot 2H_2O$, and boric acid were mixed and homogenized with H_2O at a w/s ratio of 10 according to [132]

$$4CaO \cdot Al_2O_3 \cdot \frac{(1-x)}{2} B_2O_3 \cdot xCl_2 \cdot nH_2O, \quad 0 \le x \le 1.$$

At reaction temperatures of 20 and 40 °C, chloride ions were fixed chemically in a solid solution with the composition

$$C_3A \cdot \frac{(1-x)}{2} CaHBO_3 \cdot xCaCl_2 \cdot (11.5 - 1.5x)H_2O, \quad 0.17 \le x \le 1.$$

Its lattice parameters stay constant in the range $x \le 0.2$. $C_3A \cdot CaHBO_3 \cdot 11.5H_2O$ was detected in the samples until $x \le 0.25$. The limiting solid solution contains certain amounts of OH^- ions, too. Its composition is

$$C_3A \cdot 0.17CaHBO_3 \cdot (0.83 - y)CaCl_2 \cdot yCa(OH)_2 \cdot nH_2O.$$

Because of the substitution $2Cl^- \leftrightarrow HBO_3^{2-}$ in the system, different polytypes were identified by XRD independent of temperature or the relative humidity. Pure mono borate 11.5 hydrate is monoclinic with a 2M polytype superstructure in the range $x = 0$–0.1. In the range $0.2 \le x \le 0.85$, solid solutions (6R polytypes) with the formula

$$C_3A \cdot \frac{(1-x)}{2} CaHBO_3 \cdot xCaCl_2 \cdot (11.5 - 1.5x)H_2O$$

were identified by XRD. At quite high Cl^- concentrations ($x = 0.9$–0.95) monoclinic solid solutions with a 2M superstructure crystallized.

7.5.27 The system $C_3F \cdot CaCl_2 \cdot nH_2O$–$C_3F \cdot CaSO_3 \cdot nH_2O$

In order to study the phase relations in the system, samples with the chemical compositions

$$4CaO + 2Fe(OH)_3 + (1-x)Na_2SO_3 + 2xNaCl, \quad 0 \le x \le 1,$$

were prepared and investigated [100]. In the entire system, the end members of the system are miscible at 100 and 35 % r.h., forming solid solutions with the chemical composition

$$C_3F \cdot (1-x)CaSO_3 \cdot xCaCl_2 \cdot nH_2O.$$

7.5.28 The system $C_3F \cdot Ca(NO_3)_2 \cdot nH_2O$–$C_3F \cdot CaSO_3 \cdot nH_2O$

Each sample contained solid solutions with

$$C_3F \cdot (1-x)CaSO_3 \cdot xCa(NO_3)_2 \cdot (12-2x)H_2O, \quad 0 \leq x < 0.5,$$

and

$$C_3F \cdot (1-x)CaSO_3 \cdot xCa(NO_3)_2 \cdot 16H_2O, \quad 0.5 \leq x < 1,$$

which were distinguished by XRD according to their different layer stacking sequences at 100 % r.h. [100]. At 35 % r.h., both phases are completely miscible.

7.5.29 The system $C_3A \cdot CaSO_4 \cdot nH_2O$–$C_3F \cdot CaSO_4 \cdot nH_2O$

The binary system was investigated in 1968 and 1969 [79, 114]. For its synthesis, mixtures of bayerite, iron hydroxide, CaO, and gypsum reacted in a $Ca(OH)_2$ saturated solution at 25 °C, 50 °C, and 100 °C. Just in samples stored at 100 °C, homogeneous precipitates were obtained. At lower temperatures, a broad miscibility gap and an intermediate phase $C_3A_{0.7}F_{0.3} \cdot CaSO_4 \cdot nH_2O$ was identified by XRD. The synthesis of

$$4CaO + (2x)Al(OH)_3 + (2-2x)Fe(OH)_3 + xNa_2SO_4, \quad 0 \leq x \leq 1,$$

yielded homogeneous solid solutions just at 25 °C [100]. The stacking sequence of the layers in the solid solutions were equal to 3R polytypes and had lattice parameters in the range of $a = 5.756$–5.889 Å and of $c = 26.811$–26.669 Å. Recently, solid solutions in the binary system were synthesized in order to determine the solubility product of the solid solutions and the modelling of the liquid phase [90]. Their findings demonstrated the formation of solid solutions in the range $0.0 < Al/(Al + Fe) < 0.45$ and the existence of a miscibility gap between $0.45 < Al/(Al + Fe) < 0.95$, where two different AFm phases coexisted.

7.5.30 The system $C_3A \cdot CaCO_3 \cdot nH_2O$–$C_3F \cdot CaCO_3 \cdot nH_2O$

Different samples with the targeted concentrations

$$4CaO + (2x)Al(OH)_3 + (2-2x)Fe(OH)_3 + xNa_2CO_3, \quad 0 \leq x \leq 1,$$

reacted at 25 °C, 40 °C, and 60 °C [100]. Throughout the three different temperature levels, AFm phases with variable Al/Fe ratios were identified. In all samples with $0.1 \leq x \leq 0.9$, two different phases $C_3(A_xF_{(1-x)}) \cdot CaCO_3 \cdot nH_2O$ with trigonal symmetry and R-Lattice ($C_3(A_xF_{(1-x)}) \cdot CaCO_3 \cdot 12H_2O$), and a triclinic phase $C_3(A_xF_{(1-x)}) \cdot CaCO_3 \cdot 11H_2O$ at 100 % r.h. were detected. The water content of the interlayers depended on the Al/Fe ratio of the AFm phases. The lattice parameters vary with the change of the Al/Fe ratios.

$C_3(A_xF_{(1-x)}) \cdot CaCO_3 \cdot 12H_2O$: $a = 5.80–5.91$ Å and $c' = 8.10–8.00$ Å,
$C_3(A_xF_{(1-x)}) \cdot CaCO_3 \cdot 11H_2O$: $a = 5.80–5.91$ Å and $c' = 7.60–7.50$ Å.

Crystal data on the intermediate phase with $6CaO \cdot Al_2O_3 \cdot Fe_2O_3 \cdot CaCO_3 \cdot Ca(OH)_2 \cdot 22H_2O$ is available. In contrast to these results, Dilnesa et al. did not observe any solid solution but the formation of two separate phases [106].

7.5.31 The system $C_3A \cdot CaCl_2 \cdot nH_2O–C_3F \cdot CaCl_2 \cdot nH_2O$

For the synthesis of potential solid solutions

$$C_3A_{(1-x)}F_{(x)} \cdot CaCl_2 \cdot nH_2O, \quad 0 \leq x \leq 1,$$

in the binary system, reaction mixtures with stoichiometric concentrations of $CaCl_2$, CaO, and ferrihydrite were homogenized with portlandite saturated CO_2-free water [89]. The samples reacted for six months at a constant temperature of 40 °C. Under those conditions, the precipitates were single phased containing different polytypes with a six or a three layer structure. Previous studies in this system with deviating parameters (reaction time 22 days and T = 25 °C, 50 °C, and 100 °C) achieved an incomplete miscibility in the system [114].

The iron-free sample is composed of β-monochloride, the 6R polytype. The AFm phase in sample $x = 0.1$ crystallized trigonal again as a 6R polytype but higher Fe_2O_3 contents were responsible for the stabilization of a 3R polytype, to which β-iron-monochloride belongs as well. Due to higher Fe_2O_3 concentrations, the c dimensions of the unit cells became smaller, whereas the layer distances a increased continuously (from 5.75 Å to 5.850 Å).

7.5.32 The system $C_3A \cdot CaSO_3 \cdot nH_2O–C_3F \cdot CaSO_3 \cdot nH_2O$

Samples with the chemical compositions

$$4CaO + (2x)Al(OH)_3 + (2-2x)Fe(OH)_3 + xNa_2SO_3, \quad 0 \leq x \leq 1,$$

reacted for 6 months contained single phased solids, which indicated that $C_3A \cdot CaSO_3 \cdot nH_2O$ and $C_3F \cdot CaSO_3 \cdot nH_2O$ are completely miscible in the investigated concentration area. The lattice parameters at 100 % r.h. and 35 % r.h. do not show any differences. All solid solutions had trigonal symmetry, whereby the space group changed from P$\bar{3}$ to R$\bar{3}$c [100].

7.6 Phase stabilities in the ternary system $C_3A{\cdot}CaSO_4{\cdot}nH_2O$–$C_3A{\cdot}CaCO_3{\cdot}nH_2O$–$C_3A{\cdot}Ca(OH)_2{\cdot}nH_2O$ at 35 % r.h. and 20 °C

7.6.1 Section $C_3A{\cdot}CaSO_4{\cdot}nH_2O$–$C_3A{\cdot}\frac{1}{2}CaCO_3{\cdot}\frac{1}{2}Ca(OH)_2{\cdot}nH_2O$

In order to solve the phase relations in the system, samples with the chemical compositions

$$\left(3+\frac{x}{2}\right)CaO + 2Al(OH)_3 + (1-x)CaSO_4\cdot 2H_2O + \left(\frac{x}{2}\right)CaCO_3 + (n)H_2O, \quad 0 \le x \le 1,$$

were preprepared [76]. After a reaction time of three months, the precipitates of the different samples were composed of different solid solutions with the composition $C_3A{\cdot}\frac{1}{2}CaSO_4{\cdot}\frac{1}{2}Ca(OH)_2{\cdot}nH_2O$ in equilibrium with hemicarbonate in the range $0.1 \le x \le 1$. The limiting solid solution in the system is monosulfate. The layer distances of sulfate hydroxide AFm phases decrease with the substitution of SO_4^{2-} against $2OH^-$ from the maximum value of 8.947 Å ($C_3A{\cdot}CaSO_4{\cdot}12H_2O$) to 8.748 Å ($C_3A{\cdot}0.95CaSO_4{\cdot}0.05Ca(OH)_2{\cdot}nH_2O$) at a relative humidity of 35 %. The lattice parameters a were just slightly affected. Their values were quite stable with 5.76 Å to 5.75 Å. A solid solution series whereby Afm phases contained all three anion species were not detected by XRD.

7.6.2 Section $C_3A{\cdot}\frac{1}{2}CaSO_4{\cdot}\frac{1}{2}Ca(OH)_2{\cdot}nH_2O$–$C_3A{\cdot}\frac{1}{2}CaCO_3{\cdot}\frac{1}{2}Ca(OH)_2{\cdot}nH_2O$

Different samples with the chemical compositions

$$CaO + 2Al(OH)_3 + (0.5-x)CaSO_4\cdot 2H_2O + (x)CaCO_3 + (n)H_2O, \quad 0 \le x \le 1,$$

reacted for three months [76]. Finally, XRD experiments on the dried solid matter at 35 % r.h. included two different phases with the chemical compositions $C_3A{\cdot}\frac{1}{2}CaSO_4{\cdot}\frac{1}{2}Ca(OH)_2{\cdot}12.5H_2O$ and $C_3A{\cdot}\frac{1}{2}CaCO_3{\cdot}\frac{1}{2}Ca(OH)_2{\cdot}12H_2O$ in the concentration range of $0.1 \le x \le 0.9$. The refinement of the unit cell parameters a and c plus the layer distances c' did not change for the responsible AFm phase between $0.1 \le x \le 0.9$. Therefore, a miscibility gap is present and the substitution of $(\frac{1}{2}SO_4{\cdot}OH{\cdot}6H_2O) \leftrightarrow (\frac{1}{2}CO_3{\cdot}OH{\cdot}5.5H_2O)$ is not possible.

7.6.3 Section $C_3A{\cdot}\frac{1}{2}CaSO_4{\cdot}\frac{1}{2}Ca(OH)_2{\cdot}nH_2O$–$C_3A{\cdot}CaCO_3{\cdot}nH_2O$

Precipitates with the chemical compositions of

$$\left(3+\frac{x}{2}\right)CaO + 2Al(OH)_3 + \left(\frac{x}{2}\right)CaSO_4\cdot 2H_2O + (1-x)CaCO_3 + (n)H_2O,$$
$$0.1 \le x \le 0.9,$$

were not single phased, but XRD measurements indicated the stability of three different AFm phases, monocarbonate, hemicarbonate, and solid solutions of the series monosulfate TCAH [76]. The end members of this system were monosulfate and monocarbonate. In the range $0.1 \leq x \leq 0.4$, the phase assemblages included $C_3A \cdot (1-x)CaSO_4 \cdot xCa(OH)_2 \cdot nH_2O$ and hemicarbonate. The increase of the carbonate concentrations in the range $0.5 \leq x \leq 0.9$ caused the additional crystallization of monocarbonate and sulfate rich AFm phases with the composition $C_3A \cdot (1-x)CaSO_4 \cdot xCa(OH)_2 \cdot nH_2O$. The complete substitution of $(\frac{1}{2}SO_4 \cdot OH \cdot 6H_2O) \leftrightarrow (CO_3 \cdot 5H_2O)$ was not detected.

7.6.4 The system $C_3A \cdot \frac{1}{2}CaSO_4 \cdot \frac{1}{2}Ca(OH)_2 \cdot nH_2O$–$C_3A \cdot \frac{1}{2}CaCO_3 \cdot \frac{1}{2}Ca(OH)_2 \cdot nH_2O$–$C_3A \cdot CaCO_3 \cdot nH_2O$

In this stability field, there are two different ternary AFm phases with the following chemical compositions were identified [76]:
(1) $C_3A \cdot \frac{1}{6}CaSO_4 \cdot \frac{1}{2}Ca(OH)_2 \cdot \frac{2}{6}CaCO_3 \cdot nH_2O$,
(2) $C_3A \cdot \frac{1}{6}CaSO_4 \cdot \frac{1}{6}Ca(OH)_2 \cdot \frac{4}{6}CaCO_3 \cdot nH_2O$.

At 35 % r.h., the phases contained $12H_2O$. Both ternary phases crystallized with trigonal symmetry and possible space group R*. After indexing the powder data, six formula units $[Ca_2Al(OH)_6]^+ [\frac{1}{12}SO_4 \cdot \frac{1}{2}OH \cdot \frac{2}{12}CO_3 \cdot 3H_2O]^-$ and $[Ca_2Al(OH)_6]^+ [\frac{1}{12}SO_4 \cdot \frac{2}{12}OH \cdot \frac{4}{12}CO_3 \cdot 3H_2O]^-$ exist per unit cell. The lattice parameters are for (1) $a = 5.771$ Å and $c = 49.520$ Å and for (2) $a = 5.769$ Å and $c = 49.205$ Å [207].

The single phased stability field of ternary solid solutions is marked by the boundaries $C_3A \cdot CaCO_3 \cdot nH_2O$, $C_3A \cdot 1/6CaSO_4 \cdot \frac{1}{2}Ca(OH)_2 \cdot \frac{2}{6}CaCO_3 \cdot nH_2O$, and $C_3A \cdot \frac{1}{6}CaSO_4 \cdot \frac{1}{6}Ca(OH)_2 \cdot \frac{4}{6}CaCO_3 \cdot nH_2O$. The area of single phased ternary solid solution is given by:

$$C_3A \cdot (x)CaCO_3 \cdot (y)Ca(OH)_2 \cdot zCaSO_4 \cdot nH_2O$$

with the limits $0.33 < x < 0.66$, $0.17 < y < 0.50$, $0 < z < 0.17$, and $x + y + z = 1$.

7.7 Conclusions

Calcium AFm phases with inorganic anions play an important part in cement chemistry. Due to the relevant conditions in cements, a recent work of Baquerizzo et al. summarized the most stable hydration states of the main phases to be expected in cementitious materials [82, 83]. Also, many organic molecules can enter the interlayer of these phases [208–212], forming different hydration states. The orientation of these organic molecules can vary due to the conditions of synthesis and treatment. In composite cements with blast furnace slags and other secondary cementitious materials, LDH phases with magnesium and iron can also occur. Composite cements may contain other AFm phases with varying cat- but also anions. In immobilization applications, many different cations and anions can form numerous LDH phases, which still need some further investigation.

References

[1] Pöllmann H.: Mineralogisch-Kristallographische Untersuchungen an Hydratationsprodukten hydraulischer Bindemittel, 1989, (Bd. I, 285 p.), (Bd. II, 236 p.), Habil.-Schrift

[2] Matschei T, Lothenbach B, Glasser FP. The AFm phase in Portland cement. Cement and Concrete Research. 2007; 37: 118–130.

[3] Pöllmann H, Löns JF.: Eigenschaften von lamellaren Schichtstrukturen mit Chromat, Sulfat und Chlorid. GDCH-Monographie, Siegen. 1999; 15: 171–173

[4] Bensted J. Comments on the Chemical Nomenclature System. Il Cemento. 1983; 4: 193–200.

[5] Rapin JP. Synthèse et étude cristallochimique de quelques aluminates et ferrites calciques hydratés de formule $[Ca_2M(OH)_6]^+$ $[X, nH_2O]^-$ avec X = Cl, Br, I, ClO_4, $\frac{1}{2}CO_3$ $\frac{1}{2}CrO_4$ et $\frac{1}{2}SO_4$ et M = Al et Fe. l'Université Henri Poincaré, Nancy I, 2002.

[6] Aimoz L, Wieland E, Taviot-Guého C, Dähn R, Vespa M, Churakov SV. Structural Insight into Iodide Uptake by AFm Phases. Environmental Science & Technology. 2012; 46: 3874–3881.

[7] Rapin JP, Walcarius A, Lefevre G, Francois M.: A double-layered hydroxide, $3CaO \cdot Al_2O_3 \cdot CaI_2 \cdot 10H_2O$. Acta Crystallographica Section C. 1999; 55: 1957–1959.

[8] Rapin J-P, Francois M.: Rapin JP, Noor NM, Francois M. The double layered hydroxide $3CaO \cdot Al_2O_3 \cdot CaBr_2 \cdot 10H_2O$. Acta Crystallographica Section C. 1999; 55.

[9] Rapin J-P, Francois M. The double-layered hydroxide $C_3A \cdot \frac{1}{2}CaCl_2 \cdot \frac{1}{2}CrO_4 \cdot 12H_2O$. Acta Crystallographica Section C. 2001; 57: 137–138.

[10] Rousselot I, Taviot-Guého C, Leroux F, Léone P, Palvadeau P, Besse J-P. Insights on the Structural Chemistry of Hydrocalumite and Hydrotalcite-like Materials: Investigation of the Series $Ca_2M^{3+}(OH)_6Cl \cdot 2H_2O$ (M^{3+}: Al^{3+}, Ga^{3+}, Fe^{3+}, and Sc^{3+}) by X-Ray Powder Diffraction. Journal of Solid State Chemistry. 2002; 167: 137–144.

[11] Renaudin G, Rapin JP, Elkaim E, François M. Polytypes and polymorphs in the related Friedel's salt $[Ca_2Al(OH)_6]^+$ $[X \cdot 2H_2O]^-$ halide series. Cement and Concrete Research. 2004; 34: 1845–1852.

[12] Rozov K, Berner U, Taviot-Gueho C, et al. Synthesis and characterization of the LDH hydrotalcite-pyroaurite solid-solution series. Cement and Concrete Research. 2010; 40: 1248–1254.

[13] Taylor R, Richardson IG, Brydson RMD. Composition and microstructure of 20-year-old ordinary Portland cement-ground granulated blast-furnace slag blends containing 0 to 100 % slag. Cement and Concrete Research. 2010; 40(7): 971–983.

[14] Zajac M, Bremseth SK, Whitehead M, Ben Haha M. Effect of $CaMg(CO_3)_2$ on hydrate assemblages and mechanical properties of hydrated cement pastes at 40 and 60 °C. Cement and Concrete Research. 2014; 65: 21–29.

[15] Mascolo G. Hydrotalcite Observed in Mortars Exposed to Sulfate-Solutions – a Discussion. Cement and Concrete Research. 1986; 16: 610–612.

[16] Yang Z, Fischer H, Polder R. Modified hydrotalcites as a new emerging class of smart additive of reinforced concrete for anticorrosion applications: A literature review. Materials and Corrosion. 2013; 64: 1066–1074.

[17] Lee H-S, Wang X-Y. Evaluation of the Carbon Dioxide Uptake of Slag-Blended Concrete Structures, Considering the Effect of Carbonation. Sustainability. 2016, 8: 312.

[18] Lothenbach B, Scrivener K, Hooton RD. Supplementary cementitious materials. Cement and Concrete Research. 2011; 41: 1244–1256.

[19] Escalante-Garcia JI, Sharp JH. The chemical composition and microstructure of hydration products in blended cements. Cement and Concrete Composites. 2004; 26: 967–976.

[20] Juenger MCG, Winnefeld F, Provis JL, Ideker JH. Advances in alternative cementitious binders. Cement and Concrete Research. 2011; 41: 1232–1243.

[21] De Weerdt K, Haha MB, Le Saout G, Kjellsen KO, Justnes H, Lothenbach B. Hydration mechanisms of ternary Portland cements containing limestone powder and fly ash. Cement and Concrete Research. 2011; 41: 279–291.

[22] Ben Haha M, Le Saout G, Winnefeld F, Lothenbach B. Influence of activator type on hydration kinetics, hydrate assemblage and microstructural development of alkali activated blast-furnace slags. Cement and Concrete Research. 2011; 41: 301–310.

[23] Haha MB, Lothenbach B, Le Saout G, Winnefeld F. Influence of slag chemistry on the hydration of alkali-activated blast-furnace slag – Part I: Effect of MgO. Cement and Concrete Research. 2011; 41: 955–963.

[24] Kurdowski W. Cement and Concrete Chemistry. In: Kurdowski W, editor. Kraków. New York London: Springer Dordrecht Heidelberg; 2014.

[25] Zigan, F, Rothbauer, R Neutronenbeugungsmessungen am Brucit Neues Jahrbuch für Mineralogie. Monatshefte. 1967: 137–143

[26] Catti M, Ferraris G, Hull S, Pavese A. Static compression and H disorder in brucite, $Mg(OH)_2$, to 11 GPa: a powder neutron diffraction study. Physics and Chemistry of Minerals. 1995; 22: 200–206.

[27] Greaves C, Thomas MA. Refinement of the structure of deuterated nickel hydroxide, $Ni(OD)_2$, by powder neutron diffraction and evidence for structural disorder in samples with high surface area. Acta Crystallographica Section B Structural Science. 19; 42: 51–55.

[28] Brindley GW, Kao C-C. Structural and IR relations among brucite-like divalent metal hydroxides. Physics and Chemistry of Minerals. 1984; 10: 187–191.

[29] Kuzel HJ. Synthese und Röntgenuntersuchungen des $3CaO \cdot Al_2O_3 \cdot CaSO_4 \cdot 12H_2O$ N. Jb Min Mh. 1965; 7: 193–197.

[30] Intissar M, Jumas J-C, Besse J-P, Leroux F. Reinvestigation of the Layered Double Hydroxide Containing Tetravalent Cations: Unambiguous Response Provided by XAS and Mössbauer Spectroscopies. Chemistry of Materials. 2003; 15: 4625–4632.

[31] Richardson IG. Clarification of possible ordered distributions of trivalent cations in layered double hydroxides and an explanation for the observed variation in the lower solid-solution limit. Acta Crystallogr B Struct Sci Cryst Eng Mater. 2013; 69: 629–633.

[32] Hentschel G, Kuzel HJ. Strätlingit. $2CaO–Al_2O_3–SiO_2 \cdot 8H_2O$, ein neues Mineral. N Jb Miner Mh. 1976: 326–330.

[33] Hofmeister W, Platen HV. Crystal Chemistry and Atomic Order in Brucite-related Double-layer Structures. Crystallography Reviews. 1992; 3: 3–26.

[34] Allmann R, Jepsen HP. Die Struktur des Hydrotalkits. N. Jb. Miner. Mh. 1969, 544–551.

[35] Mills SJ, Whitfield PS, Kampf AR, et al. Contribution to the crystallography of hydrotalcites: the crystal structures of woodallite and takovite. Journal of Geosciences. 2012; 57: 273–279.

[36] Drits V A, Sokolova N, Sokolova G V, Cherkashin V I. New members of the Hydrotalcite-Manasseite group. Clays and Clay Minerals. 1987; 35: 401–417.

[37] Mills SJ, Whitfield PS, Wilson SA, et al. The crystal structure of stichtite, re-examination of barbertonite, and the nature of polytypism in MgCr hydrotalcites. American Mineralogist. 2010; 96: 179–187.

[38] Rius J, Allmann R. The superstructure of the double layer mineral wermlandite $[Mg_7(Al_{0.57}Fe^{3+}{}_{0.43})(OH)_{18}]^{2+}$ $[(Ca_{0.6},Mg_{0.4})(SO_4)_2(H_2O)_{12}]^{2-}$. Z Kristallogr. 1984; 168: 133–144.
[39] Cavani F, Trifirò F, Vaccari A. Hydrotalcite-type anionic clays: Preparation, properties and applications. Catalysis Today. 1991; 11: 173–301.
[40] Khan AI, O'Hare D. Intercalation chemistry of layered double hydroxides: recent developments and applications. Journal of Materials Chemistry. 2002; 12: 3191–3198.
[41] Bourrié G, Trolard F, Refait P, Feder F. A Solid-Solution Model for Fe(ii)-Fe(iii)-Mg(ii) Green Rusts and Fougerite and Estimation of Their Gibbs Free Energies of Formation. Clays and Clay Minerals. 2004; 52: 382–394.
[42] Bellotto M, Rebours B, Clause O, Lynch J, Bazin D, Elkaïm E. A Reexamination of Hydrotalcite Crystal Chemistry. The Journal of Physical Chemistry. 1996; 100: 8527–8534.
[43] Terzis A, Filippakis S, Kuzel H-J, Burzlaff H. The crystal structure of $Ca_2Al(OH)_6Cl \cdot 2H_2O$. Zeitschrift für Kristallographie. 1987; 181(1–4): 29–34.
[44] Alberti G, Costantino U. In: Alberti G, Bein T, editors. Solid State Supramolecular Chemistry, Two- and Three-Dimensional Inorganic Networks. Comprehensive Supramolecular Chemistry 7. Oxford, UK: Pergamon and Elsevier Science; 1996.
[45] de Roy A, Forano C, Besse JP. Layered Double Hydroxides: Synthesis and Post-Synthesis Modification. In: Rives V, editor. Layered Double Hydroxides, Present and Future. New York; Nova Science Publishers: 2011. 1–39.
[46] López-Salinas E, García-Sánchez M, Montoya JA, Acosta DR, Abasolo JA, Schifter I. Structural Characterization of Synthetic Hydrotalcite-like $[Mg_{1-x}Ga_x(OH)_2](CO_3)x/2 \cdot mH_2O$. Langmuir. 1997; 13: 4748–4753.
[47] Legrand L, Abdelmoula M, Gehin A, Chausse A, Genin JMR. Electrochemical formation of a new Fe(II)-Fe(III) hydroxy-carbonate green rust: characterisation and morphology. Electrochimica Acta 2001, 46, 1815–1822.
[48] Tilley C. E, Megaw HD, Hey M. H. Hydrocalumite $(4CaO \cdot Al_2O_3 \cdot 12H_2O)$,a new mineral from Scawt Hill, Co. Antrim. Mineralogical Magazine. 1934; 23: 607–615.
[49] Inacio J, Taviot-Gueho C, Morlat-Therias S, de Roy ME, Besse JP. Synthesis and characterization of a new ruthenium containing LDH: $[Zn\text{-}Al\text{-}RuCl_5H_2O_2^-]$. Journal of Materials Chemistry. 2001; 11: 640–643.
[50] Koritnig S, Süsse P. Meixnerit. $Mg_6Al_2(OH)_{18} \cdot 4H_2O$, ein neues Magnesium-Aluminium-Hydroxid-Mineral. TMPM Tschermaks Mineralogische und Petrographische Mitteilungen. 1975; 22: 79–87.
[51] Kloprogge JT, Wharton D, Hickey L, Frost RL. Infrared and Raman study of interlayer anions CO_3^{2-}, NO_3^-, SO_4^{2-} and ClO_4^- in Mg/Al-hydrotalcite. American Mineralogist. 2002; 87: 623–629.
[52] Fogg AM, Williams GR, Chester R, O'Hare D. A novel family of layered double hydroxides-$[MAl_4(OH)_{12}](NO_3)_2 \cdot xH_2O$ (M = Co, Ni, Cu, Zn). Journal of Materials Chemistry. 2004; 14: 2369–6371.
[53] Besserguenev AV, Fogg AM, Francis RJ, et al. Synthesis and structure of the gibbsite intercalation compounds $[LiAl_2(OH)_6]X$ {X = Cl, Br, NO_3} and $[LiAl_2(OH)_6]Cl \cdot H_2O$ using synchrotron X-ray and neutron powder diffraction. Chemistry of Materials. 1997; 9: 241–247.
[54] Britto S, Kamath PV. Structure of bayerite-based lithium-aluminum layered double hydroxides (LDHs): observation of monoclinic symmetry. Inorg Chem. 2009; 48: 11646–11654.
[55] Britto S, Kamath PV. Polytypism in the lithium-aluminum layered double hydroxides: the $[LiAl_2(OH)_6]^+$ layer as a structural synthon. Inorg Chem. 2011; 50: 5619–5627.
[56] Britto S, Thomas GS, Kamath PV, Kannan S. Polymorphism and Structural Disorder in the Carbonate Containing Layered Double Hydroxide of Li with Al. Journal of Physical Chemistry C. 2008; 112: 9510–9515.

[57] Kozlova SG, Gabuda SP, Isupov VP, Chupakhina LÉ. Structural Effects of Intercalation in $[LiAl_2(OH)_6]Cl \cdot xH_2Os$ (^1H, ^7Li, ^{27}Al NMR). Journal of Structural Chemistry. 2001; 42: 189–192.
[58] Thomas GS, Kamath PV, Kannan S. Variable Temperature PXRD Studies of $LiAl_2(OH)_6X \cdot H_2O$ (X = Cl, Br): Observation of Disorder → Order Transformation in the Interlayer. The Journal of Physical Chemistry C. 2007; 111: 18980–18984.
[59] Allmann R, Jepsen H P. Die Struktur des Hydrotalkits. N Jb Miner Mh. 1969: 544–551.
[60] Taylor H F W. Crystal structures of some double hydroxide minerals. Min Mag. 1973; 30: 377–389.
[61] Olowe A. Crystal structures of pyroaurite and sjoegrenite. Advances in X-Ray Analysis. 1995; 38: 749–755.
[62] Ingram L, Taylor H F W. The crystal structure of sjögrenite and pyroaurite. Min Mag. 1967; 36: 467–469.
[63] Allmann R. Nachtrag zu den Strukturen des Pyroaurits und Sjoegrenits. N Jb Miner Mh. 1969: 55–558.
[64] Bookin AS, Drits VA. Polytype diversity of the hydrotalcite-like minerals, I, Possible polytypes and their diffraction features. Clays and Clay Minerals. 1993; 41: 551–7.
[65] Bookin AS, Cherkasin VI, Drits VA. Polytype Diversity of the Hydrotalcite-Like Minerals II. Determination of the Polytypes of Experimentally Studied Varieties. Clays and Clay Minerals. 1993; 41: 558–564.
[66] Aruja E. The unit cell and space group of $4CaO \cdot Al_2O_3 \cdot 19H_2O$ polymorphs. Acta Cryst. 1961; 14: 1213–1216.
[67] Lea F. M. The Chemistry of Cement and Concrete. London: Edward Arnold; 1970.
[68] Roberts MH. Calcium Aluminate Hydrates and related basic salts solid solution V. Intern. Symposium on the chemistry of Cements Tokio. 1968: 104–117
[69] Fischer R, Kuzel H J. Reinvestigation of the System $C_4A \cdot nH_2O–C_4A \cdot CO_2 \cdot nH_2O$. Cement and Concrete Research. 1982; 12: 517–526.
[70] Buttler FG, Dent Glasser LS, Taylor HFW. Studies on $4CaO \cdot Al_2O_3 \cdot 13H_2O$ and the Related Natural Mineral Hydrocalumite. Journal of the American Ceramic Society. 1959; 42: 121–126
[71] Sutterheim N. Proceedings of the 5th International Symposium on the Chemistry of Cement 1969, Tokyo, Cement Association of Japan.
[72] Rogers DE, Aldridge LP. Hydrates of calcium ferrites and calcium aluminoferrites. Cement and Concrete Research. 1977; 7: 399–409.
[73] Sutterheim N. In, Monograph NBoS, editor. International Symposium on the Chemistry of Cement, SCC 4symp, 1962, Washington, US Department of Commerce.
[74] Pöllmann H, Witzke T, Kohler H. Kuzelite, $[Ca_4Al_2(OH)_{12}][(SO_4) \cdot 6H_2O]$, a new mineral from Maroldsweisach/ Bavaria, Germany. N Jb Miner Mh. 1997: 423–432.
[75] Edmonds RN, Majumdar AJ. The hydration of monocalcium aluminate at different temperatures. Cement and Concrete Research. 1988; 18: 311–320.
[76] Pöllmann H. Die Kristallchemie der Neubildungen bei Einwirkung von Schadstoffen auf hydraulische Bindemittel. Erlangen, Friedrich-Alexander-Universität Erlangen-Nürnberg; 1984.
[77] Dosch W. Untersuchungen in den Alkali enthaltenden Systemen, gebildet aus den Komponenten CaO, Al_2O_3, Fe_2O_3, SiO_2, SO_3, Na_2O, K_2O, H_2O. Mainz: Universität Mainz; 1962.
[78] Kuzel H J. Röntgenuntersuchungen im system $3CaO \cdot Al_2O_3 \cdot CaSO_4 \cdot nH_2O–3CaO \cdot Al_2O_3 \cdot CaCl_2 \cdot nH_2O–H_2O$. N Jb Miner Mh. 1966: 1: 193–200.
[79] Kuzel HJ. Beitrag zur Kristallchemie der Calciumaluminathydrate. Johann Wolfgang Goethe-Universität Frankfurt; 1969.
[80] Turriziani R, Schippa G. Riconoscimento all'ATD ed ai raggi X dei solidi quaternary CaO-$Al_2O_3–CaSO_4–H_2O$. La Ricerca Scientifica. 1955: 24: 2356–2363.

[81] Pöllmann H, Neubauer J. $3CaO \cdot Al_2O_3 \cdot CaSO_4 \cdot 16H_2O$ J. Friedrich-Alexander-Universität Erlangen-Nürnberg, Germany. ICDD Grant-in-Aid 1993.

[82] Baquerizo LG, Matschei T, Scrivener KL, Saeidpour M, Wadso L. Hydration states of AFm cement phases. Cement and Concrete Research. 2015; 73: 143–157.

[83] Baquerizo LG, Matschei T, Scrivener KL, Saeidpour M, Thorell A, Wadso L. Methods to determine hydration states of minerals and cement hydrates. Cement and Concrete Research. 2014; 65: 85–95.

[84] Malquori G & Caruso E. Solfo, cloro e nitro- ferriti di calcio. Atti X Congr. Intern. Chimica. 1938; 2: 713–718.

[85] Schippa, G, I solfoferritti d calcio idrati. Ric Sei. 1958; 28: 2334–2340.

[86] Poellmann H, Gotz-Neunhoeffer F. $3CaO \cdot Fe_2O_3 \cdot CaSO_4 \cdot 16H_2O$ 00-044-0605 Friedrich-Alexander-Universität, Erlangen-Nürnberg, Germany. ICDD Grant-in-Aid 1993.

[87] Tanaka S, Machinaga O, Aoyama Y. Synthesis of Calcium Ferrite Monosulfate Hydrate and Rehydration of Its Thermally Treated Intermediates. NIPPON KAGAKU KAISHI 1998. 1998: 817–823.

[88] Ecker M, Poellmann H. $3CaO \cdot Fe_2O_3 \cdot CaSO_4 \cdot 12H_2O$ Mineralogical Institute of University, Erlangen, Germany. ICDD Grant-in-Aid 00-042-1472 1991.

[89] Götz-Neunhöffer F. Synthese und mineralogische Charakterisierung von Calciumferrathydraten-Speicherminerale für die Anwendung in Reststoffdeponien. Friedrich-Alexander-Universität Erlangen-Nürnberg; 1996.

[90] Dilnesa BZ, Lothenbach B, Renaudin G, Wichser A, Wieland E. Stability of Monosulfate in the Presence of Iron. Journal of the American Ceramic Society. 2012; 95: 3305–3316.

[91] Rietveld HM. A profile refinement method for nuclear and magnetic structures. Journal of Applied Crystallography. 1969; 2: 65–71.

[92] Allmann R. Refinement of the hybrid layer structure $[Ca_2Al(OH)_6]^+$ $[\frac{1}{2}SO_4 \cdot 3H_2O]^-$. N Jb Miner Mh. 1977; 3: 136–144.

[93] Motzet H. Der Einfluß von Calciumsulfit auf das Hydratationsverhalten von Portlandzement unter besonderer Berücksichtigung von Carbonatisierungserscheinungen der sulfithaltigen Calciumaluminathydrate. Friedrich-Alexander University; 1990.

[94] Motzet H, Pollmann H. Synthesis and characterisation of sulfite-containing AFm phases in the system $CaO\text{-}Al_2O_3\text{-}SO_2\text{-}H_2O$. Cement and Concrete Research. 1999; 29: 1005–1011.

[95] Murakami K, Tanaka H, Ikehata T, Suga F. Studies on the reactions of $CaSO_3 \cdot \frac{1}{2}H_2O$ and $3CaO \cdot Al_2O_3$, $4CaO \cdot Al_2O_3 \cdot Fe_2O_3$ and $2CaO \cdot Fe_2O_3$. J Ceram Ass Japan. 1959; 67: 56–63.

[96] Keller H. Anorganische Austauschreaktionen an $(Ca_2Al(OH)_6 \cdot ((OH) \cdot aq)$ und davon abgeleitete Mischkristalle [dissertation]. Mainz; 1971

[97] Fischer R. Über den Einbau von CO_3^{2-} in laminare Erdalkalialuminathydrate. Friedrich-Alexander Univerity, Erlangen; 1977.

[98] Yasue T, Mihara M, Arai Y. Synthesis and characteristics of new compounds in the system $CaO–Al_2O_3–CaSO_3–nH_2O$. Gypsum and Lime. 1981; 173: 139–149.

[99] Motzet H, Poellmann H, Kuzel HJ. $3CaO \cdot Al_2O_3 \cdot CaSO_3 \cdot 11H_2O$ 00-041-0477. Miner Institute of Univ, Erlangen, Germany. ICDD Grant-in-Aid 1990.

[100] Ecker M. Diadochiebeziehungen in Kalziumaluminatferraten und deren Hydratationsprodukten. 1998 Hallesches Jahrbuch für Geowissenscaften. Reihe B: ISSN 14323702.

[101] Poellmann H, Ecker M. $Ca_4Fe_2O_6(SO_3) \cdot 12H_2O$ 00-044-0448 Friedrich-Alexander Universität, Erlangen-Nürnerg, Germany. ICDD Grant-in-Aid (1993).

[102] Francois M, Renaudin G, Evrard O. A Cementitious Compound with Composition $3CaO \cdot Al_2O_3 \cdot CaCO_3 \cdot 11H_2O$. Acta Crystallographica Section C. 1998; 54: 1214–1217.

[103] Renaudin G, Francois M, Evrard O. Order and disorder in the lamellar hydrated tetracalcium monocarboaluminate compound. Cement and Concrete Research. 1999; 29: 63–69.

[104] Simon G. Untersuchungen im System $4CaO \cdot Fe_2O_3 \cdot xH_2O$–$3CaO \cdot Fe_2O_3 \cdot CaCO_3 \cdot xH_2O$ [Diplomarbeit]. Erlangen: Friedrich Alexander Universität Erlangen-Nürnberg; 1984.

[105] Poellmann H, Ecker M, Mineral. $3CaO \cdot Fe_2O_3 \cdot CaCO_3 \cdot 12H_2O$ Poellmann, H, Ecker, M, Mineral. Institut of University, Erlangen, Germany. ICDD Grant-in-Aid (1992). Institute of University, Erlangen, Germany. ICDD Grant-in-Aid 1992.

[106] Dilnesa BZ, Lothenbach B, Le Saout G, et al. Iron in carbonate containing AFm phases. Cement and Concrete Research. 2011; 41: 311–23.

[107] Friedel PM. Sur un chloro-aluminate de calcium hydraté se maclant par compression, Bull. Soc Franç Minéral. 1897; 19: 122–136.

[108] Birnin-Yauri UA, Glasser FP. Friedel's salt, $Ca_2Al(OH)_6(Cl,OH) \cdot 2H_2O$: its solid solutions and their role in chloride binding. Cement and Concrete Research. 1998; 28: 1713–1723.

[109] Renaudin G, Kubel F, Rivera JP, Francois M. Structural phase transition and high temperature phase structure of Friedels salt, $3CaO \cdot Al_2O_3 \cdot CaCl_2 \cdot 10H_2O$. Cement and Concrete Research. 1999; 29: 1937–1942.

[110] Rapin JP, Renaudin G, Elkaim E, Francois M. Structural transition of Friedel's salt $3CaO \cdot Al_2O_3 \cdot CaCl_2 \cdot 10H_2O$ studied by synchrotron powder diffraction. Cement and Concrete Research. 2002; 32: 513–519.

[111] Fischer R, Kuzel HJ, Schellhorn H. Hydrocalumit: Mischkristalle von Fridelschem' salz $3CaO \cdot Al_2O_3 \cdot 10H_2O$ und Tetracalciumaluminathydrat $3CaO \cdot Al_2O_3 \cdot Ca(OH)_2 \cdot 10H_2O$. N Jb Miner Mh. 1980; 7: 322–334.

[112] Andersen MD, Jakobsen HJ, Skibsted J. Characterization of the α–β Phase Transition in Friedels Salt ($Ca_2Al(OH)_6Cl \cdot 2H_2O$) by Variable-Temperature ^{27}Al MAS NMR Spectroscopy. The Journal of Physical Chemistry A. 2002; 106: 6676–6682.

[113] Poellmann H, Goetz-Neunhoffer F. $3CaO \cdot Fe_2O_3 \cdot CaCl_2 \cdot 10H_2O$ 00-044-0445 Friedrich-Alexander University, Erlangen-Nurnberg, Erlangen, Germany. ICDD Grant-in-Aid 1993.

[114] Kuzel HJ. Ersatz von Al^{3+} durch Cr^{3+} und Fe^{3+} in $3CaO \cdot Al_2O_3 \cdot CaCl_2 \cdot nH_2O$ und $3CaO \cdot Al_2O_3 \cdot CaSO_4 \cdot nH_2O$. ZKG. 1968; 21(12): 493–499.

[115] Shannon RD, Prewitt CT. Effective ionic radii in oxides and fluorides. Acta Crystallographica Section B. 1969; 25: 925–946.

[116] Renaudin G, Mesbah A, Dilnesa B, Francois M, Lothenbach B. Crystal Chemistry of Iron Containing Cementitious AFm Layered Hydrates. Current Inorganic Chemistry. 2015; 5: 184–193.

[117] Mesbah A, Rapin J-P, François M, Cau-dit-Coumes C, Frizon F, Leroux F, et al. Crystal Structures and Phase Transition of Cementitious Bi-Anionic AFm-(Cl^-, CO_3^{2-}) Compounds. Journal of the American Ceramic Society. 2011; 94(1): 261–268.

[118] Zivica, V. Über die korrosive Wirkung von Calciumnitratlösungen auf erhärtete Zementpasten ZKG. 1971: 175–179

[119] Ahmed SJ, Dent Glasser LS und Taylor HFW. Crystal structure and reactions of C_4AH_{12} and derived basic salts. Proc V Int Symp Chem Cem Tokio; 1968. 118–127

[120] Ahmed SJ, Taylor HFW. Crystal Structures of the Lamellar Calcium Aluminate Hydrates. Nature. 1967; 215(5101): 622–623.

[121] Kuzel HJ. Über die Hydratsstufen der Hydroxisalze $3CaO \cdot Me_2O_3 \cdot CaCl_2 \cdot nH_2O$ und $3CaO \cdot Me_2O_3 \cdot Ca(NO_3)_2 \cdot nH_2O$. N Jb Min Mh. 1970: 161–174.

[122] Feitknecht W, Buser H W. Über den Bau der plättchenförmigen Calcium-Aluminium-Hydroxysalze. Helvetica Chimica Acta. 1951; 34: 128–142.

[123] Renaudin G, Rapin JP, Humbert B, François M. Thermal behaviour of the nitrated AFm phase $Ca_4Al_2(OH)_{12}(NO_3)_2 \cdot 4H_2O$ and structure determination of the intermediate hydrate $Ca_4Al_2(OH)_{12}(NO_3)_2 \cdot 2H_2O$. Cement and Concrete Research. 2000; 30: 307–314.

[124] Kuzel HJ. Über die Hydratstufen der Hydroxisalze $3CaO \cdot Me_2O_3 \cdot CaCl_2 \cdot nH_2O$ und $3CaO \cdot Me_2O_3 \cdot Ca(NO_3)_2 \cdot nH_2O$. N Jb Miner Mh. 1970; 1: 363–374.

[125] Renaudin G, François M. The lamellar double-hydroxide (LDH) compound with composition $3CaO \cdot Al_2O_3 \cdot Ca(NO_3)_2 \cdot 10H_2O$. Acta Crystallographica Section C Crystal Structure Communications. 1999; 55: 835–838.

[126] Goetz-Neunhoeffer F, Poellmann H. $Ca_2FeO_3(NO_3) \cdot 5H_2O$ 00-048-0065 Univ of Halle, Dept of Mineralogy & Geochemistry, Germany. ICDD Grant-in-Aid; 1996.

[127] Lieber W, Effect on inorganic admixtures on the setting and hardening of portland cement. ZKG. 1973; 2: 75–79.

[128] Bensted J, IC. Callaghan, A. Lepre, Comparative study of the efficiency of various borate compounds as set retarders of class G oilwell cement. Cem Concr Res. 1991; 21: 663–668.

[129] Ramachandran VS, Lowery MS. Conduction calorimetric investigation of the effect of retarders on the hydration of portland cement. Thermochim Acta. 1992; 195: 373–387.

[130] Demirbaş A, Karslioglu S. The effect of boric acid sludges containing borogypsum on properties of cement. Cement and Concrete Research. 1995; 25: 1381–1384.

[131] Champenois J-B, Mesbah A, Cau Dit Coumes C, et al. Crystal structures of Boro-AFm and sBoro-AFt phases. Cement and Concrete Research. 2012; 42: 1362–1370.

[132] Wenda, R. (1984), Untersuchungen zur Kristallchemie der Hydratationsprodukte borathaltiger Zemente [dissertation]. Erlangen.

[133] Goetz-Neunhoeffer F, Poellmann H. $3CaO \cdot Fe_2O_3 \cdot CaHBO_3 \cdot 10H_2O$ 00-049-0013 Univ. of Halle, Dept. of Mineralogy & Geochemistry, Germany. ICDD Grant-in-Aid 1996.

[134] Dosch W, Keller H. On the crystal chemistry of tetracalcium aluminate hydrate. The 6th International Congress on the Chemistry of Cement, 1974, Moscow.

[135] Dosch W, Keller H, zur Strassen H. Additional remark on tetracalcium aluminate hydrate and its quaternary complex salts. Proceedings of the 5th International Symposium on the Chemistry of Cement, Tokyo; 1968. 72–77.

[136] Dosch W, Koestel C. Stabilized dicalcium aluminate hydrates, US Patent 4095989, 20 June 1987.

[137] Taylor H. F. W. Cement Chemistry. 2nd edition. London: Thomas Telford Publishing; 1997.

[138] Scheller T, Kuzel H.-J. Studies on dicalcium aluminate hydrates. The 6th International Congress on the Chemistry of Cement, 1974, Moscow.

[139] Richard N, Lequeux N, Boch P. Local environment of Al and Ca in CAH_{10} and C_2AH_8 by X-ray absorption spectroscopy. Eur J Solid State Inorg Chem. 1995; 32: 649–662.

[140] Ramachandran VS. Applications of Differential Thermal Analysis in Cement Chemistry. New York: Chemical Publishing Company; 1969. 206–12.

[141] George C. M. Industrial alumina cement. In: Barnes P, editor. Structure and Performance of Cement. 1983.

[142] Midgley HG. Measurement of high-alumina cement-calcium carbonate reactions using DTA. Clay Minerals. 1984; 19: 857–864.

[143] Bushnell-Watson SM, Sharp JH. The detection of the carboaluminate phase in hydrated high alumina cements by differential thermal analysis. Thermochimica Acta. 1985; 93: 613–616.

[144] Das SK, Mitra A, Das Poddar PK. Thermal analysis of hydrated calcium aluminates. Journal of thermal analysis. 1996; 47: 765–774.

[145] Guirado F, Gali S, Chinchon J S. Thermal dehydration of CAH_{10}. Cem Concr Res. 1998; 28: 381–390.

[146] Fryda H, Scrivener KL, Chanvillard G. Relevance of laboratory tests to field applications of calcium aluminate cement concretes. In: RJ Mangabhai FPG, editor. Calcium Aluminate Cements. London: IOM Communications; 2001. 227–46.

[147] Cardoso FA, Innocentini MDM, Akiyoshi MM, Pandolfelli VC. Effect of curing time on the properties of CAC bonded refractory castables. Journal of the European Ceramic Society. 2004; 24: 2073–2078.

[148] Ukrainczyk N, Matusinovic T, Kurajica S, Zimmermann B, Sipusic J. Dehydration of a layered double hydroxide – C_2AH_8. Thermochimica Acta 2007, 464, 7–15.
[149] Rives V. Layered Double Hydroxides: Present and Future. In: Rives V, editor. Nova Science Publishers; 2001. 1–493.
[150] Richard N, Lequeux N, Boch P. EXAFS Study of Refractory Cement Phases: $CaAl_2O_{14}H_2O$, $Ca_2Al_2O_{13}H_{16}$, and $Ca_3Al_2O_{12}H_{12}$. J Phys III France. 1995; 5: 1849–64.
[151] Poellmann H, Scarbath J, Goske J, Kuzel HJ. $Ca_2Al_2O_5 \cdot 8\ H_2O$ Friedrich-Alexander-Universität Erlangen-Nürnberg, Germany. ICDD Grant-in-Aid 1994.
[152] Raab B. Synthese und Charakterisierung nanoskaliger hydraulisch hochreaktiver Phasen des Portland- und Tonerdezements. 2010.
[153] Raab B, Poellmann H. $Ca_2Al_2O_5 \cdot 8.2H_2O$ Univ. of Halle, Dept. of Mineralogy and Geochemistry, Halle/Saale, Germany. ICDD Grant-in-Aid 2009.
[154] Raab B, Poellmann H. $Ca_2Al_2O_5 \cdot 8H_2O$ Univ. of Halle, Dept. of Mineralogy and Geochemistry, Halle/Saale, Germany. ICDD Grant-in-Aid 2009.
[155] Faucon P, Charpentier T, Bertrandie D, Nonat A, Virlet J, Petit JC. Characterization of Calcium Aluminate Hydrates and Related Hydrates of Cement Pastes by ^{27}Al MQ-MAS NMR. Inorganic Chemistry. 1998; 37: 3726–3733.
[156] Raab B, Poellmann $Ca_2Al_2O_5 \cdot 7.5H_2O$. Univ. of Halle, Dept. of Mineralogy and Geochemistry, Halle/Saale, Germany. ICDD Grant-in-Aid 2009.
[157] Renaudin G. I/ Etude d'un hydroxide simple d'aluminium, Le Bayerite II/ Etude d' une famille d'hydroxides doubles lammellares d'aluminium et da Calcium, Les Phases AFm (aluminates tetracalciques Hydrates). Nacy, Université H. Poincaré; 1998.
[158] Matschei T, Lothenbach B, Glasser FP. The AFm phase in Portland cement. Cement and Concrete Research. 2007; 37: 118–130.
[159] Passaglia E, Turconi B. Silicati ed Altri Minerali di Montalto di Castro(VT). Riv Mineral Ital. 1982; 4: 97–110.
[160] Stoppa F, Scordari F, Mesto E, Sharygin V, Bortolozzi G. Calcium-aluminum-silicate-hydrate "cement" phases and rare Ca-zeolite association at Colle Fabbri, Central Italy. Open Geosciences; 2010: 175.
[161] Schüller W, Betz V. Die Mineralien vom Emmelberg. Lapis. 1986: 12.
[162] Kuzel HJ. Crystallographic data and thermal decomposition of synthetic gehlenite hydrate $2CaO \cdot Al_2O_3 \cdot SiO_2 \cdot 8H_2O$. N Jb Miner Mh. 1976; 7: 319–325.
[163] Schmitt CH. Discussion of the paper, Hydration of calcium aluminates and ferrites. Proc Fourth Internat symposium on the Chemistry of Cement. Washington; 1962.
[164] Schwiete HE, Ludwig U, Wuerth KE, Grieshammer G. Neubildungen bei der Hydratation von Hochofenschlacken. ZKG. 1969; 22: 154–160.
[165] Liebau F. Silicon. In, Handbook of Geochemistry. Vol. II/3. Berlin: Springer-Verlag; 1972. 14-A-I-14-A-32.
[166] Rinaldi R, Sacerdoti M, Passaglia E. Strätlingite: crystal structure, chemistry, and a reexamination of its polytype vertumnite. European Journal of Mineralogy. 1990; 2: 841–850.
[167] Wang J. Hydration Mechanism of Cements based on Low-CO_2 Clinkers Containing Belite, Ye'elimite and Calcium Alumino-ferrite. University of Lille; 2010.
[168] Kwan S, LaRosa J, Grutzeck MW. ^{29}Si and ^{27}Al MAS NMR Study of Stratlingite. Journal of the American Ceramic Society. 1995; 78: 1921–1926.
[169] Santacruz I, Torre ÁGDl, Álvarez-Pinazo G, et al. Structure of stratlingite and effect of hydration methodology on microstructure. Advances in Cement Research. 2015; 28: 13–22.
[170] Galli E, Passaglia E. Vertumnite – Its Crystal-Structure and Relationship with Natural and Synthetic Phases. Tschermaks Mineralogische Und Petrographische Mitteilungen. 1978; 25: 33–46.

[171] Passaglia E, Galli E. Vertumnite, a new natural silicate. Tschermaks Min Petr Mitt. 1977; 24: 57–66.

[172] Cooper M. A, Hawthorne F. C. The Crystal Structure of Shigaite $[AlMn_2^{2+}(OH)_6]_3$ $(SO_4)2Na(H_2O)$ $\{H_2O\}_6$, A Hydrotalcite-Group Mineral. Can Min. 1996; 34: 91–97.

[173] Brindley G W. Motukoreaite additional data and comparison with related minerals. Min Mag. 1979; 43: 337–340.

[174] Hudson D. R, Bussell M. Mountkeithite, a new pyroaurite-related mineral with an expanded interlayer containing exchangeable $MgSO_4$. Min Mag. 1981; 44: 345–355.

[175] Rius J, Allmann R. The superstructure of the double layer mineral wermlandite $[Mg_7(Al_{0.57}Fe^{3+}{}_{0.43})(OH)_{18}]^{2+}[(Ca_{0.6},Mg_{0.4})\ (SO_4)_2(H_2O)_{12}]^{2-}$. Z Kristallogr. 1984; 168: 133–144.

[176] König Uwe. Synthese, Charakterisierung und Eigenschaften von manganhaltigen Layered Double Hydroxides (LDHs). Halle: Martin-Luther-Universität Halle-Wittenberg; 2006.

[177] König U. Synthesis and Characterization of Manganese Containing Layered Double Hydroxides (LDH's). In: Sherman IT, editor. Layered Double Hydroxides (LDHs): Synthesis, Characterization and Applications Nova Science Publisher. 2015: 51–72.

[178] Dosch W, zur Strassen H. Ein alkalihaltiges Calciumluminatsulfathydrat (Natrium-Monosulfat). ZKG. 1967; 9: 392–401.

[179] Li G, Le Bescop P, Moranville M. The U phase formation in cement-based systems containing high amounts of $Na2SO_4$. Cement and Concrete Research. 1996; 26: 27–33.

[180] Guerrero A, Goñi S, Hernández M-S. Thermodynamic Solubility Constant of Ca(OH)2 in Simulated Radioactive Sulfate Liquid Waste. Journal of the American Ceramic Society. 2000; 83: 882–888.

[181] Li G, Le Bescop P. Degradation machanisms of cement-stabilized wastes by internal sulfate associated with the formation of the U phase. In: Scrivener KL, Young JF, editors. Mechanisms of Chemical Degradation of Cement-based Systems CRC Press. 1997: 427–435.

[182] Shayan A, Quick GW, Lancucki CJ. Morphological, mineralogical and chemical features of steam-cured concretes containing densified silica fume and various alkali levels. Advances in Cement Research. 1993; 5: 151–162.

[183] Shimada Y, Young JF. Thermal stability of ettringite in alkaline solutions at 80 °C. Cement and Concrete Research. 2004; 34: 2261–2268.

[184] Post K, Poellmann H. $C_3A\cdot CaSO_4\cdot 0.5Na_2SO_4\cdot 15H_2O$ Friedrich-Alexander-Universität, Erlangen-Nürnberg, Germany. ICDD Grant-in-Aid 1992.

[185] Post K, Poellmann H. $C_3A\cdot CaSO_4\cdot 0.5K_2SO_4\cdot 15H_2O$ Friedrich-Alexander Universität, Erlangen-Nürnberg, Germany. ICDD Grant-in-Aid 1992.

[186] Li G, Le Bescop P, Moranville-Regourd M. Synthesis of the U Phase $(4CaO\cdot 0.9Al_2O_3\cdot 1.1SO_3\cdot 0.5Na_2O\cdot 16H_2O)$. Cem Concr Res. 1997; 27: 7–13.

[187] Jones FE. The Quaternary System $CaO–Al_2O_3–CaSO_4–H_2O$ at 25 °C. Equilibria with Crystalline Al_2O_3-$3H_2O$, Alumina Gel, and Solid Solution. The Journal of Physical Chemistry. 1944; 48: 311–356.

[188] D'Ans J and Eick H. Das System $CaO–Al_2O_3–CaSO_4–H_2O$ bei 20 °C ZKG. 1953; 6: 302–311

[189] Seligmann, P. and Greening, N. R. Studies of early hydration reactions of Portland cement by X-ray diffraction Tokio-Symposium. 1968: 80–10

[190] Pöllmann H. Solid Solution in the system $3CaO\cdot Al_2O_3\cdot CaSO_4\cdot aq.$-$3CaO\cdot Al_2O_3\cdot Ca(OH)_2\cdot aq$-$H_2O$ at 25 °C, 45 °C, 60 °C, 80 °C. N Jb Miner Abh. 1989; 161(1): 27–40.

[191] Pöllmann H. Mineralogical Institute of University, Erlangen, Germany. ICDD Grant-in Aid 1990.

[192] Mesbah A, François M, Cau-dit-Coumes C, et al. Crystal structure of Kuzel's salt $3CaO\cdot Al_2O_3\cdot \frac{1}{2}CaSO_4\cdot \frac{1}{2}CaCl_2\cdot 11H_2O$ determined by synchrotron powder diffraction. Cement and Concrete Research. 2011; 41: 504–509.

[193] Poellmann H, Gotz-Neunhoeffer F. $3CaO \cdot Fe_2O_3 \cdot 0.5CaSO_4 \cdot 0.5CaCl_2 \cdot 10H_2O$ 00-046-0813 Friedrich-Alexander-Universität, Erlangen-Nürnberg, Germany. ICDD Grant-in-Aid 1995.

[194] Göske J. Die mineralogisch-chemische Barriere. Fixierung der Anionen SO_4^{2-}, Cl^-, NO_3^- und CrO_4^{2-} in Speichermineralen, unter dem Aspekt verschiedener Rohstoffmischungen, Abmischungen und Modellkonzentration [dissertation]. Martin-Luther Universität Halle; 1998.

[195] Dosch W, zur Strassen H. Untersuchung von Tetracalciumaluminathydraten. I. Die verschiedenen Hydratstufen und der Einfluß von Kohlensäure. ZKG. 1965; 5: 233.

[196] Ecker M, Poellmann H. 00-045-0572 $6CaO \cdot Al_2O_3 \cdot Fe_2O_3 \cdot CaCO_3 \cdot Ca(OH)_2 \cdot 22H_2O$ Friedrich-Alexander-Universität, Erlangen-Nürnberg, Germany. ICDD Grant-in-Aid 1994.

[197] Runcevski T, Dinnebier RE, Magdysyuk OV, Pollmann H. Crystal structures of calcium hemicarboaluminate and carbonated calcium hemicarboaluminate from synchrotron powder diffraction data. Acta Crystallographica Section B. 2012; 68(5): 493–500.

[198] Ecker M, Pöllmann H. Synthesis and characterisation of CO_3-containing Ca-Al- Fe-hydrates. International symposium of chemistry of cement; Götheborg. 1997; 2ii032.

[199] Ecker M, Pöllmann H. $6CaO \cdot Al_2O_3 \cdot Fe_2O_3 \cdot CaCO_3 \cdot Ca(OH)_2 \cdot 22H_2O$ 00-045-0572 Friedrich-Alexander-Universität, Erlangen-Nürnberg, Germany. ICDD Grant-in-Aid 1994.

[200] Chatterjee S. Mechanism of the $CaCl_2$-attack on Portland cement concrete. Cem Concr Res. 1978; 8: 461–468.

[201] Turriziani R, Schippa G. Contribution to the knowledge of hydrated calcium chloroaluminates. La Ricerca Scientifica. 1955; 25: 3102–3106.

[202] Pöllmann H. Mischkristallbildung in den Systemen Monochlorid-Monocarbonat u. Monochlorid-TCAH. 1980.

[203] Pöllmann H. Solid solution series of complex calcium aluminate hydrates containing Cl^-, OH^- and CO_3^{2-}. Proc 8th Int Symp on the Chem of Cement; 1986; Rio de Janeiro. 300–306.

[204] Goetz-Neunhoeffer F, Poellmann H. $3CaO \cdot Fe_2O_3 \cdot 0.5Ca(OH)_2 \cdot 0.5CaHBO_3 \cdot 10H_2O$ 00-049-0012 Univ. of Halle, Dept. of Mineralogy & Geochemistry, Germany. ICDD Grant-in-Aid 1996.

[205] Sacerdoti M, Passaglia E. Hydrocalumite from Latium, Italy: its crystal structure and relationship with related synthetic phases. N Jb Miner Mh. 1988; 10: 462–475.

[206] Mesbah A, Cau-dit-Coumes C, Frizon F, Leroux F, Ravaux J, Renaudin G. A New Investigation of the Cl^-–CO_3^{2-} Substitution in AFm Phases. Journal of the American Ceramic Society. 2011; 94: 1901–1910.

[207] Pöllmann H. Syntheses, properties and solid solution of ternary lamellar calcium aluminate hydroxi salts (AFm-phases) containing SO_4^{2-}, CO_3^{2-} and OH^-. N Jb Miner Abh. 2006; 182: 173–181.

[208] Pöllmann H, Stöber S, Stern E. Synthesis, characterization and reaction behaviour of lamellar AFm phases with aliphatic sulfonate-anions. Cement and Concrete Research. 2006; 36: 2039–2048.

[209] Stöber S, Pöllmann H. Synthesis of a lamellar calcium aluminate hydrate (AFm Phase) containing benzenesulfonic acid ions. Cem Concr Res. 1999; 29: 1841–1845.

[210] Stöber S, Pöllmann H. Synthesis and Characterisation of Hydration Products of the Calcium-Aluminate Phase under the Presence of Sulfonic Acids. In: Mangabhai RJ, Glasser FP. Calcium Aluminate Cements 2001: Proceedings of the International Conference on Calcium Aluminate Cements (CAC) Held at Heriot-Watt University Edinburgh, Scotland, UK, 16–19 July 2001: IOM Communications; 2001.

[211] Plank J, Keller H, Andres PR, Dai Z. Novel organo-mineral phases obtained by intercalation of maleic anhydride–allyl ether copolymers into layered calcium aluminum hydrates. Inorganica Chimica Acta. 2006; 359: 4901–4908.

[212] Dosch W. Die innerkristalline Sorption von Wasser und organischen Substanzen an Tetracalciumaluminathydrat. N Jb Min Abh. 1967; 106: 200–239.

Part III: **Cementitious and binder materials**

X. Gao, B. Yuan, Q.L. Yu*, and H.J.H. Brouwers

8 Chemistry, design and application of hybrid alkali activated binders

Abstract: This chapter presents a brief overview of two typical alkali activated systems: a high calcium system and an aluminosilicates based one. Subsequently, recent progress in understanding the chemistry, design and applications of these alternative binders is reviewed. The commonly applied two alkali activated binder systems are modified by using alternatives such as sodium carbonates, slag/fly ash, limestone powder and nano-silica. The effects of those additives on reaction kinetics, gel structure, mechanical properties, porosity and shrinkage are comprehensively studied. The recipe designs of modified alkali activated binders are also investigated. The applications of these new binders in waste management, high performance products, and functional building materials are introduced. Finally, future research directions, modifications, designs and potential applications are suggested.

Keywords: alkali activation, slag, fly ash, sodium carbonate, reaction kinetics, design, application

8.1 Chemistry of alkali activated binders

8.1.1 Introduction

Although Portland cement (PC) is considered to be a very robust building material, the production of cement involves high energy-consumption and greenhouse gas-emission [1]. During the last decades, considerable research has been conducted on alkali activated materials (AAM) due to their superior properties and environmental benefits [2–4]. The AAM concept uses alkali solutions or salts to dissolve and activate waste and industrial by-products to obtain sustainable building materials. Many factors can affect the performance of AAM, e.g. alkali natures [5–7], alkali concentrations [8], reactivity and physiochemical properties of raw materials [9–12], curing methods [13, 14], and sampling methods [15, 16], etc. [17–20].

Note: X. Gao and B. Yuan are both considered as first authors

---

X. Gao, B. Yuan, H.J.H. Brouwers, Department of the Built Environment, Eindhoven University of Technology, Eindhoven, The Netherlands; State Key Lab of Silicate Materials for Architectures, Wuhan University of Technology, Wuhan 430070, PR China
***Corresponding author: Q.L. Yu,** Department of the Built Environment, Eindhoven University of Technology, Eindhoven, The Netherlands, q.yu@bwk.tue.nl

DOI 10.1515/9783110473728-009

8.1.2 Review of the chemistry of alkali activated materials

8.1.2.1 Alkali activation of low-calcium system

In the 1950s, Glukhovsky proposed a mechanism for the alkali activation of materials primarily comprising of silica and reactive alumina and defined the process as having three stages [19]. More recently, various authors have elaborated and extended the Glukhovsky theory (e.g. using zeolite synthesis routine in order to explain the geopolymerization process). The alkali activation of aluminosilicates mainly involves the dissolution of silicate and aluminate species into the solution, then the formation of aluminosilicates oligomers, and finally the growth and crystallization of zeolite-like inorganic polymers. Though presented linearly, these processes are largely coupled and occur concurrently. Despite the general acceptance of this conceptual model, the actual process of particle-to-gel conversion has not been confirmed in the highly alkaline and poorly solvated conditions prevailing during the geopolymer synthesis.

8.1.2.2 Alkali activation of high-calcium system

Kühl [23], in a patent from 1908, described a type of one-part geopolymer of slag combined with solid alkali-carbonate or alkali-sulphate as activator. Additionally, the slag also contained a percentage (2–3 %) of calcium hydroxide. This precipitated the carbonate or sulphate from the alkali carbonate/sulphate as calcium salt, thereby releasing hydroxide ions into the solution in order to accelerate dissolution of the slag. Later, Kühl [24] proposed a system of slag activated with a KOH solution. The activation of high calcium materials is a complex process including the initial decomposition of the precursor, namely the breakdown of the original T–O and Ca–O bonds, releasing Ca, Si and Al groups into the solution, followed by a polycondensation process of the reaction products. The primary distinction between the Na^+ and Ca^{2+} sites shown has a much greater extent of "damage" caused to the glass structure by the removal of a divalent cation compared to a monovalent one. It has been proposed that the first step of a glass dissolution at a moderately high pH resembles that observed under acidic conditions, being initiated by ion exchange of H^+ for Na^+ or Ca^{2+}. It should be noted that the exact mechanism in detail is still debatable and not fully understood.

The main product of sodium hydroxide activated slag is a disordered 14 Å tobermorite-like C-S-H(I) type, which presents a higher Ca/Si ratio and a more ordered structure than the C-A-S-H type gel formed in silicate-activated blast furnace slag (BFS) binders [26]. Compared to that of waterglass, a relatively low degree of crosslinking is found in the C-A-S-H type gels formed in NaOH-activated BFS. This is probably because waterglass provides extra Si which leads to a lower Ca/(Si + Al) ratio of the system and consequently the C-A-S-H gel itself. Nevertheless, the Ca content in the C-A-S-H type gel formed in the alkali activated BFS is generally lower than the Ca/Si ratio in the Portland cement system. Besides, mainly depending on the Al and Mg content in the raw materials, hydrotalcite-like phases are observed as the secondary products [11]. The amount of Al that can become substituted into the bridging sites within the C-A-S-H structure is limited, leading to the formation of the Al-rich secondary phases. A recent research also shows that Na^+ can possibly substitute some Ca^{2+} in the C-A-S-H gel forming C-(Na)-S-H type gel with a lower Ca/Si ratio than what is observed in the binding gel formed in the bulk region in both NaOH-activated and silicate activated BFS binders [27].

8.1.3 Alkali activation of slag/fly ash blends

It is known that the products of slag and fly ash based alkali activated materials are different, C-(A)-S-H gel and N-(A)-S-H gel, respectively [25, 28]. Previous research has shown that a combination of raw materials slag and fly ash leads to the formation of the coexisted C-(A)-S-H gel and N-(A)-S-H gel in certain circumstances (stable at low pH) [29]. Activation of slag normally happens at room temperature while that of fly ash requires extra thermal curing. In this case, room temperature cured alkali activated slag/fly ash blends show their advantages in field applications. A study was conducted which focused on the effects of two compositional factors: activator modulus (SiO_2/Na_2O [Ms] from 1.0 to 1.8) and slag/fly ash mass ratios (between 90/10 and 50/50) on hydration kinetics, gel characters and compressive strength [30]. The solid materials used were commercial ground granulated blast furnace slag and Class F fly ash. The alkali activator used was a mixture of sodium hydroxide pellets (analytical level) and sodium silicate solution.

8.1.3.1 Reaction kinetics and gel structure

Fig. 8.1 (a) shows the heat flow of samples with an activator modulus of 1.8, where samples with the activator moduli of 1.8, 1.4, 1.0 are assigned as A, C, E; and slag/fly ash mass ratios of 90/10, 70/30, 50/50 are represented by 1, 3, and 5. The initial dissolution peak appears at around 3-4 minutes, then all mixes exhibit an induction period lasting between 4 to 10 h before the second peak. Samples with higher fly ash contents present considerably longer induction times, attributed to the low reactivity of fly ash. The second heat evolution peak appears at around 13 h for the sample with a slag/fly ash mass ratio of 90/10, and this peak shifts to longer times with lower intensities and broader peak shapes as the fly ash content increases. Such significant changes are mainly due to the reduced total slag content that results in a smaller overall heat evolution, also partly because of the relatively low reactivity and a more moderate reaction process of fly ash under ambient temperature.

Fig. 8.1 (b) and (c) exhibit the heat evolutions of mixtures with Ms of 1.4 and 1.0, respectively. As the activator modulus decreases, higher peak flows are observed in the acceleration stage, which may indicate a higher reaction degree due to the larger covered area under the peak. It was suggested that a decrease in activator modulus

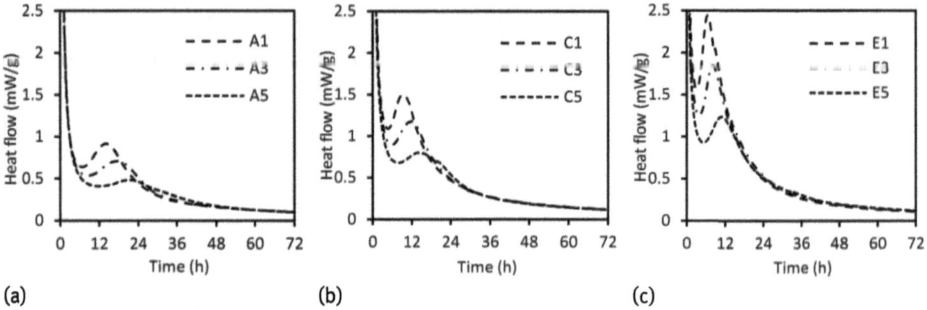

Fig. 8.1: Isothermal calorimetry results of alkali activated slag/fly ash blends up to 72 h with the waterglass moduli of (a) 1.8, (b) 1.4, and (c) 1.0 [30].

leads to a higher alkali concentration, thus enhancing the dissolution rate and reaction degree of slag [31]. A higher alkalinity also increases the solubility of silica and alumina in solution which could benefit the formation of reaction products [32, 33]. For a constant slag/fly ash mass ratio, the duration of the induction time significantly decreases when lowering the activator modulus, also the commence time of the acceleration peak is decreased. These changes demonstrate the acceleration in reaction that is caused by the decreased activator modulus.

Fig. 8.2 (left) shows the infrared spectra of the unreacted slag and fly ash. The main vibration band for slag is at around 900 cm^{-1} and about 1020 cm^{-1} for fly ash, which is associated with the asymmetric stretching vibration of Si–O–T bonds (where T = Si or Al units) [34]. Fig. 8.2 (right) shows the infrared spectra of samples with the activator moduli of 1.8, 1.4, 1.0 (represented by A, C, E) and slag/fly ash mass ratios of 90/10, 70/30, 50/50 (represented as 1, 3, and 5, respectively) after alkali activation. Regardless of the activator modulus and raw materials' relative content, all specimens generally exhibit similar absorption band positions. This includes OH groups at 1640 cm^{-1} and around 3200 cm^{-1}; O–C–O at around 1420 cm^{-1} [35]; and a main absorption band at

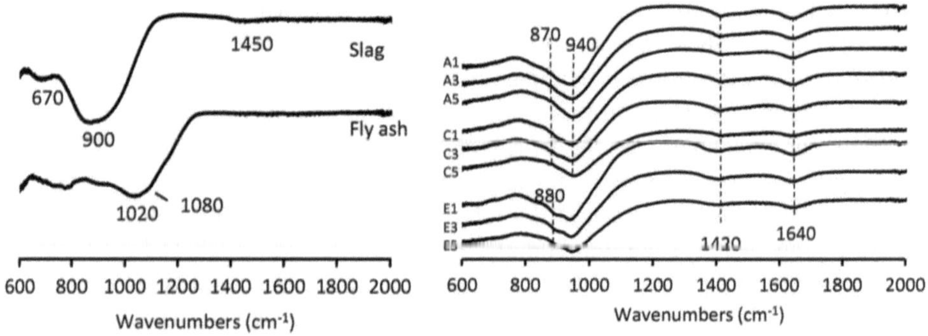

Fig. 8.2: FTIR spectra of starting materials (left) and alkali-activated slag/fly ash blends (right) [30].

940 cm^{-1}, which is assigned to the asymmetric stretching vibration of Si–O terminal bonds [36], indicating the formation of C-A-S-H type gels with short chain structures. It can be noted that the absorption bands from around 1000 to 1100 cm^{-1}, which are usually identified from alkali activated fly ash, are not significant in all mixtures.

8.1.3.2 Modification by nano-silica

Waterglass modified with sodium hydroxide activation has shown its benefits in terms of strength development and durability. However, the production of waterglass and sodium hydroxide is not environmentally friendly. Considering the high pozzolanic reactivity of nano-silica owing to its amorphous nature and high specific surface area [37], nano-silica can potentially serve as a Si source for the modification of the modulus of waterglass. On the other hand, the presence of nano-silica can also act as nuclei sites for the precipitation of reaction products, enhancing the strength development of mixtures. The influence of nano-silica on early age reaction kinetics of alkali activated slag/fly ash blends is presented.

The normalized heat flow of alkali activated slag/fly ash blends (with a slag/fly ash ratio of 70/30 and nano-silica replacement from 0 % to 3 %) within the first 72 h are shown in Fig. 8.3 (left). It can be seen that for samples without nano-silica addition, the heat release peak in the acceleration stage is located at around 8.3 h after mixing; while as the nano-silica content increases, this peak slightly shifts to longer times with lower intensities. This result is in contrast with previous studies where the presence of nano-silica accelerates the reaction process in a cement based system [50, 55–57]. Overall, the reaction process is slightly delayed with lower peak intensities when nano-silica is added.

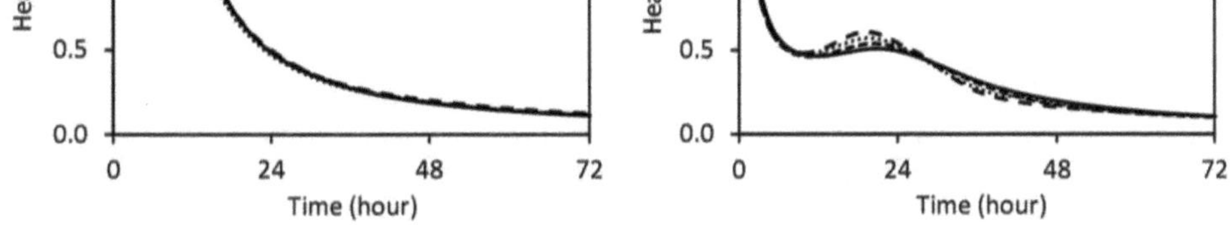

Fig. 8.3: Normalized heat flow of (left) AA slag/fly ash (70/30) and (right) AA slag/fly ash (30/70) pastes with nano-silica (NS) addition [112].

Fig. 8.3 (right) illustrates the normalized heat flow of samples with a slag/fly ash ratio of 30/70 and nano-silica replacement from 0 % to 3 %. Compared to the samples with a slag/fly ash ratio of 70/30, it can be seen that both the induction and acceleration stages are remarkably retarded. For instance, the induction period is located at around 3.7 h in mixes with a slag/fly ash ratio of 70/30, while it delays for approximately 4.5 h when the slag/fly ash ratio is changed to 30/70. Similarly, the acceleration peak shifts from about 8.3 h to 20 h, and the peak intensity decreases from about 1.5 mW/g to 0.5 mW/g. These changes demonstrate that the slag/fly ash ratio shows a dominant influence on the early age reaction. An increase in slag content can effectively accelerate the main reaction stage and result in a more intensive reaction. This is due to the intrinsic differences in the amorphous structure between slag and fly ash. Slag has a remarkable Ca content which leads to a more disordered glassy framework than an aluminosilicates dominated structure (fly ash in this case) [38]. Thus slag exhibits a much higher reactivity than fly ash under alkali activation.

8.1.3.3 Role of limestone powder

The utilization of limestone powder can further lower the environmental impact of producing alkali activated materials, but its role in AAM has not yet been systematically studied. In this study, the physical and chemical influence of limestone powder on alkali activated hybrid systems is investigated. Alkali activated slag/fly ash based mortar mixes with slag content (60, 50, and 40 %) and limestone additions from 0 % to 30 % by mass are used.

The normalized heat flow of alkali activated slag/fly ash-limestone blends (with a slag content of 60 % and limestone addition from 0 % to 30 %) within the first 72 h are shown in Fig. 8.4 [39]. The acceleration peak is located at around 16 h after mixing, which is assigned to the massive formation of reaction products from dissolved Ca, Si and Al units [40]. As the limestone powder content increases, the heat evolu-

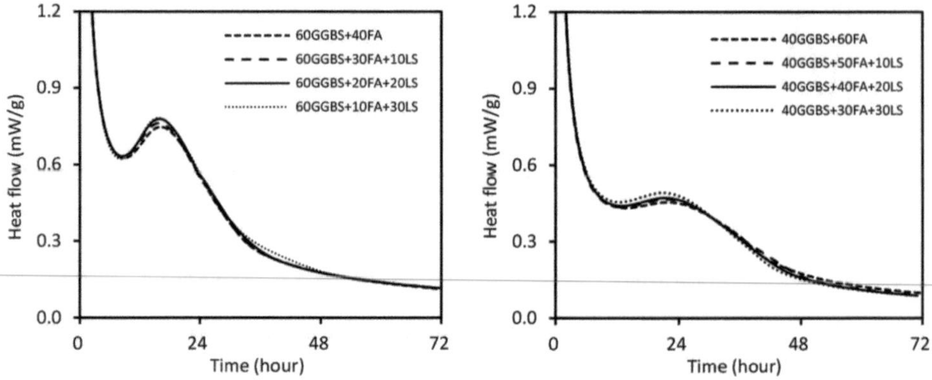

Fig. 8.4: Normalized heat flow of AA slag–fly ash–limestone pastes [113].

tion peak shifts slightly to earlier positons with higher intensities. It indicates that the reaction process is slightly accelerated in the presence of limestone powder. This is in line with previous research on the effect of limestone powder in Portland cement hydration [41, 42].

However, when compared to the mixes shown in Fig. 8.3, it can be noticed that both the induction and acceleration stages are significantly retarded. Similarly, when lowering the slag content from 60 % to 40 %, the acceleration peak shifts from about 16 to 20 h with a decreased peak intensity from about 0.8 to 0.5 mW/g.

8.1.4 Slag activated by ternary activators

Though waterglass is often reported as the most effective activator in AAS regarding strength and durability, problems [43, 44] such as fast setting and high shrinkage are also widely reported [7, 9, 45–49]. On the other hand, the prolonged setting time of sodium carbonate activated slag has often been observed. In this study, combinations of waterglass, sodium hydroxide, and sodium carbonate as the activators of slag binder were investigated and their effects on the reaction kinetics, compressive strength, and reaction products were analyzed.

8.1.4.1 Reaction kinetics

Fig. 8.5 depicts the heat release of slag activated by the ternary alkali activators (T1–9) (Tab. 8.1). When changing the proportions of waterglass and sodium carbonate, the reaction kinetics of the samples clearly differ, leading to varied reaction process and strength development. Overall, T4 shows the fastest reaction rate followed by T8 and T7, while T9 shows the slowest reaction rate.

Applying the analysis of variance (ANOVA) method, the percentage contributions of each factor on the length of time to reach reaction peak (TRRP) and on the total heat release on day 7 were calculated (more details are presented elsewhere). The order of the factors affecting the TRRP/reaction process were derived, yielding; WSR (72.6 %) > waterglass content > Na_2CO_3 content > waterglass modulus (1.8 %). For the heat release on day 7, the order was; waterglass content (71.8 %) > Na_2CO_3 content > waterglass modulus > WSR (0.7 %).

8.1.4.2 Role of ternary activators on the reaction products

The X-ray patterns of the anhydrous slag and slag activated by the ternary activators (T4–9) at the age of 28 d are shown in Fig. 8.6. Depending on the activators used, the main reaction products are calcite, hydrotalcite, C-(A)-S-H gel and gaylussite.

Tab. 8.1: Mix design of the samples activated by ternary activators [114].

Mix	SCC Na_2O wt%	WSR	WGM	WGC Na_2O wt%
T1	3	0.40	1.5	0.5
T2	3	0.45	1.3	1.5
T3	3	0.50	1.1	2.5
T4	4	0.40	1.3	2.5
T5	4	0.45	1.1	0.5
T6	4	0.50	1.5	1.5
T7	5	0.40	1.1	1.5
T8	5	0.45	1.5	2.5
T9	5	0.50	1.3	0.5
T10	5	0.40	1.3	2.5
T11	4	0.40	–	2.5 (NaOH)
T12	4	0.40	3.6	2.5

SCC = sodium carbonate, WSR = water to solid ratio, WGC = waterglass content, WGM = waterglass modulus

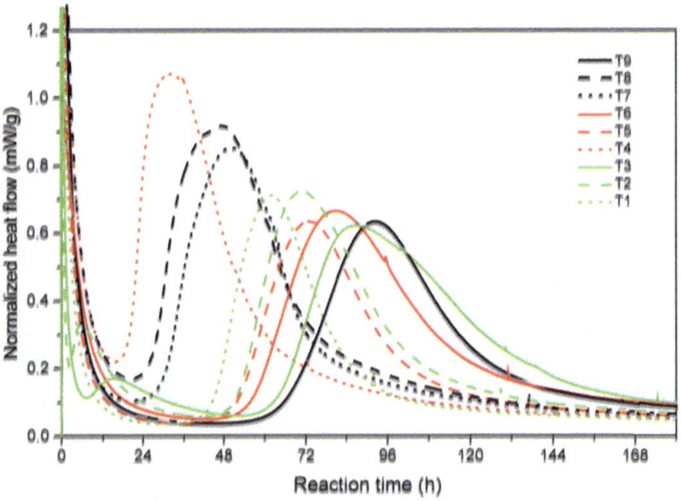

Fig. 8.5: Heat release of slag activated by the ternary activators [114].

As confirmed by the previous research, the reaction rate of slag activated by only sodium carbonate is significantly low, indicated by a long dormant period which is caused by the initial precipitation of calcium carbonate. Owing to the addition of waterglass ($Na_2O \cdot r \cdot SiO_2 \cdot n \cdot H_2O$), the reaction process is promoted. The first possible reason is related to the pH of the alkaline solution increased by the additional Na_2O content in waterglass, which has the similar effect as NaOH [50–52]. The second pos-

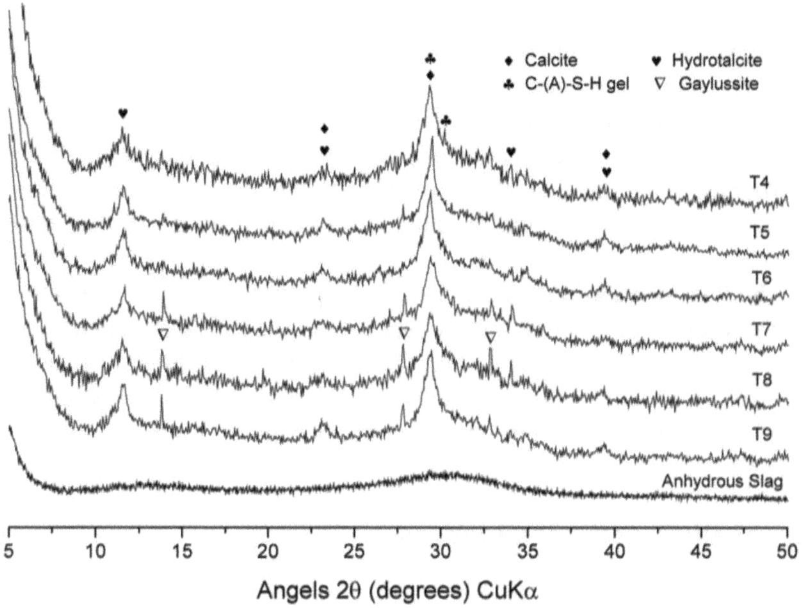

Fig. 8.6: XRD patterns of anhydrous slag and slag activated by ternary activators [114].

sibility is the additional Si element provided by waterglass, which will trigger the formation of C-(A)-S-H gel, consuming carbonate anions and promoting strength development especially at an early stage. The first mechanism is clearly observed as indicated by the dominant role of waterglass dosages (based on Na_2O wt% content) on reaction rates and strength development, while the second mechanism is not prominent since the effect of WGM on the reaction is negligible.

8.1.5 Conclusions and the future

It is known that alkali nature is a determinant of the mechanical properties and microstructure of alkali activated slag (AAS) [7, 27, 53]. Currently all activators investigated have either positive or negative aspects with respect to the setting time, strength development, durability, energy-consumption, etc. Furthermore, though waterglass is often reported as the most effective activator in AAS regarding strength and durability [43, 44], its production has a significant CO_2 footprint as it involves high energy consumptions when melting quartz and soda at about 1300 °C [54]. Problems of fast setting and high shrinkage are also widely reported [7, 9, 45–49]. Therefore, a low usage of waterglass is preferred with respect to environmental issues and fresh paste behavior. However, a diluted alkaline solution often leads to weakened mechanical properties and durability. High silica containing sources such as nano-silica have been applied as activators to modify the waterglass modulus. Compared to waterglass activated slag, the mechanical properties of slag activated by other activators such as sodium hydroxide, sodium carbonate or sodium sulphate are relatively poor [27, 49, 55]. Nevertheless, these activators can potentially provide benefits in terms of environmental, reaction kinetics and durability related properties. For example, slag activated by sodium carbonate generally shows a long setting time due to the initial precipitation of $CaCO_3$ [7] which can potentially compromise the fast setting of waterglass activation.

Until now, limited attention has been paid to the investigation of hybrid activators. This is probably because the reaction mechanism behind the hybrid system is much more complex than the sole alkali activation. However, mixes of different types of alkaline solutions as activators could potentially lead to a desirable material which resolves the problems of energy consumption, strength development and durability, etc.

8.2 Design of alkali activated binders

8.2.1 Introduction

As discussed in Section 8.1, alkali activated materials can generally be classified into two types based on their chemical compositions and reaction mechanisms, namely the calcium and silica enriched (Ca + Si) system and low calcium aluminosilicates system [38]. Each system exhibits distinguishable properties regarding rheology, setting, mechanical properties, shrinkage, durability, alongside the required alkali and curing conditions due to their intrinsic differences in the structure of the raw materials and reaction mechanisms. Thus significant differences exist in determining the key synthesizing parameters. Considering that alkali activated materials are sensitive to the key manufacturing factors, designing a certain mix with appropriate starting composition is of great importance. This section will present a brief introduction concerning the design progress of two typical alkali activated systems, as well as the design and modifications of hybrid binders based on the understanding of each individual system. Finally recommendations will be provided for potential research directions.

8.2.2 Review of the design of alkali activated binders

8.2.2.1 Alkali activated high calcium binders

For alkali activated high calcium materials, different types of slag: phosphorous, steel, blast furnace etc., can be applied as a starting precursor in alkaline activation. Amongst these, granulated blast furnace slag is particularly favored due to its high amorphous oxides contents such as SiO_2, Al_2O_3, CaO, and MgO during the quenching process [56]. The structure of slag can be regarded as an over-charge-balanced calcium aluminosilicates framework [57]. The reactivity depends mostly on the phase composition and glassy structure content. It is generally believed that the final performances of alkali activated materials are related to the type and dosage of alkali activator, fineness and glassy chemistry of the raw material and curing conditions such as temperature, time, and relative humidity (RH) [53].

The main oxides such as CaO and SiO_2 will eventually form C-A-S-H gels as main reaction products, thus the content of these reactive oxides exhibit a dominant influence. Secondary main oxides such as MgO and Al_2O_3 also play an important role. In his study, Douglas reported that as the MgO content increases, the compressive strength of alkali activated slag will be much higher [58]. A remarkable increment in strength due to the higher MgO contents was also observed [11], and more hydrotalcite type products were formed in mixes with higher MgO contents. The influences of Al content in alkali activated slags were also studied [11]. Different slags with Al content between 7 % to 17 % were chosen, and results shows that the increasing Al content has a negative influence on the early age reaction degree and early strength, but no significant influence on the long term strength.

It is generally believed that the characteristics of activators have a decisive influence on the final performance of alkali activated materials. Provis showed that in the case of alkali activated slag, silicate based activators are recommended with desirable performances [59]. Carbonate-based activators are also suitable, while alkalis and strong acid salts only give acceptable properties. Wang summarized the influences of dosage and modulus of activator on the strength of alkali activated slag [43]. By using sodium hydroxide as the activator, within the range of 1 % to 10 %, the higher dosages of activator lead to increased strength. Other researchers have found there is an optimum dosage in terms of strength. In the case of using sodium silicate, which is often linked to high mechanical properties, the usage of 2 % to 8 % equivalent Na_2O is necessary for alkali activation, the optimum content also depends on slag chemistry, activator type and curing conditions. The increase of activator dosage also leads to a higher cost; generally the recommended activator content is in the range of 3 % to 5.5 % of equivalent Na_2O by weight of slag. Another important parameter in silicate based activator is the activator modulus (SiO_2/Na_2O ratio); the silica provided by the activator will eventually participate in the hydrated gel. Generally the ideal strength can be achieved by using a SiO_2/Na_2O ratio within the range of 1.0 to 1.5, depending on slag chemistry and curing conditions [7, 9, 26, 30, 39, 53].

8.2.2.2 Alkali activated aluminosilicates binders

Typically the used precursors of aluminosilicates are class F fly ash and metakaolin. The design of these two materials presents slight differences due to the differing origins of amorphous structure but a similar design process can still be applied. Other starting precursors containing primarily amorphous silicate and aluminate can also follow the same role. The total content of silica and alumina, as well as ratios of the two play an essential role in the degree of reaction and gel structure. The starting materials such as class F fly ash and metakaolin usually contain a certain amount of crystalline phases like quartz and mullite, namely non-reactive SiO_2 and Al_2O_3, which are counted in the starting chemical composition but hardly take part in the reaction process. A desired mechanical property is usually obtained with an initial SiO_2/Al_2O_3 ratio range from 2 to 4 [60, 61], where SiO_2 and Al_2O_3 only represent the reactive phases. Thus a detailed determination concerning the starting precursors is important.

Fernandez and Palomo investigated the effect of activator Na_2O/SiO_2 ratios on alkali activated fly ash, concluding that the strength cannot simply be linked with Na_2O content or Na_2O/SiO_2 ratio [62]. A combined influence of both Na_2O content and Na_2O/SiO_2 ratio, as well as the addition of extra silica from sodium silicate contributes to the polymerization process. With this a denser matrix with a higher strength can be predicted. Palomo et al. [63] verified the importance of extra silica from activators of waterglass. When solely sodium hydroxide with a concentration from 8 to 12 M was used, the 28 d compressive strength of 35–40 MPa was obtained. However, if waterglass was incorporated into the activator, the compressive strength reached nearly 90 MPa with a SiO_2/Na_2O ratio of 1.23. An optimum modulus of 1.5 and Na_2O content of around 10 % in terms of strength was also suggested by other researchers [64–66].

Duxson and Provis studied the metakaolin based geopolymer, and analyzed the influence of total Si/Al ratio in the system [67]. Samples with Si/Al ratios between 1.15 and 2.15 were prepared. The compressive strength increased significantly from Si/Al = 1.15 to Si/Al = 1.90, then decreased with the increase of the Si/Al ratio. The improvement of strength is nearly linear in the range of $1.15 \leq Si/Al \leq 1.90$. Thus the total Si/Al ratio plays a decisive role on the compressive strength of the final products. The suggested total Si/Al ratio for activating metakaolin is around 2. Silva investigated the role of Al_2O_3 and SiO_2 on geopolymerization kinetics [68], synthesizing samples with different SiO_2/Al_2O_3 and Al_2O_3/Na_2O ratios. The strength outcomes indicate that higher strength can be obtained by increasing the total SiO_2/Al_2O_3 molar ratio with SiO_2/Al_2O_3 ratios around 3.4 to 3.8 being suitable for a high strength. Increasing the Al content, namely lowering the SiO_2/Al_2O_3 ratio, will lead to a lower strength. This is due to the formation of Na-Al-Si grains rather than amorphous N-A-S-H that is mainly responsible for strength development. Rowles studied the influence of the ratios between Si, Al, and Na on strength and microstructure of alkali activated metakaolin [69]. The results showed that samples with a Si/Al ratio around 2.5 and Na/Al around 1.25 refer to the highest compressive strength. When either the Na/Al or Si/Al is kept constant, an optimum value exists in the other one.

8.2.3 Room temperature cured alkali activated composites

8.2.3.1 Design of alkali activated slag/fly ash blended binder

Ambient temperature curing shows advantages in energy saving and convenience of application. The compressive strength of specimens after 28 days of curing is given in Fig. 8.7. The samples show compressive strengths ranging from 71.6 MPa to 100.9 MPa, which is promising for high performance applications. Higher slag contents lead to higher compressive strength in general. When the slag/fly ash mass ratio is kept constant, all mixes show a shift in the optimal Ms for compressive strength. In addition, mixes with a higher slag content exhibit a higher optimal activator modulus. The compressive strength decreases when increasing the fly ash content in general, because the glassy phases of slag are more vulnerable to alkaline attack than the aluminosilicate-enriched phases from fly ash under room temperature [70, 71]. Slag generally has a higher content of reactive phase than fly ash [21, 72].

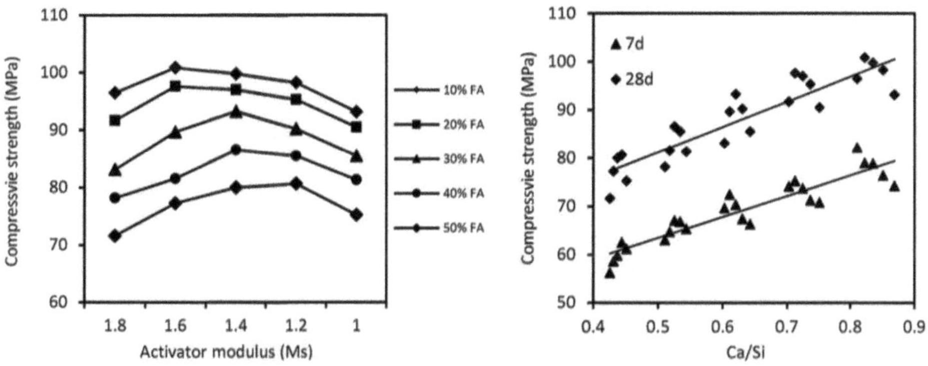

Fig. 8.7: Compressive strength of alkali activated slag/fly ash blends [30].

Fig. 8.7 shows the relationships between the total Ca/Si molar ratio and the compressive strength after 7 d and 28 d of curing, where the chemical composition of the starting materials, including solid precursors and liquid activator, are computed into the molar content. This figure shows clearly that although slag/fly ash mass ratio has a dominating influence on the total Ca/Si ratio, the activator modulus still has a non-ignorable effect. The extra silicate that is provided by the activator contains 6–9 % by mass of the total binder (from activator modulus 1.0 to 1.8). It can also be found that the calcium content is positively related to the compressive strength, which has also been confirmed [73, 74]. The strength variation caused by the activator modulus is comparable with the effects of slag/fly ash mass ratio to some extent, which reveals that silicate content is as important as calcium content in terms of compressive strength. Thus determining these two factors simultaneously in the mix design stage is of great importance.

8.2.3.2 Design of hybrid activator based alkali activated material

Fig. 8.8 shows the compressive strength of slag activated by ternary activators at different ages (7 d and 28 d) and under different curing regimes (i.e. water and RH 95 %) with their reference samples. Mix T4 provides the highest compressive strength of 63.57 MPa (7 d) and 69.10 MPa (28 d), while mix T9 gives the lowest 22.87 MPa (7 d) and 36.07 MPa (28 d). The strength gain from 7 d to 28 d of all mixtures, except T4, ranges from 10.2 MPa to 15.5 MPa, while for the references samples this is 6.9–9.9 MPa. In other words, the incorporation of waterglass not only improves the compressive strength of samples at the early age, but also has a positive effect on strength development at a later stage. Overall, the differences between the 28 d compressive strength of samples cured in water and RH 95 % are very small, indicating either of these two curing regimes is practically applicable.

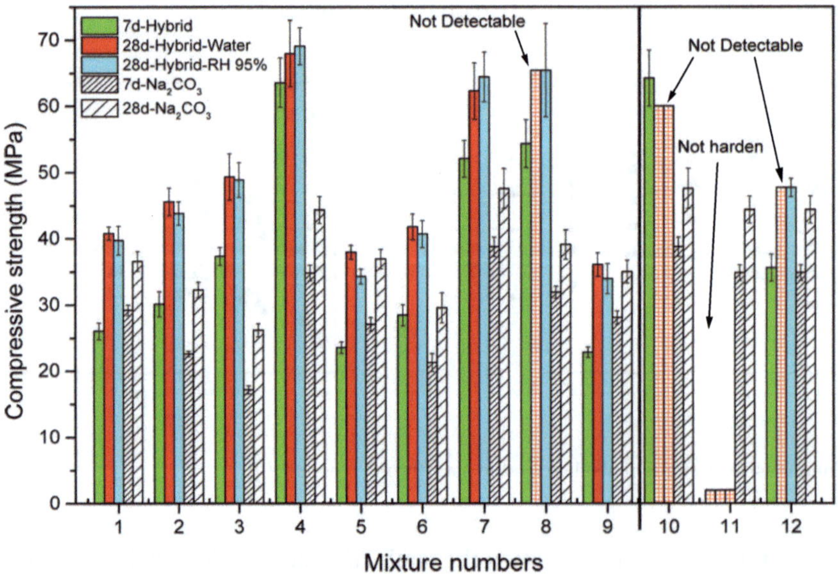

Fig. 8.8: Compressive strength of slag activated by ternary activators and sole Na_2CO_3 at different curing ages and curing conditions [114].

Furthermore, the percentage contribution of each factor on the compressive strength at different ages and different curing regimes is evaluated by applying ANOVA. The waterglass content plays a dominant role on the compressive strength, followed by water to solid ratio and sodium carbonate, while the contribution of waterglass modulus is relatively insignificant. Overall, though with lower contents, the contribution of waterglass is more significant than sodium carbonate on strength development.

8.2.4 Role of nano-silica in alkali activated slag/fly ash blends

8.2.4.1 Mechanical property

The compressive strength of samples with a slag/fly ash ratio of 70/30 and nano-silica content from 0 % to 3 % is shown in Fig. 8.9 (left). In mixes without nano-silica addition, the compressive strength is 53.5 MPa after 3 d of curing; and it increases to 66.6 MPa at 7 d and 90.2 MPa at 28 d When nano-silica is incorporated, the compressive strength is slightly increased at 3 d, and it can be seen that on increasing the nano-silica content from 0 % to 3 %, the compressive strength gradually increases from 53.5 MPa to 57.9 MPa. It should also be noted that when the nano-silica content reaches 2 %, the strength increment is no longer significant when more nano-silica is incorporated. Similarly, at the age of 7 d, the compressive strength is increased from 66.6 MPa to 71.6 MPa when increasing the nano-silica content up to 2 %, but the strength is slightly reduced to 69.5 MPa when the nano-silica content increases to 3 %. It indicates that a nano-silica content of 2 % is the optimum in terms of strength in both cases. It can be observed that the rate of increase in strength is lowered with the increase of nano-silica content. Concerning the 28 d strength; when increasing the nano-silica content from 0 % to 3 %, the compressive strength firstly increases from 90.2 MPa to 95.9 MPa, then decreases to 92.4 MPa, showing again an optimum nano-silica content of 2 %.

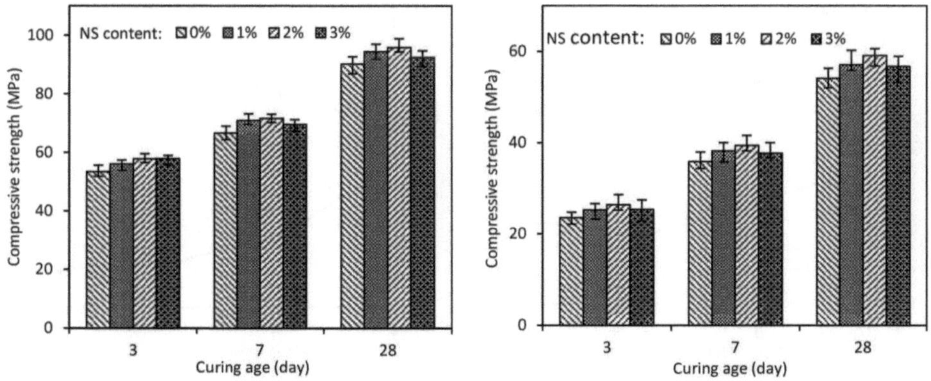

Fig. 8.9: Compressive strength of (left) AA slag/fly ash blends (70/30) and (right) AA slag/fly ash blends (30/70) with nano-silica addition [112].

Fig. 8.9 (right) depicts the compressive strength results of samples with a slag/fly ash ratio of 30/70. It is obvious that the strength is remarkably lower than those with a slag/fly ash ratio of 70/30, caused by the significant reduction of slag content as discussed elsewhere [75]. It can be seen that without the nano-silica addition, the 3, 7, and 28 d strength are 23.6 MPa, 35.9 MPa, and 54.1 MPa, respectively. The compressive strength is increased by 52.1 % and 129 % at these three typical curing ages. In samples with a slag/fly ash ratio of 70/30, these values are 24.5 % and 68.6 %, respectively, indicating that there is a difference in strength development between samples with different slag/fly ash ratios. In other words samples with a higher slag content show a higher early strength and a lower rate of increase. The compressive strength firstly increases when the nano-silica content is brought up to 2 %, then slightly reduces when the nano-silica content reaches 3 %; this tendency becomes more significant at older ages. The optimum nano-silica content of 2 % in terms of compressive strength is shown in all mixes, and the highest strength present in samples with 2 % nano-silica are 26.4 MPa at 3 d, 39.4 MPa at 7 d, and 59.1 MPa at 28 d, respectively.

8.2.4.2 Porosity

Fig. 8.10 shows the relationships between the compressive strength, nano-silica content, and porosity of alkali activated slag/fly ash blends after 28 d of curing. It can be seen that the compressive strength increases with the reduction of porosity in general, confirming again the inverse relationship between strength and porosity in materials. The porosity generally decreases with the increase of the nano-silica content. On increasing the nano-silica content up to 3 %, the reduction of porosity becomes less significant or the porosity even slightly increases.

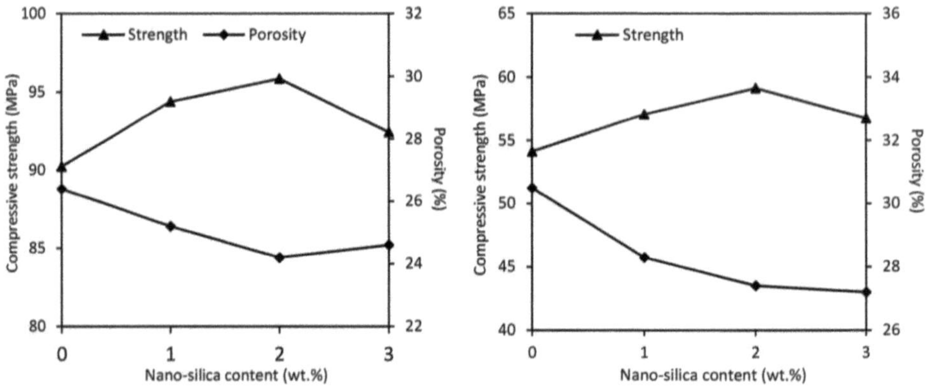

Fig. 8.10: Relations between 28 d strength, porosity and nano-silica content: (left) slag/fly ash = 70/30 and (right) slag/fly ash = 30/70 [112].

As presented in Fig. 8.10 (left), on increasing the nano-silica content from 0 % to 2 %, the compressive strength increases from 90.2 MPa to 95.9 MPa, while the porosity decreases from 26.4 % to 24.2 %. However, the further increase of nano-silica content to 3 % leads to a slight decrease of strength and increase of porosity. One possible explanation for this phenomenon is that a suitable content of nano-silica will refine the pore structure and generate more reaction products through a pozzolanic effect.

Higher contents of nano-silica lead to a significant reduction of workability (increased cohesiveness), which may result in an increase of the air content within the paste and the porosity is increased as a result. Fig. 8.10 (right) shows the results of samples with a slag/fly ash ratio of 30/70, it is obvious that higher porosities as well as lower compressive strengths are presented compared to the mixes with a slag/fly ash ratio of 70/30. This is due to the relatively low reactivity of fly ash under ambient temperature. This results in a lower content of reaction products and thus less dense matrixes and lower strengths are exhibited. Specifically, when increasing the nano-silica content from 0 % to 3 %, the porosity decreases from 30.5 % to 27.2 %, while the compressive strength firstly increases from 54.1 MPa to 59.1 MPa, then slightly decreases to 56.8 MPa at the nano-silica dosage of 3 %.

8.2.5 Limestone powder modification in alkali activated slag/fly ash blends

8.2.5.1 Mechanical property

The 7 d and 28 d compressive strengths of alkali activated slag/fly ash/limestone pastes and mortars are shown in Fig. 8.11. The water/binder ratio for paste samples is 0.35, and 0.45 for mortars. For samples with a constant slag content, both the 7 d and 28 d compressive strengths are increased when raising the limestone powder content (546 m^2/kg); the highest strength is shown in samples with the 30 % limestone powder addition. This behavior should be attributed to the filler effect of limestone powder. The addition of limestone powder leads to a higher content of fine particles within the paste, which could work as micro aggregates and reduce the total porosity, resulting in an increment of strength. Moreover, the better reaction of slag due to the presence of limestone also contributes to the improvement in strength, since the additional nuclear sites provided by the fine limestone particles promote the formation of the reaction products. It is obvious that with a constant limestone replacement, higher compressive strengths are achieved in samples with higher slag contents at both 7 d and 28 d. This should be attributed to the considerably higher reactivity of slag relative to the other components under ambient temperature. Concerning the mortar mixes, the filler effect of limestone powder and the remarkable influence of slag on strength are also obviously presented.

As shown in Fig. 8.11, based on a constant slag level, the compressive strength increases with the increasing limestone content; also based on a constant limestone addition, both the 7 d and 28 d strengths increase with the raised slag content. The highest values are obtained in samples with a 60 % slag and 30 % limestone powder, showing 7 d and 28 d compressive strengths of 51.9 MPa and 64.2 MPa, respectively. While the lowest 7 d and 28 d strength of 46.6/59.2 MPa is shown in samples with 40 % slag and 0 % limestone.

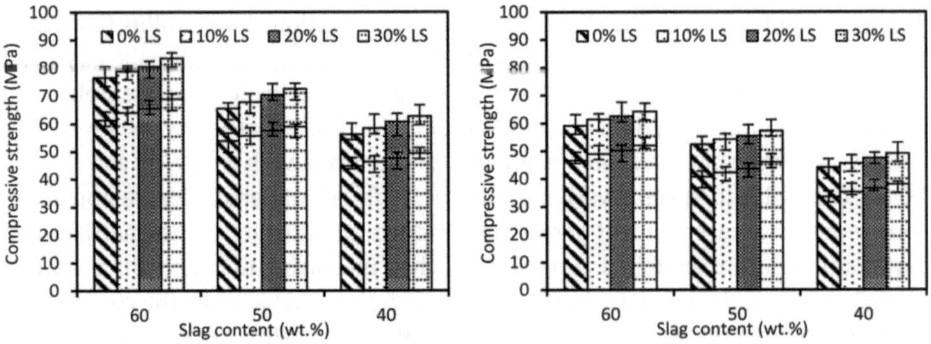

Fig. 8.11: 7 d and 28 d compressive strength of AA-slag–fly ash–limestone composites (left: pastes with $w/b = 0.35$, right: mortars with $w/b = 0.45$) [113].

8.2.5.2 Microstructure

The thermogravimetry results of samples with a slag content of 60 % and limestone additions up to 30 % are presented in Fig. 8.12. Here a calculation on physically and chemically bound water contents was carried out. The physically bound water content was calculated as the mass loss before 105 °C; the chemically bound water content in each mix was calculated by the mass loss between 105 and 1000 °C, with the exclusion of the mass loss due to the incorporated limestone. The results are presented in Tab. 8.2, where it can be seen that the addition of limestone powder leads to a slight increase in the total chemical water content. This phenomenon can be partly explained by the XRD results in this study; the addition of limestone powder does not evidently lead to the formation of additional reaction products, thus no remarkable amount of extra bound water can be observed. However, it is possible that the additional nucleation sites that are provided by the fine limestone particles promote the formation of hydrated gels and refine the pore structures, resulting in slightly increased physically and chemically bound water contents; or a slight but non-ignorable amount of Ca^{2+} is released from the fine limestone particles under alkali activation, which participates in the reaction process and leads to a slight increase in the bound water content.

Tab. 8.2: Calculation of bound water content (wt%) [39].

GGBs-FA-LS (%-%-%)	60-40-0	60-30-10	60-20-20	60-10-30
Mass loss between 105 and 1000 °C	5.09	8.11	11.46	14.53
Limestone powder incorporated	0	6.70	13.38	20.07
Mass loss of limestone powder	0	2.88	5.75	8.64
Chemical bound water	5.09	5.23	5.71	5.89
Physical bound water	16.1	16.9	16.9	16.9

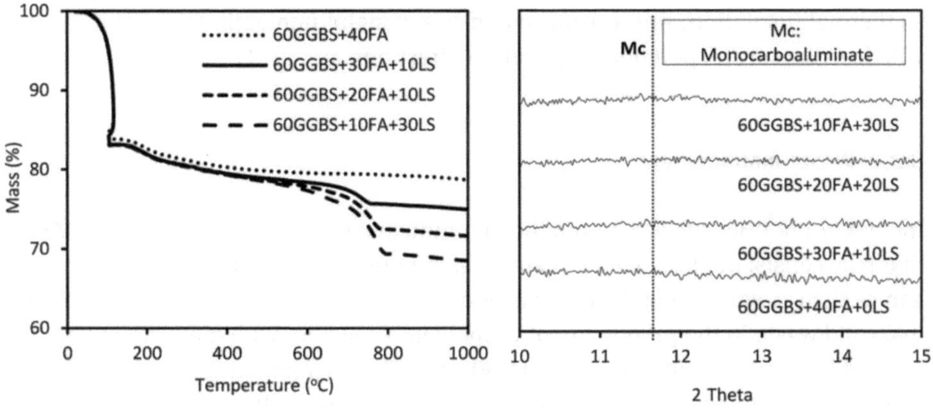

Fig. 8.12: TG and XRD analysis of alkali activated slag/fly ash/limestone pastes.

8.2.6 Conclusions and the future

From the available literature, it can be seen that the design of alkali activated individual systems are well studied. For high calcium mixes, the total Na_2O content and activator modulus are key parameters in alkali activation of slag, while in the case of alkali activated aluminosilicates, the total reactive SiO_2/Al_2O_3 ratio exhibits a dominant effect. Concerning the design of hybrid binder systems, high performance alkali activated slag/fly ash blends can be achieved under ambient temperature curing. Slag plays a dominant role in determining the reaction process and strength due to its relatively high reactivity. Also, additional silicate provided from the activator is of great importance. Modifications based on the hybrid alkali system show the potential of making this alternative binder even more environmentally friendly with comparable properties. The addition of limestone powder modifies the alkali activated slag/fly ash blends both physically and chemically. Even though the chemistry and microstructure are better understood in the hybrid systems, to determine key factors, an improved theoretical basis is still needed since there exists no guided rule in the design stage.

8.3 Applications of alkali activated binders

8.3.1 Introduction

Due to the limited understanding of alkali activated materials, there still exists limited application of them in worldwide construction up until now. Compared to existing systems, an alternative binder system requires a modified performance evaluation which also limits the large scale application of this type of material. However, attributed to the nature of alkali activated materials, excellent mechanical properties can be achieved together with durability and a number of functional properties such as thermal insulation, fire resistance and heavy metal immobilization. This chapter will describe several potential application directions.

8.3.2 Waste management

Incinerator bottom ash is a by-product of municipal solid waste incineration (MSWI) that may consist of glass, metals, ceramics, minerals, stone, brick, and unburned organics such as wood and plastics. Lancellotti studied ash from incineration of municipal solid waste in the production of geopolymers together with metakaolin [76]. The applied bottom ash content was up to 80 wt%. The microstructure analyses demonstrated that bottom ash is a suitable source material for producing geopolymers with percentages from around 50 to 70 wt%. If the bottom ash contains a certain amount of amorphous CaO, Al_2O_3, and SiO_2, then it is possible to be activate using only alkali. Yamaguchi reported a flexural strength of up to 16 MPa by using urban waste material under the curing temperature of 80 °C [77].

Alkali activated binders have also shown the potential to immobilize radioactive waste with low or intermediate levels. Cs and Sr are the two main concerns in this case because their immobilization in Portland cement system is not ideal (due to weak binds and a tendency to elute) [78, 79]. However, several studies have found that Cs can be stably bound in aluminosilicate and other zeolite-like structures and a much lower leaching is usually shown when compared with Portland cement system [80]. The use of hybrid binder system was also reported to improve the binding capacity to Cs by the specific adsorption properties of coexisting C-A-S-H and N-A-S-H structures [81]. Alkali-activated materials have also been observed to have the ability to bind other radioactive wastes containing Co and Ru [82].

Besides being activated as a binder component, applying bottom ash with bigger particle sizes in alkali activated systems also shows several benefits. The primary environmental issue of incinerated bottom ash comes from the leaching of heavy metals such as lead, copper and antimony, which could lead to severe damage to water resources and human health. Thus appropriate treatments are required before application. One study of the application of bottom ash in alkali activated slag mortars was carried out to investigate heavy metal binding behavior. The bottom ash used was incinerated urban waste from the Netherlands, with a maximum size fraction of up to 2 mm. The applied dosage was up to 50 % by volume of the fine aggregates. The results show that the alkali activated slag binder can inhibit the leaching of ions such as Cu and Se, passing Dutch regulations which limit the amount of leachable contaminants from construction materials (Soil Quality Decree).

8.3.3 High performance building materials

Even though excellent performance can be achieved from the blended alkaline systems, the relatively high shrinkage during drying, due to the nature of both the raw materials and activators, is still a remaining issue that limits large scale application. The application of steel fiber in Portland cement systems has proved its advantages in improving the flexural strength, fracture toughness, impact and fatigue resistance [83, 84]; as well as its efficiency in reducing shrinkage behavior of the brittle matrix [85, 86]. Therefore, high performance designs of alkali activated slag/fly ash composites alongside steel fibers of two different lengths are applied for strength reinforcement and to inhibit shrinkage.

All mortar mixtures are designed by applying the modified Andreasen & Andersen (A&A) particle packing model (equation (8.1)), in order to achieve optimal packing of the granular ingredients:

$$P(D) = \frac{D^q - D_{\text{min}}^q}{D_{\text{max}}^q - D_{\text{min}}^q}, \tag{8.1}$$

where $P(D)$ is a fraction of the total solid materials which are smaller than the particle size D (μm); D_{max} is the maximum particle size (μm); D_{min} is the minimum particle size (μm); and q is the distribution modulus. The distribution modulus q in the modified A&A model is used to determine the proportion between fine and coarse particles in the mixture. The value of q is fixed at 0.23 for all mixtures based on experiences in previous studies [87, 88]. The proportions of each individual material in the mix are adjusted until an optimum fit between the composed mix grading curve and the target curve is reached, using an optimization algorithm based on the least squares method (LSM), namely the deviation between the target curve and the composed mix expressed by the residual sum of squares (RSS) at defined particle sizes is minimal. Therefore, the optimized mixture will possess a compact matrix due to optimal packing. This packing model has been successfully applied to concrete designs with various properties applying different materials as binders [89–98].

Fig. 8.13 shows the 7 d and 28 d compressive strengths of mixes with a total fiber content of 1 % and different long/short fiber ratios. The strength firstly increases when lowering the long/short fiber ratio, reaching the maximum strength in mixes with long/short ratio of 60/40, followed by a gradual decrease. This result indicates the beneficial effect of using hybrid steel fiber on compressive strength: a higher strength can be achieved with the same fiber content. Their 28 d flexural strengths are presented in Fig. 8.14. As the long fiber friction increases, mixes generally show a higher energy absorption capacity and a lower stress drop rate after reaching the stress peak. This again shows the higher efficiency of long fibers in bridging the macro-cracks and therefore more stable post-peak responses. The highest flexural strength is shown in mixes with a short/long fiber ratio of 40/60; showing the positive effect of hybrid fiber on mechanical properties.

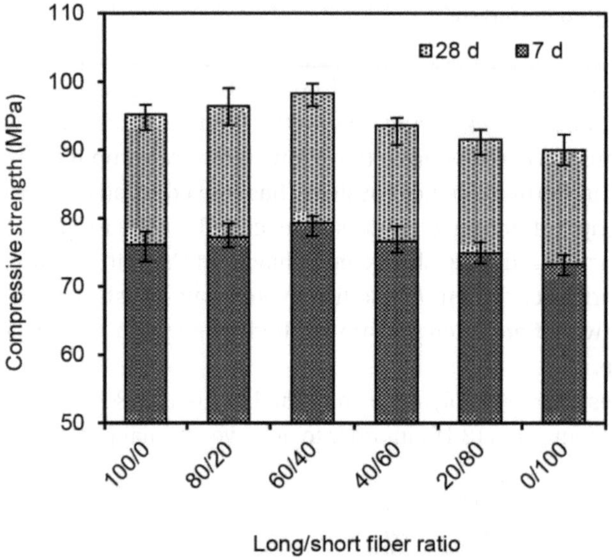

Fig. 8.13: 7 d and 28 d compressive strength of AA slag/fly ash composites with hybrid fiber addition [115].

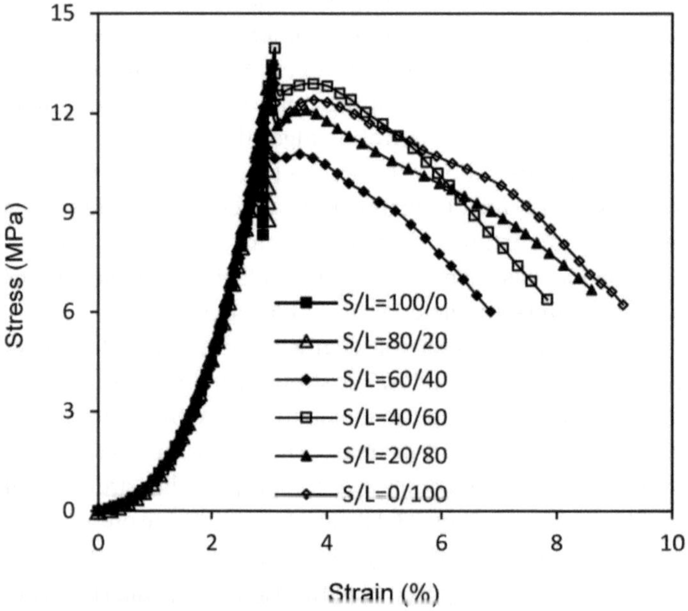

Fig. 8.14: Stress-strain curve of AA slag/fly ash composites with hybrid fiber addition [115].

Fig. 8.15 depicts the drying shrinkage results of mixes with only long or short steel fibers until 28 d. The shrinkage decreases with an increasing steel fiber content of up to 1 %. This reveals that the addition of fiber can be used as an efficient approach to inhibit the drying shrinkage of alkali activated materials. For a fixed fiber dosage, mixes with long fiber present relatively high values when compared to those with short fiber. This indicates that long fiber is slightly less effective than short fiber in inhibiting shrinkage; but a long fiber addition of 1 % still exhibits a shrinkage reduction rate of 27.6 % compared to the reference sample.

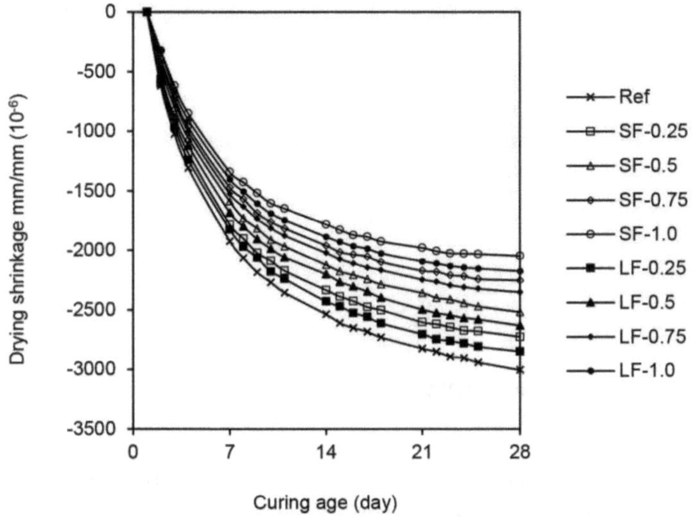

Fig. 8.15: Drying shrinkage of AA slag/fly ash mortars with different fiber lengths [115].

8.3.4 Functional building material

The concept of ultra-lightweight concrete has great potential because of the flexibility it brings in terms of architectural design, mechanical properties, thermal insulating properties and durability [99]. Recent research on blended alkaline systems has shown promising progress in understanding blended systems and the resulting modified properties indicate a promising future [30, 39]. This research addresses the development of alkali activated ultra-lightweight concrete (termed Geo-ULC), aiming towards its application in monolithic concrete façade structures in both load bearing and thermal insulating capacities. The effect of alkali activators on the properties of developed concrete is investigated. This investigation includes sodium hydroxide modified sodium silicate and sodium hydroxide modified sodium silicate further modified by sodium carbonate. Ground granulated blast furnace slag (GGBS) is used as the raw material and following that, the substitution of GGBS by fly ash and limestone powder is researched.

Fig. 8.16 shows the cut surface of the selected samples after performing the compressive strength test. It shows that the LWAs are very homogeneously and evenly distributed in the concrete matrix. This again confirms that there was no segregation in the mixtures developed in this study, indicating the suitability of the applied concrete design method as well as the sufficient viscosity of the binder paste.

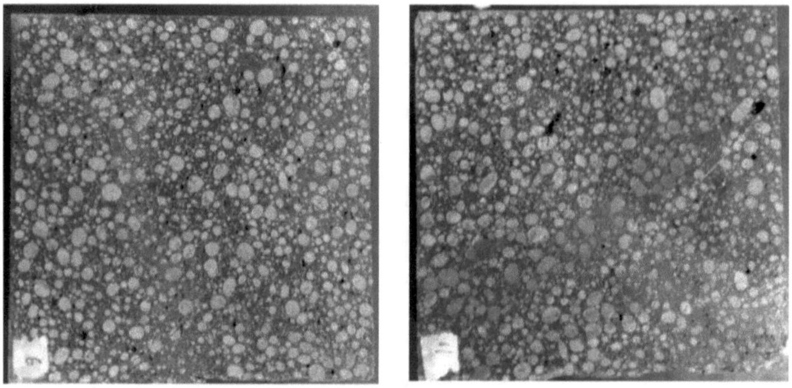

Fig. 8.16: Cross-section of the developed concrete [100].

All the developed concretes show very good thermal physical properties, reflected by very low thermal conductivities. The thermal conductivities of the three mixtures are 0.14, 0.13, and 0.13 W/(m K), respectively. These measurments were taken from the samples on 28 d under oven dry conditions. Fig. 8.17 provides a comparison between the compressive strength and thermal conductivities of different types of concrete available in literature [91, 100–105]. In Fig. 8.17 one can also see the results of the previously developed ultra-lightweight concrete based on cement. It can be clearly observed that the results show excellent thermal properties. At this thermal conductivity range, the compressive strength of the Geo-ULC is much higher than other reported data. It can also be seen that the results of this study fall into the same range of results obtained from the ultra-lightweight concrete developed in the previous study (Fig. 8.17).

Another remarkable feature of alkali activated materials is their intrinsic resistance to fire. This originates from their inorganic framework structure. Much research has been conducted on the residual mechanical properties of AAM after exposure to elevated temperatures, suggesting the advantages of AAM in high temperature applications and showing evidence of their ability to have high residual strength after exposures of up to 800 °C [73, 106–109]. With this natural benefit, AAMs are attract-

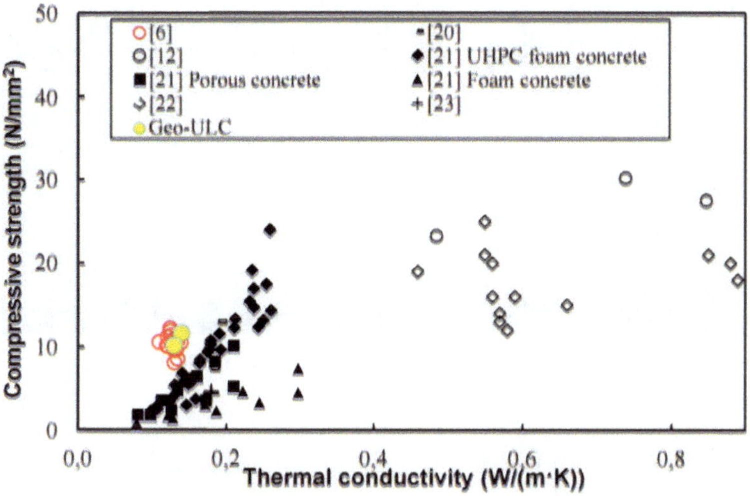

Fig. 8.17: Thermal conductivity vs. compressive strength [100].

ing increasing interest in their use in construction as ecologically friendly fireproof building materials.

Cheng and Chiu [106] studied the fire-resistant slag-metakaolinite-based AAM with different types and concentrations of alkali solutions (K^+ and Na^+) by putting a 10 mm thick panel of geopolymers into an 1100 °C flame. On increasing the amount of added metakaolinite, the fire resistance characteristics can be improved. Kong et al. [110] investigated the thermal behavior of fly ash and metakaolin-based geopolymer at exposures of up to 800 °C. As can be seen from Fig. 8.18, fly ash is known to contain a significant proportion of particles with hollow spheres and the polymer also contains many unreacted particles. The hollow cavities are possibly due to the dissolution of fly ash particles. With metakaolin-based geopolymer, there are some cracks on the surface. The SEM images show that there are nearly no cracks visible and it seems to be even more compact, mainly a layer-like structure, than prior to exposure to high temperatures. Similar results were also obtained by Prud'homme et al. [111] who investigated the properties of a metakaolin-based geopolymer with different types of activators (Na^+, K^+, and $Na^+ + K^+$) after high temperature exposure. Pan and Sanjayan studied the stress-strain behavior and abrupt loss of stiffness of geopolymer at elevated temperatures [109]. They reported that geopolymer specimens exhibited a significant increase at an elevated temperature in the range of 200–600 °C while a Portland cement based system has a decreasing trend at elevated temperature.

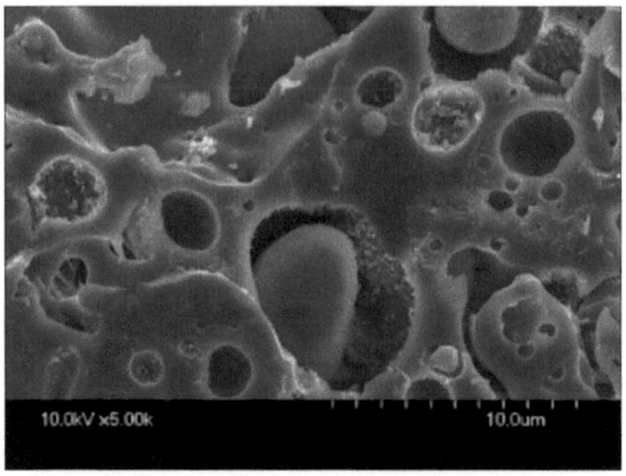

(a) Before exposure

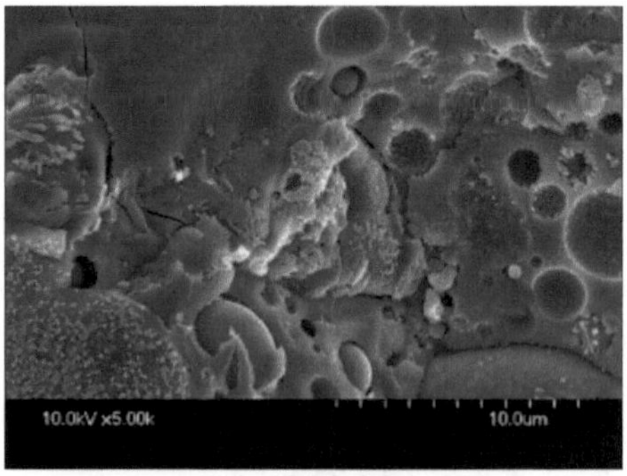

(b) After temperature exposure

(a) Before exposure

(b) After temperature exposure

Fig. 8.18: Photographs and SEM of fly ash-based geopolymer (upper) and metakaolin-based geopolymer (down) specimens: (a) before exposure, (b) after temperature exposure (up to 800 °C). Adapted from [110].

8.3.5 Conclusions and future trends

Alkali activated materials have proven their potential benefits as construction materials in terms of mechanical properties and durability as fire retardants, in heavy metal immobilization, and special applications. Clearly not all alkali activated materials will possess all the advantages listed above, however, it is possible to design a formula of alkali activated materials with appropriate concerns which will meet the requirements of a specific purpose. At the moment, the number of applications of alkali activated materials is limited. A deeper understanding of the reactions and their benefits in economic and environmental terms will attract more attention especially in special applications (e.g. waste management and high temperature applications).

References

[1] Duxson P, Lukey GC, Provis JL, van Deventer JSJ. The role of inorganic polymer technology in the development of "green concrete". Cem Concr Res. 37: 1590–1597.

[2] Pacheco-Torgal F, Castro-Gomes J, Jalali S. Alkali-activated binders, A review. Constr Build Mater. 2008; 22: 1305–1314.

[3] Pacheco-Torgal F, Abdollahnejad Z, Camoes AF, Jamshidi M, Ding Y. Durability of alkali-activated binders. A clear advantage over Portland cement or an unproven issue? Constr Build Mater. 2012; 30: 400–405.

[4] Piatak NM, Parsons MB, Seal RR. Characteristics and Environmental Aspects of Slag. A Review. Elsevier Ltd.; 2014.

[5] Puertas F, Varga C, Alonso MM. Rheology of alkali-activated slag pastes. Effect of the nature and concentration of the activating solution. Cem Concr Compos. 2014; 53: 279–288.

[6] Jimenez AF, Palomo JG, Puertas F. Alkali-activated slag mortars Mechanical strength behaviour. Cem Concr Res. 1999; 29: 1313–1321.

[7] Jimenez AF, Puertas F. Setting of alkali-activated slag cement. Influence of activator nature. Adv Cem Res. 2001; 13: 115–121.

[8] Heikal M, Nassar MY, El-Sayed G, Ibrahim SM. Physico-chemical, mechanical, microstructure and durability characteristics of alkali activated Egyptian slag. Constr Build Mater. 2014; 69: 60–72.

[9] Lee NK, Lee HK. Setting and mechanical properties of alkali-activated fly ash/slag concrete manufactured at room temperature. Constr Build Mater. 2013; 47: 1201–1209.

[10] Akcaozoglu S, Ulu C. Recycling of waste PET granules as aggregate in alkali-activated blast furnace slag/metakaolin blends. Constr Build Mater. 2014; 58: 31–37.

[11] Haha M Ben, Lothenbach B, Le Saout G, Winnefeld F. Influence of slag chemistry on the hydration of alkali-activated blast-furnace slag – Part II, Effect of Al2O3. Cem Concr Res. 2012; 42: 74–83.

[12] Rashad AM. Alkali-activated metakaolin. A short guide for civil Engineer – An overview. Constr Build Mater. 2013; 41: 751–765.

[13] Sanjayan JG, Collins F. Microcracking and strength development of alkali activated slag concrete. Cem Concr Compos. 2001; 23: 345–352.

[14] Chi M. Effects of dosage of alkali-activated solution and curing conditions on the properties and durability of alkali-activated slag concrete. Constr Build Mater. 2012; 35: 240–245.

[15] Sturm P, Greiser S, Gluth GJG, Jager C, Brouwers HJH. Degree of reaction and phase content of silica-based one-part geopolymers investigated using chemical and NMR spectroscopic methods. J Mater Sci. 2015; 50: 6768–6778.

[16] Sturm P, Gluth GJG, Simon S, Brouwers HJH, Kühne H-C. The effect of heat treatment on the mechanical and structural properties of one-part geopolymer-zeolite composites. Thermochim, Acta. 2016; 635: 41–58.

[17] Weil M, Dombrowski K., Buchawald A. Life-cycle analysis of geopolymers. In, Geopolymers, structure, processing, properties and applications. 2009; 194–210.

[18] Davidovits J. Geopolymer properties and chemistry. In, 1st European conference on soft mineralurgy. France. 1998; 25–48.

[19] Provis JL, Palomo A, Shi C. Advances in understanding alkali-activated materials. Cem Concr Res. 2015; 78: 110–125.

[20] Pacheco-Torgal F, Labrincha JA, Leonelli C, Palomo A, Chindaprasirt P. Handbook of alkali-activated cement, mortars and concretes; 2015.

[21] Duxson P, Fernandez-Jimenez A, Provis JL, Lukey GC, Palomo A, Deventer JSJ. Geopolymer technology, the current state of the art. J Mater Sci. 2006; 42: 2917–2933.

[22] Davidovits J. Geopolymer chemistry and sustainable develop- ment. The poly(sialate) terminology, a very useful and simple model for the promotion and understanding of green-chemistry. In: Proceedings of 2005 geopolymere conference. 2005; 9–15.

[23] Kuhl H. Slag cement and process of making the same; 1908.

[24] Kuhl H. Kolloide Zementerhartung I; 1930; 262–264.

[25] Duxson P, Provis JL. Designing precursors for geopolymer cements. J Am Ceram Soc. 2008; 91: 3864–3869.

[26] Provis JL, Van Deventer JSJ. Alkali Activated Materials; 2014.

[27] Haha MB, Saout GL, Winnefeld F, Lothenbach B. Influence of activator type on hydration kinetics, hydrate assemblage and microstructural development of alkali activated blast-furnace slags. Cem Concr Res. 2011; 41: 301–310.

[28] Wang S, Scrivener KL. ^{29}Si and ^{27}Al NMR study of alkali-activated slag. Cem Concr Res. 2003; 33: 769–774.

[29] Garcia-Lodeiro I, Palomo A, Fernandez-Jimenez A, MacPhee DE. Compatibility studies between N-A-S-H and C-A-S-H gels. Study in the ternary diagram Na2O-CaO-Al2O3-SiO 2-H2O. Cem Concr Res. 2011; 41(9): 923–931.

[30] Gao X, Yu QL, Brouwers HJH. Reaction kinetics, gel character and strength of ambient temperature cured alkali activated slag/fly ash blends. Constr Build Mater. 2015; 80: 105–115.

[31] Brough AR, Atkinson A. Sodium silicate-based, alkali-activated slag mortars Part I. Strength, hydration and microstructure. Cem Concr Res. 2002; 32: 865–879.

[32] Song S, Jennings HM. Pore solution chemistry of alkali-activated ground granulated blast-furance slag. Cem Concr Res. 1999; 29: 159–170.

[33] Thomas JJ, Rothstein D, Jennings HM, Christensen BJ. Effect of hydration temperature on the solubility behavior of Ca-, S-, Al-, and Si-bearing solid phases in Portland cement pastes. Cem Concr Res. 2003; 33: 2037–2047.

[34] Rovnanik P, Bayer P, Rovnaníkova P. Characterization of alkali activated slag paste after exposure to high temperatures. Constr Build Mater. 2013; 47: 1479–1487.

[35] Bernal SA, Provis JL, Rose V, Mejia De Gutierrez R. Evolution of binder structure in sodium silicate-activated slag-metakaolin blends. Cem Concr Compos. 2011; 33: 46–54.

[36] Zhang Z, Wang H, Provis JL, Bullen F, Reid A, Zhu Y. Quantitative kinetic and structural analysis of geopolymers. Part 1. the activation of metakaolin with sodium hydroxide. Thermochim Acta. 2012; 539: 23–33.

[37] Qing Y, Zenan Z, Deyu K, Rongshen C. Influence of nano-SiO2 addition on properties of hardened cement paste as compared with silica fume. Constr Build Mater. 2007; 21: 539–545.

[38] Li C, Sun H, Li L. A review, The comparison between alkali-activated slag (Si+Ca) and metakaolin (Si+Al) cements. Cem Concr Res. 2010; 40: 1341–1349.

[39] Gao X, Yu QL, Brouwers HJH. Properties of alkali activated slag–fly ash blends with limestone addition. Cem Concr Compos. 2015; 59: 119–128.

[40] Shi C, Day RL. A calorimetric study of early hydration of alkali-slag cements. Cem Concr Res. 1995; 25: 1333–1346.

[41] Tydlitat V, Matas T, CernY R. Effect of w/c and temperature on the early-stage hydration heat development in Portland-limestone cement. Constr Build Mater. 2014; 50: 140–147.

[42] Zajac M, Rossberg A, Le Saout G, Lothenbach B. Influence of limestone and anhydrite on the hydration of Portland cements. Cem Concr Compos. 2014; 46: 99–108.

[43] Wang S, Scrivener KL, Pratt PL. Factors affecting the strength of alkali-activated slag. Cem Concr Res. 1994; 24: 1033–1043.

[44] Bakharev T, Sanjayan JG, Cheng YB. Alkali activation of Australian slag cements. Cem Concr Res. 1999; 29: 113–120

[45] Palacios M, Puertas F. Effect of shrinkage-reducing admixtures on the properties of alkali-activated slag mortars and pastes. 2007; 37: 691–702.
[46] Neto AAM, Cincotto MA, Repette W. Drying and autogenous shrinkage of pastes and mortars with activated slag cement. Cem Concr Res. 2008; 38: 565–574.
[47] Gong C, Yang N. Effect of phosphate on the hydration of alkali-activated red mud–slag cementitious material. Cem Concr Res. 2000; 30: 1013–1016.
[48] Chang JJ. A study on the setting characteristics of sodium silicate-activated slag pastes. Cem Concr Res. 2003; 33: 1005–1011.
[49] Duran Atis C, Bilim C, Celik O, Karahan O. Influence of activator on the strength and drying shrinkage of alkali-activated slag mortar. Constr Build Mater. 2009; 23: 548–555.
[50] F. Collins, J.G. Sanjayan. Early age strength and workability of slag pastes activated by NaOH and Na2CO3. Cem Concr Res. 1998; 28: 655–664.
[51] Bakharev T, Sanjayan JG, Cheng YB. Effect of admixtures on properties of alkali-activated slag concrete. Cem Concr Res. 2000; 30: 1367–1374.
[52] Puertas F, Carrasco MT. Use of glass waste as an activator in the preparation of alkali-activated slag. Mechanical strength and paste characterisation. Cem Concr Res. 2014; 57: 95–104.
[53] Rashad AM. A comprehensive overview about the influence of different additives on the properties of alkali-activated slag – A guide for Civil Engineer. Constr Build Mater. 2013; 47: 29–55.
[54] European Commission. Reference document on best available techniques for the manufacture of large volume inorganic chemicals – solids and other industry; 2007.
[55] Rashad AM, Bai Y, Basheer PAM, Milestone NB, Collier NC. Hydration and properties of sodium sulfate activated slag. Cem Concr Compos. 2013; 37: 20–29.
[56] Palomo A, Shi C, Jimenez a. F. New cements for the 21st century, The pursuit of an alternative to Portland cement. Cem Concr Res. 2011; 41: 750–763.
[57] Granizo ML, Alonso S, Blanco-Varela MT, Palomo A. Alkaline Activation of Metakaolin, Effect of Calcium Hydroxide in the Products of Reaction. J Am Ceram Soc. 2002; 85: 225–231.
[58] Douglas E, Brandstetr J. A preliminary study on the alkali activation of ground granulated blast-furnace slag. Cem Concr Res. 1990; 20: 746–756.
[59] Provis JL, van Deventer JSJ. Alkali Activated Materials, State-of-the-Art Report, RILEM TC. 224-AAM.
[60] Garcia-Lodeiro I, Fernandez-Jimenez A, Pena P, Palomo A. Alkaline activation of synthetic aluminosilicate glass. Ceram Int. 2014; 40: 5547–5558.
[61] Chindaprasirt P, De Silva P, Sagoe-Crentsil K, Hanjitsuwan S. Effect of SiO 2 and Al 2O 3on the setting and hardening of high calcium fly ash-based geopolymer systems. J Mater Sci. 2012; 47: 4876–4883.
[62] Fernandez-Jimenez A, Palomo A. Composition and microstructure of alkali activated fly ash binder, Effect of the activator. Cem Concr Res. 2005; 35: 1984–1992.
[63] Palomo A, Grutzeck MW, Blanco MT. Alkali-activated fly ashes. A cement for the future. 1999; 29: 1323–1329.
[64] Komljenovic M, Bascarevic Z, Bradic V. Mechanical and microstructural properties of alkali-activated fly ash geopolymers. J Hazard, Mater. 2010; 181: 35–42.
[65] Criado M, Fernandez-Jimenez A, de la Torre AG, Aranda MAG, Palomo A. An XRD study of the effect of the SiO2/Na2O ratio on the alkali activation of fly ash. Cem Concr Res. 2007; 37: 671–679.
[66] Criado M, Fernandez-Jimenez A, Palomo A. Alkali activation of fly ash, Effect of the SiO2/Na2O ratio. Microporous Mesoporous Mater. 2007; 106: 180–191.

[67] Lukey GC, Provis JL, Kriven WM, van Deventer JSJ, Mallicoat SW, Duxson P. Understanding the relationship between geopolymer composition, microstructure and mechanical properties. Colloids Surfaces A Physicochem Eng Asp. 2005; 269: 47–58.

[68] Silva P De, Sagoe-Crenstil K, Sirivivatnanon V. Kinetics of geopolymerization, Role of Al2O3 and SiO2. Cem Concr Res. 2007; 37: 512–518.

[69] Rowles M, O'Connor B. Chemical optimisation of the compressive strength of aluminosilicate geopolymers synthesised by sodium silicate activation of metakaolinite. J Mater Chem. 2003; 13: 1161–1165.

[70] Ismail I, Bernal SA, Provis JL, San Nicolas R, Hamdan S, van Deventer JSJ. Modification of phase evolution in alkali-activated blast furnace slag by the incorporation of fly ash. Cem Concr Compos. 2014; 45: 125–135.

[71] Fan Y, Yin S, Wen Z, Zhong J. Activation of fly ash and its effects on cement properties. Cem Concr Res. 1999; 29: 467–472.

[72] Fernandez-Jimenez A, De La Torre AG, Palomo A, Lopez-Olmo G, Alonso MM, Aranda MAG. Quantitative determination of phases in the alkali activation of fly ash. Part I. Potential ash reactivity. Fuel. 2006; 85: 625–634.

[73] Xu H, Van Deventer JSJ. The geopolymerisation of alumino-silicate minerals. Int J Miner Process. 2000; 59: 247–266.

[74] Yip CK, Lukey GC, van Deventer JSJ. The coexistence of geopolymeric gel and calcium silicate hydrate at the early stage of alkaline activation. Cem Concr Res. 2005; 35: 1688–1697.

[75] Kumar S, Kumar R, Mehrotra SP. Influence of granulated blast furnace slag on the reaction, structure and properties of fly ash based geopolymer. J Mater Sci. 2009; 45: 607–615.

[76] Lancellotti I, Ponzoni C, Barbieri L, Leonelli C. Alkali activation processes for incinerator residues management. Waste Manag. 2013; 33: 1740–1749.

[77] Yamaguchi N, Nagaishi M, Kisu K, Nakamura Y, Ikeda K. Preparation of monolithic geopolymer materials from urban waste incineration slags. J Ceram Soc Japan. 2013; 121: 847–854.

[78] El-Kamash AM, El-Dakroury AM, Aly HF. Leaching kinetics of 137Cs and 60 Co radionuclides fixed in cement and cement-based materials. Cem Concr Res. 2002; 32: 1797–1803.

[79] Tits J, Wieland E, Muller CJ, Landesman C, Bradbury MH. Strontium binding by calcium silicate hydrates. J Colloid Interface Sci. 2006; 300: 78–87.

[80] Xuequan W, Sheng Y, Xiaodong S, Mingshu T, Liji Y. Alkali-activated slag cement based radioactive waste forms. Cem Concr Res. 1991; 21: 16–20.

[81] Qian G, Sun DD, Tay JH. New aluminium-rich alkali slag matrix with clay minerals for immobilizing simulated radioactive Sr and Cs waste. J Nucl Mater. 2001; 299: 199–204.

[82] Zosin AP, Priimak TI, Avsaragov KB. Geopolymer materials based on magnesia-iron slags for normalization and storage of radioactive wastes. At Energy. 1998; 85: 78–82.

[83] Sahmaran M, Yaman IO. Hybrid fiber reinforced self-compacting concrete with a high-volume coarse fly ash. Constr Build Mater. 2007; 21: 150–156.

[84] Bencardino F, Rizzuti L, Spadea G, Swamy RN. Experimental evaluation of fiber reinforced concrete fracture properties. Compos Part B Eng. 2010; 41: 17–24.

[85] Corinaldesi V, Moriconi G. Durable fiber reinforced self-compacting concrete. Cem Concr Res. 2004; 34: 249–254.

[86] Bernal S, De Gutierrez R, Delvasto S, Rodriguez E. Performance of an alkali-activated slag concrete reinforced with steel fibers. Constr Build Mater. 2010; 24: 208–214.

[87] Martin Hunger. An integral design concept for ecological Self-Compacting Concrete. 2010.

[88] Yu QL, Brouwers HJH. Development of a self-compacting gypsum-based lightweight composite. Cem Concr Compos. 2012; 34: 1033–1043.

[89] Yu R, Spiesz P, Brouwers HJH. Mix design and properties assessment of Ultra-High Performance Fibre Reinforced Concrete (UHPFRC). Cem Concr Res. 2014; 56: 29–39.

[90] Ng TS, Foster SJ. Development of a mix design methodology for high-performance geopolymer mortars. Struct Concr. 2013; 14: 148–156.

[91] Yu QL, Spiesz P, Brouwers HJH. Development of cement-based lightweight composites – Part 1, Mix design methodology and hardened properties. Cem Concr Compos. 2013; 44: 17–29.

[92] Brouwers HJH, Radix HJ. Self-compacting concrete, Theoretical and experimental study. Cem Concr Res. 2005; 35: 2116–2136.

[93] Hunger M, Entrop AG, Mandilaras I, Brouwers HJH, Founti M. The behavior of self-compacting concrete containing micro-encapsulated Phase Change Materials. Cem Concr Compos. 2009; 31: 731–743

[94] Yu R, Spiesz P, Brouwers HJH. Effect of nano-silica on the hydration and microstructure development of Ultra-High Performance Concrete (UHPC) with a low binder amount. Constr Build Mater. 2014; 65: 140–150.

[95] Quercia G, Husken G, Brouwers HJH. Water demand of amorphous nano silica and its impact on the workability of cement paste. Cem Concr Res. 2012; 42: 344–357.

[96] Husken G, Brouwers HJH. On the early-age behavior of zero-slump concrete. Cem Concr Res. 2012; 42: 501–510.

[97] Mikanovic N, Jolicoeur C. Influence of superplasticizers on the rheology and stability of limestone and cement pastes. Cem Concr Res. 2008; 38: 907–919.

[98] Husken G, Brouwers HJH. A new mix design concept for earth-moist concrete, A theoretical and experimental study. Cem Concr Res. 2008; 38: 1246–1259.

[99] Yu QL, Spiesz P, Brouwers HJH. Ultra-lightweight concrete, Conceptual design and performance evaluation. Cem Concr Compos. 2015; 61: 18–28.

[100] Yu QL, Yuan B, Gao X, Brouwers HJH. Alkali activated ultra-lightweight concrete, an innovation on functionality and sustainability. In, H.-B. Fischer, ed. 19. Internationale Baustofftagung. Weimar, Bundesrepublik Deutschland: 2015; 2468–2477.

[101] Kralj D. Experimental study of recycling lightweight concrete with aggregates containing expanded glass. Process Saf Environ Prot. 2009; 87: 267–273.

[102] Topcu IB, Uygunoglu T. Effect of aggregate type on properties of hardened self-consolidating lightweight concrete (SCLC). Constr Build Mater. 2010; 24: 1286–1295.

[103] Ahmed M, El M, Ingenieurwissenschaften D Der. Conceptual and Structural Design of Infra-Lightweight Concrete. 2010; 105.

[104] Schauerte M, Trettin R. Neue Schaumbetone mit gesteigerten mechanischen ind physikalischen Eigenschaften. In, Proceedings of the 18th Ibausil, International Conference on Building Materials. Weimar, Germany, 2012; 2-66-2-0072.

[105] Zareef Me. Conceptual and structural design of buildings made of lightweight and infra-lightweight concrete. 2010.

[106] Cheng TW, Chiu JP. Fire-resistant geopolymer produced by granulated blast furnace slag. Miner Eng. 2003; 16, 205–210.

[107] Lyon RE. geopolymere.. 1997; 21: 67–73.

[108] Davidovits J, Davidovics M, Balaguru PN, Foden A, Sorathia U, Lyon RE. Geopolymer Composites.

[109] Pan Z, Sanjayan JG. Stress–strain behaviour and abrupt loss of stiffness of geopolymer at elevated temperatures. Cem Concr Compos. 2010; 32: 657–664.

[110] Kong DLY, Sanjayan JG, Sagoe-Crentsil K. Comparative performance of geopolymers made with metakaolin and fly ash after exposure to elevated temperatures. Cem Concr Res. 2007; 37: 1583–1589.

[111] Prud'homme E, Michaud P, Joussein E, et al. Structural characterization of geomaterial foams – Thermal behavior. J Non Cryst Solids. 2011; 357: 3637–3647.

[112] Gao X, Yu QL, Brouwers HJH. Characterization of alkali activated slag–fly ash blends containing nano-silica. Constr Build Mater. 2015; 98: 397–406

[113] Gao X, Yu QL, Brouwers HJHJH. Properties of alkali activated slag/fly ash blends with limestone addition. In, Fischer HB, Fischer H-B, Neugebauer M, Boden C, editors. 19. Internationale Baustofftagung. Weimar, Bundesrepublik Deutschland. 2015; 119–128.

[114] Yuan B, Yu QL, Brouwers HJH. Reaction kinetics, reaction products and compressive strength of ternary activators activated slag designed by Taguchi method. Mater Des. 2015; 86: 878–886.

[115] Gao X, Yu QL, Brouwers HJH. Properties of hybrid steel fiber reinforced alkali activated slag/fly ash composites. In, Jones M., Newlands M., Halliday JE, et al., editors. 9th international concrete conference. Dundee, Scotland, UK. 2016; 625–634.

Christian Pritzel*, Torsten Kowald, Yilmaz Sakalli, and Reinhard Trettin

9 Binding materials based on calcium sulphates

Abstract: Calcium sulphate based binding materials attract a lot of attention due to the multiple fields of their application in industry and housing. Exemplified, they are used in the building industry in plasterboard, screed, stucco or gypsum fireboards. In cements, calcium sulphate retards the hydration of tricalcium-aluminate and thus influences the period of workability of cement paste. In ceramics, calcium sulphate binders are utilized in the fabrication of negative forms for ceramic casting. Furthermore, these materials are used in the medical sector for dental castings. Calcium sulphate based binding materials are mainly produced by the formation of hemihydrate or anhydrite through the dehydration of natural gypsum or byproduct gypsums (calcium sulphate dihydrate). A variety of gypsum sources are discussed in this study besides the typical natural gypsum crystal morphologies. To improve certain of the material's characteristics, the subhydrates produced are mixed with additives and transported to the point of use, where they are mixed with water for setting. The hemihydrate reacts with water to form gypsum crystals, thereby developing the strength and hardness of the gypsum based binding material. The mechanism of the hydration process is debated in literature. Yet, it is known that it is a process with several reaction steps. The hydration process includes combined interactions in the liquid phase and in the solid phase from hemihydrate or anhydrite and dihydrate. For this study, both phases, liquid and solid, were investigated separately. The dehydration and rehydration process form part of this study.

The advantages of calcium sulphate based binding materials compared to other typical binding materials is their eudermic pH-value, their small energy consumption during production, their compatibility for recycling as well as their good resistance to fire. The main disadvantage is the significant decrease in strength of gypsum based binders in the presence of moisture. Major reasons for this decrease in strength and also on the strength development of calcium sulphate based binding materials are discussed in this study.

For all given technical applications, information about different material properties are needed for a better understanding and in order to improve material properties. The strength of the binding material is one of these properties. This is alongside its high accuracy in uses as a casting material, its special porosity for screed, plaster

***Corresponding author: Christian Pritzel,** Universität Siegen, Naturwissenschaftlich-Technische Fakultät, Department Chemie-Biologie, Institut für Bau- u. Werkstoffchemie, Siegen, Germany, pritzel@chemie.uni-siegen.de
Torsten Kowald, Yilmaz Sakalli, Reinhard Trettin, Universität Siegen, Naturwissenschaftlich-Technische Fakultät, Department Chemie-Biologie, Institut für Bau- u. Werkstoffchemie, Siegen, Germany
DOI 10.1515/9783110473728-010

bondage or ceramic casts, its high resistance to fire for plasterboard or stucco or its low expansion during hydration. All these technical properties are strongly influenced by the crystal morphology of the gypsum crystals which are created and adjusted by different additives. The influence of different additives on technical properties and the morphology of the thus created gypsum crystals are explained.

Additionally, the historical use of gypsum materials and some methods for investigation of their properties are included in this study. An example of a frequently used method for investigation is optical microscopy, which can provide information about gypsum morphology and the influence of additives thereon. It can also help to check for possible side effects when using different additives at once, as is typical in technical products.

Keywords: gypsum based binding materials, additives for the hydration of hemihydrate, creation of hemihydrate, gypsum morphologies

9.1 Introduction

Several inorganic binders are well known with regard to their application in construction materials. The most frequently used are Portland cement based concretes, followed by limestone and calcium sulphate (i.e. gypsum) based binding materials to which this study is dedicated. Calcium sulphate based binders are suitable for a wide range of different technical applications; for floor screed, mortar, plaster, gypsum plaster board, gypsum fiber board, gypsum blocks, and stucco. Furthermore, they are applied as a material for ceramic, metal, and dental casting, in plaster bandage, in the arts for plaster casting and more recently for 3D printouts. However, calcium sulphate is predominantly used in the cement industry to retard the hydration of the highly reactive calcium aluminate phase during the setting of the cement paste. Some of the applications mentioned above have been utilized by humankind since very early on. With regard to each application, the technical properties are adjusted to achieve the performance demanded of the material. For floor screed, mortar, and plaster bandage a high porosity is required so that the material is fast drying. For ceramic casting an open porosity is necessary to ensure the intrinsic water is transported out of the ceramic body during sintering. In the case of dental casting material, a low porosity and fine grained material is required for the formation of small crystals through which a high accuracy in casting can be ensured. Furthermore, the hydration time of the hemihydrate is of importance for the technical applicability of the binder. For plaster bandage or 3D printing it has to be within the range of 5 minutes or even less, whilst in contrast, for floor screed the hydration time has to be much longer. To improve the respective technical properties, various types of calcium sulphates and additives are applied. In the course of this study, all modifications in the system of calcium sulphates and water are presented. Historical and recent applications of the material are discussed. Regarding the respective technical applications, the feasibility of raw materials is explained considering the hydration process of the hemihydrate. Additionally, the influence and manipulation of different crystal morphologies of gypsum is pointed out with regard to the technical requirements of the final material.

9.2 Historical use of gypsum based binders

Humankind started using gypsum as a binding material very early on in history; it was first described during the Assyrian empire [1]. The word *gypsum* is based on the ancient Greek word for burned earth. The first use of gypsum was for artworks like statues or for the conserving of fruits or dead bodies. Gypsum based mortars were found in the pyramid of Chephren [2, 3]. The Roman Empire [4] and ancient Greeks used gypsum as a construction or casting material for death masks. It was also used in stucco. During medieval times, gypsum was used in stucco and screed. In romanticism, gypsum based binders were applied with horsehair and straw as fiber reinforcement to increase flexural strength. Later on gypsum was used as flame retarding material. Ludwig XIV. introduced a law that construction wood in houses had to be covered by gypsum due to its resistance to fire. In 1894, Austin Sacket got the patent for gypsum plaster board production [5]. Nowadays gypsum is used in many different fields as described in the introduction.

9.3 Principles

9.3.1 Phase transformation in the system of calcium sulphate and water

The system of calcium sulphate and water has five main phases with different amounts of chemically bounded water and different crystal structures. Calcium sulphate dehydrate (known under its mineral name *gypsum*), has the highest content of crystal water in the system with two moles of crystal water. It is followed by hemihydrate (also known as mineral *bassanite*), with half a mole of water per mole calcium sulphate and *anhydrite* without chemically bound water. There are three known anhydrite phases which differ in their crystal structures: anhydrite III (AIII), anhydrite II (AII), and anhydrite I (AI) [6, 7]. The interrelation of all phases, which are partly transformable by changing the water content and temperature, is summed up in Fig. 9.1.

Anhydrite and gypsum can be found in nature and could be used to produce inorganic binding materials alongside some calcium sulphate resources which are produced as byproducts in technical processes.

Prior to its application as a binding material, processing the respective materials is necessary. Anhydrite has to be ground and mixed with an activator such as potassium sulphate, while gypsum can be transformed to hemihydrate or anhydrate by

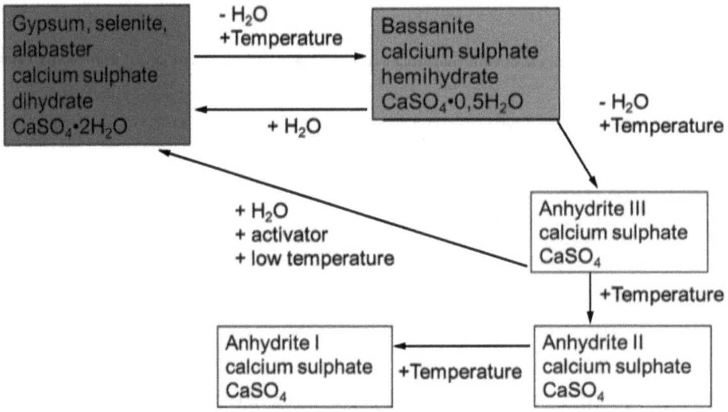

Fig. 9.1: System of calcium sulphate and water.

Fig. 9.2: Different hemihydrates; wet autoclave alpha (left), steam autoclave alpha (middle), beta hemihydrate (right).

thermal treatment. The process of dehydration of gypsum can be conducted in different ways to obtain hemihydrate or anhydrite with different properties. One way to produce hemihydrate is to bring dihydrate in an autoclave with water and additives at a higher temperature [8, 9]. During this process dihydrate undergoes a recrystallization into large hemihydrate crystals with only a small number of defects. The product is called alpha hemihydrate (depicted in Fig. 9.2, left). It is also possible to produce alpha hemihydrate in a steam autoclave. For this process, dihydrate is pressed into blocks and stored in an autoclave with steam. With this process large hemihydrate crystals with a higher concentration of defects are produced. The resulting crystals are shown in Fig. 9.2 (middle).

Beta hemihydrate can be obtained in the form of small crystals with many defects as pointed out in Fig. 9.2 (right), by heating dehydrate in a rotary kiln, an oven or in a gypsum kettle [10].

By thermal processing it is also possible to produce anhydrite III but then a higher temperature must be applied. Anhydrite II is produced by heating up AIII. If it is heated up to 1190 °C, AI is created. AI is only stable above 1180 °C and decomposes at around 1200 °C. In addition, a sintering starting at around 500 °C could be observed.

The processes of weight loss and deformation of dihydrate crystals during decomposition dehydration of dihydrate and the following transformation phase of anhydrite could be observed by simultaneous thermal analysis and dilatometry as a function of temperature. The result is given in Fig. 9.3. With the thermal gravimetric measurement the dehydration could be detected around from 100 °C to 180 °C with a weight loss of 21 %. This process is endothermic as the DSC curve shows. The phase transformation of anhydrite III to anhydrite II is exothermic (340 °C). The dilatometry starts with the thermal expansion of dehydrate. When around 6 % water is lost there is a small shrinkage due to the creation of hemihydrate. During the creation of hemihydrate, the crystal bursts along the single cleavage planes of dehydrate. There is a small expansion during that process on the red curve. At around 340 °C the phase transformation of anhydrate takes place. This is the reason for shrinkage at that temperature. At higher temperatures, shrinkage is created by a sintering process. The temperature of dehydration influences the reactivity of anhydrite based binding materials [11].

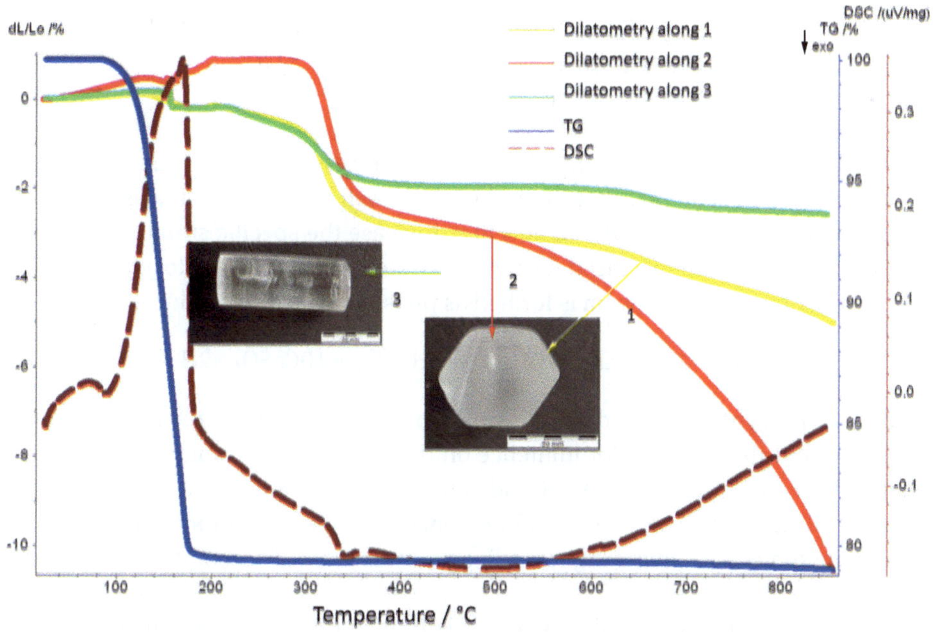

Fig. 9.3: Simultaneous thermal analysis and dilatometry of gypsum crystal. Arrow shows the measured direction in dilatometry.

9.3.2 Raw materials and their properties

As raw material for the processes above, different gypsum resources were tested. Natural gypsum can have varying compositions such as fiber gypsum, alabaster, gypsum stone or gypsum crystals. Natural gypsum has some typical impurities; for example limestone ($CaCO_3$), quartz (SiO_2), talc ($Mg_3[Si_4O_{10}(OH)_2]$), feldspar or mica, soluble salts like halite (NaCl), sylvine (KCl) or magnesium chloride ($MgCl_2$) and claystone, sulfur, iron, iron oxide or copper oxide. Some impurities can cause problems during processing [12]. Quartz, for example, is very hard and could damage the machines used in production. Claystone could intercalate additives like superplasticizers in their layer structure, so that they will become ineffective. But claystone also improves resistance to fire through endothermic release of its chemical bound water and it changes the rheological properties of the binding material.

In addition to natural gypsum, different byproduct gypsum materials could be used as a resource for the production of hemihydrate. Byproduct gypsum is formed during the flue gas desulfurization (FGD) process, when energy is generated in coal-fired plants. Thereby, flue gas is induced in lime milk and gypsum is formed from the sulfur trioxide in the coal flue gas following the reaction equation (9.1).

$$\begin{aligned} 2SO_2 + 2Ca(OH)_2 &\rightarrow 2CaSO_3 \cdot 0.5H_2O + H_2O \\ 2CaSO_3 \cdot 0.5H_2O + O_2 + 3H_2O &\rightarrow 2CaSO_4 \cdot 2H_2O \\ 2SO_2 + Ca(OH)_2 + O_2 + 2H_2O &\rightarrow 2CaSO_4 \cdot 2H_2O \end{aligned} \tag{9.1}$$

Typical impurities of gypsum formed during this process are limestone, soluble salts, metals, especially mercury, fly ash, and coal, as well as moisture. Fly ash and coal influence the color of the material and they increase the specific surface area of the particles, leading to an increased demand for water upon hydration to gypsum.

Another byproduct-gypsum is formed as phosphoric acid is produced

$$[Ca_3(PO_4)_2]_3CaF_2 + 10H_2SO_4 + 20H_2O \rightarrow 6H_3PO_4 + 10CaSO_4 \cdot 2H_2O + 2HF. \tag{9.2}$$

Typically, the gypsum formed during this process only contain a small amount of impurities but they have major influence on the material. For example, phosphorus pentoxide (P_2O_5) decreases the pH and thus works as a retarder for the hydration of hemihydrate. It may also cause efflorescence through a higher amount of alkalis. Phosphate sometimes causes higher radioactivity because of accompanying elements [10, 13, 14].

The most common byproduct gypsum is formed during the production of citric acid, oxalic acid and lactic acid. With fermentation, saccharose is converted to citric acid monohydrate, which is transferred to calcium citrate with lime milk. The calcium citrate is then converted to citric acid and gypsum with sulfuric acid. The produced gypsum could include a remainder of citric acid which works as a retarder in the hydration of hemihydrate.

9.3.3 Morphology and mineralogical properties of gypsum crystals

There are many different morphologies of gypsum crystals described in literature which can be found in nature and in technical calcium sulphate based binding materials [15–17]. Needle like structures with more or less branches of dove tailed twins are typical. Some examples for natural gypsum crystals are depicted in Fig. 9.4.

These morphologies can also be found in different technical used binding materials after the hydration of hemihydrate. Fig. 9.5 shows some examples for hemihydrate 1 to 3. In Fig. 9.6, the morphology and the crystal lattice of a typical gypsum crystal are shown.

Twinning lines are typical for gypsum crystal (Fig. 9.7) as are cleavage planes which both could be observed by the naked eye in the crystal depicted in Fig. 9.8. The density of gypsum is 2.2 to 2.4 g/cm^3; it has a perfect cleavage along the lattice plane [010] and a good cleavage along [100]. Gypsum includes 20.93 % of crystalline water and has an atomic mass of 172.17 g/mol [18].

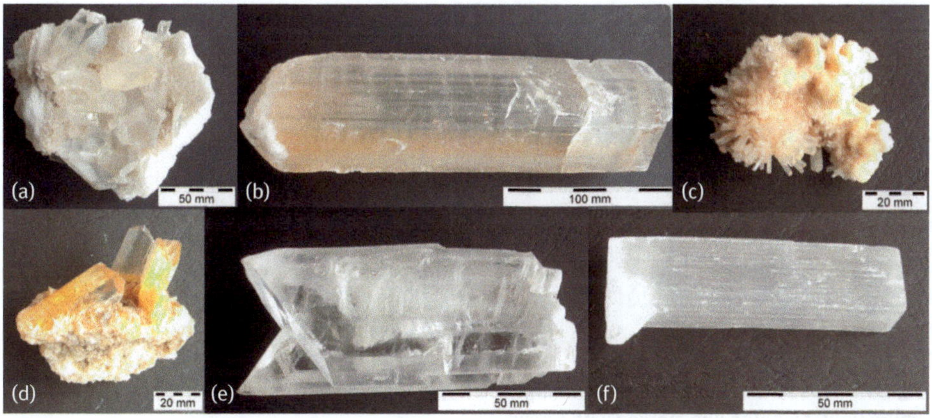

Fig. 9.4: Different natural gypsum samples from author's collection. Crystals are found in (a) Mexico, (b) Ohio, USA, (c) Whyalla, Australia, (d) Peru, (e) Eisleben, Germany, (f) selenite from Marocco.

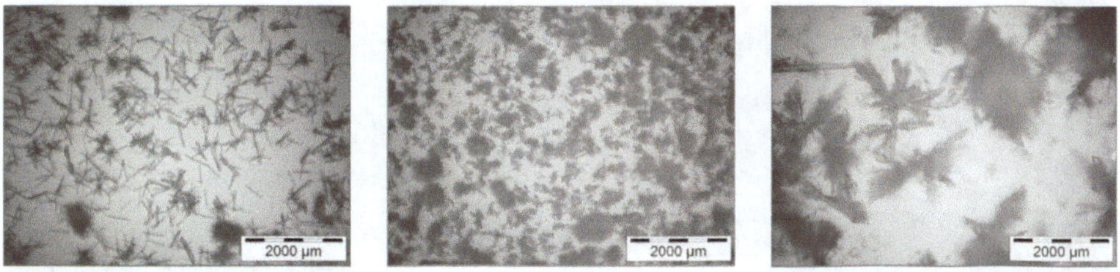

Fig. 9.5: Gypsum crystals created by hemihydrate 1, 2, and 3. Hemihydrate 1 creates longer crystals with some branches, hemihydrate 2 very small crystals with some branches, and hemihydrate 3 longer crystals with many branches, optical microscopy after 5 hours in water.

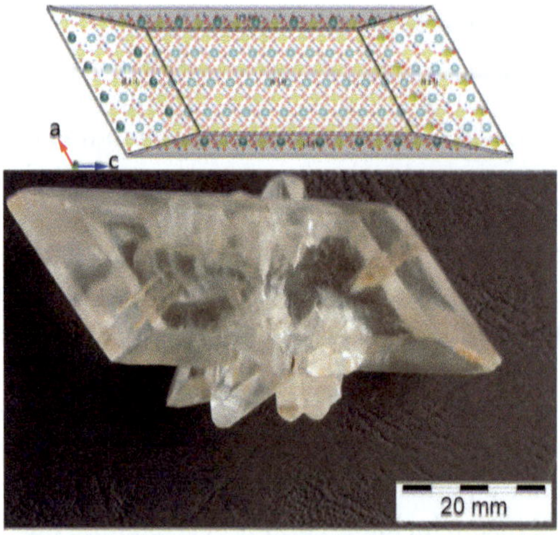

Fig. 9.6: Crystal lattice and surface of a gypsum crystal compared to a natural gypsum crystal.

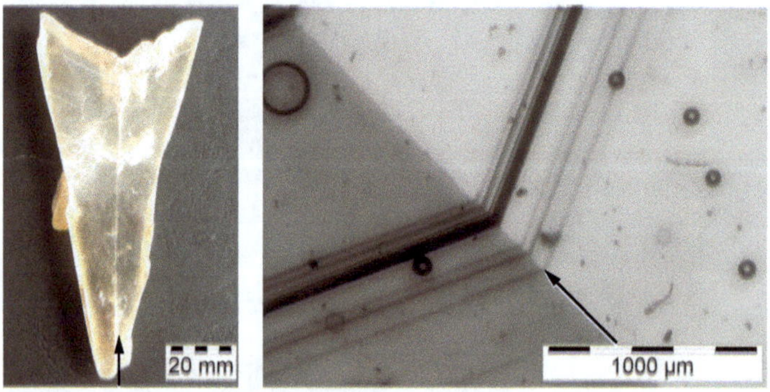

Fig. 9.7: Gypsum crystal dove tailed twin figure left, polarization microscopy of gypsum crystal twin figure right, twinning line marked with arrow.

Fig. 9.8: Gypsum crystal with cleavage planes marked with arrows (left), SEM figure of cleavage planes in a gypsum crystal (right).

9.4 Materials and Methods

9.4.1 Materials

To investigate the hydration of hemihydrate and the influence of additives [19, 20], alpha hemihydrates of differing technical quality were used. These materials are applied in special industrial sectors. Hemihydrate 1 is present in casting gypsum in the ceramics industry, hemihydrate 2 in casting gypsum for use in dentistry and hemihydrate 3 is used in building materials. A special alpha hemihydrate from a steam autoclave for building materials was chosen for the hydration process and to investigate the influence of additives on the hydration of hemihydrates.

In the investigation of the influence of different storage conditions on the strength of gypsum, two alpha hemihydrates with different dihydrate morphologies were selected. All the measurements were done with a variety of hemihydrates (including some variety of beta hemihydrates).

9.4.2 Morphology with optical microscopy

Optical microscopy was used to investigate the hydration process of hemihydrate and the morphology of the created dihydrate, looking especially at the influence of additives. A special measuring cell was used for this investigation. This consists of a glass slide with a fixed rubber ring and a piece of covering glass. The individual parts are glued together with water insoluble grease. The volume of such a cell was checked with pure water. For each measurement it was filled with 0.387 ml of deionized water and 0.01 g of hemihydrate. The additives were mixed with water and the amount of additives was calculated depending on the amount of hemihydrate. If nothing else is specified, the photos were taken five hours after the hydration was completed. The pictures were taken with an Olympus BX61 Microscope, a ColorViewII camera, and the software AnalySIS five in reflection mode.

9.4.3 Morphology with scanning electron microscopy (SEM)

Scanning electron microscopy imaging was used to check the morphology of the dihydrate created through the hydration of hemihydrate from mixtures with a lower water to gypsum ratio than the those used with optical microscopy. A mixture of hemihydrate and water with a respective water to gypsum ratio of 0.9 was applied to a SEM sample holder and stored in a closed polymer box. To avoid the sample drying out, wet paper was placed at the bottom of the box. After five hours the hydration solution was sucked off with paper in order to minimize drying of the artifacts. The measurements were done in low vacuum mode without sputtering, at a pressure of 110 Pa using a FEI Quanta FEG 250 environmental scanning electron microscope.

9.4.4 Thermal behavior with simultaneous thermal analysis and dilatometry

Simultaneous thermal analysis and dilatometry were used to investigate the thermal properties of dihydrate single crystals. A dihydrate crystal was cut into lengths of 20 mm along the different crystal surfaces with a slow moving crystal saw. The samples were stored in a Netzsch DIL 403 dilatometer and heated up to 900 °C with a slow heating rate of 0.5 K/min. The simultaneous thermal analysis of a single gypsum crystal with a mass of around 40 mg was done with a Netzsch 449 C Jupiter STA using the same heating rate as during the dilatometry measurements.

9.4.5 Reaction heat with isothermal heat flow calorimetry

Isothermal heat flow calorimetry was used to measure the development of the reaction heat of the hydration of the various hemihydrates. For each measurement, 2 g of hemihydrate was put into a test tube and 2 g of water (with or without additive), was stored in a syringe. The syringe was plunged into the test tube and both were connected by sealing film. These samples were measured in a ToniCalHexa calorimeter. The water to gypsum ratio was chosen in order to ensure direct contact of all the hemihydrate with water.

9.4.6 Liquid reaction phase

The calcium concentration during five hours of hydration was monitored by measurement with an ion selective electrode from Mettler Toledo. For each sample, water was added to a beaker. Stirring and measurement commenced using a measurement interval of 30 s, prior to the addition of the respective hemihydrate.

9.4.7 Mechanical properties

Flexural and compressive strength were measured according to DIN 1168, but with smaller samples of 15 mm × 15 mm × 60 mm. Samples were prepared by adjusting the water to gypsum ratio to achieve the same workability for each hemihydrate used. The influences of different storage conditions on mechanical properties were examined by applying a two-step curing approach. All samples were firstly dried at 35 °C. Three samples of each hemihydrate were stored at 25 °C in a dryer (≈ 0 % humidity), in the lab (room climate), in an over super saturated sodium chloride solution (75 %

humidity), in an over-super saturated potassium nitrate solution (98 % humidity), in deionized water, in saturated gypsum solution and in an oven at 60, 100, 400, and 800 °C, respectively. The conditions were chosen as the differing availability of water under each of them should influence the layer of water between individual crystals of gypsum. Samples stored in liquid water should result in a maximum water layer and storage in a gypsum solution should avoid the dissolution of gypsum. Different temperatures in the oven were selected according to the main steps obtained from the STA results. Expansion during hydration was also measured according to DIN 1168 with shrinkage drainage.

9.4.8 Porosity with mercury intrusion measurement

The porosity of the samples was measured by mercury intrusion. Samples were prepared like those for measuring strength; dried in a dryer at 35 °C until reaching a constant weight after the storage conditions described above and crushed into small pieces with diameters of between 2 and 4 mm. The measurements were performed with an Autopore II 9220 from Micromeritics.

9.5 Experiments

9.5.1 Hydration of hemihydrate

It is typical to use calcium sulphate hemihydrate or anhydrite for calcium sulphate based binding materials which are mixed with water and sometimes with additives and aggregates in order to set the material. The reaction is described by the chemical equations (9.3) and (9.4).

$$CaSO_4 + 2H_2O\ (+\text{activator}) \rightarrow CaSO_4 \cdot 2H_2O \quad (\text{activator: } K_2SO_4 \text{ for example}) \tag{9.3}$$

$$2CaSO_4 \cdot 0.5H_2O + 3H_2O \rightarrow 2CaSO_4 \cdot 2H_2O \tag{9.4}$$

These equations give information about the molar ratio but none about the reaction progress and phenomena taking place during the hydration process. From the equation it could be calculated that 18.6 % water is needed for the hydration of hemihydrate and 26.5 % water in the case of anhydrate. Normally a higher amount of water is used due to workability. There are different theories discussed in literature regarding the phenomena taking place during hydration. Lavoisier and Le Chatelier explained hydration by the dissolution of hemihydrate, causing the creation of a super saturated solution followed by the subsequent crystallization of dihydrate from said solution [21]. This approach is comprehensive because the “solubility” of hemihydrate is higher than the one of dihydrate. According to literature, the solubility of hemihydrate is highly dependent on the type of hemihydrate and the temperature is in the range of

8 to 10 g/l at 25 °C [22, 23]. The solubility is lowered with increasing temperatures but for dihydrate it is nearly constant (around 2 g/l between 20 and 100 °C). The solubility of anhydrite also depends on temperature and it is only a little higher compared to dihydrate at low temperatures. The reaction must have four steps: the dissolving of hemihydrate followed by the creation of a super saturated solution, the creation of dihydrate seed crystals and the growing of these seeds.

A second theory was published by Cacazzi and Traube, who discussed a gel-like-interface which is converted stepwise into a crystalline phase [24].

Later Fiedler explained the hydration of hemihydrate through water retention in the hemihydrate and a spontaneous turn of the crystal lattice from hemihydrate to dehydrate [25]. Perededij and Eipeltauer described an inner hydration of the hemihydrate alongside crystallization from a super saturated solution [26–29]. The hydration progress was investigated with optical microscopy, heat flow calorimetry and the measurement of the calcium ion concentration in the liquid phase. With these investigations it could be proven that at first the hemihydrate starts to dissolve and a super saturated solution is created. The next step is the formation of seed crystals. This step takes places at different speeds depending on the degree of supersaturation. In case of a faster seed crystal formation caused by a higher super saturation, more seed crystals with more branches were observed. During the seed crystal formation and the following crystal growth, the calcium concentration decreases until it reaches a plateau where as much hemihydrate gets dissolved as ions are needed for the crystal growth. At the end of the reaction, most of the hemihydrate is dissolved and the ion concentration in the solution gets lower until it reaches the solubility of dihydrate whereby the remaining hemihydrate is dissolved. Until that time the gypsum crystals grow. Normally there is only one point in the reaction where seed crystal formation takes place, leading to only one generation of gypsum crystals. Sometimes, if the drying of the redundant water is very fast, there may be a second generation of very small crystals. A small amount of dihydrate seeds added as an accelerator could also lead to dihydrate crystals of different sizes. In the next step hemihydrate continues to dissolve and the dihydrate crystals grow. During the last reaction step most of the hemihydrate is dissolved, the growing of dihydrate crystals decelerates until all the hemihydrate is dissolved and then dihydrate stops growing. Fig. 9.9 shows the hydration progress measured using different techniques. Fig. 9.10 shows the morphological course of hydration measured by optical microscopy following different hydration times (chosen depending on the different reaction steps as described above).

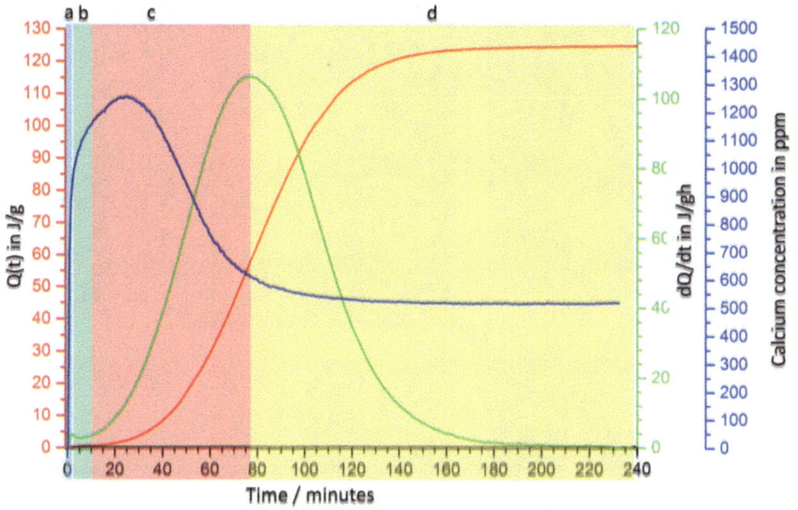

Fig. 9.9: Reaction energy measured by heat flow calorimetry vs. calcium concentration measured by ion chromatography, α-hemihydrate, (a) dissolving of hemihydrate and formation of super saturated solution, (b) seed crystal formation and dissolving of hemihydrate, (c) growing of dihydrate crystals and dissolving of hemihydrate, (d) most hemihydrate dissolved, growing of dihydrate crystals and end of reaction.

9.5.2 Different morphologies of dihydrate created by the hydration of hemihydrate and its influence on technical properties

As described in Section 9.3.3, gypsum can adopt various morphologies. Frequently one can observe longer or shorter needle like crystals with different amounts of branches alongside disk like crystals (normally look like dove tailed twins). When additives are present, xenomorphic plates can also be obtained. SEM figures of hemihydrate 1 to 3 are shown in Fig. 9.11.

The short and less branched crystals (depicted in Fig. 9.11), are used in highly accurate castings with binding materials. In this case, the value of flexural and compressive strength decreases. Porosity and expansion during hydration also have the same trend. With screed long crystals with more branches are required for more strength and a higher porosity (a higher porosity causing the material to dry faster). Depending on the porosity, gypsum can also influence room climate; creating a humidity between 45 to 90 % according to its deliquescence. The properties of gypsum based binders created from hemihydrate 1 to 3 are shown in the following figures. Fig. 9.12 shows the flexural strength, Fig. 9.13 the compressive strength, Tab. 9.1 enlists the porosity, and Fig. 9.14 shows expansion during hydration from hemihydrate 1 to 3 [30, 31]. Reaction temperature also influences the crystal morphology of the created dihydrate crystals. Higher temperatures lead to longer or larger crystals as the ion concentration in the reaction solution is lower influencing seed crystal formation [32, 33].

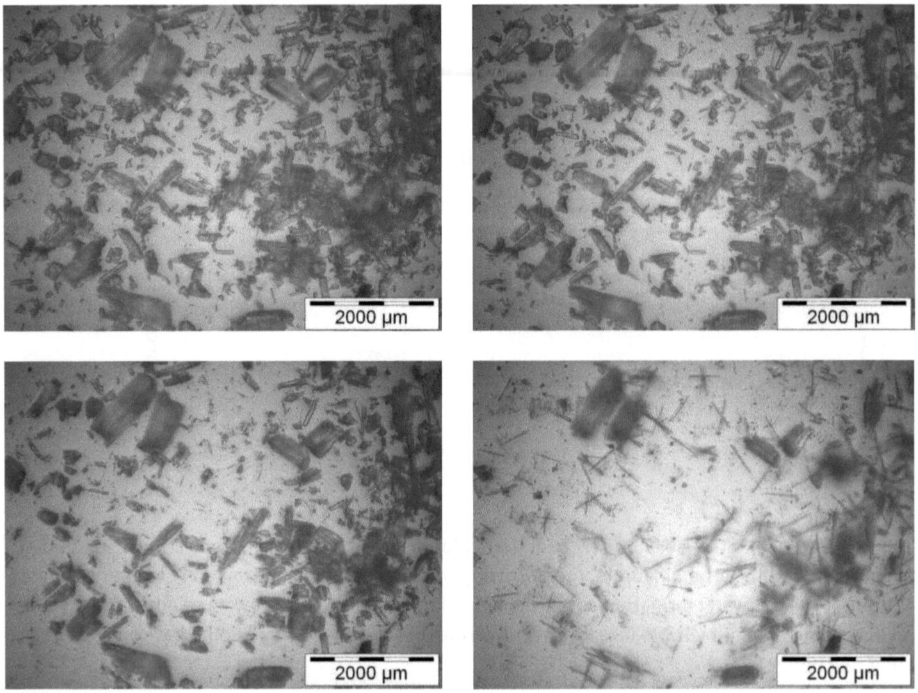

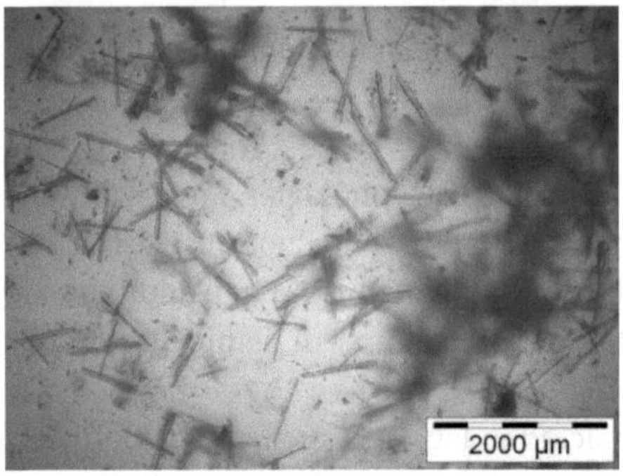

Fig. 9.10: α-hemihydrate reacting with water after different reaction times: after 1 (upper left), 5 (upper right), 20 (middle left), 80 (middle right), and 240 minutes (lower right).

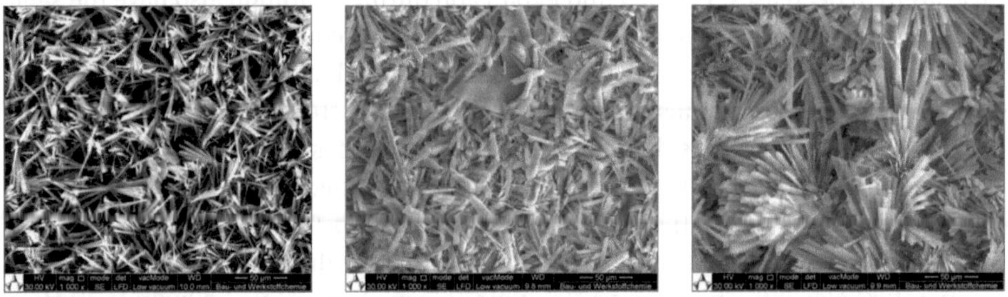

Fig. 9.11: Gypsum crystals created by hemihydrate 1, 2, and 3. Hemihydrate 1 creates longer crystals with some branches, hemihydrate 2 very small crystals with some branches, and hemihydrate 3 longer crystals with many branches; SEM low vacuum mode after 5 hours.

Tab. 9.1: Porosity of the gypsum stone created by hemihydrate 1 to 3 measured by mercury intrusion porosimetry.

Sample	Porosity (vol.%)	> 10 μm Air pores (vol.%)	10–0.03 μm Capillary pores (vol.%)	< 0.03 μm Gel pores (vol.%)
Hemihydrate 1	16.01	2.05	13.96	0.00
Hemihydrate 2	14.34	2.47	11.88	0.00
Hemihydrate 3	25.98	2.65	23.30	0.03

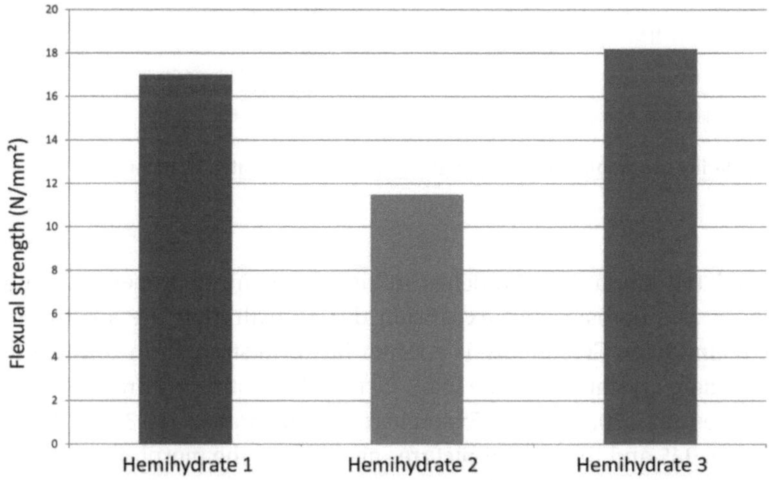

Fig. 9.12: Flexural strength of gypsum stone created by hemihydrate 1 to 3.

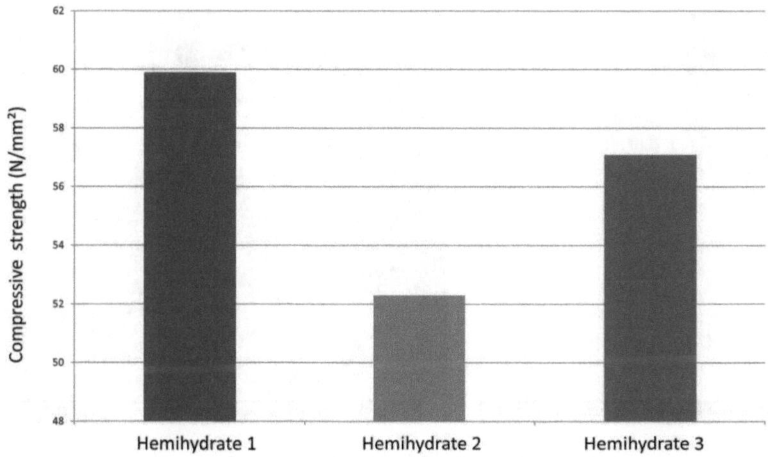

Fig. 9.13: Compressive strength of the gypsum stone created by hemihydrate 1 to 3.

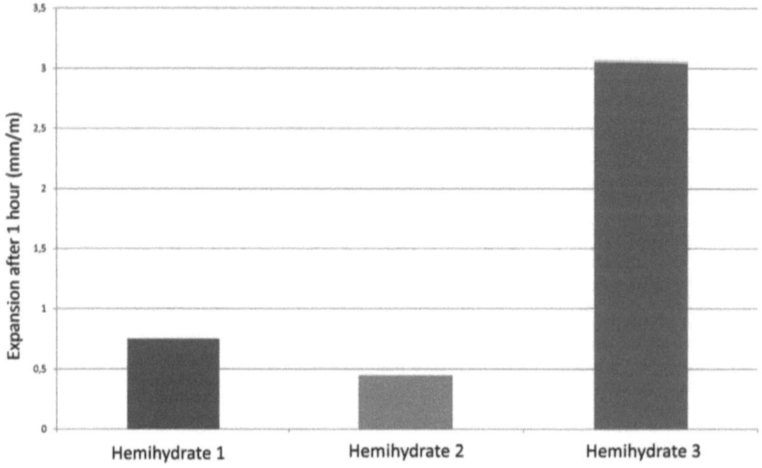

Fig. 9.14: Expansion during hydration after 1 hour of the gypsum stone created by hemihydrate 1 to 3.

Crystal length and the amount of branches influence technical properties. With regards to strength, the porosity and expansion during hydration the amount of branches is the main factor. Crystal size is a minor factor in strength in contrast to casting accuracy where crystal size is the main factor. The casting accuracy is measured by casting lines of 25, 50, or 75 μm. Examples of hemihydrate 2 and 3 for a 25 μm line are given in Fig. 9.15 and the optical pictures again show the morphology of the two hemihydrates. The sizes measured from these crystals in the optical microscopy pictures are collected in Tab. 9.2.

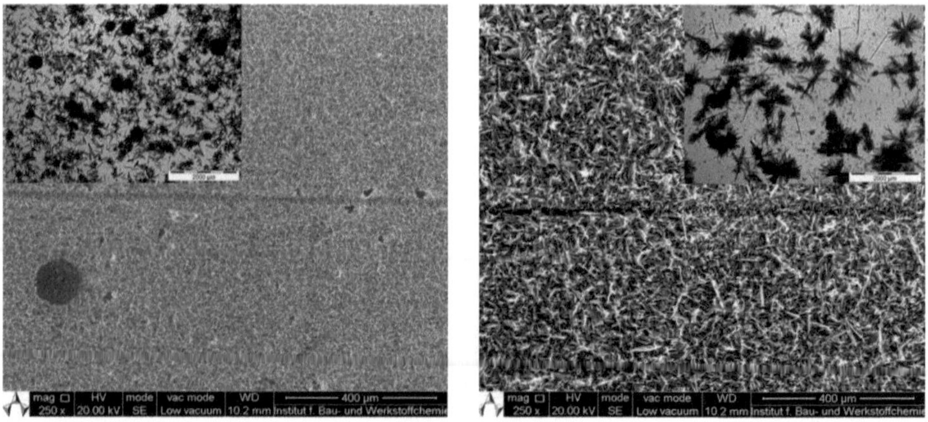

Fig. 9.15: Casting accuracy measured for hemihydrate 2 (left) and 3 (right) 25 micrometer lines. Optical microscopy pictures show the hemihydrate morphology.

Tab. 9.2: Length measurement of the gypsum crystals used in Fig. 9.15, optical microscopy pictures.

	Crystal length (μm), Fig. 9.15	
	Left hemihydrate	Right hemihydrate
Number	25	25
Minimum	109	570
Maximum	501	1098
Standard deviation	81	148
Average	206	934

9.5.3 Influence of additives on the morphology of the created dihydrate crystals

Normally different additives are used to influence technical properties like hardening time, porosity, and workability or to create special properties like hydrophobic surfaces, air pores, better workability, or others. Mixtures of additives are frequently used [34–46].

As accelerators gypsum seeds, potassium sulphate, other sulphates, aluminates, sodium chloride or alum can be used. These additives are capable of adjusting the required hardening time. Possible side effects are efflorescence, larger porosity, and changes in the final gypsum morphology. Polycarboxylate ethers, lignin sulphonates, melamine resin, casein, or copolymers with phosphate groups can be used as liquidizers to decrease the demand for water for workability. Thereby, the porosity is lowered. Some additives also decrease the degree of hydration as well as compressive strength and flexural strength. As adhesives methylcellulose, cellulose paste, polysaccharides, or bentonite are feasible. They lead to a change in viscosity, increase stickiness and are water retarding agents (yet some of them also retard the hydration process). Typical retarders are citric acid or other fruit acids, alcohols, sugars, phosphates, or polyphosphates. These additives normally influence the morphology of the created dihydrate. Citric acid for example, decreases the growing speed of the c-axis forming shorter crystals with a larger diameter. Supplementary fibers (e.g. glass, ceramic, polymer, wood, or metal fibers) enhance flexural strength and fire resistance.

The influence of additives to the morphology of formed dihydrate is illustrated in the following Fig. 9.16.

The first micrograph shows the dihydrate crystals which are created through hydration of a typical alpha hemihydrate. The influence of citric acid on the morphology can be observed in Fig. 9.16 in the lower right: Shorter crystals with less branches and larger diameters form. If potassium sulphate and citric acid are used, the citric acid dominates the influence on morphology (shown in Fig. 9.16, lower left). With potassium sulphate, the reaction is accelerated and longer crystals are grown, but the amount of branches decreases as shown in Fig. 9.16 in the upper right. A reference crystal, formed upon hydration without additives is given in Fig. 9.16 (upper left).

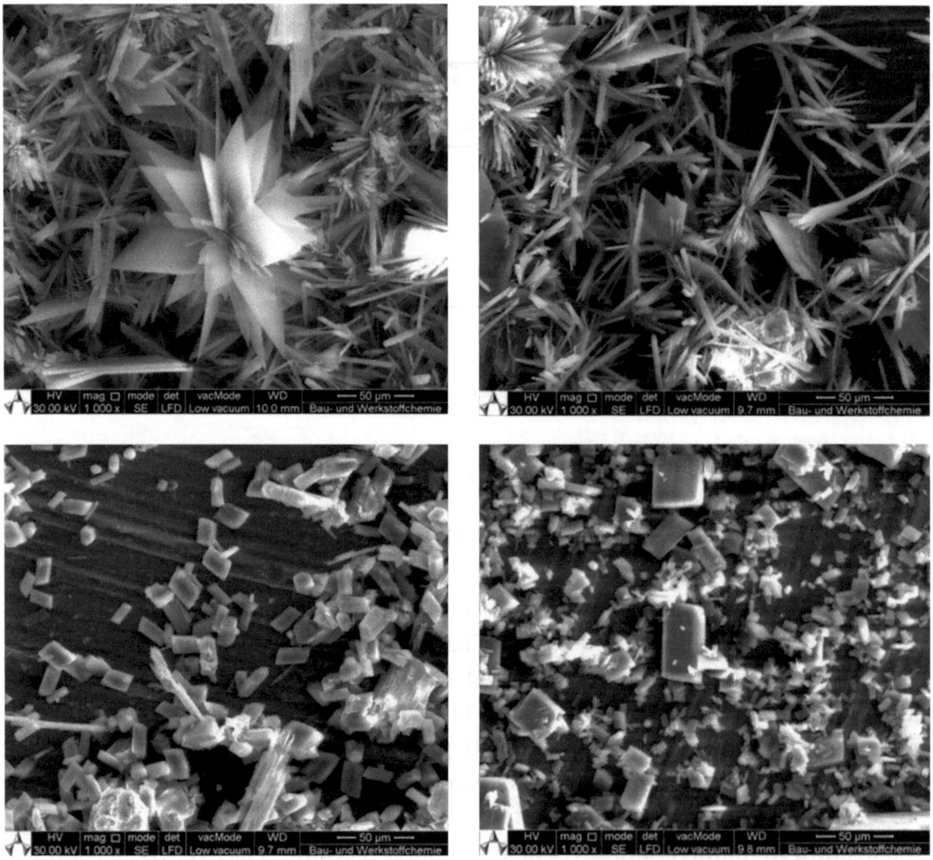

Fig. 9.16: Influence of different additives on the crystal morphology of dihydrate: without additive (upper left), with 2 % of potassium sulphate (upper right), with 2 % of potassium sulphate and 0.2 % of citric acid (lower left), and with 0.2 % of citric acid and 2 % of gypsum seed crystals (lower right); SEM after 5 hours reaction with water.

9.5.4 Strength development and strength decrease in presence of moisture

Different theories about the development of strength in gypsum stone are discussed in literature. The first theory explains that dihydrate creates strength in gypsum stone by felting its crystals [47]. Another theory explains strength development through single gypsum crystals growing together and thereby creating a network [48]. A third approach explains that the surfaces of the single crystals are glued together by van der Waals forces of the adsorbed water. All of these theories have been verified by single experiments, but none of them can explain all phenomena observed. Depending on these theories, the decrease of strength in the presence of moisture is explained differently. If the strength of gypsum stone is measured in the presence of moisture, its value is sometimes only one fifth of those of dry samples. In case of felting of single crystals it is explained by the dissolution of those crystals. In the network theory, the starting points of dissolving crystals overlap the points of crystals growing together, so breaking the network and decreasing the strength. If the strength development is described by van der Waals forces, its decrease is explained by an increase in diameter of the water layer [49]. Porosity also influences the strength decrease and the strength of gypsum [50]. In order to prove these theories, different experiments were performed, which are presented in this contribution. Single gypsum crystals could show a plastic deformation if pressure is orthogonal to the cleavage planes [51].

The theory of felting crystals as a mechanism for hardening explains that larger crystals with more branches create higher strength than smaller ones with fewer branches. At the same time, small crystals without branches give a gypsum stone with good strength properties as well. This cannot be explained with the felting crystal theory. The creation of a gypsum crystal network by an intergrowth of individual crystals has never been observed under optical microscopy. Instead, growing crystals push each other away. This theory for hardening cannot be the only mechanism for strength creation in calcium sulphate dihydrate stones. In nature however, some interpenetration of twins could be observed. Gypsum stones stored at higher temperatures are free of water, so there should not be any strength if the third theory about the hardening of gypsum is true. However, there are still high measurable strengths. At lower humidity the strength increases a little so it can be concluded that van der Waals forces influence the strength of gypsum stone because thin water layers could work as glue between the single crystals.

The decreasing strength of gypsum stone in the presence of moisture is not primarily caused by dissolving particles (it is nearly the same in a saturated gypsum solution where no dissolution can take place). The interesting point is that due to recrystallization, the strength of gypsum stone stored in gypsum solution might be a little higher or lower compared to that stored in water. This depends on the morphology of the dihydrate crystals. Smaller crystals with fewer branches create less porosity, so that newly created crystals are not able to grow into the pores and thus create a crystallization pressure leading to lower strength. In the case of highly ramified crystals, more pores could be detected in the gypsum stone. During the recrystallization process these pores get smaller leading to a small increase of strength. Results for flexural strength and compressive strength are given in Fig. 9.17 and 9.18.

The combination of the results described leads to a new theory about the hardening of gypsum stone and the strength decrease in the presence of moisture. Regarding this theory, the main factor in the hardening of gypsum stone is the friction between the single gypsum crystals. Smaller factors are van der Waals forces and felting of crystals. Branches lead to felting causing a stopper effect on the gliding of gypsum crystals. Thinner water layers could work as glue between the planes of individual crystals which explains the slight increase of strength at lower humidity. Thicker water layers, like in liquid water or at higher humidity, could change the friction mechanism from a solid-solid friction into a solid-liquid friction, resulting in a change of the co-

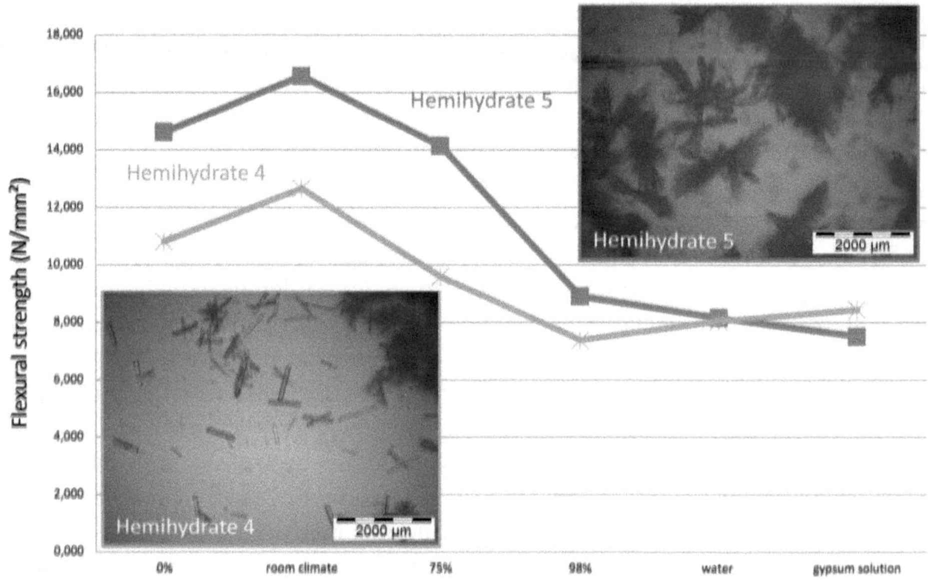

Fig. 9.17: Flexural strength of gypsum based binding material at different humidities. The optical microscopy pictures show the morphology of two different gypsum materials.

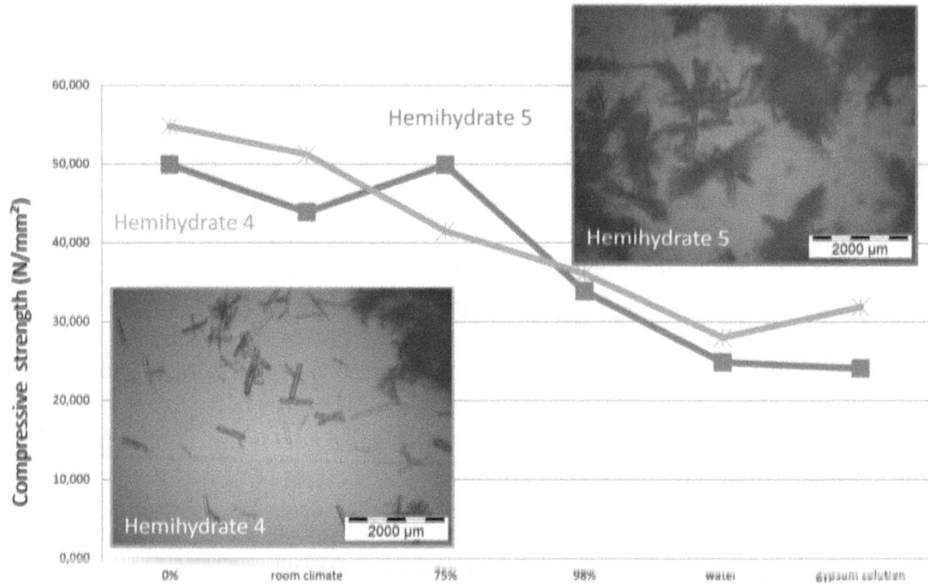

Fig. 9.18: Compressive strength of gypsum based binding material at different humidity. The optical microscopy pictures show the morphology of two different gypsum materials.

efficient of friction of a factor to about 10. It can be assumed that there is a mixture of solid-solid friction and solid-liquid friction in any case, so that an 80 % strength decrease for gypsum stone is realistic. This theory could explain all experimental results but it is still a subject of research.

According to computational simulations, the binding energy between HOH···OSO_3 is in the range of 35 kJ/mol and for water to water about 19 kJ/mol for two water molecules (Dr. Funk [52]). The binding energy for water to water in liquid water is given by literature at 7.9 kJ/mol. This explains why small water layers could work as glue and larger ones as a lubricant. In the case of single gypsum crystal flexural strength test, the chemical bounded water in the gypsum crystal also acts as a lubricant. The cleavage planes are the sites where crystalline water is stored in the gypsum crystal. Pressure from outside the binding force will be exceeded and the single layers could glide on each other.

The changes in strengths in the samples which were stored at different temperatures could be explained by the previous theory. At 100 °C hemihydrate is formed and the volume changes, thus more pores are created and the gluing effect of water disappears with the physically bound water. At 400 °C anhydrite with a higher sur-

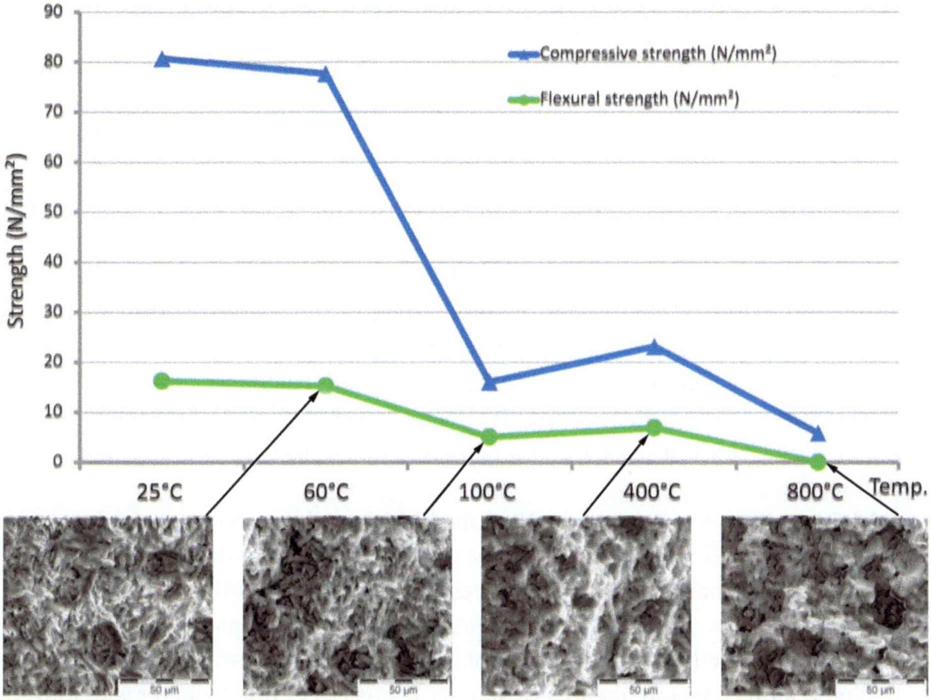

Fig. 9.19: Bending tensile and compressive strength of gypsum based binding material stored at different temperatures. Samples were stored at different temperatures. Pictures show the structure at a given temperature.

face roughness is formed. This results in a higher friction which could be seen as explanation of the increase in strength. At temperatures above 600 °C a sinter process accompanied by larger shrinkage starts, which includes the creation of small cracks in the crystals and therefore reduced strength. Results are given in Fig. 9.19.

9.6 Summary

Calcium sulphate based binders have many advantages: e.g. eudermic pH values, good fire resistance, high strength, low imbedded energy, convenient handling, recyclability, and in the creation of comfortable room climates. The main disadvantage of calcium sulphate based binding materials is their extreme decrease in strength in the presence of moisture. These materials are therefore not applicable for use in outdoor construction.

The different morphologies of dihydrate crystals observed in natural gypsum could also be investigated in calcium sulphate based binding materials. These morphologies influence the technical properties of the material. Additives are feasible in the adjustments of technical properties by influencing the dihydrate morphology but they also sometimes cause side effects.

Calcium sulphate based binding materials are used for many different technical applications in a variety of fields. They are mainly used for interior fittings like plaster board, between wall tiles and gypsum fiber board, as stucco and screed building materials. Additionally they can be used as a casting material for dental fillings due to fast hydration and good casting accuracy. Furthermore, they are utilized for metal casts due to resistance to high temperatures and in ceramic casts for slush due to their porosity which ensures suitable water transport.

Optical microscopy is a capable method to observe the influence of different additives and the influence of mixtures of additives on the morphology of the hydration products of hemihydrate. These morphologies influence important technical properties e.g. porosity, strength and strength development, casting accuracy, expansion during hydration, and surface properties. By understanding these interrelations, the creation of new calcium sulphate based binding materials with special properties is within reach.

Additionally, a new theory was developed which explains the creation of strength in gypsum based materials and the decrease of said strength in the presence of moisture. This theory combines former theories like gypsum crystal felting and van der Waals forces with friction between crystals. The strength decrease in the presence of moisture is explained through a change of the friction mechanism from solid-solid to solid-liquid alongside a weakening of van der Waals forces due to larger water layers and the dissolution of gypsum crystals.

References

[1] Kirca Ö. Ancient binding materials, mortars and concrete technology: History and durability aspects. In: Modena Lourenco & Roca, editors. Structural Analysis of Historical Constructions. London: Taylor & Francis Group; 2005. ISBN 0415363799.

[2] Lucas A, Harris JR. Ancient Egyptian Materials and Industries. 4th edition. Dover Publications Inc; 2005. ISBN 978-0486404462.

[3] Aitcin P-C. Binders for durable and sustainable concrete. Taylor & Francis Group; 2008. ISBN 978-0-415-38588-6.

[4] Lancaster LC. Innovative Vaulting in the Architecture of the Roman Empire: 1st to 4th Centuries, CE. Cambridge University Press; 2015. ISBN 978-1107059351.

[5] Gypsum association Gypsum through the ages [cited 2016 Jun 20]. Available from: https://www.gypsum.org/about/gypsum-101/history-gypsum.

[6] Bundesverband der Gipsindustrie e.V. GIPS-Datenbuch Bundesverband der Gipsindustrie e.V. Berlin; 2013. Available from: www.gips.de.

[7] Wirsching F. Die Phasen des Systems $CaSO_4$–$CaSO_4 \cdot 2H_2O$. ZKG. 1966; 10.

[8] Jäger R, Brosig A. Verfahren zur Herstellung von Alpha-Calciumsulfat-Halbhydrat aus Calciumsulfat-Dihydrat. Patent publication no. WO2008141627A1, patent no. PCT/DE2008/000854; 2008.

[9] Förster HJ. Aufbereitung von Abfallgips aus der Phosphorsäure-Produktion nach dem Giulini-Verfahren. Chem Ing Techn. 2005; 44: 969–972.

[10] Schoch EP, Cuningham WA. Production of gypsum plaster by wet methods. Chem Engin. 1941; 37(8): 1–18.

[11] Leskeviciene V, Sarlauskaite I, Nizeviciene D, Kybartiene H. Influence of the Gypsum Dehydration Temperature and Alkali Additives on the Properties of Anhydrite Cement Science of Sintering. 2010; 42: 233–243.

[12] Wang H, Zhang W, Dong G, Zhang J. Influence of impurities on hydration, hardening and morphology of flue gas desulphurization gypsum plaster. Procedia Engineering. Chinese Materials Conference. 2011; 27: 384–393.

[13] Kogel JE, Trivedi NC, Barker JM. Industrial Minerals & Rocks. 7th edition. Soc for Mining Metallurgy; 2006. ISBN 978-0873352338.

[14] Euro Gypsum Factsheet on: What is Gypsum? Euro Gypsum; 2007. Available from: www.eurogypsum.org.

[15] Goldschmidt Victor M. Atlas der Kystallformen. Book IV, Tables 63 to 73. Heidelberg: Winters; 1923.

[16] Offermann E. Kristalle und ihre Formen. Band 1, Theoretische Kristallmorphologie. Achberg: KristalloGrafik Verlag; 2004. ISBN 3-00-008112-7.

[17] Offermann E. Kristalle und ihre Formen. Band 2, Praktische Kristallmorphologie. Achberg: KristalloGrafik Verlag; 2004. ISBN 3-00-008112-7.

[18] Schorn S, et al. Mineralienatlas – Fossilienatlas Gypsum MineralDate [cited 2016 Jun 19]. Available from: https://www.mineralienatlas.de/lexikon/index.php/MineralData?mineral=Gypsum.

[19] Badens E, Veesler S, Boistelle R. Crystallization of gypsum from hemihydrate in presence of additives. Journal of Crystal Growth. 1999; 198/199: 704–709.

[20] Muryanto S. The Role of Impurities and Additives in the Crystallization of Gypsum [PhD thesis]. Curtin Universtity of Technology; 2002.

[21] Le Chatelier MH. Crystalloids against colloids in the theory of cements. Trans Faraday Soc. 1919; 14.

[22] Christoffersen J, Christoffersen M. The Kinetics of Dissolution of calcium sulphate dihydrate in water. Journal of Crystal Growth. 1976; 35: 79–88.

[23] Autorenkollektiv Der Baustoff Gips. 1st edition. Berlin: VEB Verlag Bauwesen; 1977.

[24] Cavazzi A. Das gelatinöse Calciumsulfat und das Abbinden des Gipses. Kolloid-Zeitschr. 1912; 11.

[25] Becherer G, Fiedler H. Über röntgenografische Untersuchungen des Abbindevorgangs bei Gips. Silikattechnik. 1955; 7(6).

[26] Perederij IA. Theorie der Bildung, Erhärtung und Festigkeit von normalem Gips und hochfestem Gips. GP Chem Techn. 1956; 8.

[27] Eipeltauer E. Erzeugung von kriechfesten Hartgipsen. Zem.-Kalk-Gips. 1960; 6.

[28] Pritzel C, Sakalli Y, Trettin R. In situ optical microscopy and in situ ultra sonic measurements for quality assurance in gypsum plants. Global Gypsum Magazine. 2010; September/October. ISSN: 1463-9661.

[29] Follner S, et al. The Setting Behaviour of α- and β-$CaSO_4 \cdot ½H_2O$ as a Function of Crystal Structure and Morphology. Cryst Res Technol. 2002; 37(10): 1075–1087.

[30] Pritzel C, Trettin R. Influencing the morphology of gypsum. In: Broekmans ATM, editor. Proceedings of the 10th International Congress for Applied Mineralogy (ICAM). Berlin: Springer; 2012. Pages 541–548, ISBN 978-3-642-27681-1, e-ISBN 978-3-642-27682-8.

[31] Gartner EM. Cohesion and expansion in polycrystalline solids formed by hydration reactions – The case of gypsum. Cement and Concrete Research. 2009; 39: 289–295.

[32] Ridge MJ. Effect of Temperature on the Structure of Set Gypsum Plaster. Nature. 1958; 192: 1224–1225.

[33] Pritzel C, Trettin R. Einfluss der Hydratationstemperatur auf die Kristallmorphologie von Calciumsulfathydraten. GDCh-Monographien Band 39; 2009. Pages 359–360, ISBN: 978-936028-54-6.

[34] Van Der Leeden M, Van Rosmalen G, De Vreugd K, Witkamp G. Einfluß von Additiven und Verunreinigungen auf Kristallisationsprozesse. Chem Ing Tech. 1989; 61(5).

[35] Mallon T. Die Verzögerungswirkung von Gipsverzögerern verschiedener chemischer Zusammensetzung in Abhängigkeit vom pH-Wert des Gipses. Zement-Kalk-Gips. 1988; 41(6).

[36] Chatterji S, Jeffrey W. Crystal growth during hydration of $CaSO_4 \cdot ½H_2O$. Nature. 1993; 200: 463–464.

[37] Komarov VF, Severin AV, Melikhov IV. Fluctuations Growth rate of gypsum crystals. Crystallography reports. 2000; 45(2): 329–335.

[38] Sangwal K. Additives and Crystallization Processes, From Fundamentals to Applications. Wiley; 2007. ISBN: 978-0-470-06153-4.

[39] Koslowski T. Zitronensäure ein Verzögerer für Gips [dissertation]. Aachen; 1983.

[40] Pritzel C, Trettin R. The Phase transformation of Calciumsulfate subhydrates reacting with water. 12th International Congress on the Chemistry of Cement, Montreal; 2007.

[41] Combe EC, Smith DC. The Effects of some organic acids and salts on the setting of gypsum plaster I. acetates. J Appl Chem. 1964; 14.

[42] Combe, EC, Smith DC. The Effects of some organic acids and salts on the setting of gypsum plaster II. tatrates. J Appl Chem. 1965; 15.

[43] Bosbach D, Hochella MF, Jr. Gypsum Growth in the Presence of Growth Inhibitors: A Scanning Force Microscopy Study. Chem Geol. 1996; 132.

[44] Müller M, Fischer H-B, Hummel H-U, Scheller L. Acceleration of the setting of hemihydrate plaster with calcium sulfate dehydrate. ZKG. 2009; 62(3): 47–53.

[45] Hill J-R, Plank J. Retardation of Setting of Plaster of Paris by Organic Acids: Understanding the Mechanism through Molecular Modeling. Wiley, InterScience; 2004.

[46] Peng J, Qu J, Zhang J, Chen M, Wan T. Adsorption characteristics of water-reducing agents on gypsum surface and its effect on the rheology of gypsum plaster. Cement and Concrete Research. 2005; 35: 527–531.

[47] Krenkler, K. Chemie des Bauwesens. Band 1, Anorganische Chemie. Springer; 1980. ISBN 978-3642814761.

[48] Müller M. Die Abbindebeschleunigung von Stuckgips durch Calciumsulfatdihydrat. Kassel University Press; 2007. ISBN 978-3899583281.

[49] Pellenq Roland J-M, Van Damme H. Why does concrete set? The Nature of cohesion forces in hardened cement-based materials. MRS Bulletin: 319–323.

[50] Singh M, Grag M. Relationship between mechanical properties and porosity of water-resistant gypsum binder. Cement and Concrete Research. 1996; 26(3): 449–456.

[51] Craker WE, Schiller KK. Plastic Deformation of Gypsum. Nature. 1962; 193: 672–673.

[52] Pritzel C, Trettin R, Sakalli Y. Beitrag zur Festigkeitsentwicklung von Gipsstein und zum Festigkeitsabfall bei Durchfeuchtung. In: Finger FA. Tagungsbericht, 2. Weimarer Gipstagung 26.–27. März 2014 in Weimar. Weimar: Institut für Baustoffkunde Bauhaus-Universität Weimar; 2014. Pages 313–318, ISBN 978-3-00-045359-5.

[53] Okrusch M, Matthes S. Mineralogie: Eine Einführung in die spezielle Mineralogie, Petrologie und Lagerstättenkunde. 7th edition. Berlin: Springer; 2005.

[54] Förthner S. Kristallisation und Morphologiebeeinflussung bei Gipsen: Untersuchungen zur Steuerung der makrophysikalischen Eigenschaften. Südwestdeutscher Verlag für Hochschulschriften; 2011. ISBN 978-3838123691.

Daniela Freyer

10 Magnesia building material (Sorel cement) – from basics to application

Abstract: Magnesia building material is a special cementitious material made up of magnesia (magnesium oxide, MgO) as a binder and a concentrated magnesium chloride solution as the mixing liquid. Admixtures of different aggregates modify the resulting mechanical properties alongside pigment additives which modify the color. The special features of this material are based on specific material properties and a selective range of applications, in comparison to the more extensively used ordinary Portland cement (OPC). The starting material, MgO, is also more expensive than Portland cement. On the other hand, MgO has a smaller carbon dioxide footprint than OPC due to the lower burning temperature used in its production.

In the following subsections, the application of the magnesia building material (also called magnesia cement, magnesia oxychloride cement (MOC), or Sorel cement) is described. Starting with its historical development and continuing up to recently reached, state-of-the-art, science and technology.

Keywords: magnesia cement, magnesium oxychloride, setting reaction mechanism, application, building material in salt formations

10.1 History and application

The discovery of magnesia cement dates back to the Frenchman Stanislas Sorel in 1867 [1], who had previously reported in 1855, the discovery of zink oxychloride cement [2]. The material is named Sorel cement after him. Sorel justified the hardening reaction with the formation of a basic magnesium oxychloride. The first investigation to determine the constitution was carried out in 1870 by Bender [3]. By 2010 [4], the composition of all relevant basic magnesium oxychloride phases had been determined accurately.

During this period of more than 100 years, numerous investigations addressing the characterization of material composition, preparation and properties were published [5–30].

Since the beginning of the 19th century magnesia cement has been used in the manufacture of special cement flooring (also known as stone wood floor) in living spaces as well as in highly resilient industrial floors. Numerous patents appeared and protected the increasingly improved processing methods and properties of this floor

Daniela Freyer, TU Bergakademie Freiberg, Institut für Anorganische Chemie, Leipziger Str. 29, 09596 Freiberg, Germany, daniela.freyer@chemie.tu-freiberg.de

DOI 10.1515/9783110473728-011

material. Its production became a major industry [31]. Prepared as plates, the material is known under the name "Xylolithe", and the plates are used in flooring or as wallboards [32, 33]. In combination with various aggregates such as wood flour, wood shavings, cork, asbestos, grog, and kieselgur, a different type of stone wood floor results. Depending on these aggregates, the stone wood floor and Xylolithe panels show properties similar to that of wood or stone. The resulting high mechanical strength, lower alkalinity and absence of shrinking (in comparison with OPC products) and its high elasticity, low thermal conductivity, fire resistance, resistance to oil, grease, gasoline, benzene and boron water, and the possibility to prepare it in any color by pigment admixture, still makes it a unique flooring material today (Fig. 10.1). It is found in machine factories, assembly plants, textile and shoe factories, or under certain conditions, in aesthetically demanding areas such as homes offices, studios, lofts, and schoolhouses [34].

Fig. 10.1: Examples of magnesia cement floor applications (copyright by courtesy of Walo Bertschinger AG [34]).

In addition, magnesia cement is also used for some other applications, such as bonding agents in grinding wheel production, artificial stone manufacturing (lithographic stone), as artificial ivory (e.g. billiard balls) and in decorative interior plasters with embedded stone aggregate [35–37].

However, magnesia cement is not resistant to acids and alkalis and is corrosive on contact with metal, which must be taken into account for potential applications. Even though it is known that magnesia cement is not resistant to water and to permanent humidity changes (outdoor situation), there are providers who offer panels or wall boards made of magnesia cement for outdoor applications. They describe the panels as being resistant to humid environments and due to the resulting corrosion (decomposition) there are customer complaints.

A special feature of magnesia cement is its resistance to salt and salt solutions, which is why the material has been used for many years as a construction material in salt formations.

All the known and specific properties of magnesia building material mainly result from experiences collected while using it and while searching its applications. Thus, 'From Basics to Application' is rather the currently reached route for different constructive applications. Scientific knowledge about phase formation in the magnesia binder system has improved considerably during the last decade due to its special importance as a building material in salt formations. Its application in barrier constructions, as part of plug and sealing systems in current concepts for toxic and nuclear waste repositories in salt rock formations [38, 39], requires proof of the materials' long-term stability. Therefore, knowledge about the thermodynamic solubility equilibria in the binder phase–salt solution system is essential.

10.2 Magnesia cement composition

Magnesia cement is produced through the reaction of caustic magnesium oxide (MgO) with concentrated magnesium chloride solution. Depending on the MgO content (and purity) and the $MgCl_2$ concentration in the mixing liquid, various magnesium chloride hydroxide hydrates (Mg-oxychloride) with the general composition $Mg_k(OH)_lCl_m . nH_2O$ (alternatively written as *x-y-z* phases according the double salt hydrate notation: $xMg(OH)_2 \cdot yMgCl_2 \cdot zH_2O$) are formed during hydration, setting and hardening of the cement according the reactions (10.1) and (10.2).

$$3MgO + MgCl_2 + 11H_2O \rightarrow 2[Mg_2(OH)_3Cl \cdot 4H_2O] = 3Mg(OH)_2 \cdot MgCl_2 \cdot 8H_2O = 3\text{-}1\text{-}8 \text{ phase} \quad (10.1)$$

$$5MgO + MgCl_2 + 13H_2O \rightarrow 2[Mg_3(OH)_5Cl \cdot 4H_2O] = 5Mg(OH)_2 \cdot MgCl_2 \cdot 8H_2O = 5\text{-}1\text{-}8 \text{ phase} \quad (10.2)$$

The 3-1-8 phase was also found as a mineral in veinlets in dolomitic marble and named after the locality as korshunovskite [40, 41].

The two phases, 3-1-8 and 5-1-8, have long been known as the relevant binder phases of magnesia cement. Their stoichiometry were finally determined in the 1950's [16, 18, 42]. Alongside these two phases, more have been unambiguously characterized over the last years. These are the 9-1-4, 2-1-4, 2-1-2, and 3-1-0 phases [4, 43–47], of which only the 9-1-4, a high temperature phase, has importance as binder phase in magnesia cement.

Beside the Mg-oxychloride system, magnesium oxysulfate and magnesium phosphate cements are mentioned. However, their importance is of minor significance for applications and so will not be considered further.

10.3 Phase formation, stability, and properties of Mg-oxychloride phases

The formation and stability of basic magnesium chloride hydrates (Mg-oxychloride) must be considered from different perspectives. On the one hand, phase formation during setting and hardening of magnesia cement must be considered (a kinetical controlled process). On the other hand, the phase formation and stability under equilibrium conditions in the solid-liquid-system, as in the system $Mg(OH)_2$–$MgCl_2$–H_2O at definite temperatures, must also be understood.

The setting reaction depends on the detailed formulation (the ratio of MgO to $MgCl_2$ in solution, as well as their concentration) and the temperature that develops during the exothermic setting reaction as well as which phase or phases are formed and are the resulting phase or phases after the completed setting reaction (Section 10.3.4). These facts and details can only be understood and properly analyzed if the temperature dependent solid-liquid equilibria in the system $Mg(OH)_2$–$MgCl_2$–H_2O are known [48].

10.3.1 Solubility equilibria in the system $Mg(OH)_2$–$MgCl_2$–H_2O

The ternary system $Mg(OH)_2$–$MgCl_2$–H_2O represents the scientific basis of magnesia building materials. From the temperature dependent solubility data of the system, one can derive the formation conditions and the ranges of stability of Mg-oxychloride phases in the presence of different $MgCl_2$ solution concentrations.

As found in reference materials, appropriate solubility data are given either as OH^- [4, 49–53] or as H^+ molalities [54, 55] dependent on $MgCl_2$ molalities in equilibrium with the solid phase. Conversion is only possible with knowledge of the ionic product of water in dilute to concentrated $MgCl_2$ solutions. A more direct way to quantify the solubility equilibria is with determination of the OH^- solution concentration according to the solution reaction of basic magnesium chloride hydrates, e.g.

$$3Mg(OH)_2 \cdot MgCl_2 \cdot 8H_2O \rightarrow 4Mg^{2+} + 6OH^- + 2Cl^- + 8H_2O.$$

However, in view of the further use of $Mg(OH)_2$ and Mg-oxychloride to buffer the pH value of brines in evaporating marine environments [56], the knowledge of the H^+ molalities (representative of pH value) is necessary. In relation to radioactive waste repositories, the application of $Mg(OH)_2$ and Mg-oxychloride will buffer a near neutral pH to favorably influence the geochemical conditions in the zone near the waste repository (radionuclide retention) – a special scientific field of research [57, 58].

Current solubility data ($Mg(OH)_2$ molalities vs. $MgCl_2$ molalities) in the system $Mg(OH)_2$–$MgCl_2$–H_2O from 25 °C to 120 °C [48, 59, 60] are displayed in Fig. 10.2. According to this data, there are four thermodynamically stable Mg-oxychloride hydrates

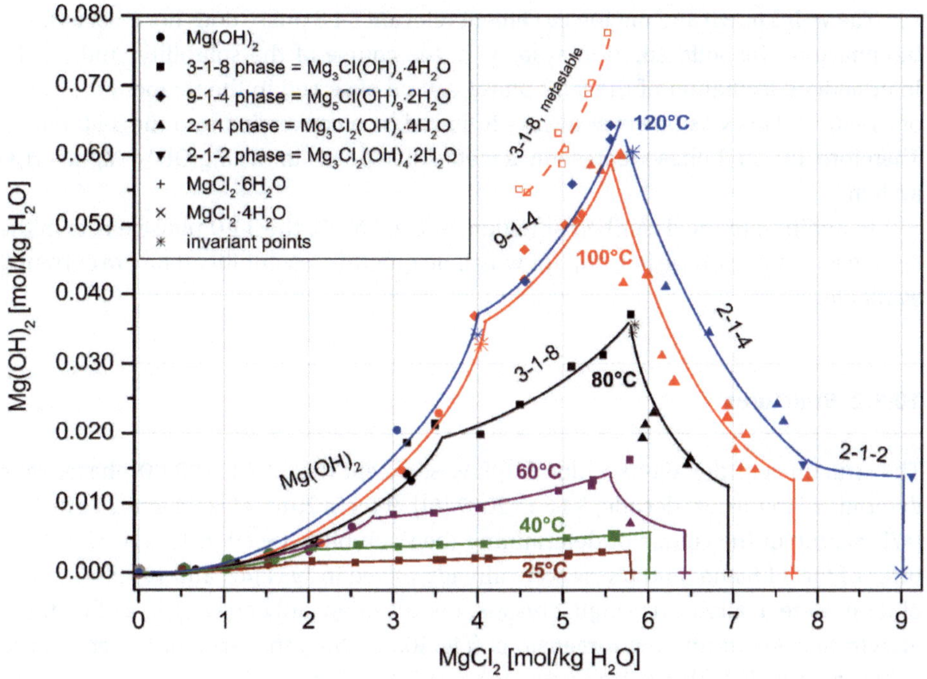

Fig. 10.2: Solubility equilibria in the System Mg(OH)–$MgCl_2$–H_2O (isotherms from 25 °C to 120 °C) [48, 59, 60].

(3-1-8, 9-1-4, 2-1-4, and 2-1-2 phases) in the system between 25 and 120 °C. Their crystallization fields are limited by $Mg(OH)_2$ and by the magnesium chloride hydrates, $MgCl_2 \cdot 6H_2O$ (bischofite) and $MgCl_2 \cdot 4H_2O$. $Mg(OH)_2$ occurs in low concentrated $MgCl_2$ solutions throughout the entire temperature range. Its stability field increases at higher $MgCl_2$ concentrations with increasing temperatures. The first occurring Mg-oxychloride is the 3-1-8 phase following the brucite range. Above 80 °C, instead of the 3-1-8 phase, the 9-1-4 phase occurs in the presence of appropriate $MgCl_2$ solution concentrations (≈ 4–5.6 mol $MgCl_2$/kg H_2O) as the thermodynamically stable phase. Even if a crystallization field of the 3-1-8 phase is found at 100 °C, due to the higher solubility in comparison to the 9-1-4 phase, the 3-1-8 phase is metastable at this temperature and transforms, dependent on time, into the 9-1-4 phase. At higher $MgCl_2$ solution concentrations (> 5.6 molal) the 2-1-4 phase occurs from 60 °C onwards. Because the $MgCl_2$ solution concentration reaches higher values with temperature, the crystallization field of the 2-1-4 phase expanded up to about 8 molal $MgCl_2$ at 120 °C. Above this concentration the 2-1-4 phase transforms into the 2-1-2 phase.

Generally, the solubility of the Mg-oxychloride phases increases with temperature. Maximum values range between 5–6 $MgCl_2$-molal concentrated solutions (at the appropriate two salt points).

The well-known and magnesia cement relevant 5-1-8 phase does not exist as a stable phase in the solid solution system. In the course of the solubility studies, the intermediate formation of the 5-1-8 phase can be observed in the temperature range of 25–40 °C. This was however always followed by a conversion into the 3-1-8 phase. Therefore, the 5-1-8 phase represents a metastable phase in the $Mg(OH)_2$–$MgCl_2$–H_2O system.

In addition to the described situation, above 120 °C, the 3-1-0 phase exists in the system as a stable phase [43, 46], for which no quantified solubility data are currently available.

10.3.2 Structures

The structures of the Mg-oxychloride phases, except the 3-1-8 and 3-1-0 phase, were determined in the last decade, 5-1-8 in 2007 [61], 9-1-4 in 2010 [4], 2-1-4 and 2-1-2 in 2012 [47], mainly in the course of determining their solubility (Section 10.3.1). The structures of the 3-1-8 and 3-1-0 phase were already solved in 1953 [42] and 1954 [44]. None of them were available as single crystals, because they only crystallize as fine intergrowth needles in nm–μm dimensions (Fig. 10.3). Thus, the structures were solved using exclusively high resolved powder X-ray diffraction data.

The structures of the Mg-oxychloride hydrates (3-1-0 phase is not considered) are closely related to the crystal structure of brucite, $Mg(OH)_2$. The Mg ion is coordinated by six ligands in an octahedral form. The octahedra are edge-linked and form definite layers. In the $Mg(OH)_2$ structure there are only hydroxide ions within the coordination sphere of magnesium (Fig. 10.4 (a)). In view of a setting reaction, brucite is formed from MgO in contact with water, however, the presence of appropriate chloride ions (with concentrated $MgCl_2$ solution as the mixing liquid) seems to open the closed octahedra layer sequence through the intrusion of chloride ions and water molecules. Thus forming (dependent on temperature and magnesium chloride solution concentration) the appropriate new edge-linked octahedra of the respective Mg-oxychloride hydrates. There is a clear trend visible in the condensation of the MgO_6 octahedra with respect to the $Mg(OH)_2$ content of the oxychloride phase.

The 9-1-4 phase (monoclinic, space group I2/m [4]), with the highest magnesium hydroxide content, consists of corrugated layers of edge- and corner-linked distorted MgO_6 octahedra, which are separated by interstitial one-dimensional zigzag chains of disordered Cl^- ions and H_2O molecules. Their atom positions are half occupied, thus every second position is vacant. Each MgO_6 octahedron, with only hydroxide ions within the coordination sphere of magnesium according to brucite, is connected to eight neighboring octahedra, to six through corner-sharing, and to two through edge-sharing (Fig. 10.4 (b)).

The crystal structure of the monoclinic 5-1-8 phase (space group P2/m [61]), with the second highest magnesium hydroxide content, consists of infinite triple chains

Fig. 10.3: SEM images of the Mg-oxychlorides, (a) and (b) 3-1-8 phase, (c) 9-1-4 phase, and (d) 2-1-4 phase.

of MgO_6 octahedra and intercalated occupationally disordered chlorine ions and water molecules (Fig. 10.4 (c)). The magnesium is exclusively coordinated with hydroxide and water molecules according to the structure of brucite and the 9-1-4 phase.

The crystal structure of the triclinic 3-1-8 phase (space group $P\bar{1}$ [42]), consists of double chains of distorted $Mg(OH)_4(H_2O)_2$ octahedra, which run parallel to alternately ordered chains of intercalating chlorine atoms and water molecules (Fig. 10.4 (d)). Both positions are occupied with the factor one, meaning no disordered occupancy.

The crystal structures of the 2-1-2 and the 2-1-4 phases (with the lowest magnesium hydroxide content) are similar (monoclinic, space group C2/m [47]). The main building blocks of both structures are infinite triple chains of edge-linked distorted octahedra in which firstly the magnesium is also coordinated with chlorine beside hydroxide ions and water molecules (Fig. 10.4 (e) and (f)). The phases crystallize in the presence of very highly concentrated $MgCl_2$ solutions at high temperatures (Fig. 10.2), so the magnesium in the solid phase structure cannot keep the preferred coordination of hydroxide and water molecules exclusively.

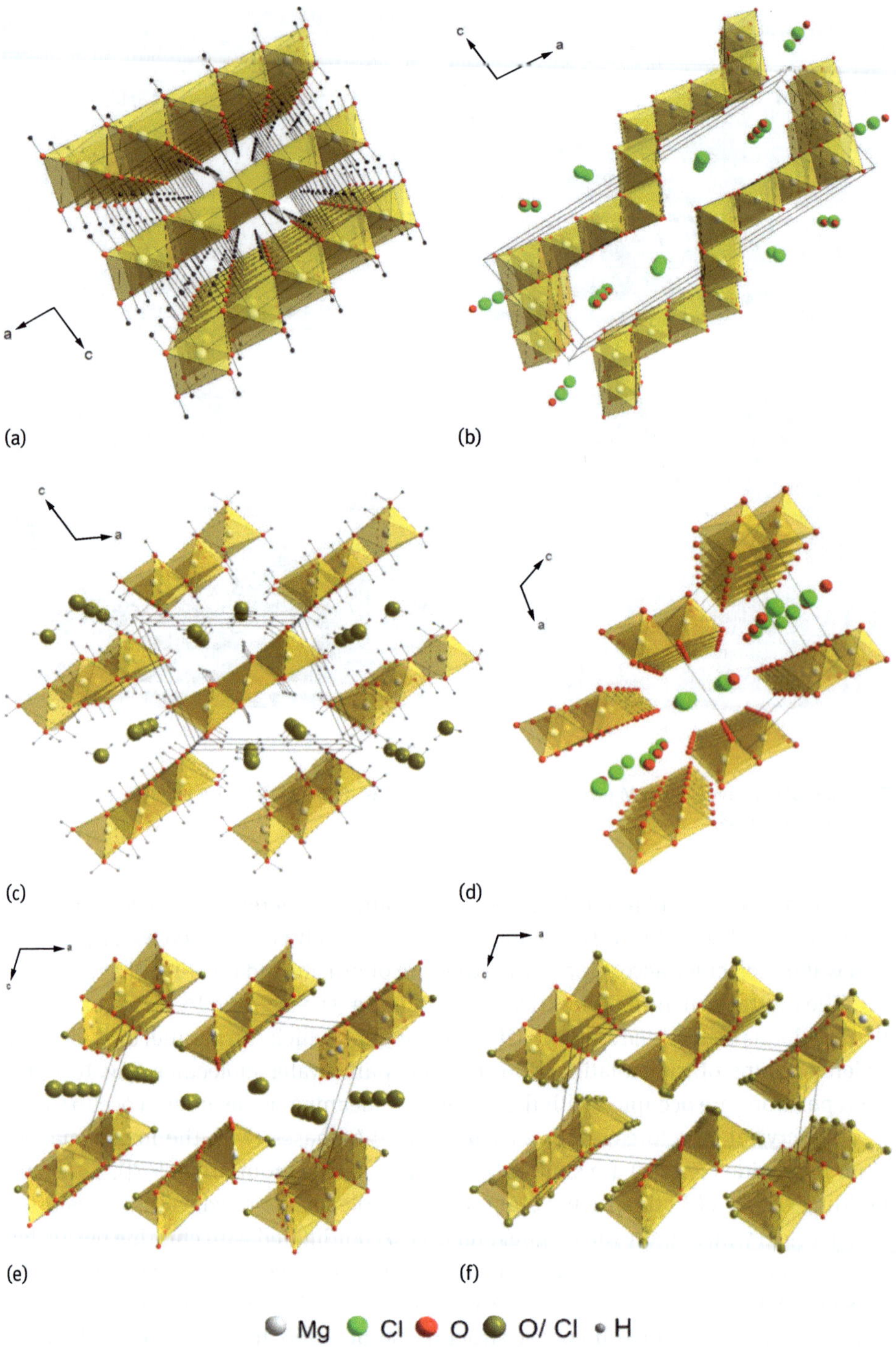

Fig. 10.4: Crystal structure motives of (a) $Mg(OH)_2$ and the Mg-oxychlorides: (b) 9-1-4 phase, (c) 5-1-8 phase, (d) 3-1-8 phase, (e) 2-1-4 phase, and (f) 2-1-2 phase.

The main difference between the triple chains of the 2-1-2 and 2-1-4 phases is found in the occupancy of the octahedra sites. Whereas disordered chloride and hydroxide ions/water molecules in the structure of the 2-1-4 phase only occupy one site of the outer octahedra of triple chains, there are two sites occupied in the outer octahedra of the 2-1-2 phase structure. Additionally, in the 2-1-4 phase structure, the octahedra chains are interrupted by an interstitial zigzag chain of disordered Cl^- ions and water molecules, which no longer exist in the structure of the 2-1-2 phase. This changed situation is caused by conditions during formation. The 2-1-2 phase crystallizes in the presence of the highest concentration of $MgCl_2$ solution, meaning, in presence of the lowest water activity in the system. Thus, in comparison to the 2-1-4 phase (with the least possible water content which can leave the structure), it is replaced in the magnesium coordination sphere of the 2-1-2 phase by the originally interstitial chlorine ions of the 2-1-4 phase (Fig. 10.4 (e) and (f)).

All crystal structures are stabilized by H-bonds formed between the OH and H_2O groups of the octahedra layers and the interstitial Cl^-/H_2O. In the case of the 2-1-2 phase, only H-bonds exist between the neighboring OH/H_2O groups of the octahedra units.

In addition, the different Mg-oxychloride can be distinguished by their Raman as well as IR-spectra [47, 62, 63].

10.3.3 Thermal behavior

The thermal behavior of Mg-oxychloride was characterized by TG/DTA measurements. In general, temperature increases causes decomposition starting with dehydration followed by the formation of MgO under hydrogen chlorine development as the last step. The dehydration already starts below 100 °C depending on the type of temperature exposure and the condition of materials: compact or finely dispersed, on air or in a closed system. During the steps of dehydration and decomposition, there occur different intermediate lower hydrates with imperfections and pronounced structural stacking faults. Following further dehydration, different intermediate solid solutions form [15, 47, 64, 65]. All the intermediate phases (lower hydrates and solid solutions) formed during dehydration and decomposition of the Mg-oxychloride binder are insignificant for the application of the magnesia building material. They do however contribute to the fundamental understanding of material chemistry. Noteworthy is the fact the development of HCl (hydrogen chlorine) above about 400 °C [65] in view of the reference to the material as a fire-retardant.

10.3.4 Binder phase formation during setting reaction

To apply as a building material, a cured material is formed following mixing and processing. Because of different applications and experiences in processing, different formulas with respect to the MgO solution ratio have been handled. If the desired consistency is not achieved during processing (e.g. the paste tends to 'bleed' or sedimentation occurs), additional MgO is admitted. From some applications some negative properties were also reported: such as the 'sweating' and 'swelling' [31] of the hardened material. Changes on the surface due to the possible carbonation of the binder phase(s) post-setting are visually problematic.

What are the reasons of these properties? They are directly associated with the phase formation process during the setting reaction and will be explained below in context.

In the reaction (see equations (10.1) and (10.2)), the binder phase 3-1-8 or 5-1-8 is formed if the amount of MgO and the concentration of the $MgCl_2$ solution is adjusted to the phase stoichiometry. That means, e.g. 3 mol MgO + $MgCl_2$ solution from 1 mol $MgCl_2$ and 11 mol water for a 3-1-8 phase formation or 5 mol MgO + $MgCl_2$ solution from 1 mol $MgCl_2$ and 13 mol water for a 5-1-8 phase formation. Values in between lead to the formation of a 3-1-8/5-1-8 phase mixture. According to the stoichiometric setting reaction, all mixing liquid is completely consumed for solid phase formation. If a higher MgO amount is used than necessary for the 5-1-8 phase formation, some MgO remains in the cured cement.

In case of the stoichiometric adjusted setting reaction for the 3-1-8 phase formation is the amount of solution in the formulation higher than in case of a "5-1-8 phase formulation" and leads to a more flowing paste or suspension. In detail, the flowing behavior at one formulation depends on the initial MgO reactivity.

By using a more reactive MgO (the burning temperature, specific surface and particle size determine the reactivity of MgO, i.e. the rate of reaction and formation of hydroxide in the suspension after mixing MgO with a solution. One way to measure this is the citric acid test-value [48]), the setting reaction begins earlier and the hardening process is completed earlier. This affects the material's workability. The reason is the quicker formation of a super saturated hydroxide concentration in the mixing liquid – the driving force of the formation of basic magnesium chloride hydrates (binder phases) – accompanied by the lowering of the OH^- concentration [19, 48] and the equal consumption of the solution. Thus, the use of more reactive MgO yields a higher setting temperature value than the use of less reactive MgO. This is because the heat of the exothermic setting reaction will be released over shorter and longer times, respectively.

Formation of the 5-1-8 phase is usually observed in all formulas, since this phase is kinetically preferred with respect to the formation of the 3-1-8 phase. For structural reasons, the high ordered structure of the 3-1-8 phase requires a longer time for crystallization. Over time, the conversion of the 5-1-8 phase through the consumption of still remaining pore solution in the 3-1-8 phase occurs and approaches solid-liquid equilibrium (Section 10.3.1). This is binder phase formation of a "3-1-8 phase formulation", as shown in equations (10.3) and (10.4). The lower the increase in temperature during the setting reaction, the less pronounced is the formation of the 5-1-8 phase.

$$3MgO + \{MgCl_2 + 11H_2O\}_{\text{mixing liquid}} \rightarrow 0.6[5Mg(OH)_2 \cdot MgCl_2 \cdot 8H_2O] + \{0.4MgCl_2 + 6.2H_2O\}_{\text{pore solution}} \quad (10.3)$$

$$0.6[5Mg(OH)_2 \cdot MgCl_2 \cdot 8H_2O] + \{0.4MgCl_2 + 6.2H_2O\}_{\text{pore solution}} \rightarrow 3Mg(OH)_2 \cdot MgCl_2 \cdot 8H_2O \quad (10.4)$$

Processing large amounts of material can effect high setting temperatures up to about 100 °C. During a longer period with temperatures of more than 80 °C within the setting process, the 9-1-4 phase (high temperature phase) occurs as a binder phase. With cooling and time, the phase formation always tends to reach equilibrium with the formation of a 3-1-8 phase. This is because pore solution is still present alongside the primary formed 9-1-4 phase (equations (10.5) and (10.6)).

$$3MgO + \{MgCl_2 + 11H_2O\}_{\text{mixing liquid}} \xrightarrow{T \geq 80\,°C,t} \tfrac{1}{3}[9Mg(OH)_2 \cdot MgCl_2 \cdot 4H_2O] + \left\{\tfrac{2}{3}MgCl_2 + \tfrac{20}{3}H_2O\right\}_{\text{pore solution}} \quad (10.5)$$

$$\tfrac{1}{3}[9Mg(OH)_2 \cdot MgCl_2 \cdot 4H_2O] + \left\{\tfrac{2}{3}MgCl_2 + \tfrac{20}{3}H_2O\right\}_{\text{pore solution}} \xrightarrow{T<80\,°C,t} 3Mg(OH)_2 \cdot MgCl_2 \cdot 8H_2O \quad (10.6)$$

These two-phase setting reactions, over primary and secondary binder phase formation, continue for a long time (months to years; Fig. 10.5 (a), (b), (d)). The final properties of the material are reached after this time. Additional over this longer time period, an amorphous solid phase formation accompanied the 3-1-8 phase formation and crystallized as a 3-1-8 phase.

Fig. 10.6 shows the microstructure of a hardened "3-1-8 formulation" in the period of secondary phase formation (at a setting time when the temperature has dropped below 80 °C). One can observe that there are crystals of the 3-1-8 phase in the pores where previously there was pore solution. Amorphous phase quantities are also visible [48, 66].

In case of a "5-1-8 phase formulation", the mixing liquid is completely consumed for a 5-1-8 phase formation (equation (10.2), Fig. 10.5 (c)). As there is no more solution in the system, why do no further reactions take place in "equilibrium direction". A partial 3-1-8 phase formation is possible if a very low reactive magnesium oxide is used, because then the setting reaction is slow and the mixing liquid is present longer in the forming microstructure.

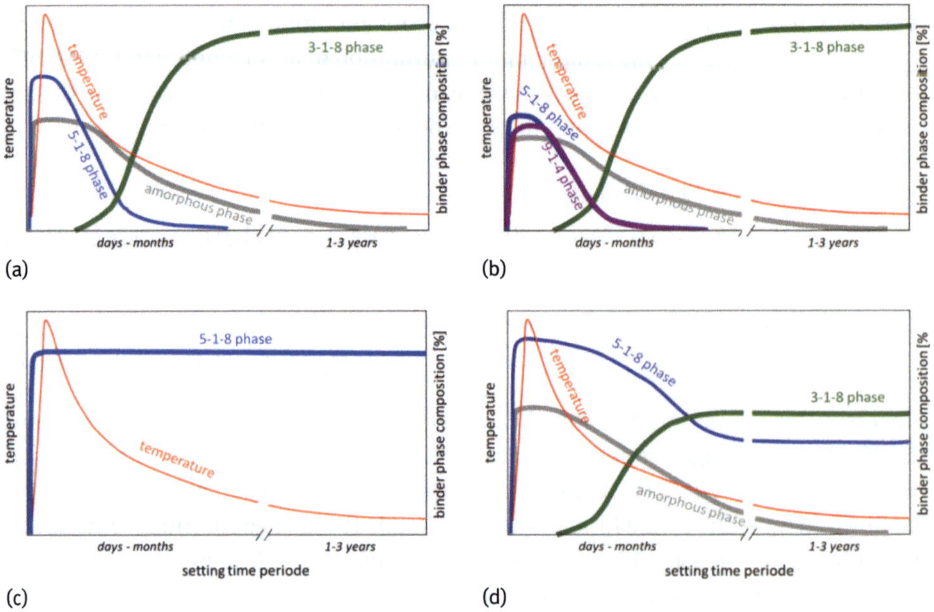

Fig. 10.5: Possible binder phase formation development of magnesia cement formulations, e.g.: (a) 3-1-8 formulation, T_{max} setting < 90 °C, (b) 3-1-8 formulation, T_{max} setting > 90 °C, (c) 5-1-8 formulation, T_{max} setting < 90 °C, and (d) a formulation between 3-1-8 and 5-1-8 formulation, T_{max} setting < 90 °C [66].

In the case of using a too highly concentrated $MgCl_2$ solutions (> 5 molal) as the mixing liquid, an excess of $MgCl_2$ (not used for the stoichiometric binder phase formation) would remain in the microstructure. $MgCl_2$ crystallizes at ambient conditions as $MgCl_2 \cdot 6H_2O$ (bischofite). This hydrate attracts water (is hygroscopic). The result is a "sweating" of the material.

In the case using a too lowly concentrated $MgCl_2$ solution (< 4 molal) as the mixing liquid, not enough $MgCl_2$ is available for the stoichiometric Mg-oxychloride formation, so appropriate portions of $Mg(OH)_2$ are formed within the microstructure.

In summary, the phase formation during the setting reaction of magnesia cement is a kinetically controlled process with a tendency to reach equilibrium by forming equilibrium phases (as a function of temperature) so long as one (*or* any) pore solution is present in the primary formed microstructure.

Secondary $MgCl_2$ solution accessing the 5-1-8 binder phase, or in general, water access or high humidity (causing water adsorption) cause reactions approaching solid-liquid equilibrium. This can mostly be observed as "swelling" (Fig. 10.7, left). With a late $MgCl_2$ solution accessing a "5-1-8 phase" material, the formation of 3-1-8 phase begins (as in equation (10.4)). The material's surface is completely overgrown with secondary crystallite (Fig. 10.8).

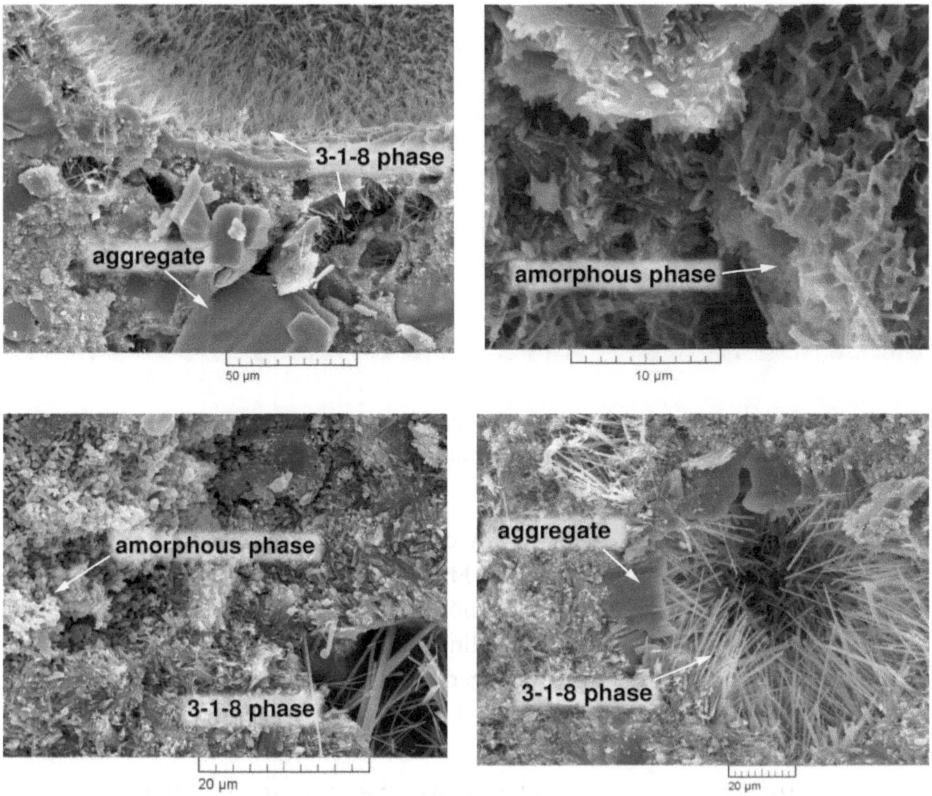

Fig. 10.6: SEM images of the microstructure of a "3-1-8 phase formulation" at the period of secondary phase formation: crystallization of long 3-1-8 phase needles in the pore space (beside the whole matrix of 3-1-8 phase and aggregates) and still existing aggregates of amorphous phase.

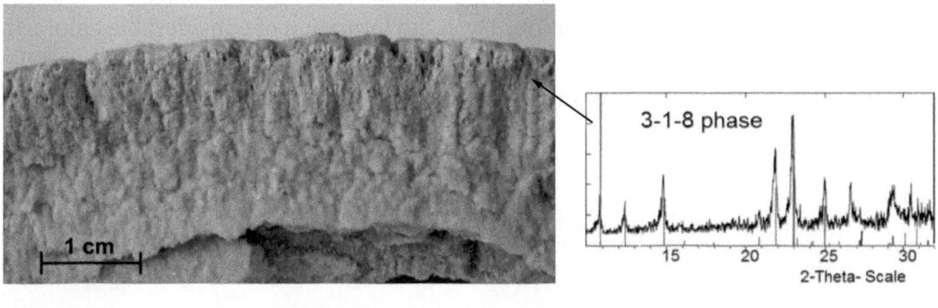

Fig. 10.7: Sample of "swelled" magnesia cement material (right), consisting of 3-1-8 phase indicated by X-ray diffraction (left).

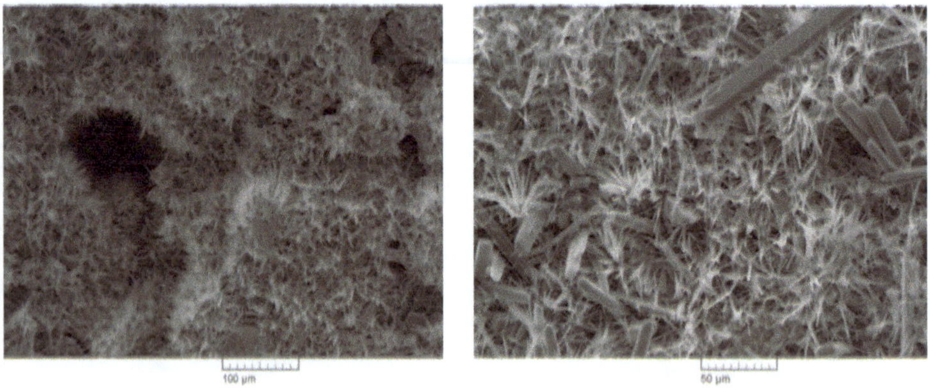

Fig. 10.8: SEM pictures of magnesia cement microstructure of the "swelled" sample zone (Fig. 10.7), originally consisting of 5-1-8 phase, after contact with solution, 3-1-8-phase-overgrown.

Water access or adsorption leads, in any case, to the decomposition of the binder phase. According to the solubility data in Fig. 10.2, only $Mg(OH)_2$ is stable in the presence of pure water. As such, the Mg-oxychloride phase decomposes into $Mg(OH)_2$ and the water is enriched with $MgCl_2$. Depending on the extent of the access of water, the original material microstructure is damaged in this way at a higher or lower rate.

Another possible reaction of hardened magnesia cement is carbonation, the formation of chlorartinite [67]:

$$Mg_2(CO_3)(OH)Cl \cdot 2H_2O = Mg(OH)_2 \cdot 2MgCO_3 \cdot MgCl_2 \cdot 4H_2O.$$

This is only possible due to an excess of carbon dioxide. Sufficient CO_2 contact should not be possible after the surface is correctly sealed. This is the usual procedure for the qualitative magnesia floor preparation [31], as very small amounts of chlorartinite are usually found in magnesia floors. However, carbonation occurs to a greater extent in the presence of humidity or a solution. Thus, the formation of considerable chlorartinite amounts (Fig. 10.9, right) can indicate bricolage in the surface sealing or that the surface sealing was ineffective due to the presence of pore solution in the magnesia floor.

Fig. 10.9: Magnesia cement material with very small (left) and considerable (right) amounts of chlorartinite (white facing) on surface.

Generally, it depends on the application of magnesia cement whether carbonation is problematic or not.

10.4 Mechanical properties

The setting reaction of magnesia cement depends on the formulation; a single- or two-phase process. The material characteristics such as the mechanical properties strength and creeping (the latter is of special significance in view of the material's use as a building material in salt formations), depend in turn on the mechanism of the setting reaction.

The single-phase reaction, in the case of a “5-1-8 phase formulation” (Section 10.3.4), yields the kinetically controlled formation of the 5-1-8 binder phase. Their spontaneous crystallization under intensive intergrowth of needle-like crystals results in a high material strength (up to the order of ultra-high strength concrete [48]).

With the two-phase setting reaction process (“3-1-8 phase formulation”), a material of relatively lower strength results. The reason for this is the slow crystallization of the secondary phase (3-1-8) under complete transformation of the primary formed binder phase (5-1-8 or/and high temperature phase 9-1-4). By the slow secondary crystal growth of the 3-1-8 phase, the growth of the needle-like crystals dodge external constraints. Therefore, the crystallization phase does not undergo such an intense intergrowth as the crystals of the primary 5-1-8 phase. Finally, the strength is lower than for a “5-1-8 magnesia cement material”. The addition of different types of inert aggregates does not effect this principle difference [48].

Independent from the type of formulation, setting and curing occurs without shrinkage (apart from thermally induced deformations).

10.5 Application as a building material in salt formations

Long-term storage of toxic or radioactive waste in deep geological formations, based on their isolation from the biosphere with multi-functional barrier systems [39], consists of the geological formation itself, suitable backfill materials and geotechnical constructions (e.g. dams, shaft seals, etc.). Geotechnical barriers can provide additional shields around waste and act against the mobilization of waste components through accidental penetration by brines. The chemical stability of barrier materials against these brines is one of the first prerequisites for long-term safety [68].

Whereas ordinary Portland cement is corroded by calcium-magnesium exchange [69] in such brines, magnesia cement is a stable material. For scientifically reliable proof, thermodynamic solubility data for the system $Mg(OH)_2$–$MgCl_2$–H_2O (Section 10.3.1) are now available, along with data of sodium chloride saturation (not included in this thesis).

(a)

(b)

(c)

Fig. 10.10: Constructions of a drift seal: (a) site concrete processing, (b) shotcrete, and (c) shaft seal in pilot plant scale for testing of technological aspects (material handling, workability of large quantities), of temperature and pressure development during setting time period and for follow-up investigations of the resulting geomechanical properties [70].

Together with fundamental investigations in the last years concerning the geochemical and geomechanical properties of magnesia building materials (mainly supported by the German Federal Ministry of Economic Affairs and Energy and the Federal Office for Radiation Protection), its complex material behavior depending on formulation has been thoroughly studied [48, 70].

Specifically, magnesia cement has long-term stability in the presence of salt and salt solutions if small amounts of $MgCl_2$ components or concentrations are present (a typical situation of salt rock sequences). This is because the Mg-oxychloride phase 3-1-8, $3Mg(OH)_2 \cdot MgCl_2 \cdot 8H_2O$, is the thermodynamically stable phase in the presence of $MgCl_2$-containig solutions (Fig. 10.2). The long-term safety of a binder material consisting of the 5-1-8 phase in the case of access to it by a secondary solution can also be proven due to the hydro-mechanical integrity of the sealing construction material itself.

Magnesia building material can be processed as site concrete or as shotcrete (Fig. 10.10). For shotcrete technology, the effect of thermally induced deformations is of minor importance due to layer concreting [48, 71].

The material can be formed with consideration given to various strength demands, i.e. "5-1-8 formulations" are characterized by very high strengths (uniaxial strength > 70 MPa) in contrast to relatively lower values for "3-1-8 formulations" (varies between 30 to 38 MPa). Accordingly, the compaction behavior of a "3-1-8" type is significantly higher, and the stress relaxation behavior is more pronounced compared to a "5-1-8" material. Since the "5-1-8" type is generally characterized by low relaxation effects under high loadings and strength properties, it may therefore act in sealing systems as "stiff" abutment. The "3-1-8 material" may be in turn seen as a "weak" inclusion. These properties should be considered alongside the demands on a barrier construction in salt rock and the properties and behavior of the salt rock itself in appropriate geological formations [72].

10.6 Summary

Magnesia cement can be used in special applications according to specific chemical and mechanical properties. Current state-of-the-art science and technology allows for adjustment of the material's properties and formulation according to demands and regulations. The key issue is the mechanism of the setting reaction (process of binder phase(s) formation). The selective formulation adjustment takes place via the MgO–solution ratio. Here, the MgO characteristics (reactivity, purity, particle size) are of decisive importance, as well as the concentration of the $MgCl_2$ solution mixing liquid. Thus, tailor-made magnesia building material products can be prepared both for interior application (floor material, artificial stone, plaster, etc.) as well as for the construction of long-term stable geotechnical barriers in salt formations.

References

[1] Sorel S. Sur un nouveau ciment magnésien. (Note de monsieur Sorel présentée par monsieur Dumas) C. R. Acad. Sci. 1867; 65: 102–104.

[2] Sorel S. Procédé pour la formation d'un ciment très-solide par l'action d'un chlorure sur l'oxyde de zinc. C. R. Acad. Sci. 1855; 41: 784–785.

[3] Bender C. Hydrate des Magnesiaoxychlorids. Ber. d. chem. Ges. Bd. 1870; 3: 932–934.

[4] Dinnebier RE, Freyer D, Bette S, Oestreich M. 9 $Mg(OH)_2 \cdot MgCl_2 \cdot 4\ H_2O$, a high temperature phase of the magnesia binder system. Inorg Chem. 2010; 49: 9770–9776.

[5] Krause O. Ueber Magnesiumoxychlorid. Liebig's Ann. 1873; 165: 38–44.

[6] Krieger A. Über Magnesiazement. Chemiker-Zeitung. 1910; 29: 246–247.

[7] Lahrmann D. Das Magnesiumoxychlorid im Magnesiazement. Tonindustrie-Zeitung. 1911; 35: 265–266.

[8] Krieger A. Erhärtung der Sorelzemente. Tonindustrie-Zeitung. 1930; 54: 577–578.

[9] Kallauner O. Studien über Sorelzement: I. Die Konstitutionsfrage. Chemiker Zeitung. 1913; 105: 1045–1056.

[10] Bury C R, Davies ERH. The system magnesium oxide – magnesium chloride – Water. J Chem Soc. 1932; 2008–2015.

[11] Feitknecht W. Zur Kenntnis der Magnesiumoxyd-Zemente I. Helv Chim Acta. 1926; 9: 1018–1049.

[12] Feitknecht W. Zur Kenntnis der Magnesiumoxyd-Zemente II. Helv Chim Acta. 1927; 10: 140–167.

[13] Feitknecht W. Zur Kenntnis der Magnesiumoxyd-Zemente III. Helv Chim Acta. 1930; 13: 1380–1390.

[14] Lukens HS. The composition of magnesium oxychloride. J Am Ceram Soc. 1932; 54: 2372–2380.

[15] Feitknecht W, Held F. Über die Hydroxychloride des Magnesiums. Helv Chim Acta. 1944; 27: 1480–1495.

[16] de Wolff PM, Walter-Levy L. Structure et formules de quelques constituants du ciment Sorel. C. R. Acad. Sci. 1949; 229: 1232–1234.

[17] Newman ES, Gilfrich JV, Wells S. Heat Generation in the Setting of Magnesium Oxychloride Cements. J Res Nat Bur Stand. 1952; 49(6): 377–383.

[18] Cole WF, Demediuk T. X-ray, thermal, and dehydration studies on magnesium oxychlorides. Austr J Chem. 1955; 8: 234–251.

[19] Newman ES. A Study of the System Magnesium Oxide-Magnesium Chloride-Water and the Heat of Formation of Magnesium Oxychloride. J Res Nat Bur Stand. 1955; 54(6): 347–355.

[20] Kaßner B. Ein Beitrag zur Kenntnis der Bildung von Magnesiumhydroxychloriden und ihres thermischen Verhaltens. Tonindustrie-Zeitung und Keramische Rundschau. 1958; 28: 290–291.

[21] Dietzel H. Physikalisch-chemische Grundlagen des Systems $MgCl_2$-MgO-H_2O. Freiberg. Forschungsh. A. 1960; 171: 5–24.

[22] Smirnov BI, Solv'eva ES, Segalova EE. Study on chemical interactions of magnesium oxide with magnesium chloride solutions of various concentrations (in Russian). Zh Prikl Khim. 1967; 40: 505–515.

[23] Ved, EI, Zharov EF, van Phong H. Mechanism of magnesium oxychloride formation during the hardening of magnesia cement (in Russian). Zh Prikl Khim. 1976; 49(10): 2154–2158.

[24] Sorrell CA, Armstrong CR. Reaction and Equilibria in Magnesium Oxychloride Cements. J Am Ceram Soc. 1976; 59(1/2): 51–54.

[25] Ball MC. Reactions of compounds occurring in Sorel's cement. Cem Concr Res. 1977; 7(5): 575–584.

[26] Matkovice B, Popovic S, Rogic V, Zunic T, Young JF. Reaction Products in Magnesium Oxychloride Cement Pastes. System MgO-$MgCl_2$-H_2O. J Am Ceram Soc. 1977; 60(11/12): 504–507.

[27] Urwongse L, Sorrell CA. The System MgO-$MgCl_2$-H_2O at 23 °C. J Am Ceram Soc. 1980; 63(9/10): 501–504.

[28] Bilinski H, Matkovic B, Mazuranic C, Zunic TB. The formation of magnesium oxychloride phases in the systems MgO-$MgCl_2$-H_2O and $NaOH$-$MgCl_2$-H_2O. J Am Ceram Soc. 1984; 67(4): 266–269.

[29] Norlund Christensen A, Norby P, Hanson JC. Chemical Reactions in the System MgO-$MgCl_2$-H_2O followed by Time-Resolved Synchrotron X-Ray Diffraction. J Solid State Chem. 1995; 114: 556–559.

[30] Bensted J. Sorel and related cements. Part 1 – Sorel cement, also known as magnesium oxychloride cement (MOC). Cement Wapno Beton. 2006; 5: 297–316.

[31] Scherer R. Die künslichen Fußböden, Wandbeläge und Deckenverkleidungen. Wien, Leipzig: A. Bartleben's Verlag; 1922.

[32] Stade F. Die Steinkonstruktionen. Salzwasser-Verlag GmbH; 2013. Reprint of the original from 1907.

[33] Dorsch KE. Chemie der Zemente. Chemie der hydraulischen Bindemittel. Berlin, Heidelberg: Springer-Verlag; 1932.

[34] Walo Bretschinger AG. Available from: www.walo.ch/de/produkte/bodenbelaege/fama-hartsteinholzbelaege/.

[35] Holleman AF, Wiberg N. Lehrbuch der Anorganischen Chemie. 102nd edition. De Gruyter; 2007. ISBN 978-3-11-017770-1.

[36] grinding wheels – Master Abrasives. www.master-abrasives.co.uk, ISO 9001:2008. Available from: www.master-abrasives.co.uk/downloads/content/Master%20Grinding%20Wheels.pdf.

[37] Grecian Magnesite. Abrasives. Available from: www.grecianmagnesite.com/markets/construction/abrasives.

[38] Nuclear Waste Management. Available from: www.bfs.de/EN/topics/nwm/repositories/repositories_node.html.

[39] Müller-Hoeppe N, Buhmann D, Czaikowski O, Engelhardt H-J, Herbert H-J, Lerch C, Linkamp M, Wieczorek K, Xi M. Vorläufige Sicherheitsanalyse für den Standort Gorleben (VSG): Integrität geotechnischer Barrieren, Teil 1 Vorbemessung – AP 9.2, GRS 287, 2012 (in German). ISBN 978–3-939355-63-2.

[40] Malinko SV, Lisitsyn AE, Purusova SP, Fitsev BP, Khruleva TA. Korshunovskite, $Mg_2Cl(OH)_3 \cdot nH_2O$, a new hydrous magnesium chloride. Zapiski Vserossiyskogo Mineralogicheskogo Obshchestva (in Russian). 1982; 111: 324–329.

[41] Malinko SV. Korshunovskite $Mg_2Cl(OH)_3 \cdot nH_2O$ a new hydrous magnesium chloride. International Geology Review. 1983; 25: 1105–1110.

[42] de Wolff PM, Walter-Levy L. The crystal structure of $Mg_2(OH)_3(Cl,Br) \cdot 4H_2O$. Acta Cryst. 1953; 6: 40–44.

[43] Bianco Y. The formation of basic magnesium chlorides at 50–175 °C by aqueous methods. (in French) C. R. Acad. Sci. Paris. 1951; 232: 1108–1110.

[44] de Wolff PM, Kortlandt D. Crystal structure determination from an x-ray powder diffraction pattern of β-$Mg_2(OH)_3Cl$. Appl Sci Res. 1954; 3: 400–408.

[45] Demediuk T, Cole WF, Hueber HV. Studies on magnesium and calcium oxychlorides. Austr J Chem. 1955; 8(2): 215–233.

[46] Bianco Y. The basic chlorides and bromidesof magnesium (in French). Ann de Chim. 1958; 13: 370–404.

[47] Dinnebier RE, Oestreich M, Bette S, Freyer D. $2Mg(OH)_2 \cdot MgCl_2 \cdot 2H_2O$ and $2Mg(OH)_2 \cdot MgCl_2 \cdot 4H_2O$, high temperature phases of the magnesia cement system. Z Anorg Allg Chem. 2012; 638(3/4): 628–633.

[48] Freyer D, Matthias Gruner M, Popp T. Zusammenhang von Chemismus und mechanischen Eigenschaften des MgO-Baustoffs. Freiberg. Forschungsh. E15 – Naturwissenschaften. 1st edition. Freiberg: Verlag der TU Bergakademie; 2015. ISBN 978-386012-516-8.

[49] Robinson WO, Waggaman WH. Basic Magnesium Chlorides. J Phys Chem. 1909; 13: 673–678.

[50] Gjaldbaek JK. Untersuchungen über die Löslichkeit des Magnesiumhydroxids. II. Die Löslichkeitsprodukte und die Dissoziationskonstante der Magnesiumhydroxide. Z Anorg Allg Chem. 1925; 144: 269–288.

[51] D'Ans J, Katz W. Magnesiumhydroxyd-Löslichkeiten, pH-Zahlen und Pufferung im System H_2O-$MgCl_2$-$Mg(OH)_2$. Kali-Zeitschrift für Kali, Steinsalz- und Erdölindustrie sowie Salinenwesen. 1941; 35: 37–41.

[52] Nakayama M. A new basic triple salt containing magnesium hydroxide. IV. The quinary system KCl-K_2SO_4-$MgCl_2$-$MgSO_4$-$Mg(OH)_2$-H_2O at 50 °C. Bull Agr Chem Soc Japan. 1960; 24: 362–371.

[53] Nakayama M. A new basic triple salt containing magnesium hydroxide. Part II. The quinary system KCl-$MgCl_2$-$Mg(OH)_2$-H_2O at 100 °C. Bull Agr Chem Soc Japan. 1959; 23: 46-48.

[54] Altmaier M, Metz V, Neck V, Müller R, Fanghänel T. Solid-liquid equilibria of $Mg(OH)_2$(cr) and $Mg_2(OH)_3Cl\cdot 4H_2O$(cr) in the system Mg-Na-H-OH-Cl-H_2O at 25 °C. Geochim Cosmochim Acta. 2003; 67: 3595–3601.

[55] Xiong Y, Deng H, Nemer M, Johnsen S. Experimental determination of the solubility constant for magnesium chloride hydroxide hydrate ($Mg_3Cl(OH)_4\cdot 4H_2O$, phase 5) at room temperature, and its importance to nuclear waste isolation in geological repositories in salt formations. Geochim Cosmochim Acta. 2010; 74: 4605–4611.

[56] Bodine MW. Magnesium hydroxychloride: a possible pH buffer in marine evaporite brines? Geology. 1976; 4: 76–80.

[57] Metz V, Schüßler W, Kienzler B, Fanghänel T. Geochemically derived non-gaseous radionuclide source term for the Asse salt mine – assessment for the use of a $Mg(OH)_2$-based backfill material. Radiochimica Acta. 2004; 92: 819–825.

[58] Awwad NS, Daifullah AAM. Preconcentration of U(VI) from aqueous solutions after sorption using Sorel's cement in dynamic mode. J Radioanal Nucl Chem. 2005; 264: 623–628.

[59] Pannach M, Freyer D, Voigt W. Temperaturabhängige Lösegleichgewichte der Sorelphasen. Endlagerforschung, Fachgespräch Verschlusssysteme – In-situ-Bauwerke aus Magnesiabaustoff und dessen chemisch- mechanische Eigenschaften im Hinblick auf HAW-Endlager, Freiberg: 28.–29.4.2015. Available from: https://www.ptka.kit.edu/downloads/ptka-wte-e/FG_Verschluss_2015_Vortraege_Web-Version.pdf.

[60] Pannach M, Bette S, Freyer D. Solubility Equilibria in the System $Mg(OH)_2$-$MgCl_2$-H_2O from 298 K to 393 K. Chem Ing. Data. 2017; 62: 1384–1396.

[61] Sugimoto K, Dinnebier RE, Schlecht T. Structure determination of $Mg_3(OH)_5Cl\cdot 4H_2O$ (F5 phase) from laboratory powder diffraction data and its impact on the analysis of problematic magnesia floors. Acta Cryst. 2007; B63: 805–811.

[62] Kanesaka I, Aoyama S. Vibrational spectra of magnesia cement, phase 3. J Raman Spectrosc. 2001; 32: 361–367.

[63] Bette S, Dinnebier RE, Freyer D. $Ni_3Cl_{2.1}(OH)_{3.9}\cdot 4H_2O$, the Ni analogue to $Mg_3Cl_2(OH)_4\cdot 4H_2O$. Inorg Chem. 2014; 53: 4316–4324.

[64] Dinnebier RE, Halasz I, Freyer D, Hanson JC. The crystal structures of two anhydrous magnesium hydroxychloride phases from in situ synchrotron powder diffraction data. Z Anorg Allg Ch. 2011; 637: 1458–1462.

[65] Rünčevski T, Dinnebier RE, Freyer D. Dehydration of the Sorel cement phase $3Mg(OH)_2\cdot MgCl_2\cdot 8H_2O$ studied by in situ Synchrotron X-ray Powder Diffraction and Thermal Analyses. Z Anorg Allg Chem. 2014; 640(1): 100–105.

[66] Paschke I, Freyer D. Das rezepturbedingte Abbindeverhalten des Magnesiabaustoffs (Phasenbildung). Endlagerforschung, Fachgespräch Verschlusssysteme – In-situ-Bauwerke aus Magnesiabaustoff und dessen chemisch- mechanische Eigenschaften im Hinblick auf HAW-Endlager, Freiberg, 28.–29.4.2015. Available from: https://www.ptka.kit.edu/downloads/ptka-wte-e/FG_Verschluss_2015_Vortraege_Web-Version.pdf.

[67] Sugimoto K, Dinnebier RE, Schlecht T. Chlorartinite, a volcanic exhalation product also found in industrial magnesia screed. J Appl Crystallogr. 2006; 39: 739–744.

[68] Storck R, Aschenbach J, Hirsekorn RP, Nies A, Stelte N. PAGIS–Performance Assessment of Geological Isolation Systems for Radioactive Waste: Disposal in Salt Formations. München: Commission of the European Communities; 1988.

[69] Bube C, Metz V, Bohnert E, Garbev K, Schild D, Kienzler B. Long-term cement corrosion in chloride-rich solutions relevant to radioactive waste disposal in rock salt – Leaching experiments and thermodynamic simulations. Physics and Chemistry of the Earth, Parts A/B/C. 2013; 64: 87–94.

[70] Endlagerforschung, Fachgespräch Verschlusssysteme – In-situ-Bauwerke aus Magnesiabaustoff und dessen chemisch- mechanische Eigenschaften im Hinblick auf HAW-Endlager, Freiberg, 28.–29.4.2015. Available from: https://www.ptka.kit.edu/downloads/ptka-wte-e/FG_Verschluss_2015_Vortraege_Web-Version.pdf.

[71] Entwicklung eines Grundkonzeptes für langzeitstabile Streckendämme im leichtlöslichen Salzgestein (Carnallitit) für UTD/ UTV, Teil 2: Erprobung von Funktionselementen, Teilbericht TB 7: Errichtung und Test von Funktionselementen: Laufzeit: 1.11.2004 bis 30.6.2010. DOI: 10.2314/GBV:666744882. Available from: https://www.tib.eu/suchen/id/TIBKAT:666744882/.

[72] Kreienmeyer M, Lerch Ch, Polster M, Tholen M. Nachweiskonzept zur Integrität der einschlusswirksamen technischen Barrieren. Peine: DBE TECHNOLOGY GmbH; 2008. Available from: https://www.dbe-technology.de/fileadmin/user_upload/unterlagen/f_e_berichte/ISIBEL_4_-_AP_5_Integritaet_einschl._Barrieren.pdf

Peter Stemmermann

11 New CO_2-reduced cementitious systems

Abstract: Since the beginning of the century, the global cement industry and universities have developed more and more elaborated solutions to reduce the high CO_2 emissions of cement production. This article gives an overview of new and newly discovered strategies and technologies based on globally available raw materials, and discusses their most important properties and possible applications. In addition, it is shown that all of these new technologies face common technological and non-technological problems that will make their way into the market a long one. Finally, an outlook is given on a future circular economy in concrete construction.

Keywords: new cementitious materials, raw materials, processing, carbon footprint, properties, common challenges

11.1 Introduction

First mentioned as a scientific fact in the 1980s, climate change became an objective of European politics in the 1990s. Up to that point, the technological development of cement production was mainly focused on energy and resource efficiency. In 2000, the first European Climate Change Programme started, which quickly resulted in the proposal to install an emission trading system for greenhouse gases [1]. As two thirds of the CO_2 from OPC production are from the raw materials, beginning in 2000 alternative technological approaches with a lower CO_2 footprint came more and more into the focus of the stakeholders, with an intensity that has ever since been coupled to the changing legislative and political climate activities. It therefore seems reasonable to designate all manufacturing technologies for cementitious materials available in 2000 as "old" and all subsequently developed cementitious systems as "new". This is also reflected by a strong increase in publication numbers (Fig. 11.1). In fact, following this definition, only a very limited number of "new" CO_2-reduced cementitious systems have been proposed. Most of the approaches that have been pushed forward in order to reduce the CO_2 emissions of cement production were developed more than 40 years ago but were newly spotted as a response to climate change.

Peter Stemmermann, Karlsruher Institut für Technologie (KIT), Institut für Technische Chemie, Bereich TTC-TAB, Abt. Feststoffe und Analytik, Eggenstein-Leopoldshafen, Germany, peter.stemmermann@kit.edu

DOI 10.1515/9783110473728-012

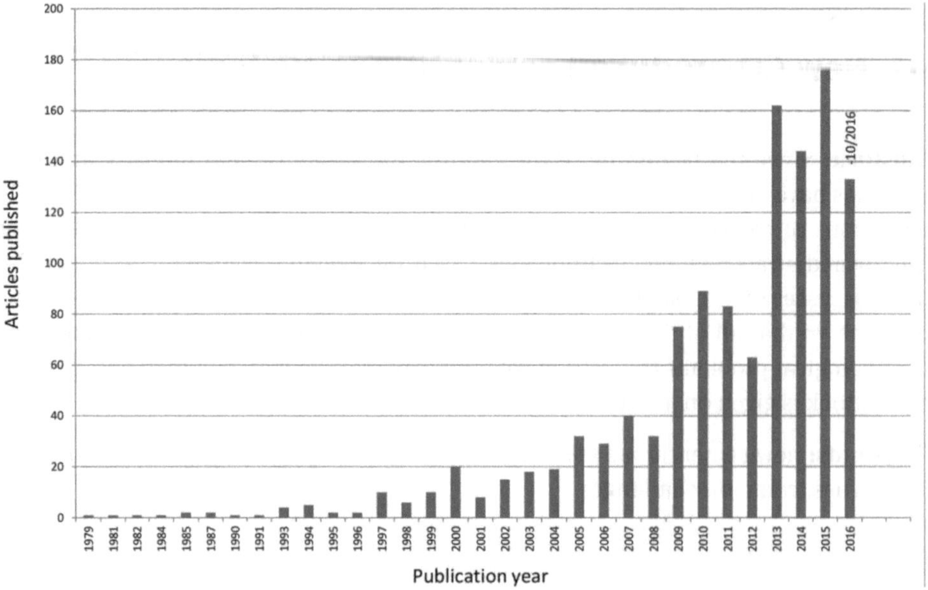

Fig. 11.1: Annual number of publications which contain "cement", "CO_2", and "emission" in title, abstract, or keywords (search performed on scopus.com on Oct 14, 2016).

This article first briefly summarizes the industrial ecosystem in which new CO_2-reduced cementitious systems have to perform as innovative. In order to understand why innovation in the cement sector is a long-term business and why the expected game changer – the emission trading system – did not change that much, one has to take a closer look at the resources that are available on a global scale, at the motivations, and at the stakeholders in the development of CO_2-reduced cements and at possible strategies and obstacles to innovations. In the second chapter, selected examples of new or newly spotted cementitious systems are presented that represent approaches using different raw materials or yield different functionalities, e.g. hydraulic binder or supplementary cementitious material (SCM). Third, the production, properties, and application of the different approaches are looked at. Finally, technological problems that are common to all new CO_2-reduced materials are discussed.

11.2 Development of CO_2-reduced cements

Published numbers from 2000 suggest that 1.66 billion tons of cement were produced globally [2], with an estimated CO_2 emission of roughly 1.3 billion tons. This equates to a 5 % share of the total 24 billion tons CO_2 emitted from fossil fuel and cement production in that year [3]. Production numbers for 2015 suggest that there was a global cement production of 4.10 billion tons [4], an increase of nearly 250 % in the 15 years after the start of active climate policies. Moderate reductions in specific CO_2 emissions have been "overcompensated by the sharp absolute growth in cement production" [5]. Most projections assume that no quick turnaround is imminent [6]. Thus, new CO_2-reduced cementitious materials are urgently needed as "the options available today (BAT, alternative fuels, and clinker substitutes) are not sufficient to achieve a meaningful reduction of CO_2 emissions" [7].

11.2.1 Globally available raw materials

As already stated, roughly two thirds of the CO_2 from OPC production originates from the raw materials. Thus, the search for alternatives to ordinary cement production may start with the identification of different raw materials which can be used for the production of hydraulic cements and posses a lower CO_2-burden. Critical preconditions that have to be met by these materials are their global availability in huge amounts, low costs in mining and processing, and good physical and chemical compatibility with the material and technologies actually used in construction. From a geochemical point of view, the earth's continental crust is mainly composed out of the 7 oxides SiO_2 (61.5 wt%), Al_2O_3 (15.1 wt%), CaO (5.5 wt%), MgO (3.7 wt%), Na_2O (3.2 wt%), Fe_2O_3 (6.3 wt%), and K_2O (2.4 wt%), which comprise more than 97 % of it [8]. In addition, this small number of elements is present in a very limited number of rock-forming minerals. Only quartz, limestones, and clays are accessible in globally distributed deposits. Other potential sources are secondary materials (coal fly ashes, slags, gypsum from desulphurization) or evaporates from sea water (NaCl, anhydrite, gypsum, etc.).

> In reality, other sources [...] may be available, but any particular source may be much less widespread. For example, bauxite is readily available in some locations but not very widespread [...] The same is true for magnesite and for phosphate ores. Thus, alternative cements that are very rich in Al, Mg or P seem less likely to be viable on a global basis, although they might be useful for local or specialized applications [9].

A hitherto largely ignored source of potential secondary raw materials, which is present on the global scale in enormous quantities, is the processed residues of metal mining. Essentially, these residues come from concentration processes, for example flotation. Due to their average composition they could possibly be used as a source of SiO_2 and Al_2O_3, but often problematic components like sulfides and heavy metals are present in high concentrations. Simple addition of these materials to the raw meal is limited. On the other hand, these substances are finely ground (average particle size 20–100 μ) and represent a waste disposal problem.

11.2.2 What is new? Strategies in cement production to lower CO_2 emissions

There are several possibilities to decrease the specific emission of CO_2 per ton of cement. The most obvious is the replacement of cement clinker with supplementary material. The use of supplementary material has already been established decades ago in order to recycle secondary materials from steel production (bfs) or coal power plants (cfa) and increase the energy efficiency. This resulted in the definition of 27 cement types in the European standard. However, the available quantities of SCMs in Europe are already being used. Additional reserves are not available. On the other hand, at least in Germany the amount of coal fly ash will probably decrease by the increased use of renewable energy. Therefore, intensive research on the industrial production of SCMs is under way [10].

Another strategy is to reduce the clinker factor by adding nonreactive materials, e.g. limestone. The reduced amount of hydraulically reactive adhesive must be compensated for, in this case by an optimized particle size distribution. Carbon capture during cement processing (CCS/CCU) is a technological option, which increases energy demand, is very cost-intensive, and lacks the infrastructure for the use or storage of CO_2 [11].

Besides these more traditional approaches, another strategy is focused on the replacement of OPC clinker by other hydraulic materials. One way is to adapt known but not yet established technologies that cause generally lower CO_2 emissions to the requirements of the mass market. The most prominent example is belit-sulfoaluminate cement. The other way is to invest in completely different production processes (e.g. calcium hydrosilicates, Celitement) or even different chemical systems (carbonates, magnesium based cements e.g. Calera, Novacem, Solidia). Most of these technologies were developed on the laboratory scale at universities.

11.2.3 Obstacles

Concrete made from cement ranks second in mass in the list of the world's processed materials after drinking water. The vast majority of the material is produced in plants with capacities of up to 4 million tons/year and is processed in mass market applications with very low margins. Thus, any truly "new" cementitious material faces a very competitive environment. CO_2 reduction as a unique selling point is hardly conceived as a strategy by incumbents [5]. Market entry is therefore more likely in niche markets based on additional properties which allow for a high added value.

In addition, the reduction of CO_2 emissions is a widely accepted task, but when it comes to technical solutions one realizes that it may be in conflict with other important objectives. Lowering CO_2 emissions by CCS may greatly increase energy consumption. Using special raw materials to reduce CO_2 emissions is in conflict with resource efficiency. Separating these materials from lower grade resources is again not energy efficient. Another critical obstacle for new cementitious materials is standardization and regulation. Standards in general can be classified as either prescriptive or performance-based [12]. Prescriptive standards for truly new cementitious materials require a separate set of rules for every cement type. It is predictable that such an approach is extremely time consuming and expensive – especially for start-ups or small companies. Performance-based standards, on the other hand, require test methods for a variety of properties that cannot be checked for all cements in the same way. These test methods have to provide comparability and security under variable conditions. The development of such regulations and test methods has just begun. An overview of the international situation is provided by Juenger et al. (2009) [12]. The objective is further complicated by different national regulations and interests. Finally, even if all performance issues, the process technology, the availability of raw materials, and the energy consumption of a specific approach were successfully solved, the greatest obstacle to reduce future greenhouse gas emissions by applying this solution is the long period of time needed for an in-service track record to make these systems globally adopted on the mass market.

11.3 New cementitious materials classified according to the raw materials used

It is not within the scope of this article to give a full review of all interesting approaches to new cementitious materials. Instead, a selection which focusses on important raw materials and new technologies is presented. Further information can be found elsewhere [13].

11.3.1 Magnesia-based cements

Walling and Provis recently reviewed magnesia based cements grouped according to basic chemistry as magnesium carbonate and reactive magnesia, magnesium phosphate, magnesium-silicate-hydrate, and magnesium oxysalt (both chloride and sulfate) cements [14]. They concluded that some of these cements are ideally suited to specialist applications and mentioned especially precast construction, road repair, and nuclear waste immobilization. One major obstacle for global use in mass construction is the availability of raw materials. The most abundant magmatic minerals with high magnesium content are magnesium silicates (forsterite, high magnesia pyroxenes). Rocks made up in high concentrations out of these minerals are globally distributed but not easily accessible. The energy needed to concentrate MgO from these raw materials is quite high and thus makes MgO-based cements more costly to produce than Portland cement [14].

11.3.1.1 Novacem

The development of Novacem is closely related to studies which investigated peridotites and their metamorphic relatives serpentinites as raw materials for carbon sequestration [15]. Both rock types contain high concentrations of magnesium silicates. Carbon dioxide is sequestered according to the exemplary reaction of forsterite (11.1):

$$Mg_2SiO_4 + 2CO_2 + 2H_2O \rightarrow 2MgCO_3 + H_4SiO_4. \qquad (11.1)$$

The idea of Novacem was born when a group of scientists from Imperial college, London extended this concept to a 2-step process. In the first step, MgO is separated from the silica and partly carbonized in an endothermic reaction (11.2).

$$4Mg_2SiO_4 + 6CO_2 + 11H_2O \rightarrow 5MgO + 3Mg(HCO_3)_2 + 4H_4SiO_4. \qquad (11.2)$$

The properties of MgO and $Mg(HCO_3)_2$ are adapted to the requirements of a cementitious material. The new cement mainly hardens due to carbonation, according to equation (11.3). Some M-S-H (magnesium silicate hydrates) are also formed.

$$\begin{aligned} &5MgO + 3Mg(HCO_3)_2 + 4H_4SiO_4 \\ &\quad \rightarrow 6MgCO_3 + M_2{-}S_{2x}{-}H_y + (2-x)H_4SiO_4 + (7-2x+y)H_2O. \end{aligned} \qquad (11.3)$$

Carbonation of MgO results in a carbon-negative mass balance if less CO_2 is emitted during the production of MgO then can be fixed by carbonation. A specialty of the solution is the combination of MgO with magnesium hydroxycarbonate and hygroscopic salts, which guarantees fast hardening [14].

Processing of MgO-based cementitious materials from magnesium silicates starts with the crushing and milling of the raw materials. The powderous magnesium silicates are then transferred in an autoclave and carbonated in solution with supercritical CO_2 at temperatures of 180 °C and a maximum pressure of 150 bar. The intended carbonation product is $MgCO_3$, together with silica and other residues, especially iron hydroxides. After drying and possibly the separation of by-products, the $MgCO_3$ is calcined at around 700 °C. Magnesium hydroxycarbonate is processed from MgO and formulated together with hygroscopic salt to the final product Novacem. CO_2 from calcination is recirculated in the autoclave and again used for carbonation of magnesium silicates. A detailed discussion of energy and mass balances is given elsewhere [16]. Novacem company was liquidated in 2012.

11.3.1.2 M-S-H

For more than a hundred years, magnesium-silicate-hydrates have been postulated and investigated as a cementitious phase. Several patents have been granted. In general, a mixture of MgO or $Mg(OH)_2$ and a highly reactive silica source is used as starting material. The resulting M-S-H gel is not yet well characterized. Most structural features resemble magnesium phyllosilicates like talc and serpentine. The MgO to SiO_2 ratio determines the relative amounts of these structural elements with medium-range order. A good overview over the literature is given in Walling and Provis (2016) [14]. The pH value of these systems varies in the range of 9.5–10.5, which makes them interesting for the deposition of nuclear wastes. Using special dispersive additives allows for the formulation of mortars with a maximum compressive strength of up to 70 MPa [17]. Fundamental research on durability and other physical properties is still missing.

11.3.2 Assisted carbonation

Solidia

The carbonation of lime, e.g. in non-hydraulic lime mortars, is a very old process for a binder of limited strength. In order to investigate possible routes for carbon sequestration, artificial carbonation of wollastonite ($CaSiO_3$) and rankinite ($Ca_3Si_2O_7$) have been investigated [15]. As in the case of Novacem, research at Rutgers University, USA resulted in the finding that a controlled carbonation of precast wollastonite or rankinite water mixtures at around 60 °C and optimized humidity results in high strength structural elements which are possibly suitable for construction (11.4), (11.5).

$$CaSiO_3 + CO_2 + 2H_2O \rightarrow CaCO_3 + H_4SiO_4 \tag{11.4}$$

$$Ca_3Si_2O_7 + 3CO_2 + 4H_2O \rightarrow 3CaCO_3 + 2H_4SiO_4 \tag{11.5}$$

Curing is typically done at atmospheric pressure in a sealed chamber with gas circulation, using CO_2 concentrations of 60–90 % [13]. As the reaction is not a hydration reaction and as it is thus not possible to use mortars or ready-mixed concrete on the construction site, the material is not a true cement, and rather exhibits the features of ceramics or tile.

As wollastonite and rankinite deposits are relatively rare in nature, they have to be synthesized. Processing ground limestone and silica raw meal with a CaO/SiO_2 ratio between 1 and 1.5 is done in a rotary kiln at 1200 °C. Milling of the obtained clinker needs less energy compared to OPC.

Calera

Calera is another example of a new CO_2-reduced cementitious material which has been developed outside the industry at Stanford University based on carbon sequestration [18]. The general idea is to use electrochemically generated alkalinity from e.g. the electrolysis of NaCl, combine it with dissolved calcium ions from brines or technical waste streams, and produce $Ca(OH)_2$ containing solutions in a proprietary process [19].

In a second step, up to 90 % of CO_2 from the off-gas of power plants can be carbon captured and sequestered using these solutions. Stable calcareous material and bicarbonate solutions are produced [18]. Subsequently, the solution and the calcareous material is further processed and dried, probably forming metastable vaterite and aragonite [19].

11.3.3 Geopolymer

The basic strength-developing concept of geopolymers or alkali-activated binders is the hydrolysis of Si–O–Si or Si–O–Al bonds in alkaline solutions, followed by condensation under the release of water. Condensation results in the formation of a three-dimensional network which incorporates water and various cations for charge equilibration. As hydrolysis is a slow process, it is necessary to use raw materials which are far away from thermodynamic equilibrium. Thus, amorphous natural pozzolanes or secondary materials from industry, e.g. fly ash or blast furnace slags are used. Another option is the use of technically processed calcined clays, e.g. metakaolin [12]. The first one who noted the reaction between amorphous alumosilicates and alkali hydroxide solutions was Purdon in 1940 [20], during the development of a test method for the reactivity of blast furnace slags in PC. Processing of so called two-part geopolymers is as simple as mixing solid alumosilicate powders with an alkaline solution [21]. However, meeting standardized properties demands very complex formulations of several components. In the last years, dry ready-mixed binders have been developed to avoid handling of caustic alkaline solutions. The next objective will be to make so-called one-part geopolymer cements available. The idea is to melt an ideal recipe of network forming, network modifying, and network breaking oxides together in order to get a cement which reacts just as OPC by simply adding water [21].

11.3.4 Calcined clays

Another very old cementitious material, which essentially goes back to the Romans, is closely interlinked to the concept of geopolymers: the use of calcined clays as a supplementary cementitious material (SCM). Most supplementary cementitious materials are in general amorphous alumosilicates, which are alkaline activated by the roughly 30 % of portlandite that result from the hydration of OPC. Roman cements in fact just use the activation of natural pozzolanes or crushed bricks, which are in effect calcined clays by caustic lime [22]. Most studies on the use of calcined clays in binder systems investigate metakaolin, a resource of limited availability, which is in addition an important raw material for the ceramic industry. New approaches focus on low quality clays and have proposed combining the obtained calcined clays and finely ground limestone in a magnitude of up to 45 % as SCMs with ordinary Portland cement [23]. Due to the high amount of calcium present in these OPC-based systems, hydration results in traditional phases known from the hydration of OPC, mainly C-S-H and calcium aluminate hydrates.

The processing of calcined clays starts with drying and pulverization, followed by calcination in a rotary kiln, fluidized bed or flash calcination at temperatures between 600–800 °C. Crystallization must be avoided [22].

11.3.5 Calcium hydrosilicates

Calcium hydrosilicates or prehydrated calciumsilicates are amorphous mineral binders that contain small silicate groups and isolated silicate tetrahedra as their main building units [24]. They are synthesized in the system $CaO–SiO_2–H_2O$ with a molar ratio of CaO/SiO_2 in the range between roughly 1 and 2. Their nominal water content fluctuates between approximately 5 and 10 wt%. One of their main characteristics is the presence of silanol and hydroxyl groups together with varying amounts of molecular water.

This new class of binders has been developed at Karlsruhe Institute of Technology, Germany, based on the assumption that during the early hydration of OPC the main clinker phases C_3S and C_2S react to a partly hydrolyzed intermediate phase. The intermediate phase is only present in a thin layer at the beginning of the hydration reaction. Structural elements of the intermediate phase are written as $Ca_3[HSiO_4]_2$ and $Ca_2[H_2Si_2O_7]$, respectively, which after nucleation of C-S-H completely transforms to the known C-S-H gel phase during the progress of hydration [25].

Proposing the existence of these highly reactive intermediates, the synthesis of cements with similar structural elements has been investigated [26]. Several minerals have been identified, e.g. afwillite ($Ca_3[HSiO_4]_2 \cdot 2H_2O$), α-$Ca_2[(HSiO_4)(OH)]$, and hydrothermally synthesized C-S-H, which are built out of the proposed structural elements of the intermediate phases and can be synthesized on the industrial scale. These minerals do not react with water as they are stabilized by strong hydrogen bonds. In order to use these materials as hydraulic cementitious binders, the hydrogen bonds have to be destabilized, which is realized in a technical mill by applying mechanical energy ("mechanical activation"). Hydration of the new binder phases results in C-S-H as the only product in accordance with the general equation (11.6).

$$Ca_{1.5x+2y+0.5z}[(HSiO_4)_x(H_2Si_2O_7)_y(OH)_z] \cdot nH_2O + mH_2O \rightarrow \text{C-S-H}, \qquad (11.6)$$

with $CaO/SiO_2 \in [1, \ldots, 2]$; $m/n \approx 3$.

In a first pilot, plant-slaked lime and milled sand are used as natural raw materials. Secondary raw materials, e.g. crushed concrete, have been successfully tested. After milling the raw materials, the obtained raw meal is autoclaved at 190 °C in saturated steam followed by cooling and drying. A nonreactive powder with a high specific surface (up to 80 m^2/g) results. The main mineral phases are C-S-H and α-C_2SH.

Autoclave technology has long been established in the industrial production of autoclaved aerated concrete. For drying and milling, conventionally available technology is applied. Milling energy is mainly used for the adjustment of the particle size distribution. Mechanical activation only adds a minor contribution. A flow sheet of materials in the process is given in Fig. 11.2. The so called "Celitement" process and the materials derived are not yet market-ready. Another variation on the process, which results in calcium hydrosilicates with an even lower water content has been presented elsewhere [27].

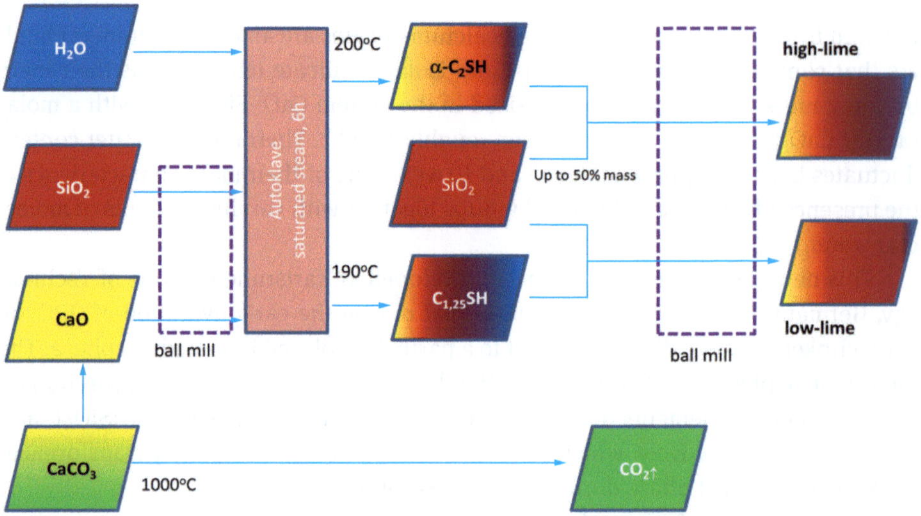

Fig. 11.2: Materials flow-sheet for the production of calium hydrosilicates with a variable CaO to SiO_2 ratio between 1 and 2.

Other types of cements formed from hydrothermal precursors are C_2S cements that can be produced e.g. from crushed concrete [28] or coal fly ash [29]. Due to the presence of χC_2S and X-ray amorphous C_2S crystals, these cements possess a high reactivity. A special variant has recently been established [25].

11.3.6 Calcium sulfoaluminate belite cements

Calcium sulfoaluminate belite cements (C$AB) have been known about for more than 40 years. In the last three decades, they have been standardized and produced, mainly in China [30]. Their standard raw materials are limestone, bauxite, and calcium sulfate (gypsum or anhydrite). As high aluminous raw material is indispensable, their use in general is restricted to special applications with a high added value or to special locations with high availability of alumina-rich residues or wastes. The main clinker minerals are ye'elimite ($C_4A_3\$$) and belite (C_2S) ± C_4AF. The clinker is interground with roughly 15 % of calcium sulfate. The amount of added $CaSO_4$ controls the formation of Afm and Aft phases and thus its reaction kinetics and properties. In order to reduce the necessary amount of high aluminous raw material, the belite content has largely been increased so that ye'elimite becomes a minor phase, which is only responsible for early strength [25]. These belite calcium sulfoaluminate cements (BC$A) face the problem of the highly different reaction kinetics of ye'elimite and belite. To obtain reaction kinetics that are comparable to OPC, the reactivity of belite has been increased by the stabilization of the high temperature phase $\alpha'C_2S$, e.g. with B_2O_3 or by adding further phases with increased reactivity, e.g. ternesite ($C_5S_2\$$).

The clinkering process of BC$A cements resembles those of OPC. Rotary kilns are used at a temperature which is, with 1250–1350 °C, lower than the one used in the clinkering of OPC. Control of SO_3 emissions and melt formation is critical. The clinker obtained is easier to mill, which further lowers energy costs.

11.4 Properties of new CO_2-reduced cementitious systems

Most important in the assessment of new technological developments is to define the objectives which have to be solved and compare these objectives with the (potential) results obtained. From the viewpoint of civil engineering, the first priority in any cement research is to look at materials properties, e.g. strength, durability, etc. Two of the approaches presented in the previous section differ in this respect from the others. Novacem and Calera set their highest priority in CO_2 sequestration. Unfortunately, being carbon-negative is an asset which counts for little on the present global cement market. The other technologies focus on cement or cementitious materials, which are at least in the long run conceptually competitive to OPC and should be able to enter the global market. In the following, the chemical principles, resulting materials properties, and potential applications of selected CO_2-reduced cements are discussed.

Magnesia-based cements: Novacem

Novacem cement consists of MgO, Magnesium hydroxycarbonate, and hygroscopic salt [14]. The cementing process is a mixture of hydration and carbonation. Thus, a coordinated reaction of the different phases present is of highest importance. Novacem claims that the hydration of MgO in the presence of magnesium hydroxycarbonates produces $Mg(OH)_2$ crystals with larger surface area and higher early strength. The carbonates additionally increase the rate of MgO hydration and carbonation over time. Carbonation takes place under atmospheric conditions, assisted by the high humidity induced in the presence of the hygroscopic salt [14]. "The examples provided showed a maximum compressive strength of 25 MPa, and the example applications were limited to use as mortars, masonry block replacements, roof tiles, and bricks, which do not require high strengths and are sufficiently porous to allow carbonation" [14]. In addition, the presence of hygroscopic chlorides prohibits the use of mild steel as reinforcement and may cause efflorescence.

Assisted carbonation: Solidia

Carbonation of calcium silicates, especially wollastonite, at standard conditions is extremely slow. The carbonation of precast wollastonite water mixtures in the Solidia process is sped up by humidity at around 60 °C. The presence of water is necessary for kinetic reasons, but to proceed further on the water in the already carbonated regions of the sample has to be evaporated. The carbonation reaction is highly exothermic, so that just a limited amount of additional heat is necessary to remove the excessive water. Curing is completed in a period of a few hours or up to one day, depending on the volume and shape of the sample [11].

The mechanical properties of concrete made from Solidia cement are reportedly similar to those of Portland cement-based concrete [31]. The pH generated in the solution is below 10, as can be assumed in equilibrium with $CaCO_3$. Thus, if reinforced elements are produced, other materials e.g. glass fibers have to replace mild steel [13].

Geopolymer and calcined clays

The main components of geopolymers are SCMs, e.g. calcined clays. These SCMs are amorphous alumosilicates, which are activated by alkali metal hydroxides, silicates, or carbonates and transformed during hydration into one or more highly disordered hydrate phases. This provides the analytical problem that neither the amount nor structural changes of either the starting materials or of hydration products can be traced by XRD. Only statistical information about structural changes in the short-range order can be deduced as a function of time, mainly by using spectroscopic methods like MIR, Raman spectroscopy, and NMR [32]. In addition, calorimetry and thermal analysis may help to characterize the hydration kinetics. Thermodynamic modelling is difficult because of the statistical nature of the reactions and their unknown fundamental thermodynamic data [12]. An extreme complexity arises in modern formulations, which are often composed out of several amorphous precursors and react to several amorphous products. In addition, amorphous alumosilicate does not react at all. Investigations of simple basic systems show that a miscibility gap exists between low calcium-high sodium alumosilicate gels of zeolithic character and C-(A)-S-H phases, with structural elements close to 14Å-tobermorite. In the alumina-free system, the miscibility gap encloses the CaO to SiO_2 ratios from approximately 0.25 to 0.66 [33].

Zeolithic gels tend to be highly permeable. A mixture of both gel types is reported to be beneficial. The formation of low-calcium gel phases is described in two phases [12]: first, solidification and hardening and, second, evolution, which is characterized by an increasing cross-link density and the reduction of the amount of bound water.

Geopolymer systems can yield good early and high final strength comparable to OPC, if SCMs as e.g. granulated blast furnace slag (ggbfs) and class F fly ashes are used [34]. The heat of hydration is low. This reduces the costs of reinforcement and potential cracking issues when the material is placed in large volumes [35]. The main challenges for geopolymers are secure durability and carbonation. Durable systems have been demonstrated but "it appears that this is strongly dependent on the application of adequate curing regimes" [12]. Geopolymers have already been tested in major infrastructure applications, e.g. road repair and in niche applications. Another issue is workability, as construction chemicals designed for OPC application do not work properly in many cases. With respect to calcined clays, rheology is especially challenging. Geopolymer mixes that use slags and ashes offer far better rheological properties and lower water demand compared to mixes based on calcined clays. This seems to be due to particle size, grading, and shape [32]. Metakaolin shows good pozzolanic performance, but the pozzolanic activity of calcined clays processed from mixed clay mineral resources is lower and variable. The systematic investigation of technologies to calcine low quality clays is in an early stage.

Calcium hydrosilicate

After mixing the calcium hydrosilicate binder ("Celitement") with water, calcium hydrosilicate is dissolved until the nucleation of C-S-H starts. The pH value rises to approximately 12.5. Usually, a water to solid ratio of 0.4 is applied. The C-S-H phase formed is similar to the one that originates from OPC hydration, but possesses a higher ordering in c-direction. No other phases are necessarily present. The induction period is less marked than in the case of OPC. The heat maximum occurs after 5 to 20 h. The addition of alkaline sources, e.g. Na(OH) or $Ca(OH)_2$, tends to retard the hydration due to the so called "common ion" effect. Compared to OPC, the heat of hydration is generally lower. In typical experiments, it amounts to 150 J/g. After 7 days, hydration is usually completed. Fig. 11.3 gives a hydration sequence from 5 h to 7 days. Details about the hydration reaction of calcium hydrogen based Celitement are given elsewhere [36]. Until now, roughly 30 t of calcium hydrosilicate cements with different properties have been processed. After one day, a compressive strength up to 40 MPa has been obtained. The final compressive strength varies between 50 and 60 MPa in standard mortars. Chemical resistance and resistance to freeze-thaw cycles are expected to be superior to OPC. Due to the high C-S-H content, chemical shrinkage increases OPC. Celitement mortars possess a lower buffer capacity because of the low calcium content. On the other hand, they show an extremely low capillary pore con-

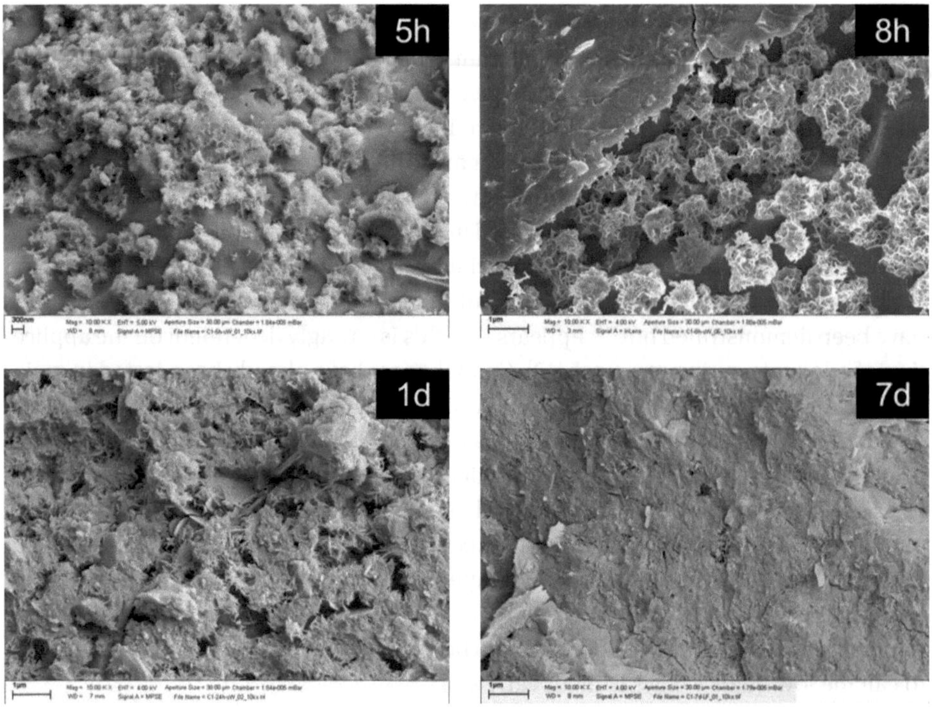

Fig. 11.3: Proceeding hydration of a calcium hydrosilicate binder 5 h, 8 h, 1 d, and 7 d after the addition of water as pictured in the Cryo-SEM. Glassy regions are amorphous ice.

tent, which should impede carbonation [36]. No significant carbonation has been observed in test samples. In general, calcium hydrosilicate based cements have a low need for additional chemically bound water due to the fact that they already contain some chemically bound water in the unreacted material.

Calcium sulfoaluminate belite cement

The early hydration of C$A and BC$A cements is dominated by the reaction of ye'elimite and $CaSO_4$. The first hydration products are AFt and, after consumption of the gypsum or anhydrite, AFm. In addition, amorphous aluminum hydroxide is usually present. Later on, belite and minor components, e.g. C_4AF hydrate, form C-S-H and additional hydration products, such as strätlingite and C_4AH_{10} [12]. The chemically bound water in the hydration product is higher than in OPC, which increases the necessary water to binder ratio. The heat of hydration of C$A cements is, with 400 J/g after 72 h, higher than for OPC [12]. Higher belite contents may reduce this value.

Due to the expansive character of ettringite formation, calcium sulfoaluminate based cements can be tailored as shrinking compensated, expansive or self-stressing cements. The formation of ettringite additionally results in a very dense microstructure. The initial pH value between 10 and 11 is quite low, but after consumption of calcium sulfates, the pH value increases to about 12.5.

The early and late strength of hydrated products is usually higher than that of OPC. The chemical shrinkage also exceeds OPC due to the higher water demand, which is also a reason for the potential of self-desiccation. Durability studies have found contradictory results [25]. High resistance against chemical attack due to the dense microstructure is reported. Carbonation seems to be faster due to the decomposition of ettringite, which may cause elution of sulfate. The protection of mild steel from corrosion seems to be provided [12]. In general, additional investigations are necessary. Due to the wide variability in composition and process technologies investigated, potential applications are widespread as long as the supply with raw materials is guaranteed. If produced from raw materials with high iron content, the color may be an issue.

11.5 Common technical issues to be solved

Construction chemistry

The performance of concrete has dramatically increased during the last 20 years. This is due in great part to modern construction chemicals, which have been developed for standard and special OPC-based systems. As most of the new CO_2-reduced cementitious binders differ from OPC with respect to pH, specific surface, particle distribution, and chemical water demand, a redesign of these construction chemicals is necessary. For example, many of the superplasticizing admixtures, which are usually used in concretes based on Portland cement, are not effective in enhancing the flow behavior of alkali-activated binders, as they are often degraded by highly alkaline activators [12].

Water/binder ratio

The application and testing of cementitious materials is generally based on the assumption that the chemically incorporated water amounts to about 20 % of the binder's mass [37]. This value is practicable for OPC, but differs greatly for other binders. Geopolymers possess very low Celitement with approximately 10 % a lower and BC$H with roughly 30 % a higher value. Thus the performance of these new binder types is underestimated in standard performance tests. In addition, existing applications have to be redesigned.

Reinforcement

The necessity of self-generating a high pH value to passivate mild steel reinforcement is an argument often heard, which may preclude the use of some CO_2-reduced cement types [14]. In general, there are three ways to get around this problem: A guaranteed pH value of 10.0 is sufficient for passivation [38]. Most systems initially generate a higher pH value, but carbonation tends to lower the basicity. The best strategy (which is also true for OPC) is to produce dense structures, as the case of UHPC shows. By adding reactive silica, the pH value of the UHPC system drops with respect to OPC, but due to their low permeability, these systems are known to be extremely robust with respect to carbonation. The second strategy is to add alkaline components as a buffer. The high initial pH of OPC (pH = 13) is due to alkali metal hydroxides. Portlandite generates a pH value of 12.5. A third strategy would be to replace steel bars by other materials with a high bending strength, lower specific weight, and less sensitivity to acid corrosion – glass, cellulose, carbon, polymer fibers, and textiles are the most promising candidates.

Compatibility with other construction materials

Compatibility issues are a common problem in construction with OPC based materials. Solutions for most problems, e.g. separating agents, have been developed. Especially for systems, which rely on a completely different chemistry, e.g. magnesium hydroxides, such solutions have to be investigated.

CO_2 footprint

Deriving rules for calculating a CO_2 footprint for OPC based cements, which are listed in the European standard has been a time-consuming process [39]. The published procedure uses indicators which can only be applied if the material under investigation is standardized and its performance therefore guaranteed. Thus, in analogy to the question of standardization of new cements, the question remains if the reporting standard, which is based on CO_2 emissions relative to mass, is applicable for cementitious systems with not fully known properties.

A good example is the CO_2 footprint of Novacem. The company claimed that a net absorption of 0.59 tonnes of CO_2 per tonne of MgO can be achieved [14]. In contrast, a scientific evaluation estimated a net emission of 0.25 to 0.5 tonnes of CO_2 [16]. Without standardized performance indicators, these values are of little significance. Nevertheless, general trends can be given, assuming that the new systems perform similarly with respect to durability, strength, etc. to standardized cements. The Solidia technology in principle avoids CO_2 emissions from the raw materials, if $CaCO_3$ is used as a calcium source. Other sources like class c fly ash in principle face the problem that they are limited. Thus, the use of such fly ash in the Solidia process or in geopolymers reduces the specific carbon footprint of the respective technology, but increases that of ordinary cement producers, which lose a source of SCMs. With respect to climate change, this is a zero-sum situation. The statement that making geopolymers using alkali activated slag and ash cement causes 80 % less CO_2 emissions than OPC cement is thus again of limited significance [35]. For calcium hydrosilicate binders, the reported reduction potential in CO_2 emission and energy demand is 50 %, as long as only primary raw materials are used [26]. For CSA cements, a reduction of CO_2 of 30 % has been reported.

Natural and secondary raw materials versus recycling

A general difficulty in establishing new cement technologies is ensuring the continuous availability of raw materials. Even secondary materials are an inadequate approach because there are not enough secondary materials available on a global scale.

11.6 Conclusion and outlook

In summary, several solutions to reduce the CO_2 emissions of global cement production have been developed. The future acceptance of these new technologies will probably differ in different global regions and for different purposes and applications. In the long run, one of the most important issues in future industrial concrete production is the need for a truly circular concrete economy. Traditionally, the cement industry's role has been the final utilization and proper disposal of process residues like blast furnace slags, fly ashes, or waste fuels. Further requirements of the material are covered by primary commodities. Thus, a linear economy is in practice. There have been approaches to reuse crushed concrete for cement production, which until now were not successful mainly for the reason that a secure supply chain was lacking and that there is a difficulty in generating economically relevant amounts of materials. The lesson to be learnt is that if one has to use decentralized sources of raw materials for cement and concrete production, one needs decentralized small cement plants. Another question which has to be addressed in the future is whether it is possible to use the cement process to extract valuables (especially metals) out of primary or secondary sources and recirculate them. On the other hand, with respect to energy efficiency and economy, plant size matters. Thus, a prerequisite of a circular economy in cement and concrete production would be to supply cheap renewable energy and possibly an adapted legislation and regulation.

References

[1] European Comission. Final Report WG1-Flexible Mechanisms. Brussels; 2001 [cited 2016 Jul 20]. Available from: http://ec.europa.eu/clima/policies/eccp/first/docs/final_report_en.pdf.

[2] Kelly TD, Matos GR, comps. Historical statistics for mineral and material commodities in the United States (2016 version): U.S. Geological Survey Data Series 140; 2014 [cited 2016 Jul 22]. Available from: http://minerals.usgs.gov/minerals/pubs/historical-statistics/.

[3] Friedlingstein P, Houghton RA, Marland G, et al. Update on CO2 emissions. Nature Geosci. 2010; 3: 811–812.

[4] U.S. Geological Survey Mineral commodity summaries 2016: U.S. Geological Survey; 2016. 202 pages. [cited Jul 22, 2016]. Available from: http://minerals.usgs.gov/minerals/pubs/commodity/cement/mcs-2016-cemen.pdf.

[5] Dewald U, Achternbosch M. Why more sustainable cements failed so far? Disruptive innovations and their barriers in a basic industry. Environmental Innovation and Societal Transitions. 2016; 19: 15–30.

[6] Achternbosch M, Kupsch C, Nieke E, Sardemann G. Climate-friendly production of cement: A utopian vision? GAIA. 2011; 20: 31–40.

[7] European Commission. 2013 Technology Map of the European Strategic Energy Technology Plan, Technology Descriptions. Luxembourg: Publications Office of the European Union; 2014. JRC Science and Policy Reports [cited 2016 Jul 22]. Available from: https://setis.ec.europa.eu/sites/default/files/2013TechnologyMap.pdf.

[8] Wedepohl HK. The composition of the continental crust. Geochimica et Cosmochimica Acta. 1995; 59: 1217–1232.

[9] Gartner EM. Industrially interesting approaches to low-CO2 cements. Cem Concr Res. 2004; 34: 1489–1498.

[10] Schneider M, Romer M, Tschudin M, Bolio H. Sustainable cement production – present and future. Special Issue: 13th International Congress on the Chemistry of Cement. Cem Concr Res. 2011; 41: 642–650.

[11] European Cement Research Academy GmbH. Technical Report TR 044/2007 Carbon Capture Technology – Options and Potentials for the Cement Industry; 2007 [cited 2016 Jul 20]. Available from: www.nrmca.org/taskforce/item_2_talkingpoints/sustainability/sustainability/sn3022[1].pdf.

[12] Juenger M, Winnefeld F, Provis JL, Ideker JH. Advances in alternative cementitious binders. Conferences Special: Cement Hydration Kinetics and Modeling, Quebec City, 2009 & CONMOD10, Lausanne, 2010. 2011. 41: 1232–1243.

[13] Gartner E, Hirao H. A review of alternative approaches to the reduction of CO2 emissions associated with the manufacture of the binder phase in concrete. Cement and Concrete Research. 2015; 78: 126–142.

[14] Walling SA, Provis JL. Magnesia-Based Cements: A Journey of 150 Years, and Cements for the Future? Chem Rev. 2016; 116: 4170–4204.

[15] Sanna A, Uibu M, Caramanna G, Kuusik R, Maroto-Valer MM. A review of mineral carbonation technologies to sequester CO2. Chem Soc Rev. 2014; 43: 8049–8080.

[16] Achternbosch M. Sind „Green Cements“ die Zukunft? Erste systemanalytische Abschätzungen zu innovativen Bindemittel. Teil 1: Novacem. Karlsruhe: KIT Scientific Publishing; 2011. Online-Resource. KIT scientific reports; vol. 7589. ISBN 3866446837 [cited 2016 Jul 22]. Available from: www.itas.kit.edu/pub/v/2011/acua11a.pdf.

[17] Zhang T, Vandeperre LJ, Cheeseman CR. Formation of magnesium silicate hydrate (M-S-H) cement pastes using sodium hexametaphosphate. Cement and Concrete Research. 2014; 65: 8–14.

[18] Zaelke D, Young O, Andersen SO. Scientific Synthesis of Calera Carbon Sequestration and Carbonaceous By-Product Applications: Donald Bren School of University of California, Santa Barbara; 2011 [cited 2016 Jul 20]. Available from: www.bren.ucsb.edu/news/documents/Calera_Carbon_Capture.pdf.
[19] Achternbosch M, Dewald U, Kupsch C, Nieke E, Sardemann G. New polymorphous CaCO3-based cementitious materials Part 1: Calera – availability of resources. ZKG International. 2013; 66: 50–62.
[20] Purdon AO. The action of alkalis on blast-furnace slag. Journal of the Society of Chemical Industry. 1940; 59: 191–202.
[21] Duxson P, Provis JL. Designing Precursors for Geopolymer Cements. J Am Ceram Soc. 2008; 91: 3864–3869.
[22] Sabir B, Wild S, Bai J. Metakaolin and calcined clays as pozzolans for concrete, A review. Cement and Concrete Composites. 2001; 23: 441–454.
[23] Scrivener KL. Options for the future of cement. Indian Concr J. 2014; 88: 11–21.
[24] Gartner EM, Macphee DE. A physico-chemical basis for novel cementitious binders, Special Issue: 13th International Congress on the Chemistry of Cement. Cem Concr Res. 2011; 41: 736–749.
[25] Ludwig H-M, Zhang W. Research review of cement clinker chemistry. Cement and Concrete Research. 2015; 78: 24–37.
[26] Stemmermann P, Achternbosch M. Dekarbonisierung im Baustoffsektor. In: Hacker J, editor. Rolle der Wissenschaft im globalen Wandel. Stuttgart; 2013. 313–332.
[27] Garbev K, Beuchle G, Schweike U, Merz D, Dregert O, Stemmermann P. Preparation of a Novel Cementitious Material from Hydrothermally Synthesized C-S-H phases. J Am Ceram Soc. 2014; 97: 2298–2307.
[28] Stemmermann P, Garbev K, Schweike U, Beuchle G, inventors; Forschungszentrum Karlsruhe GmbH, assignee. Verfahren zur Herstellung von Belit-Bindemittel. Deutschland 10 2005 037 771.8. 2007 Feb 22.
[29] Guerrero A, Goni S, Moragues A, Dolado JS. Microstructure and Mechanical Performance of Belite Cements from High Calcium Coal Fly Ash. J American Ceramic Society. 2005; 88: 1845–1853.
[30] Wang Y, Su M. The third cement series in China, World Cem. 1994; 25(8): 6–10.
[31] Ashraf W, Jeong H, Olek J, Jain J. An Experimental Investigation of the Selected Properties of Calcium Silicate Based Carbonated Concrete (CSCC) Systems, presented at the ACI fall convention, 2014.
[32] Provis JL, Duxson P, van Deventer JS. The role of particle technology in developing sustainable construction materials. Advanced Powder Technology. 2010; 21: 2–7.
[33] Bornefeld M. Einfluss von Kalium auf die Struktur von nanokristallinen und kristallinen C-S-H Phasen mit einem CaO/SiO2-Verhältnis < 1 [dissertation]. Heidelberg; 2012. 151 pages.
[34] Nath P, Sarker PK. Effect of GGBFS on setting, workability and early strength properties of fly ash geopolymer concrete cured in ambient condition. Construction and Building Materials. 2014; 66: 163–171.
[35] van Deventer JS, Provis JL, Duxson P. Technical and commercial progress in the adoption of geopolymer cement. Minerals Engineering; 2011.
[36] Garbev K, Beuchle G, Schweike U, Stemmermann P. Hydration Behavior of Celitement®: Kinetics, Phase Composition, Microstructure and Mechanical Properties. In: Palomo Á, Zaragoza A, Agüí J, editors. Proceedings of the 13th International Congress on the Chemistry of Cement (ICCC), Madrid; 2011.
[37] Brouwers H. The work of Powers and Brownyard revisited, Part 1. Cement and Concrete Research. 2004; 34: 1697–1716.

[38] Huet B, L'Hostis V, Miserque F, Idrissi H. Electrochemical behavior of mild steel in concrete: Influence of pH and carbonate content of concrete pore solution. Electrochimica Acta. 2005; 51: 172–180.

[39] World Business Council for Sustainable Development (WBCSD). CO2 Accounting and Reporting Standard for the Cement Industry, The Cement CO2 Protocol, 2005. [cited 2007 Oct 30]. Available from: www.resourcesaver.org/file/toolmanager/Custom016C45F64628.pdf.

Thomas A. Bier

12 Composition and properties of ternary binders

Abstract: Ternary binders are used to a large extent in dry mortar technology. In this field, mortars are composed, in addition to ternary binders, of mineral and organic fillers such as redispersible powders produced by spray-drying from emulsions, and most importantly admixtures.

In this article, ternary binders are defined and described by their chemical and mineralogical composition. Microstructure development as a function of hydration is characterized by SEM, calorimetry, X-ray diffraction, and the development of pore size distribution as measured by mercury intrusion porosimetry (MIP).

Technological properties such as workability and rheology as well as mechanical properties are shown for different mineralogical compositions. Special attention is paid to early shrinkage behavior and its possible impact on durability.

Keywords: mineralogy, hydration, microstructure, rheological properties, mechanical properties, early shrinkage, long-term behavior

12.1 Introduction

12.1.1 General description

In ternary systems, different binders are combined to achieve specific technological properties. For about five decades, calcium aluminate cement (CAC) has been used to improve OPC-based high-performance mortars such as self-levelling underlayments (SLU), tile adhesives (TA), or grouting mortars (GM). The improvement or optimization concerns specific properties or functions such as rapid set, early strength, rapid drying, sulfate, or corrosion resistance and shrinkage compensation.

Such high-performance mortars usually consist of three reactive materials, which are CAC, OPC, and $C\bar{S}$. The $C\bar{S}$ used may be in the form of gypsum, anhydrite, or hemihydrate. Besides these complex binders, the mortars are composed of mineral and organic fillers, and most importantly admixtures. Amongst the organic fillers, redispersible powders produced by spray drying from emulsions are added to improve flexibility, adhesion, and water resistance.

Thomas A. Bier, TU Bergakademie Freiberg, Institut für Keramik, Glas- u. Baustofftechnik, Freiberg, Germany, Thomas.Bier@ikgb.tu-freiberg.de

DOI 10.1515/9783110473728-013

12.1.2 Terminology

In the following, chemical compositions and/or mineralogical, anhydrous, and hydrate phases are given as oxides, which means that cement nomenclature is used (C = CaO, S = SiO_2, A = Al_2O_3, etc.). A non-metallic, inorganic binder is defined as a finely ground powder which forms a paste when mixed with water and hardens to become rock-like as a function of time. Hardening happens as a hydraulic, hydratic, latent hydraulic, or pozzolanic reaction with water to transform anhydrous phases to hydrates. Throughout this article, the notion of ternary binder refers to the specific combination of CAC, OPC, and $C\bar{S}$ [1].

In most recent publications, the notion of quaternary binders is used for OPC combined with three mineral additives, such as Limestone powder, fly ash, slag, metakaolin, etc. [2, 3]. In a few cases, pozzolanic materials are added to a ternary binder, as defined above. In the context of self compacting concrete (SCC), even pentanary systems or mixes are mentioned which refer to the combination of OPC, sand, aggregate, fine powders, and admixtures.

Mineral raw materials or constituents of ternary binders and high performance mortars, which are technically used as industrial or chemical products, are listed below – together with the abbreviations used throughout the text:
- OPC: ordinary Portland cement
- CAC: calcium aluminate cement
- $C\bar{S}$: calcium sulfate used as
  - α/β-hemihydrate
  - anhydrite
  - gypsum
- QP: quartz powder (or flower)
- LSP: limestone powder
- GGBFS: ground granulated blast furnace slag

For the mortars, additionally, organic and inorganic admixtures are used, such as:
- SP: super plasticizer
- Acc: accelerators
- Ret: retarders
- VMA: viscosity modifying agents
- RP: redispersible powders

12.2 Chemistry and mineralogy

12.2.1 Composition

Typical combinations of CAC-OPC-C$\bar{S}$ reflecting binders of special products of the European market were depicted by Bier and Mathieu in a mixing triangle or ternary diagram [4]. The logic presented has been used by several studies involving ternary binders [5–10]. In Tab. 12.1 and Fig. 12.1, the compositions of specific ternary binders used in recent studies are presented [11–15].

Additionally, areas of typical compositions for OPC-dominated mortars as well as for CAC-dominated mortars together with a representative mix are depicted. CAC-dominated mixes are based on mixtures of CAC and C$\bar{S}$, possibly with an addition of a lime source such as OPC or calcium hydroxide (CH) (typically represented by binder 5), while OPC-dominated mixes consist of OPC and CAC, possibly with an addition of C$\bar{S}$ (typically represented by binder 3).

Tab. 12.1: Composition of ternary mortars (all ingredients in grams).

Sample	OPC	CAC	C$\bar{S}$	Sand	LSP	Water	SP
1	350	0	0	380	270	192.5	2
2	280	45.5	24.5	380	270	192.5	3.5
3	175	108.5	66.5	380	270	192.5	10
4	108.5	171.5	70	380	270	192.5	2.5
5	45.5	225.75	78.75	380	270	192.5	2.3
6	0	245	105	380	270	192.5	1.1
7	0	350	0	380	270	192.5	1

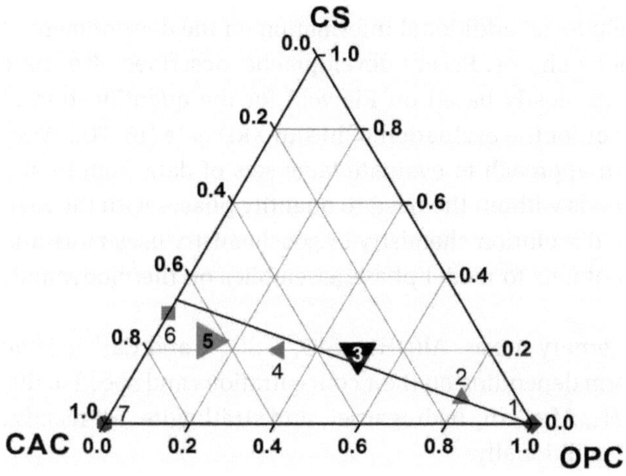

Fig. 12.1: Composition and classification of ternary binders.

Additionally, it should be kept in mind that OPC already represents a ternary binder since it contains C_3A, C_3S, and C_2S, as well as $C\bar{S}$.

In terms of (re)active mineralogical phases, a ternary binder therefore might be composed of the following phases – depending on the position in the ternary diagram:
- Tricalciumsilicate: C_3S,
- Dicalciumsilicate: C_2S,
- Tetracalciumaluminaferrite: C_4AF
- Tricalciumaluminate: C_3A
- Monocalciumaluminate: CA
- Calciumdialuminate: CA_2, in the case of high alumina cement
- Mayenite: $C_{12}A_7$ or $C_{11}A_7CaF_2$
- Anhydrite: $C\bar{S}$
- Hemihydrate: $C\bar{S}H_{0,5}$
- Gypsum: $C\bar{S}H_2$
- Ye'elimite: $C_4A_3\bar{S}$, when using CSA cements

Besides mineral compounds, resins are often added as so-called redispersible powders to improve abrasion resistance or cracking through a control of shrinkage or an expansion of the mortar. Research results concerning the influence of different resins on the rheological and shrinkage behavior of ternary self-levelling underlayments have been published recently [6].

12.2.2 Hydration and microstructure

12.2.2.1 Phase development

The phase development during hydration is classically characterized by XRD for crystalline phases. NMR could help to get additional information on the development of poorly crystalline or amorphous phases. Recent developments described elsewhere present calculation procedures mostly based on Rietveld for the quantification of amorphous phases from XRD or for the evaluation of in situ XRD data [16–20]. Westphal describes and applies an approach to evaluate large sets of data from in situ XRD based on statistical methods without the need to quantify phases with the Rietveld method [21]. In addition, dissolution chemistry or geochemistry uses more and more data on the dissolution of ions to model phase assemblies on thermodynamic equilibria [22, 23].

During the hydration of ternary mixes, $Al(OH)_4^-$, SiO_4^{4-}, SO_4^{2-}, and Ca^{2+} ions go into the solution and could form depending on their concentrations and speed of dissolution C-S-H, C-A-H, CH, AH_3, Afm, Aft, hydrogarnet, and strätlingite. All reactive phases release Ca^{2+} ions and, additionally:

- Calcium aluminate phases serve as a source of $Al(OH)_4^-$ ions
- Calcium silicate phases serve as a source of SiO_4^{4-} ions
- Calcium sulfate phases serve as a source of SO_4^{2-} ions

Besides the setting times, the morphology of ettringite is influenced by the ion concentration in the solution [24]. The phase development and phase assembly will also be influenced by the use of mineral additives, secondary cementitious materials, redispersible powders, and admixtures. Specifically, with the use of limestone powder, the formation of hemicarbonate has been observed [12, 13].

The formation of ettringite is one of the first reactions in ternary binders, and its precipitation from the dissolved ions can be described according to equation (12.1):

$$6Ca^{2+} + 2Al(OH)_4^- + 3SO_4^{2-} + 4OH^- + 26H_2O \rightarrow 3CaOAl_2O_3 \cdot 3CaSO_4 \cdot 32H_2O. \quad (12.1)$$

The coefficient of solubility for ettringite has, with $K_{ett} = 4.9 \times 10^{-44}$, quite a low value. The rate of nucleation and crystal growth from ions in a solution depends on several parameters amongst which the saturation coefficient β can be considered a major parameter, because it depends largely on the Ca^{2+} concentration and the OH^- concentration [24].

$$\beta = (a_{Ca^{2+}})^6 \cdot (a_{Al(OH)^{4-}})^2 \cdot (a_{SO_4^{2-}})^3 \cdot (a_{OH^-})^4 / K_{ett}, \quad (12.2)$$

where a_i represents the ion activity (ion concentration C_i multiplied with the coefficient of ion activity y_i) [24]. In order for precipitation to happen, the coefficient of saturation β needs to be larger than 1. Above this critical saturation β_c, the nucleation rate is drastically increased with an almost spontaneous nucleation. Below β_c, the nucleation rate is very low.

The ionic composition of the (pore) solution depends in the beginning on the dissolution period from the rate of dissolution of the active ingredients of the binder. For a mixture with significant amounts of C_3A (in OPC for example), ettringite will be formed according to equation (12.3).

$$C_3A + 3C\bar{S} + 32H \rightarrow C_3A \cdot 3C\bar{S} \cdot 32H. \quad (12.3)$$

For a CAC/$C\bar{S}$ mix, the components will dissolve upon contact with water until a critical supersaturation with respect to ettringite is reached. Besides ettringite, AH_3 will also precipitate according to equation (12.4).

$$3CA + 3C\bar{S} + 38H \rightarrow C_3A \cdot 3C\bar{S} \cdot 32H + 2AH_3. \quad (12.4)$$

For ternary binders, CAC-OPC-$C\bar{S}$ with significant additions of a lime source such as OPC or CH, ettringite could be formed as a unique hydrate according to equation (12.5).

$$CA + 3C\bar{S} + 2C + 32H \rightarrow C_3A \cdot 3C\bar{S} \cdot 32H. \quad (12.5)$$

Lamberet investigated comparable compositions as given in Tab. 12.1 and depicted them in a ternary diagram [10]. In the area close to the OPC corner (OPC-dominated mixes), C-S-H and CH are the major phases. In the area of CAC and C$\bar{S}$ (CAC-dominated mixes), ettringite, AH_3, and monosulfate are the major phases. In the intermediate area (around composition 4 from Fig. 12.1), strätlingite, hydrogarnet, and Afm phases are mostly present.

For comparable compositions based on mixes involving iron rich calciumsulfoaluminate cement (CSA) and based on thermodynamic modelling, Pelletier et al. calculated phase development as a function of time and found ettringite, strätlingite, and AH_3 in addition to the remaining anhydrous constituents [25]. For hydration times > 2 months, the appearance of monosulfoaluminatehydrate (Afm) has also been observed.

12.2.2.2 Scanning electron microscopy (SEM)

Morphological images obtained for fracture surfaces for different ternary mortars are presented in Fig. 12.2. Specifically, ettringite can clearly be identified as a component for compositions where additional sulfate has been added to the binder (number 2–6). In the absence of a rapidly dissolving lime source, as for mixture 6, AH_3 also becomes quite clearly visible. Binders only based on CAC form only C-A-H phases which are visible as hexagonal plates. For OPC-dominated mixtures and for OPC as the unique binder, calcium hydroxide (CH) is besides ettringite the most easily observable hydrate phase.

12.2.2.3 X-ray diffraction

Fig. 12.2 depicts data from several XRD measurements during the first 28 days for the ternary mixes 1–7 in a principal component plot (PCA diagram) based on a cluster analysis. It becomes obvious that the different compositions yield different phase assemblies which stay stable over the investigated time of 28 days. Three different families can be observed and are indicated in the diagram with reference to the ternary phase diagram. Thus, the classification indicated in Fig. 12.1 and to a certain extent presented by Lamberet [10] can be confirmed. The first family represents OPC and OPC-dominated mixtures showing ettringite and portlandite as the major crystalline phases. It also confirms the ternary character of OPC. The second family represents CAC-dominated binders with ettringite as the major phase and Afm type cubic phases and AH_3 as minor phases. The same features can be observed by SEM. The third "family" are calcium aluminate cements, which are actually single binders exhibiting only C-A-H phases.

Fig. 12.2: Morphological features of selected ternary binders corresponding to compositions 1, 3, 6, and 7 from Fig. 12.1.

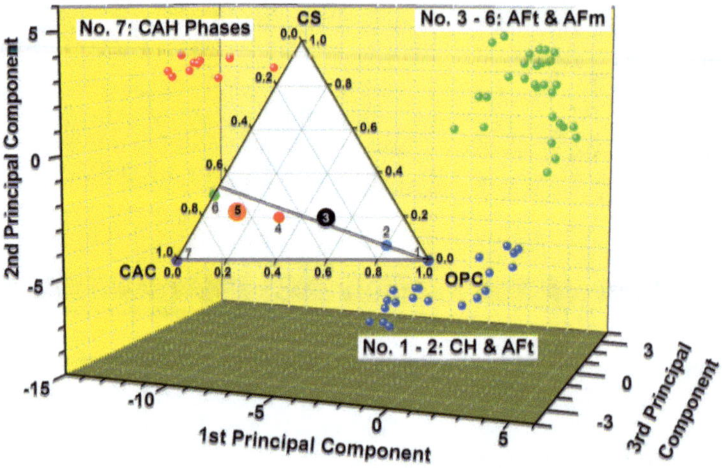

Fig. 12.3: Phase development of ternary mortars 1–7 during the first 28 days.

While Fig. 12.3 demonstrates statically the phase assembly as a stable condition after a definite time of hydration, in situ XRD allows one to observe the consumption of anhydrous phases and the formation of hydrate phases. Fig. 12.4 shows phase development during hydration for the first 24 hours of a binder corresponding to system B. While monocalciumaluminate and sulfate hemihydrate decrease, a maximum of about 20 % ettringite is formed. Due to the fast dissolution of hemihydrate, intermediate gypsum can be detected as well.

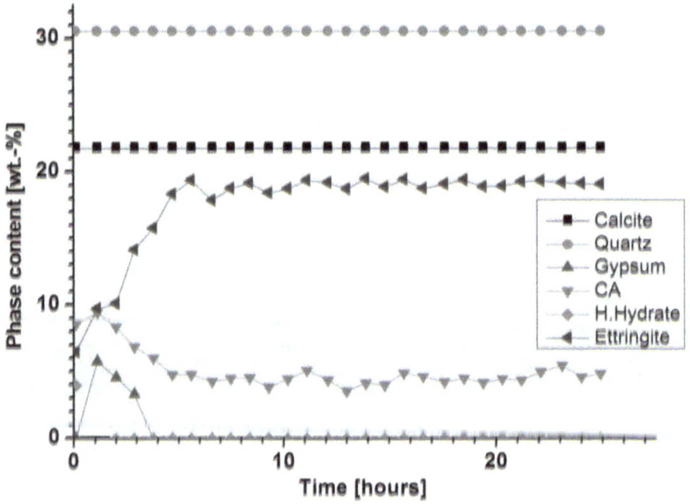

Fig. 12.4: Phase development of a CAC rich ternary mortar (binder 5) during the first 24 hours; in situ XRD measurement in transmission mode [26].

The same sequence is detected by cluster analysis, which divides the measured time series of X-ray patterns into groups or families of similar pattern. This is shown in Fig. 12.5. Five significantly different groups of X-ray patterns are distinguished which are represented by a single typical or representative diagram. The changes in patterns are observed at up to 6 hours of hydration. The 24 hours' pattern does therefore not show significant changes compared to the 6 h pattern.

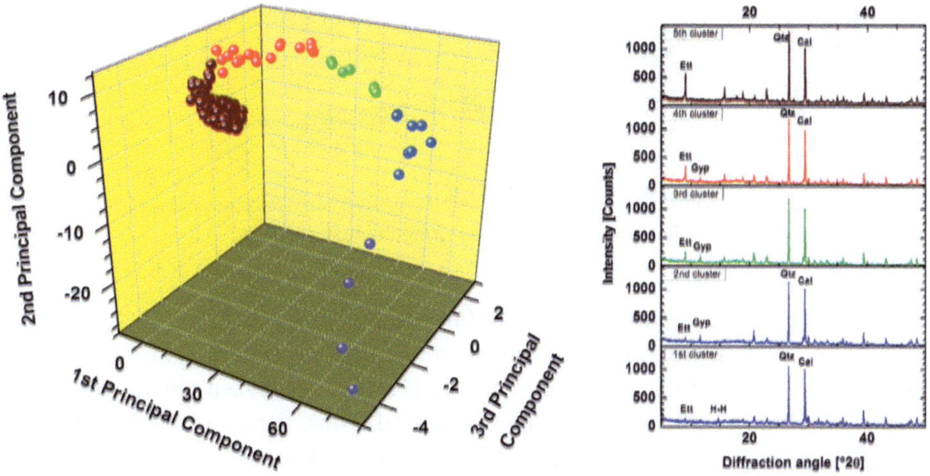

Fig. 12.5: Principal component analysis (right) with individual X-ray patterns representing the different stages of hydration for a CAC rich mortar (binder 5) during the first 24 hours.

12.2.2.4 Heat flow calorimetry

In his thesis, Kighelmann published two heat flow curves for ternary binders with compositions close to system A and B, respectively (Fig. 12.1) [6]. The OPC-dominated composition showed a heat flow curve with three pronounced maxima besides the dissolution peak whereas the CAC-dominated composition showed one pronounced maxima. The OPC-dominated composition seems to be similar to an OPC (mix 1), while the CAC-dominated mixture seems to be close to a CAC mortar (mix 7).

Heat flow curves for the first 24 hours and mixes 1–7 from Tab. 12.1 are presented in Fig. 12.6. A similar observation can be made. Compositions with significant amounts of Portland cement exhibit heat flow curves similar to Portland cement. With the addition of sulfate and alumina, some peaks can be more pronounced or additional peaks may appear. For compositions with CAC as the major compound, the heat released increases, and peaks become more pronounced and appear earlier. The curves are to a certain extent comparable to CAC. Also, due to the fast dissolution of calcium aluminates and hemihydrate, the initial dissolution is much more pronounced.

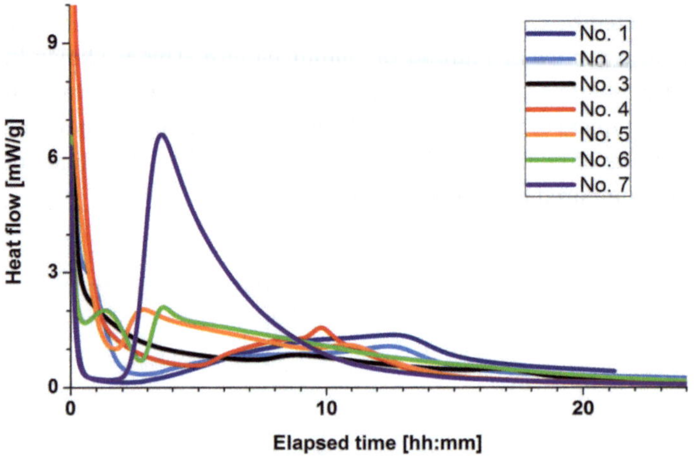

Fig. 12.6: Heat flow curves for ternary mortars 1–7 during the first 24 hours.

12.2.2.5 Mercury intrusion porosimetry (MIP)

Cumulative pore size distributions from mercury intrusion porosimetry (MIP) for hardened mixes 1–7 from Tab. 12.1 are presented in Fig. 12.7. The measurements were done on samples hydrated for 120 days. In very general terms, the measured pore volume and the threshold radius decrease as the composition changes from OPC-dominated mixes to CAC-dominated mixes. However, within the OPC-dominated mixes 1–3, the pure OPC composition exhibits the mist dense microstructure while the addition of sulfate and CAC lead to an increase in porosity and threshold radius. For the CAC-dominated mixes 5–7 the addition of sulfate and small amount of OPC lead (mix 5) lead to a decrease in threshold radius and an increase in pore volume. Mix 4 is in between, but shows a pronounced bi-modal distribution with a second pore population around sizes of about 100–200 nm. For this composition, a very pronounced expansion has also been observed (Fig. 12.11) in experiments with unrestrained shrinkage and/or with the addition polymer as redispersible powders [6, 13, 15].

In Fig. 12.8, results from MIP measurements on samples taken after restrained and free shrinkage experiments are presented. The shrinkage measurements lasted 24 hours and were realized in a shrinkage cone [27] (free shrinkage) and as a volume shrinkage measured on a sealed sample under water [28]. The pore volume from the high-pressure part of the MIP experiment (radius < 5000 nm) is plotted as a function of the water cement ratio for mixes where workability was established by super plasticizers and the W/C ratio was kept constant and for series where the workability was controlled by different W/C ratios. For the series where the W/C ratio is different, a regression line can be calculated and good coefficients of regression are obtained. However, porosities determined after the volume and after the cone shrinkage test follow a different trend indicating higher porosities in the cone measurement where the expanding phases could create additional porosity. The two straight lines cross somewhere in the area of a W/C ratio of 0.33 – a value probably close to the water demand of the binder. For system III, the use of polymer powder also increased the porosity. This can be explained by a more pronounced expansion due to polymer film formation. This effect has been reported elsewhere [6].

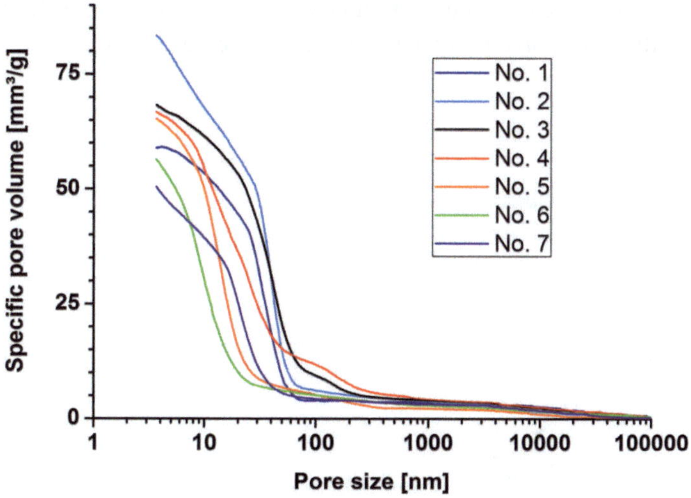

Fig. 12.7: Cumulative pore size distributions from MIP measurements for ternary mortars after 120 days hydration.

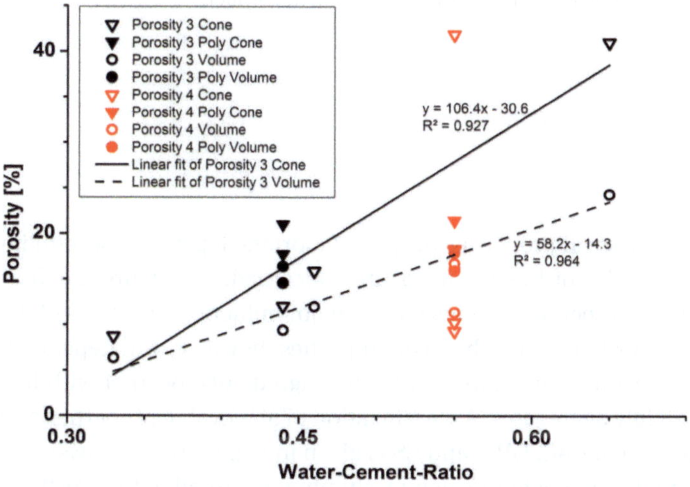

Fig. 12.8: Total porosity as a function of W/C ratio measured by MIP after 24 hours hydration in free and confined shrinkage conditions for mortars based on ternary binders.

The effect of polymer addition is demonstrated in Fig. 12.9. Cumulative pore size distributions are shown for composition 5, which is part of the CAC-dominated area and close to system B, which has shown pronounced expansion in several studies. The mixtures have been measured in free and restrained shrinkage conditions and samples with 0 % by mass, 2 % by mass and 6 % by mass polymer addition are compared. The pronounced formation of ettringite in this type of composition leads, in the case that no restraint is exercised on the material, to an increase in pore volume as well as threshold radius. The addition of polymer increases the two observed effects.

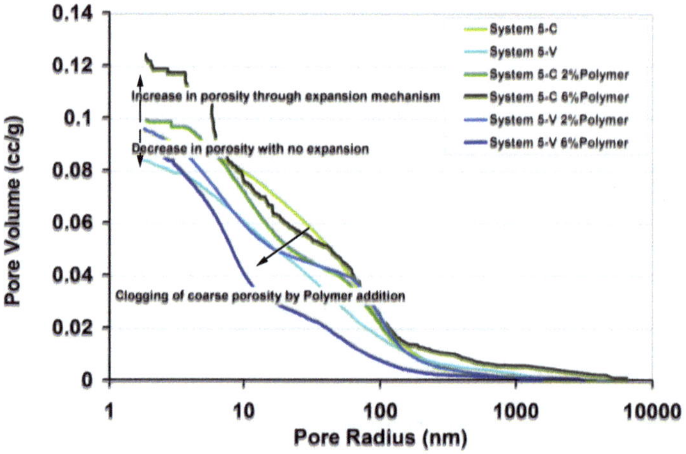

Fig. 12.9: Cumulative pore size distribution from MIP for a CAC-rich mortar (binder 5) after 24 hours hydration in free and confined shrinkage conditions with and without polymer.

12.3 Properties

12.3.1 General

The engineering or technological properties of special mortars depend on several influences. Where the properties of fresh mortars are concerned, admixtures such as retarders, accelerators, and super plasticizers have a major influence regardless of the specific mixture chosen. Most of the mechanical properties, however, will depend on the so called mineral base – a composition of reactive ingredients together with fine powders, sand, and possibly aggregates. A classification of achievable properties as a function of the composition of CAC, OPC, and CS is given in Fig. 12.10 as follows:

(1) This area represents binders where calcium aluminates are added to Portland cement in order to shorten the initial setting time. The setting time in OPC is a balance between sulfate and C_3A content and the speed of dissolution. The addition of calcium aluminate, phases will disturb this equilibrium by consuming

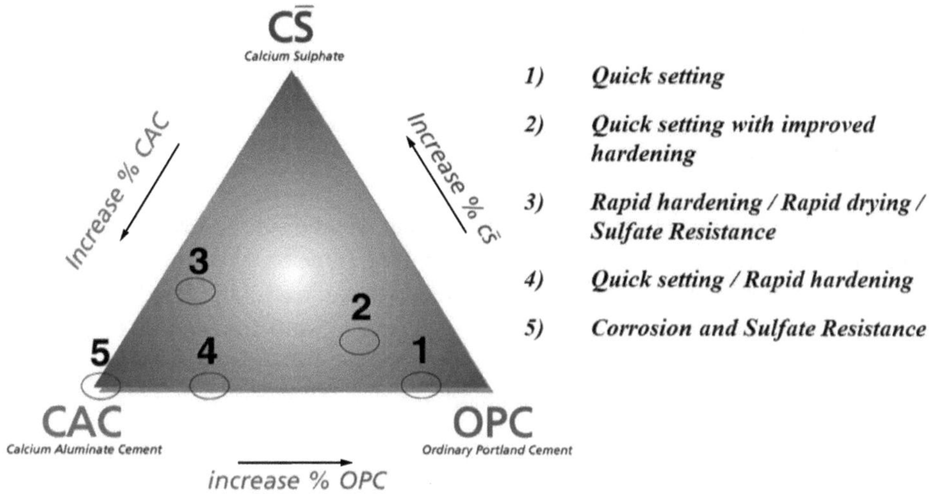

Fig. 12.10: Basic properties for ternary mixes as a result of their composition.

sulfate ions. As a consequence, lime-rich calcium aluminate phases precipitate as C-A-H phases rather rapidly. Strength values, on the other hand, will stay below strength values for OPC without the addition of CAC.

(2) In order to overcome the loss in strength performance in area 1, small amounts of calcium sulfate, preferably anhydrite, are added. The corresponding mortars still exhibit a rather quick initial set combined with acceptable strength performance.

(3) In area 3, a CAC-rich mixture where ettringite is the main hydrate phase is depicted. Ettringite forms quickly and combines 32 molecules of water. Therefore, quick setting, rapid hardening and internal drying is achieved. There is definitively no CH or soluble calcium and the sulfate resistance is increased.

(4) The setting of calcium aluminate cement, which normally consists of monocalciumaluminate or calciumdialuminate, can be accelerated by the addition of OPC. OPC releases additional Ca ions which allow, through an increased C/A ratio in the solution, a rapid or even spontaneous precipitation of C-A-H phases associated with a setting reaction.

(5) CAC has been developed as a cement with a high sulfate resistance due to the absence of portlandite. Investigations on samples subject to biogenic attack [29] showed that CAC also inhibits bacteria growth in the biofilm on the mortar surface and as a consequence higher pH values result, leading to a milder acidic attack.

Lamberet published a similar diagram and comes to the same statements as above [10]. Additionally, she identifies the area of corrosion protection which is associated with the presence of portlandite (OPC-dominated binders). The upper part of the compositional triangle above the line of mixes 1–6 presented in Fig. 12.1 is associated with expansion due to the formation of sulfate carrying phases and due to the addition of higher amounts of C$\bar{S}$.

12.3.2 Rheology

Rheological properties characterized by the measurement of flow values as well as flow times are presented in Fig. 12.11. The flow time (time to reach the flow value) is depicted as a function of the flow value for mortars made with binders 1–7. For these mortars, flow values of 300 and 370 mm, respectively, were adjusted by the quantities of superplasticizer. Additionally, to the set of samples with 370 mm flow values, redispersible polymer powders were added.

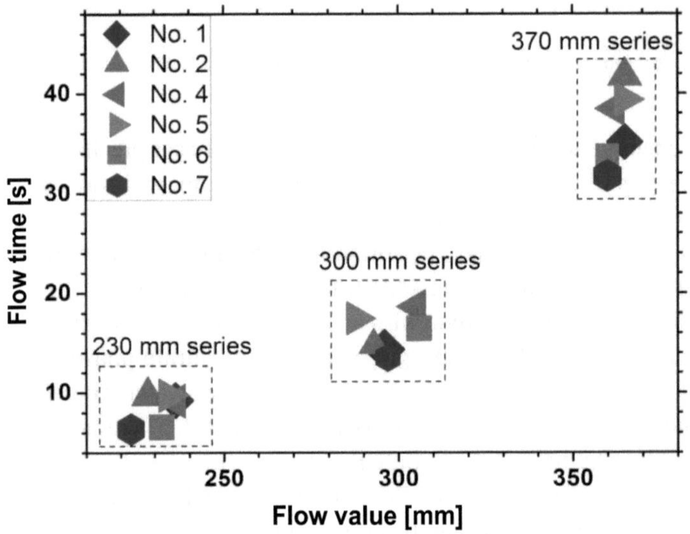

Fig. 12.11: Workability characterized by flow time and flow value for different ternary mixes with adjustment by superplasticizers.

Three populations can be distinguished concerning workability. The major parameter to control workability is the superplasticizer. The composition of the ternary binder plays only a minor role. The redispersible powders can play a role, specifically since the powder used has been adapted to SLUs through admixtures.

12.3.3 Strength development

The strength development and development of mechanical (elastic) properties during the first 28 days is shown in Fig. 12.12, where Young's modulus is given as a function of the compressive strength for the series 1–7. Young's modulus has been determined by ultra-sonic measurement based on the resonant frequency method.

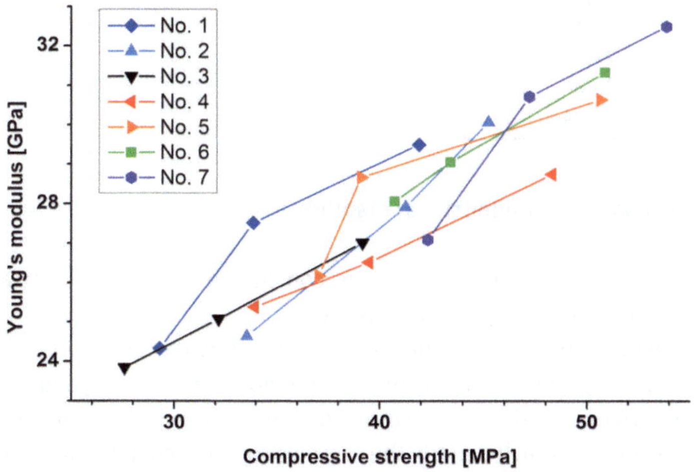

Fig. 12.12: Young's modulus as a function of compressive strength development during the first 28 days for ternary mixes 1–7.

As expected, the strength as well as Young's modulus increase with time as hardening through hydration takes place. Also, a quite clear relationship between the composition of the binder and the strength values can be observed, showing increased strength as the composition changes from OPC-dominated binders to CAC-dominated binders. For the OPC-dominated binders, the acceleration of setting time through additions of CAC and/or $C\bar{S}$ leads to a slightly lower strength as compared to the pure OPC binder. For CAC-dominated compositions, the strength increases as the amount of CAC increases in the mixture. The measured strength behavior corresponds to the general properties associated with different compositions, as shown in Fig. 12.10.

With increasing strength, Young's modulus increases, too. The fracture toughness as determined on notched samples in a bending test, however, decreases as shown in Tab. 12.2. The material becomes quite brittle.

Tab. 12.2: Fracture toughness and Young's modulus as determined by US resonance method.

Binder	K_{IC} (MPa mm$^{1/2}$)		Young's Modulus E (GPa)	
	$\bar{X}$	σ	$\bar{X}$	σ
1	0.274	0.028	27.45	0.32
2	0.225	0.056	28.71	0.15
3	0.225	0.008	24.96	0.23
4	0.228	0.006	26.47	0.06
5	0.215	0.010	27.97	0.70
6	0.196	0.024	28.84	0.32
7	0.156	0.019	31.14	0.11

12.3.4 Shrinkage compensation or dimensional stability

The characterization of shrinkage and expansion is considered in terms of early shrinkage and in terms of the classical drying shrinkage.

- Early shrinkage measurements started after ten minutes of water addition and lasted for the first 24 hours. The mortars were cast in a mold representing a modified version of the German classical "Schwindrinne", meaning shrinkage drain apparatus measuring $4 \times 6 \times 25\,cm^3$; usually the samples were tested in uncovered conditions and could lose moisture through the top surface. In uncovered conditions, the top surface was sealed with a plastic wrap and no drying occurred.
- Drying shrinkage on hardened samples between 1 day and 180 days has been characterized according to DIN 52450.

A comparison of the early shrinkage and expansion behavior for the ternary mixes 1–7 is given in Fig. 12.13. The top part of the figure represents mixes without plasticizer, the lower portion those with superplasticizer. For mixes 1, 2, and 7, without the addition of a plasticizer only shrinkage is observed. Expansion after an initial pronounced shrinkage is observed for mixes 3–6, which contain sulfate and form additional ettringite upon hydration. The expansion is most pronounced for sample 4, where the lime content is highest. Sample 7 does not show any expansion since the hydration products are mostly plate-like crystals from calcium-aluminum-hydrates. For the series with SP, the two cements, OPC and CAC, as single binders show only shrinkage. All other samples show expansion after an initial shrinkage. The extent of expansion, however, is lower than for samples without SP.

This behavior has been confirmed for many repeated measurements with different admixtures, specifically SP additions, where the extent of dimensional changes depends largely from the admixtures. Fig. 12.14 shows the shrinkage behavior for comparable compositions as in Fig. 12.13 but with higher flow values, which means higher SP dosage. In the early shrinkage curves, initial and final set times marked with * are also always found in the area where the first (plastic/chemical) shrinkage occurs. The final set always lies before the material expands or shrinkage is arrested. The comparison supports the idea that shrinkage depends not only on ettringite formation, but also on the visco-elastic properties of the matrix.

In an analogues fashion, the drying shrinkage behavior over the first 180 days is shown in Fig. 12.15 for the ternary mixes 1–7 with the top part representing mixes without plasticizer, the lower portion those with superplasticizer. For the series without SP, mix 4 does not get stabilized and shows expansion over the time considered. The other mixes reach stable expansion and shrinkage values after 1–7 days. For the series with SP, mix 4, the shrinkage and all series seem stable after about 1–7 days. The shrinkage and expansion values show higher absolute values for mixes without su-

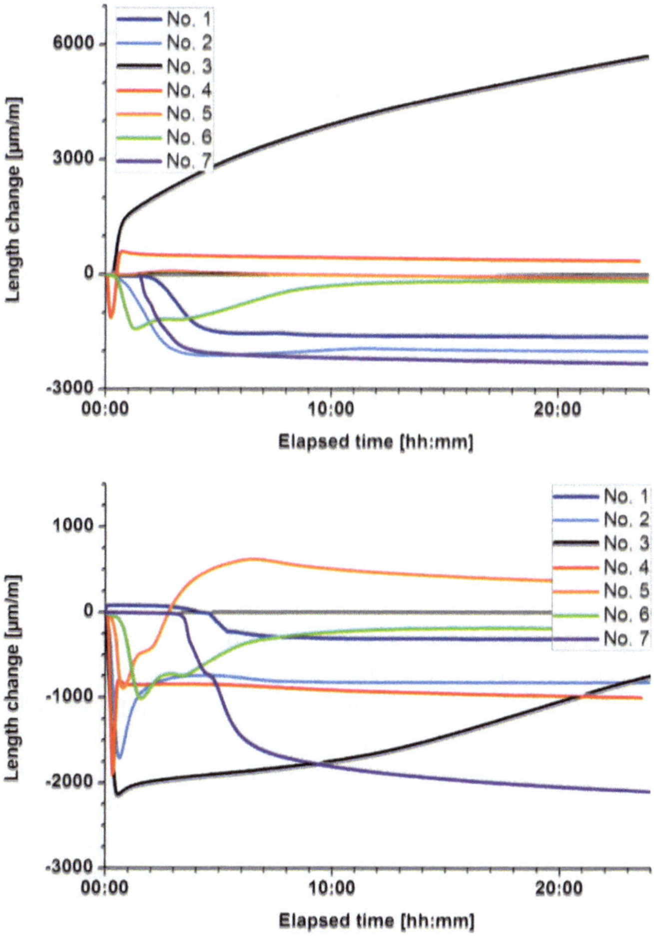

Fig. 12.13: Dimensional variation during the first 24 hours for ternary mixes 1–7 without (top) and with (below) superplasticizer addition.

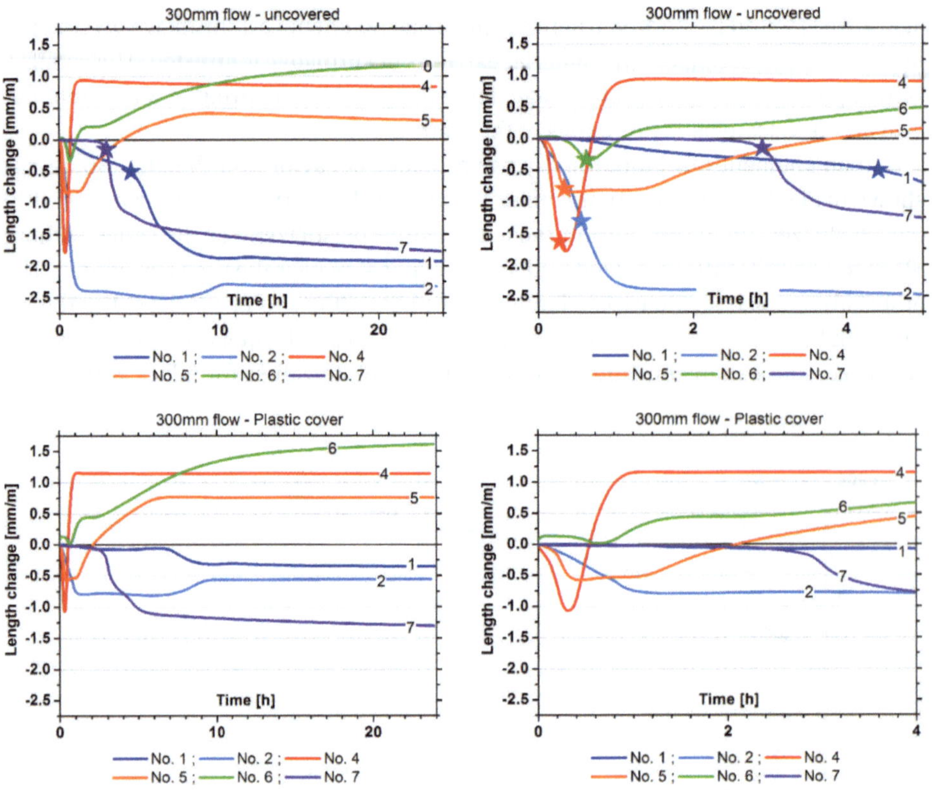

Fig. 12.14: Dimensional variation during the first 24 hours for ternary mixes 1–7 in covered and uncovered conditions.

perplasticizer. The most pronounced expansion values are observed for mixes which are between the OPC- and CAC-dominated mixes – probably due to a pronounced potential for ettringite formation because of the significant availability of Ca and sulfate ions in the solution.

When the samples are stored underwater, hydration reactions can continue and the subsequent dimensional changes are more pronounced than in drying conditions. This is demonstrated in Fig. 12.16, where dimensional changes are shown for samples stored underwater. Mix 4 shows comparable behavior, as observed in Fig. 12.14. Mix 3, which is close to mix 4 where composition is concerned but with a slowly dissolving anhydrite instead of hemihydrate, shows expansion with and without SP. With the exception of samples 3 and 4, all ternary binders show fairly stable values after 7 days either after an initial expansion or shrinkage. Absolute values show slightly higher values for underwater storage as compared to the standard climate of 65 % relative humidity at 20 °C.

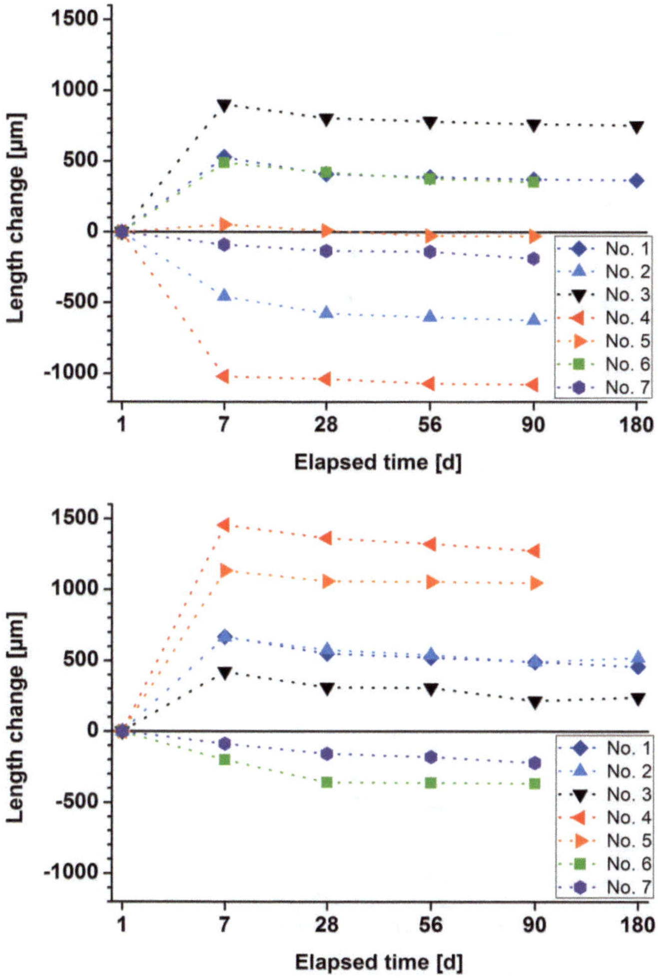

Fig. 12.15: Drying shrinkage at 20 °C and 65 % r.h. during 180 days for ternary mixes 1–7 without (top) and with (below) superplasticizer addition.

Fig. 12.17 shows a closer look at the development of shrinkage and expansion as a function of time for the ternary binder 4 during the first 2.5 hours. This composition shows very short setting times even in the case that plasticizers, retarders, or redispersible powders are used. The different admixtures were used to adjust flow values. Some measurements were also done with a plastic cover and with some redispersible powder added. Nevertheless, all curves show a very similar pattern, which can be divided into four phases:

(1) During the first minutes, there is little change in dimensional variation.

(2) Thereafter, very pronounced shrinkage is observed, followed by

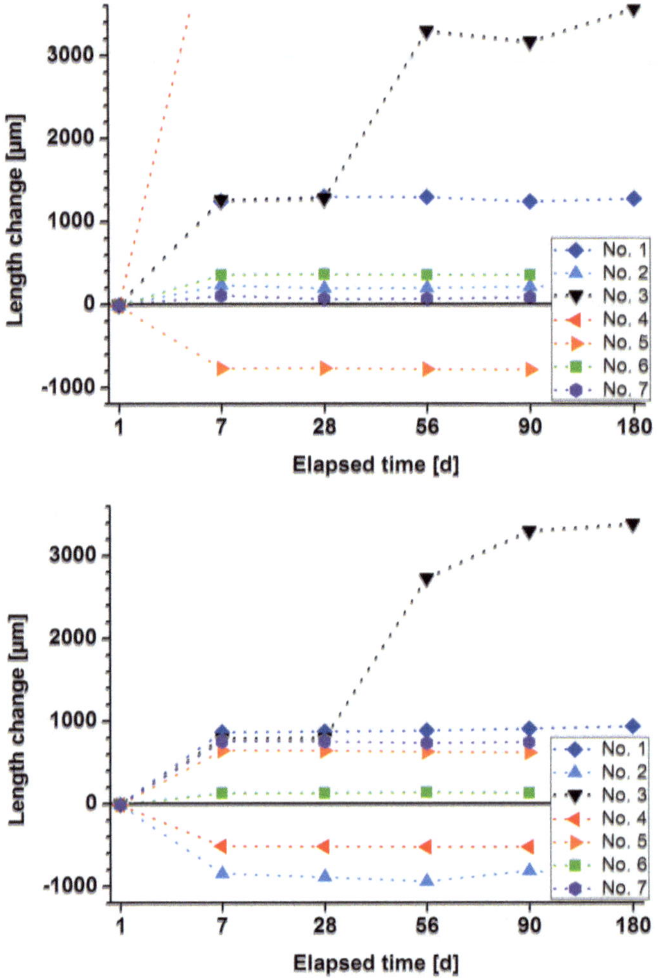

Fig. 12.16: Dimensional changes during 180 days under water at 20 °C for ternary mixes 1–7 without (top) and with (below) the addition of superplasticizer.

(3) a pronounced expansion, which
(4) finishes after about one hour.

During the shrinkage and expansion phases, the formation of ettringite could be detected by XRD. The small sketches inserted into the diagram symbolize the idea of structural changes during hydration. Firstly, ettringite is formed on individual particle surfaces comparable to the mechanisms of set control in OPC through the addition of sulfate. Particles can still move against each other freely. As ettringite forms more pronouncedly, water is consumed and a strong chemical shrinkage is observed. The ettringite grows and the particles are more and more hindered in their movement.

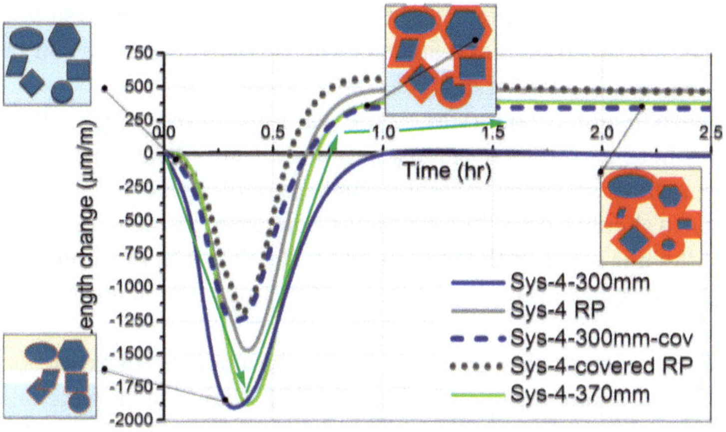

Fig. 12.17: Shrinkage and expansion for the first 2.5 hours after water addition to binder mix 4.

Macroscopically, the material stiffens. During this phase, the initial and final set occur. After the final set, the matrix is stiff enough to allow for significant expansion due to the growing ettringite particles. After the raw materials calcium aluminate, calcium sulfate, and/or water are depleted, the ettringite formation stops.

12.3.5 Long-term behavior

Long-term behavior or durability could classically be evaluated through different characteristics or tests, such as reinforcement corrosion, sulfate attack, freeze-thaw damage, or the stability of ettringite as a function of time. Few published results are available on this issue for ternary binders. One reason might be the fact that ternary binders are used for special mortars and not for reinforced concrete or larger concrete parts or structures. It is possible, however, to draw some conclusions.

The protection of reinforcement bars is assured by maintaining an alkaline environment with pH values above 10. In most cases, pure OPC will be the binder of choice for ordinary reinforced concrete. Portlandite is associated with alkaline environments and therefore mixes containing sufficient CH will provide such protection. These mixes, according to Lamberet [10], are situated in Fig. 12.1 in the lower right hand corner close to OPC.

In contrast to those mixes, compositions containing less calcium which do not form CH upon hydration like CAC-based binders 5–7, show a high sulfate resistance, as CAC was invented to create a cement with high sulfate resistance.

The freeze thaw resistance is primarily influenced by the availability of water in capillary pores and the use of air entraining admixtures. In this sense, good freeze thaw resistance can also be achieved with materials based on ternary binders as long as the pore structure is sufficiently dense for test regimes employing salt solution as testing fluid (cf. CDF test according to Setzer) [31]. Unpublished results on rapid setting grouting mortars could demonstrate that a sufficient freeze thaw resistance can be achieved for example through low water to binder ratios.

The major challenge in terms of durability is the performance of ettringite. If ettringite stays stable without converting to monosulfate and forming delayed ettringite at later stages, the materials should stay stable. Lamberet indicated a pronounced swelling potential which should exclude these compositions for mixes containing too much sulfate (above the line between OPC and mix 6 in Fig. 12.1) [10]. However, this swelling potential also depends on the mechanical and microstructural make-up of the cementitious matrix.

For OPC-based mixes, specifically 3 and 4, long-term measurements of expansion in drying conditions as well as underwater (Figures 3.6 and 3.7) show that the materials start to expand.

CAC-based mixes stay rather stable if the composition allows a consumption of sulfates and aluminates during the first days of hydration. Nevertheless, long term studies should be performed in the future.

Acknowledgment: The author acknowledges the help of his co-workers at the Institute for Ceramics, Glass and Construction Technology, TU Bergakademie Freiberg. Specifically, Dr. T. Westphal, as well as the PhD students A. Bajrami and E. Qoku and the diploma, master, and guest students F. Dlugosch, A. Qorllari, and K. Telhaj. Fig. 12.10 has been provided courtesy of Kerneos, Aluminate Technologies.

References

[1] Li G, et al. Study on a high strength ternary binder cured under different conditions. Construction and Building Materials. 2016; 107: 385–393.
[2] Dave N, et al. Experimental analysis of strength and durability properties of quaternary cement binder and mortar. Construction and Building Materials. 2016; 107: 117–124.
[3] Makhloufi Z, et al. The strength of limestone mortars with quaternary binders: Leaching effect by demineralized water. Construction and Building Materials. 2012; 36: 171–181.
[4] Bier TA, Amathieu L. Calcium Aluminate Cement (CAC) in Building Chemistry Formulations. CONCHEM Kongress; 1997.
[5] Emoto T, Bier TA. Rheological behaviour as influenced by plasticizers and hydration kinetics. Cement and Concrete Research. 2007; 37: 647–654.
[6] Bier TA, Nukita M. Influence of Redispersible Powder on Expansion in Self Levelling Underlayments, 6th ASPIC, Shanghai, China; 2009.
[7] Onishi K, Bier TA. Investigation into relations among technological properties, hydration kinetics and early age hydration of self-levelling underlayments. Cement and Concrete Research. 2010.
[8] Scrivener K. Calcium aluminate cements. In: Newman J, editor. Advanced Concrete Technology – Constituent Materials. London: Department of Civil Engineering, Imperial College.

[9] Kighelman J, Hydration and structure development of ternary binder systems as used in self-levelling compounds [dissertation]. Switzerland: École Polytechnique Fédérale de Lausanne; 2007.

[10] Lamberet S. Durability of ternary binders based on Portland cement, calcium aluminate cement and calcium sulphate, Thèse N° 3151. Lausanne; 2005.

[11] Dlugosch F. Einfluss der Sulfatzugabe in Mischbindern auf Dauerhaftigkeit und Phasenentwicklung [dissertation]. Freiberg; 2010.

[12] Westphal T, Bier TA, Dlugosch F. Influence of Admixtures on Phase Development in Ternary Binder Systems; XIII International Congress on the Chemistry of Cement, Madrid; 2011.

[13] Bier TA, Bajrami A. Influence of hydrate phase development on early shrinkage in ternary binders, 33rd International Conference on Cement Microscopy; 2011.

[14] Bier TA. Bindemittel, Spezialzemente und Ternäre Systeme. GDCh Aktuelle Wochenschau. 2011; 6.

[15] Bier TA, Bajrami A. Influence of Polymer Addition on Early Microstructure Development in Ternary Binders. Restoration of Buildings and Monuments. 2013; 19(2/3): 109–116.

[16] O'Connor BH, Raven MD. Application of the Rietveld Refinement procedure in Assaying Powdered Mixtures. Powder Diffraction. 1988; 3: 2–6.

[17] Jansen D, Goetz-Neunhoeffer F, Stabler C, Neubauer J. A remastered external standard method applied to the quantification of early OPC hydration. Cement and Concrete Research. 2011; 41: 602–608.

[18] Scarlett NVY, Madsen IC. Quantification of phases with partial or no known crystal structure. Powder Diffraction. 2006; 21: 278-284.

[19] De la Torre AG, Santacruz I, León-Reina L, Cuesta A, Aranda MAG. Diffraction and crystallography applied to anhydrous cements. In: Pöllmann H, editor. Cementitious Materials Composition, Properties, Application. Berlin, Boston: De Gruyter; 2017.

[20] Aranda MAG, De la Torre AG. Diffraction and crystallography applied to hydrating cements. In: Pöllmann H, editor. Cementitious Materials Composition, Properties, Application. Berlin, Boston: De Gruyter; 2017.

[21] Westphal T, Bier TA. Correlating XRD data with technological properties, In: Pöllmann H, editor. Cementitious Materials Composition, Properties, Application. Berlin, Boston: De Gruyter; 2017.

[22] Lothenbach B, Winnefeld F. Thermodynamic modelling of the hydration of Portland cement. Cement and Concrete Research. 2006; 36: 209–226.

[23] Lothenbach B, Matschei T, Möschner G, Glasser FP. Thermodynamic modeling of the effect of temperature on the hydration and porosity of Portland cement. Cement and Concrete Research. 2008; 38: 1–18.

[24] Amathieu L, Bier TA, Scrivener KL. Mechanisms of set acceleration of Portland Cement through CAC addition. Proceedings of the International Conference on Calcium Aluminate Cements (CAC). Edinburgh, Scotland; 2001.

[25] Pelletier L, Winnefeld F, Lothenbach B. The ternary system Portland cement–calcium sulphoaluminate clinker–anhydrite: Hydration mechanism and mortar properties. Cement & Concrete Composites. 2010; 32: 497–507.

[26] Qoku E, Westphal T, Bier TA, Dilo T. In-situ X-ray diffraction analysis of early hydration of cementitious systems and microstructural investigation with SEM. 36th International Conference on Cement Microscopy, Milan, Italy; 2014.

[27] Schleibinger Geräte Teubert. Schleibinger testing materials. Buchbach, Germany: Schleibinger Geräte Teubert [cited 2017 Mar 13]. Available from: www.schleibinger.com.

[28] Esping O. Early age properties of self-compacting concrete – Effects of fine aggregate and limestone filler [dissertation]. Göteborg, Sweden; 2007. ISBN 978-91-7291-890-0.

[29] Ehrich S, Helard L, Letourneux R, Willocq J, Bock E. Biogenic and chemical sulfuric acid corrosion of mortars. J Mater Civ Eng. 1999; 11: 340–344.
[30] Bier TA, Bajrami A, Westphal T, Qoku E, Qorllari A. Influence of Redispersible Powders on Very Early Shrinkage in Functional Mortars. Advanced Materials Research. 2015; 1129: 77–85.
[31] Setzer MJ. Zum Frost-Tausalz-Widerstand von Beton, Vom Werkstoff zur Konstruktion. Institut für Massivbau und Baustofftechnologie, Universität Karlsruhe; 1990.

Part IV: **Measurement and properties**

Christiane Rößler*, Bernd Möser, and Horst-Michael Ludwig

13 Characterization of microstructural properties of Portland cements by analytical scanning electron microscopy

Abstract: For decades, electron microscopy has been a major analytical technique for the characterization of cementitious materials. During the last twenty years, new developments in scanning electron microscopy (SEM) have significantly improved the resolution of imaging and analytical possibilities.

The current contribution aims to give an overview of the latest developments in analytical SEM, which offers parallel chemical-crystallographic characterization and SEM imaging. Crystallographic information is thereby obtained by collecting electron backscatter diffraction (EBSD) patterns and, in the same region of interest, energy-dispersive X-ray (EDX) spectroscopy delivers chemical composition. This new quality of spatially resolved combined information is unique and only possible when the corresponding requirements of sample preparation are achieved.

The results of the present investigation show how elemental mapping data obtained by EDX spectroscopy can be used to deduce phase distribution maps and how these phase maps can be further analyzed to reveal the chemical composition of individual clinker phases including minor and trace element contents. For the first time, it has been shown how EBSD can be applied to characterize cement clinkers. First, EBSD patterns of alite in cement clinkers could be obtained by introducing argon broad ion beam polishing as a final preparation step. Combined EDX and EBSD mapping data can be used to extract the phase distribution maps, grain size, and grain boundary analysis of cement clinkers. EBSD analysis can further be used to identify areas with amorphous or low (nano) crystalline structure.

The second part of the investigation shows how EDX phase mapping in combination with low voltage high resolution SEM imaging can be used to characterize hydrating cements.

Keywords: Portland cement, clinker, cement hydration, microstructure, SEM, EDX, EBSD, high resolution SEM imaging

***Corresponding author: Christiane Rößler,** Bauhaus-Universität Weimar, 99423 Weimar, Germany, christiane.roessler@uni-weimar.de
Bernd Möser, Horst-Michael Ludwig, Bauhaus-Universität Weimar, 99423 Weimar, Germany

DOI 10.1515/9783110473728-014

13.1 Introduction

Microstructural investigations provide basic insights into properties of hydrating cements as well as a basic understanding of the underlying processes. Furthermore, the results of microstructural investigations are required for modelling the macroscopic properties of cement and concrete. This understanding is the basis for a targeted optimization of cement composition in order to obtain sustainable and durable concretes.

Generally, microstructural investigations comprise qualitative and quantitative information obtained by analytical methods such as electron microscopy, X-ray diffraction, and others [1]. Modern imaging techniques of scanning electron microscopy (SEM) are valuable because the presence or absence of nanoscaled hydration products can be evaluated without extensive sample preparation. Thereby the sample is fractured, inserted into the microscope, and imaged. The drawback of SEM imaging is that the quantification of results is challenging. Traditionally, phase identification by SEM imaging is supported by chemical analysis provided by energy or wavelength dispersive X-ray spectroscopy (EDX/WDX).

13.1.1 SEM imaging and EDX spectroscopy

Backscattered electrons (BSE) are often used for the differentiation and quantification of phases by SEM imaging [1, 2]. The image reflects the fact that regions with different mean atomic number backscatter electrons at different rates [3]. This results in grey value images that can be used to identify and quantify phases. Due to similar mean atomic numbers of relevant phases, the method does not differentiate between all phases present in cement clinker. For hydrated cements, the resolution of BSE imaging is also limited mostly due to the effects of sample preparation (see Section 13.4.2).

Recently, Stutzman [4], Kocaba [5], and Yio [6] have shown that a combination of SEM-BSE imaging with energy dispersive X-ray spectroscopy (EDX) is a valuable tool to quantify phase assemblages in cement clinkers and the hydration of slag in hydrating Portland cement. The protocol applied by Stutzman [4] and Kocaba [5] includes sample embedding, polishing, coating, and time consuming EDX mapping followed by image analysis. But the gain in information obtained justified the effort. Other attempts to quantify slag fraction, degree of hydration, porosity, etc. from SEM-BSE imaging [6] or EDX point counting method [7] have been successfully applied. Coupled SEM-BSE and EDX point counting differentiates more phases than BSE alone but precision depends mainly on the selection of points for analysis (as well as many other factors) [7].

For analyzing the amount and composition of fly ash in hydrating cement, it was shown that phase analyses of EDX mappings offer new insights into quali- and quantitative microstructural characterization [8]. EDX mapping in combination with phase analysis (EDX phase mapping) is a new way of combining individual elemental maps to a conclusive image of phase distribution [9, 10]. This is a big step forward compared to phase differentiations based on BSE grey level images or on single EDX elemental distribution maps.

EDX phase mapping not only allows drawing phase maps but also suming up all spectra assigned to a specific phase. The EDX sum spectra of each phase comprise counts in higher orders of magnitude than a normal single spot spectrum. Thus the quantification of these spectra for analyzing phase composition is straight forward. Quantification of EDX spectra can be carried out both with and without standards [3]. Both approaches have specific advantages and disadvantages [3, 11, 12]. Standards-based EDX analysis provides a high degree of accuracy and the detection limits can be in the same order of magnitude as for the quantification of WDX spectra [3, 12, 13]. One major disadvantage of standardless quantification is that all quantified elements are summed up to 100 percent [3, 12]. Mistakes in peak identification therefore lead to erroneous results. However, previous studies have shown that the precision of standardless EDX analysis is comparable to that of standards-based analysis [3, 11, 14]. Especially for elements that are present at concentrations of 10 wt% and above, standardless quantification can be applied [14]. For improving accuracy, it is necessary to regularly perform tests on standard materials containing the elements of interest [3]. Accuracy, precision, and detection limit are improved by increasing the number of detected characteristic X-rays (i.e. improved peak to background ratio) [3, 11, 14]. Often, the limits of EDX quantification are dictated by the sample itself, i.e. sample homogeneity, surface roughness, sputtering, etc. [3].

Over the last years, the accuracy and precision of main, minor, and trace elements analysis by EDX spectroscopy has advanced significantly [12]. A prerequisite to taking advantage of this is that certain sample, hardware, and software parameters are set up correctly. Overall, a standardless EDX quantification with a degree of analytical error comparable to standards-based quantification is feasible if some sample information is available and if the method is validated for the class of materials investigated [14].

The present study aims to adapt EDX phase mapping to cementitious materials and to establish the reliability of standardless EDX quantification if some precautions are followed.

13.1.2 Electron backscatter diffraction (EBSD) in the SEM

A new analytical method in the field of microstructural characterization of cement and concrete research is the detection of electron backscatter diffraction (EBSD) patterns. The basics for this techniques were established during the last decades, starting with the first contribution on the topic by S. Kikuchi 1928 [15], after whom the diffraction bands detected in the SEM are named. The analysis of the EBSD data for various applications/materials is still developing. A comprehensive overview on the application of EBSD in materials science is given elsewhere [16]. The present contribution aims to give an introduction to how EBSD analysis can be used to characterize cement clinkers.

The chemical-mineralogical characterization of Portland cement clinker is an important prerequisite for specifying the hydration processes and thus the properties of concretes. Especially for the development of new, sustainable, and application-specific cements, the extension of existing analytical methods for product characterization is essential.

Clinker characterization in cement standards is usually based on the bulk chemical composition. On the basis of this data and by assuming the stoichiometric composition of phases, the mineral clinker phase fractions are calculated (using Bogue's calculation modified by Taylor) [17, 18]. A further approach is to use finely ground clinker samples for quantitative X-ray diffraction (QXRD) analysis [19]. Crystal modifications and their proportions can thereby also be determined, but often the peak overlap limits the precision of quantification [1, 19]. A disadvantage of all these methods is that they are carried out on bulk powders or the dissolution residues thereof. A separate analysis of individual clinker phases is only possible by microscopic analysis [20–22]. Currently, for a technical multi-phase mixture such as Portland cement, it is difficult to demonstrate changes in the crystal modifications of single clinker phases in dependence on the chemical composition. The reason is that a spatially resolved chemical and crystallographic analysis is not available. The accuracy of the determination of the clinker phase fractions is also still unsatisfactory (3–5 % deviations are typical) [1, 19].

The advantage of the characterization of Portland cement clinker by means of analytical techniques applied on polished sections in the SEM is that it is spatially resolved. If the SEM crystallographic (EBSD) information is combined with the chemical (EDX spectroscopy) analysis, a spatially resolved chemical-crystallographic characterization of the clinker phases is obtained. Furthermore, it is also possible to get a quantitative evaluation if these analyses are recorded as a mapping. This allows, for the first time, a clear quantitative chemical-mineralogical characterization of Portland cement clinker.

In the following, it will be shown how imaging (SE and BSE) and analytical SEM (EDX and EBSD) can be successfully applied to characterize cement clinker and hydrated cements.

13.2 Materials

Pure monoclinic alite (71.7 wt% CaO, 25.9 wt% SiO_2, 1.8 wt% MgO, and 0.6 wt% Al_2O_3) was provided by C. Naber (University of Erlangen-Nürnberg, Germany). A detailed characterization of this material can be found elsewhere [23].

For investigations with cement clinker, an unground technical Portland cement clinker has been used. The chemical composition of clinker (Tab. 13.1) is determined by X-ray fluorescence spectrometry (Orbis PC, AMETEK, USA) on a molten flat disc. According to powder X-ray diffraction combined with quantitative Rietveld analysis (software Topas 4.2; Bruker, Germany), the clinker contains approximately: 60 wt% alite, 19 wt% belite, 5 wt% cubic and 2 wt% orthorhombic aluminate, 11 wt% ferrite, and 1 wt% free lime, periclase, and arcanite.

Cement of type CEM I 42.5 R and lab prepared mixtures of this cement with 36 (CEM III/A) respectively 66 (CEM III/B) wt% ground granulated blast furnace slag (GGBFS) were used for hydration investigations. The chemical composition of the cement (CEM I 42.5 R) and GGBFS are given in Tab. 13.1.

Tab. 13.1: Chemical composition of clinker, cement (CEM I 42.5 R), and GGBFS (given in wt%).

	CaO	SiO_2	Al_2O_3	Fe_2O_3	MgO	TiO_2	K_2O	Na_2O	SO_3	P_2O_5 (S)
Clinker	65.4	21.1	5.8	3.2	2.0	0.3	1.0	0.1	0.6	0.3
CEM I	63.5	20.0	4.4	2.4	1.5	0.2	1.3	0.3	3.7	–
GGBFS	41.1	35.4	11.8	0.7	7.4	1.15	0.5	0.2	0.6	(1.3)

Specific surface area of CEM I 42.5 R and GGBFS were determined by the Blaine method (i.e. 460 and 400 m^2/kg). Cement and GGBFS have been homogenized in a TURBULA mixer (WAB AG, Switzerland).

The compositions of all investigated binders are given in Tab. 13.2. For hydration investigation, cementitious material and water have been mixed ($w/c = 0.5$) and stored at room temperature in sealed plastic containers.

Tab. 13.2: Composition of binders (given in wt%).

Sample	**CEM I 42.5 R**	**GGBFS**
CEM I 42.5 R	100	–
CEM III/A	64	36
CEM III/B	34	66

Hydration has been stopped by cutting/fracturing samples and immersing them for 6 h in 2-propanol to remove residual water. For drying, samples were placed in a high vacuum or at 35 °C for approximately 12 h.

Alpha $Ca_2[SiO_3(OH)](OH)$ (C_2SH as abbreviated by cement chemist notation) was provided by T. Link, Bauhaus-Universität Weimar (Germany). The details of the synthesis are described elsewhere [24].

13.3 Methods

13.3.1 Sample preparation

For SEM-EBSD, investigations samples were cut and embedded in epoxy resin (vacuum impregnation). Using standard mechanical polishing, sections were polished in incremental steps with diamond oil slurries of particle sizes 15, 3, 1, and 0.25 μm.

For EBSD analysis, the last preparation step was done by ion etching using an argon Broad Ion Beam (BIB) mill (TIC 3X, Leica, Germany).

For EDX spectroscopy and BSE imaging, the polished sections were coated with carbon (approx. 15 nm).

13.3.2 SEM imaging and analysis

SEM investigations were carried out by using a Nova NanoSEM 230 (FEI, Netherlands) equipped with a field emission gun. An EBSD camera (DigiView III, EDAX/AMETEK, USA) and an EDX spectrometer (Silicon Drift Detector, Apollo 40, EDAX/AMETEK) are attached to the SEM.

13.3.2.1 SEM imaging

All secondary electron (SE) imaging on fractured surfaces were acquired in a high vacuum and at low accelerating voltage (2 kV) or in a water vapor environment and with a high accelerating voltage (20 kV), thus no electric coating was necessary. Under high vacuum conditions, the through the lens detector (TLD) and, under gaseous environment, the low vacuum detector (LVD) were used for SE imaging.

BSE imaging has been used for imaging polished, carbon coated clinker surfaces at an accelerating voltage of 12 kV.

13.3.2.2 SEM-EDX analysis

In the following, conditions for SEM-EDX mappings with untilted sample and high vacuum conditions are described (i.e. combined EDX-EBSD parameters are given in Section 13.4.1.2.3).

For high resolution SEM-EDX mappings, the following conditions were applied: acceleration voltage 12.0 kV, max. 9000 cps, 512 × 400 and 1024 × 800 pixel mapping resolution (≈ 0.08 and 0.63 μm/pixel, depending on mapped area), dead time of EDX detector was always below 30 %. With respect to the limitations of EDX resolution for hydrated cements (see Section 13.4.2), the mapping resolution of 0.63 μm/pixel was taken as appropriate. The number of 9000 X-ray counts per seconds (cps) was deliberately set low (i.e. up to 100 000 cps is possible) by choosing a medium beam current

(≈ 1 nA) for the applied 12 keV accelerating voltage. This was done to limit the sample (especially for hydrated cements) deterioration under electron bombardment. To get a high total count number per mapping, 123–256 frames were recorded. This leads to a maximum mapping time of up to 13.7 h and a total count number of up to 440 million counts per map.

Phase analysis of EDX mappings was carried out with software package Genesis 6.35 (EDAX/AMETEK, USA). The word phase thereby refers to an area of the sample with similar chemical composition. The principal workflow of phase analysis consists of:

(1) acquiring of EDX elemental mapping of the sample,
(2) definition of phases of interest (by using a self made EDX phase library),
(3) setting up match parameters (chi square goodness of fit) for phase identification,
(4) recalculation of phase map (assigning a phase to each pixel, summing up of pixels identified as phase to larger areas).

The allowed variability for phase composition was set to 80–90 % for the applied chi square fit. Phase maps were obtained by comparing average spectra of the defined EDX pixel size with surrounding EDX spectra. If a certain fit (chi square) of a measured spectra with the phase spectrum from a library is detected, the analyzed pixel is colored in the respective color of phase. In a first run, an automated phase cluster analysis was carried out to check which phases can be differentiated. The obtained phases were compared with the estimated phase list. If necessary, new phases were added to the existing library and the phase map was recalculated. In this way, we assured that all chemically different phases were identified. Because the software Genesis 6.35 offers at most seven different colors for phases, it was partly necessary to combine phases.

Phase maps can be used to sum up all EDX spectra of individual phases. These spectra contain a very good count statistic (i.e. up to 10–200 million counts). This is essential for quantitative analysis of the mean phase composition, i.e. for determining major, minor, and trace elemental concentrations of the respective phases.

Software Genesis 6.35 and TEAM 4.3 (EDAX/AMETEK) were used to quantify EDX spectra by deconvolution (applying ZAF and background correction, standardless and standards-based analysis). User defined SEC (standardless element coefficients) for light elements have been included.

For comparison, standards-based EDX quantification was carried out using a glass standard (K-0412, SPI Supplies, USA).

13.3.2.3 SEM-EBSD

The software package "OIM Data Collection 6.1" (EDAX/AMETEK, USA) has been used for recording EBSD maps. In order to avoid electrical coating and charging during SEM-EBSD investigation, samples were investigated in water vapor atmosphere at

0.1–0.7 mbar. The optimum quality of EBSD patterns was achieved for higher voltages (20 kV) and high beam currents (8–16 nA).

For orientation image analysis of the mappings, the software "OIM Analysis 6.2" (EDAX/AMETEK, USA) was used. Proper indexing of EBSD patterns (using interband angles) is indicated by the value of confidence index [24], a reliability parameter. At a confidence index (CI) greater than 0.1, approximately 95 % of indexing solutions are correct if at least 6 Kikuchi bands are used for indexing [24]. In the present study, 6–14 Kikuchi bands have been used for indexing and the average CI for analyzed phases was always above 0.1. It is possible to improve the amount of properly indexed EBSD patterns by applying so-called clean up routines to the recorded data. The applied clean up routines are described as follows: firstly correction of pseudo-symmetric misindexing, followed by confidence index (CI) standardization (grain tolerance angle 5, minimum grain size 4, CI > 0.1), secondly neighbor orientation correlation (clean up level 4, single iteration), and finally grain dilation (minimum grain size 4, single iteration). Details of the applied clean up functions can be found at Humphreys [25], Nowell & Wright [26], and Wright [27]. Recorded EBSD data are often displayed as inverse pole figure maps [28]. After applying clean up functions, the amount of EBSD patterns with a CI value below 0.1 was reduced to less than 10 % for the good pattern quality of alite, aluminate, and ferrite. Only the belite EBSD pattern quality is lower, as discussed in Section 13.4.1.2.

13.4 Results

13.4.1 Characterization of unhydrated clinker materials

13.4.1.1 Energy dispersive X-ray spectroscopy (EDX)

13.4.1.1.1 Standardless vs. standards-based quantification of EDX spectra using a glass standard material (K-0412, SPI Supplies, USA)

Quantitative measures for studying the composition of clinker phases are needed to predict the properties of cements. EDX spectroscopy is a spatial resolved analysis that can analyze the composition of individual phases in multiphase materials such as cements. The quantification of EDX spectra can be carried out with or without standards. Usually, standardless analysis is referred to as semi-quantitative [3, 12]. But the technique is developing and most users perform standardless EDX analysis. This fact is based on the far simpler and faster way to get results with standardless EDX analysis as compared with standards-based methods. Thus in the following some principle investigations on the accuracy and precision of standardless and standards-based EDX analysis of materials with composition similar to cements are carried out.

Standards-based EDX analysis requires standard materials of known chemical composition and measurements of beam current used for analysis. The standard needs to be of a similar composition as the analyzed material [3]. EDX spectra of

Tab. 13.3: Results of EDX standards-based and standardless quantification of SPI glass standard.

	SPI Glass (wt%)		EDX spot spectra (wt%)				EDX sum spectra (wt%)	
			Std-bas.		Std-less			
	Mean	Std dev.	Mean	Std dev.	Mean	Std dev.	Std-bas.*	Std-less*
MgO	19.33	0.2	19.9	0.1	20.5	0.1	19.5	18.8
Al_2O_3	9.27	0.2	9.3	0.1	9.7	0.1	9.5	9.7
SiO_2	45.35	0.2	45.3	0.2	44.0	0.2	45.5	46.5
CaO	15.25	0.2	15.4	0.1	16.0	0.1	15.7	15.0
FeO	9.96	0.2	10.0	0.2	9.8	0.2	9.9	10.0
CaO/SiO_2	0.33		0.3		0.4		0.3	0.3

	Relative error (%)		
	Spot		Sum
	Std-bas.	Std-less	Std-less*
MgO	3.0	5.9	2.6
Al_2O_3	0.8	5.0	4.1
SiO_2	0.1	2.9	2.7
CaO	1.2	4.8	1.9
FeO	0.3	1.7	0.6

standard and sample were recorded at the same, defined SEM and EDX operating conditions. Thus for the current investigation a glass standard (Glass standard K-0412, SPI Supplies, USA) containing magnesium, aluminium, silicon, calcium, and iron was used. To evaluate the accuracy and precision of EDX quantification, the standardless and standards-based EDX analysis is compared to results given by the glass standard supplier (Tab. 13.3). All results are given as oxides in weight percent. Mean values and standard deviations are calculated from 10 EDX analyses. As expected, the standards-based analysis closely resembles the composition of the glass standard, i.e. except for magnesium oxide within the standard deviation given by the supplier. The relative errors of standardless analysis are also acceptable (Tab. 13.3). The results in Tab. 13.3 depict that the given standard deviations of standards-based and standardless EDX quantification are often small compared to the absolute deviation from the target value (composition given by glass standard supplier).

That means that for EDX analysis the precision is good but the accuracy can be low.

In addition to EDX quantification of spot spectra, the sum spectrum obtained on a mapping area is also quantified using a standards-based and standardless approach (Tab. 13.3). The advantage of this sum spectrum is that the number of counts is relatively high, the background is well defined and in principal trace elements can also be identified. This results in slightly decreased relative errors of standardless quantification on sum spectra as compared to standardless quantification of EDX spot analyses (Tab. 13.3, last two rows).

In conclusion, it is shown that compared to standards-based EDX quantification, the relative error of standardless analysis is slightly increased. The values of errors are in an acceptable range. For comparative studies between clinkers, knowing the relative error makes standardless analysis more useful.

13.4.1.1.2 EDX phase mapping of cement clinkers; deducing clinker phase composition including minor components

In the following, it is discussed how BSE-imaging and EDX phase mapping can be applied to characterize cement clinkers.

Fig. 13.1 shows a typical BSE image of a polished clinker surface coated with a few nanometres of carbon. The image contrast is related to the mean atomic number of individual phases. Thus these images are helpful to differentiate some phases contained in cement clinkers. More phases can be differentiated, if the chemical information obtained by EDX spectroscopy is analyzed in parallel. Traditionally, EDX mappings deliver distribution maps of elements of interest (Fig. 13.2, 13.3). It is thereby important to set up the mapping grit of spot analyses as such, so that the smallest expected phases can be differentiated. The resolution of these maps is determined by the acceleration voltage needed to excite X-rays from elements of interest and the resulting excitation volume for the phases of interest. Normally, for cement clinkers the heaviest major element of interest is iron. Thus 12 keV acceleration voltages are sufficient to excite characteristic X-rays from iron. For unhydrated calcium silicates, this acceleration voltage causes excitation of X-rays from a volume with approx. 1.8 μm diameters (calculated with software Electron Flight Simulator, Small World LCC, USA). This means that in principal the EDX quantification of calcium silicates spectra requires homogeneous areas with a diameter larger than 1.8 μm.

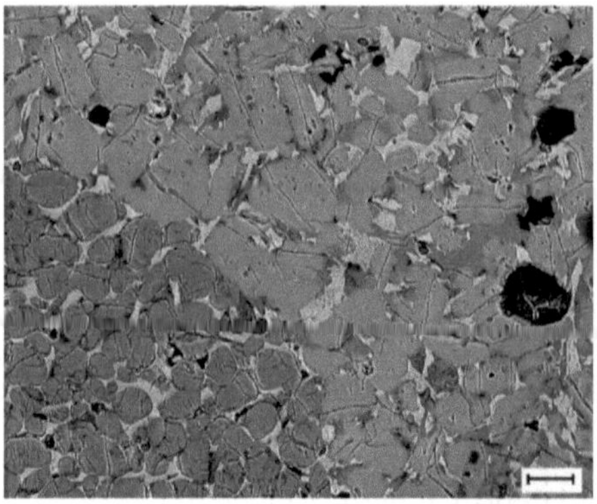

Fig. 13.1: BSE image of cement clinker; light grey: ferrite, medium grey: alite, darker grey: belite and aluminate, black: pores, CaO, and MgO (scale bar = 20 μm).

Fig. 13.2: EDX element distribution map of silicon (light color indicates increased silicon concentration); differentiation of alite and belite is obvious (compare with Fig. 13.1).

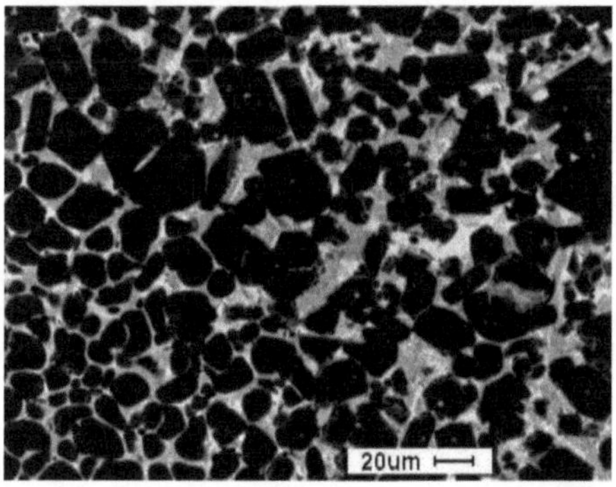

Fig. 13.3: EDX element distribution map of aluminium (light color indicates increased aluminium concentration); differentiation of ferrite and aluminate is obvious (compare with Fig. 13.1).

EDX element distribution maps are further helpful for identifying clinker phases on the basis of chemical composition. Using modern software from EDX suppliers, it is also possible to search automatically for phases or to search for user defined phases.

Fig. 13.4 shows an EDX phase map of Portland cement clinker with user defined phases of interest. These user defined phases are saved as EDX spectra of expected phases that are collected at the same SEM presets (i.e. accelerating voltage) at the same or pure samples. For phase identification, a chi square fit of 80–90 % was defined, i.e. spectra of unknown sample that are identified as phases must not deviate more than 10–20 % from given phase.

Obviously, the results in Fig. 13.4 show that all major clinker phases such as alite, belite, ferrite, and aluminate can be differentiated by this phase analysis procedure. Minor phases such as MgO, CaO, and calcium-potassium-sulphates are also easily identified. In addition, a magnesium rich aluminate (Fig. 13.4, inset) was identified by comparing the elemental distribution map of magnesium and aluminium with the obtained phase map.

Another benefit of EDX phase maps is that all collected point spectra that belong to an identified phase can be summed up. This EDX sum spectrum has very good count statistics (i.e. the number of counts per spectrum is suitable for quantitative analysis, Fig. 13.5).

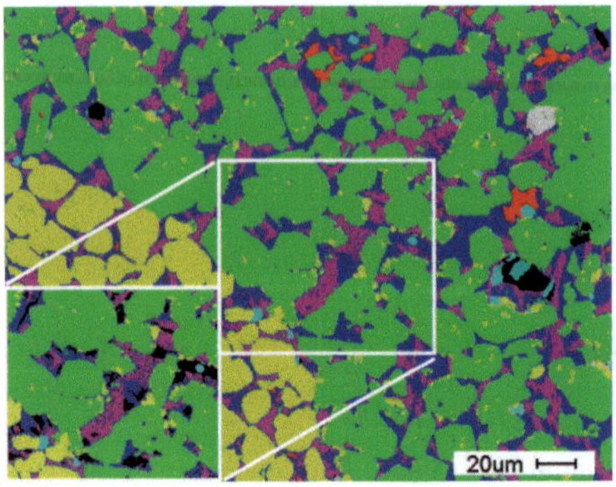

Fig. 13.4: Phase distribution map deduced from EDX mapping; green: alite, yellow: belite, magenta: ferrite, blue: aluminate, red: MgO, cyan: potassium sulphate, grey: pores, black: CaO; lower left side inset – black: aluminate, blue: magnesium rich aluminate (other colors as above).

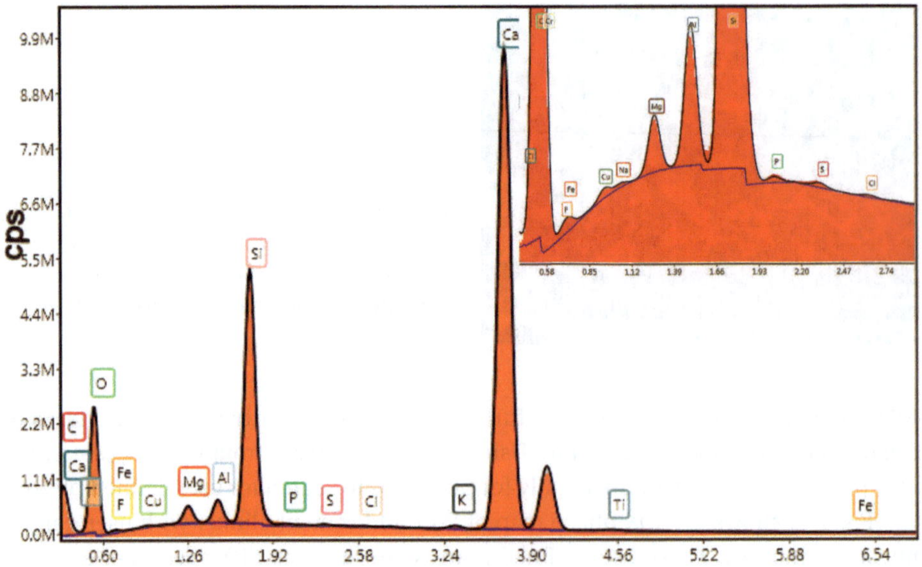

Fig. 13.5: Left side: sum spectra of alite obtained from phase map in Fig. 13.4. Right side: enlarged spectra of belite showing clear peaks of minor and trace elements.

The X-ray intensities of minor and trace elements can thus also be identified and quantified [12]. According to previous studies the limit of detection for EDX analysis (with modern silicon drift detectors as spectrometers) can be as low as 0.05 wt% [12]. The preconditions are that the sample is flat and electric conductive, that the count number of spectrum is very high and that the detector dead time is low [12]. The relative error for trace elements concentrations is given to be at least 25 %, if standards-based quantification using k-ratio measurements is carried out [12].

Tab. 13.4: Composition of clinker phases deduced from standardless EDX analysis (given in wt%).

Oxide	Aluminate	Mg-aluminate	Ferrite	Alite	Belite	Sulphate	Arcanite (spot)
CuO	–	0.2	–	0.0	0.5	0.3	–
Na_2O	0.2	0.1	0.0	0.4	0.1	0.1	–
MgO	1.7	7.7	2.4	1.4	0.7	0.9	–
Al_2O_3	27.8	21.3	19.6	2.0	1.8	4.5	–
SiO_2	5.1	9.7	5.8	25.9	31.3	6.0	–
P_2O_5	0.1	0.1	0.1	0.2	0.1	–	–
SO_3	0.1	0.5	0.2	0.1	0.1	25.9	44.0
K_2O	0.9	2.8	0.6	0.3	1.2	32.5	52.3
CaO	55.3	49.0	52.5	71.5	61.3	26.2	3.7
TiO_2	0.7	0.5	1.4	0.3	0.2	0.8	–
Fe_2O_3	8.1	8.2	17.4	0.7	2.2	2.1	–

Compositions of phases identified by EDX mapping were calculated using standardless EDX quantification (Tab. 13.4). Obviously, the two aluminate phases show a significantly different composition, i.e. the magnesium free aluminate can be considered as low alkali aluminate, whereas the magnesium and potassium content of the magnesium rich aluminate is increased. This finding is in line with results from X-ray phase analysis which shows that the clinker contains merely alkali-poor, cubic aluminate (see Section 13.2). Further inspection of EDX phase analysis reveals that most of the potassium contained in the clinker is bound to sulphate (Tab. 13.4). Apparently, the composition of phases in Tab. 13.4 shows that also aluminium and silicon are contained in the sulphate phase. To determine whether this aluminium and silicon content arises because the potassium sulphate inclusions are of small size, an EDX mapping at increased magnification was recorded (Fig. 13.6–13.8). The elemental distribution maps of aluminium, iron, sulphate, and potassium are mixed in an RGB overlay (Fig. 13.6). This map shows that aluminium and iron are only detected on the rim of the potassium and sulphur phase. Quantification of EDX spot spectra in the middle of the yellow areas in Fig. 13.6 reveal analyses of a potassium-calcium sulphate (Tab. 13.4). This finding is in line with the X-ray diffraction results of the clinker, indicating that alkali sulphates are present at approx. 1 %.

Inspecting the phase mapping in Fig. 13.4 and the results of phase composition given in Tab. 13.4 raises the question of whether the identified alkali-poor aluminate can be further differentiated into alkali-enriched aluminate and alkali-poor aluminate. Therefore, RGB overlays of elemental distribution maps are inspected in detail (Fig. 13.7 and 13.8). The overlay of calcium, potassium, and aluminium elemental distribution maps (Fig. 13.7) reveals a light-colored phase in the aluminate area that, according to the legend shown in Fig. 13.6, possesses an increased amount of potassium (the color would plot in the middle of the triangle in Fig. 13.6). In Fig. 13.8, magnesium is added to potassium corner of triangle (green side of triangle in Fig. 13.6). The

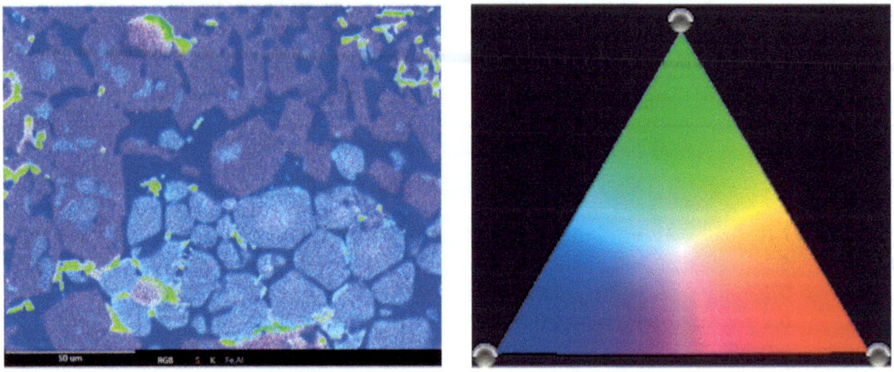

Fig. 13.6: Left: RGB overlay of sulphur (red), potassium (green), and iron/aluminium (blue) EDX element maps. Right: color legend.

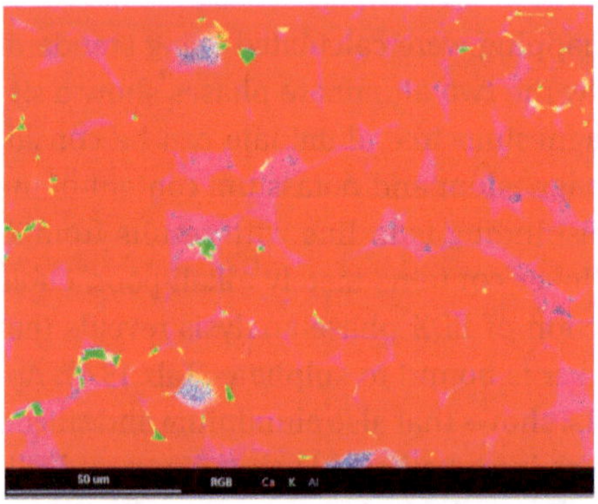

Fig. 13.7: RGB overlay of calcium (red), potassium (green), and aluminium (blue) element maps. Color legend: see Fig. 13.6 (right).

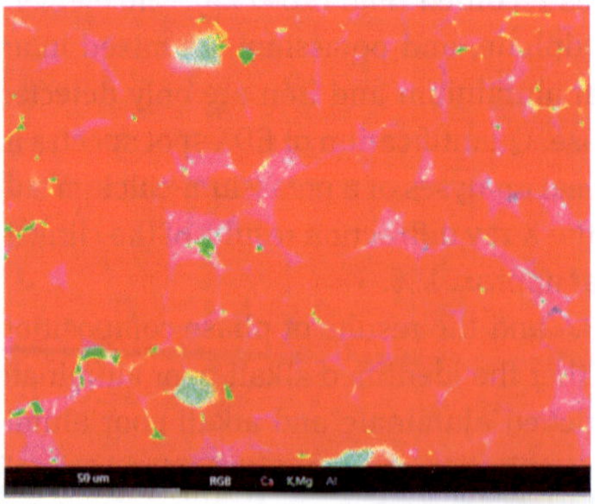

Fig. 13.8: RGB overlay as in Fig. 13.7 but magnesium is added to potassium (green). Color legend: see Fig. 13.6 (right).

effect is that the potassium-rich areas in the aluminate phase appear even lighter in color. This clearly indicates that the increased potassium content is associated with increased magnesium content. Thus, as has previously been found by EDX phase analysis, potassium content is increased in magnesium-rich aluminate phase. In conclusion, it is not clear whether this magnesium and potassium rich aluminate is the expected orthorhombic aluminate, which is indicated by X-ray diffraction results (2 % orthorhombic C_3A as given in Section 13.2) because the magnesium content is far higher than described in previous studies [17].

Comparing the compositions of alite and belite (Tab. 13.4) deduced from EDX sum spectra reveals that alite contains increased magnesium but decreased iron and potassium content. The aluminium content in both phases appears to be similar (i.e. deviates less than the expected standard deviation for the result).

Whether these are significant findings or not can be identified by looking at the EDX phase analysis shown in Fig. 13.9 and comparing it to the elemental distribution maps shown in Fig. 13.10–13.13. Thereby it is clearly shown that magnesium is enriched in alite (Fig. 13.10) and potassium is enriched in belite (Fig. 13.11). The difference in iron content in alite and belite is less marked but still recognizable (Fig. 13.12). Surprisingly, aluminate content in alite and belite is not as similar as depicted by the results in Tab. 13.4. Rather, in belite an increase in aluminium content from the center

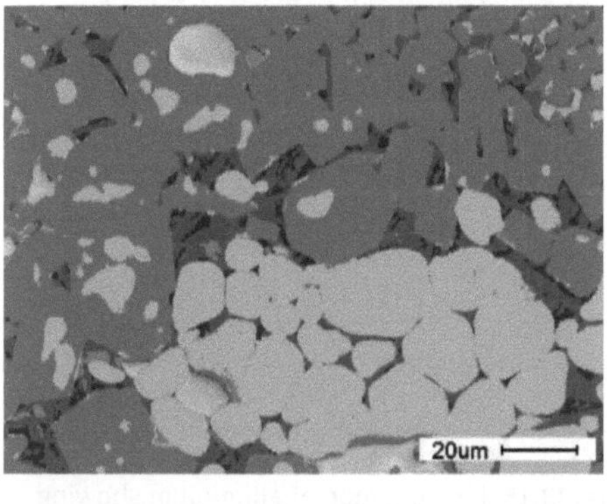

Fig. 13.9: EDX phase map of clinker at increased magnification; green: alite, yellow: belite, magenta: ferrite, blue: aluminate, red: magnesium rich aluminate, cyan: potassium sulphate, grey: pores.

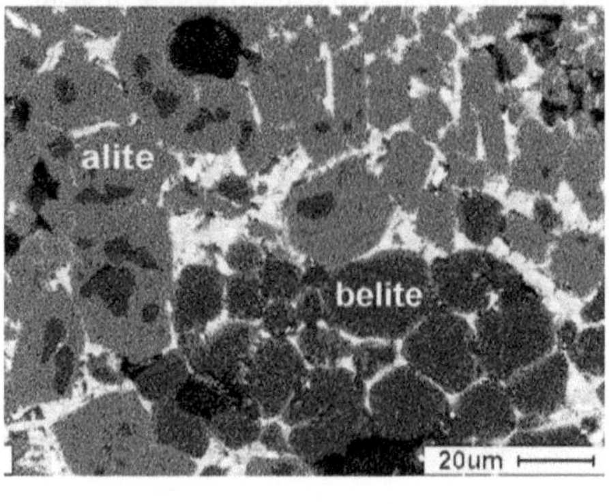

Fig. 13.10: Magnesium element map showing enrichment in alite (light grey areas) as compared to belite (dark grey area, compare with phase map in Fig. 13.9).

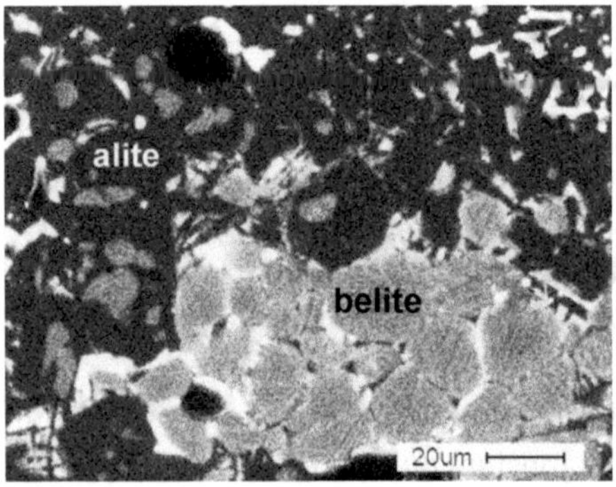

Fig. 13.11: Potassium element map showing enrichment in belite (light grey areas) as compared with alite (dark grey area, compare with phase map in Fig. 13.9).

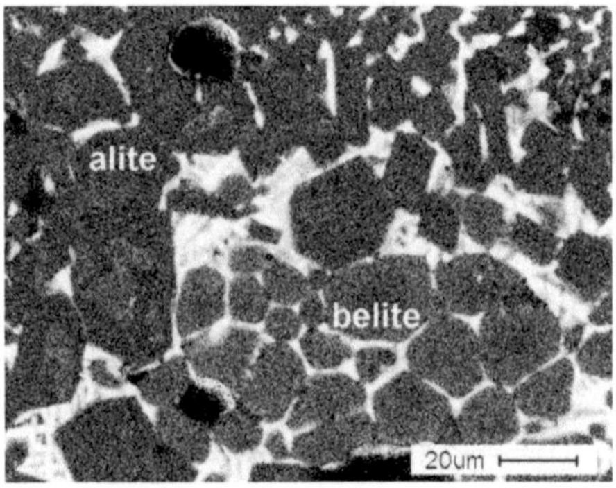

Fig. 13.12: Element map of iron showing slight enrichment of iron in belite as compared to alite.

Fig. 13.13: Element map of aluminium showing zoning of aluminium in belite (i.e. core of belite contains less aluminium than rim).

to the rim of the grains is observable (Fig. 13.13). In conclusion, EDX element distribution maps largely support the results of standardless EDX phase analysis.

The prerequisite is that element distributions in identified phases are homogeneous. To depict a gradual distribution of elements within identified phases, looking at the element distribution maps is indispensable.

13.4.1.2 Coupled SEM EBSD-EDX analysis for clinker characterization

13.4.1.2.1 Sample preparation

In investigations using electron microscopy, sample preparation is often the biggest challenge. Over the last 20 years, technical novelties in the microscope setup have introduced the possibility to investigate samples at low pressure in the microscope chamber. This opened the opportunity not only for imaging samples that are unstable under high vacuum conditions but also this development has paved the way for investigating samples that are electrically non-conductive without the necessity to coat them with a conductive layer. In parallel, the analytical techniques attached to the electron microscope have developed greatly. One recently enforced technique opens crystallographic characterization in the SEM by detecting and analysing electron back scatter diffraction (EBSD) patterns [16]. EBSD patterns of electrically nonconductive samples can now be recorded. The only prerequisite is that the sample surface is flat and free of defects introduced by sample preparation (depth where the EBSD signal origins is only about 50–100 nm). Thus, for EBSD analysis on cement clinkers, an undisturbed crystal lattice at the sample surface is required. Conventional polishing techniques applied for EBSD on metals, rocks, and ceramics fail for building materials. The reason is that unhydrated cement is highly reactive in water and humid environments and all polishing exposes the material surface to humidity. One very promising method for sample preparation that has been found is argon broad ion beam (BIB) polishing.

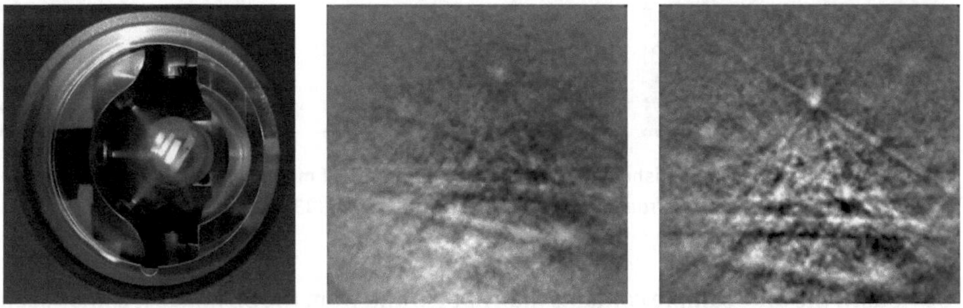

Fig. 13.14: Left: three argon BIBs (coming from left side) that polish an alite sample (Ø 2.5 cm). Middle: EBSD pattern with poor quality. Right: EBSD pattern after argon BIB.

In the present study, samples used for argon BIB polishing have been epoxy embedded and mechanically polished, as described in Section 13.2. The applied argon BIB polisher rotates the sample and uses three argon ion beams to polish the sample (Fig. 13.14, left). The sample's tilt, the voltage (and current) of the guns, and the duration of the process can be adjusted to vary the rate of material removal. After subsequent ion polishing, the EBSD pattern quality was assessed (Fig. 13.14, middle and right). Optimum pattern quality was obtained after two stages of ion polishing.

The first step uses a flat tilt angle, approx. 4 kV accelerating voltage and is applied for 20–30 min. The second step is carried out at a higher sample tilt (up to 30°), increased accelerating voltage (6 kV), and for up to 45 min. The precise time of argon polishing is very much dependent on the sample's properties. Extended ion polishing creates a rough surface, which is unsuitable for further EBSD analysis. Thus, it is more advisable to start polishing with shorter periods of time and repeat until the desired EBSD pattern quality is obtained.

13.4.1.2.2 EBSD investigations on synthetic alite and C_3S (Ca_3SiO_5)

The first EBSD investigations were carried out on synthetic alite and C_3S. Working with pure clinker phases allows for showing the principal workflow of EBSD analysis for cement clinker. To get an overview of the sample surface, it is essential to record an SE and/or BSE image of the sample. Because EBSD pattern are recorded at a high tilt angle (70°) of the sample, regular BSE detectors are not effective. Thus, either a forescatterd electron detector (FSD) might be used for imaging or the regular low vacuum SE detector is applied (Fig. 13.15, left).

Fig. 13.15: Left: SE image of polished alite. Middle: EBSD pattern of monoclinic alite. Right: crystal orientation (IPF) map deduced from EBSD mapping (legend see Fig. 13.17, right).

Obviously, the alite sample contains many pores and from the SE image it can be seen that the sample is sintered very well, so that large alite crystals might be present. Recorded EBSD patterns confirm the very good pattern quality (Fig. 13.15, middle), suitable to carry out EBSD mappings (Fig. 13.15, right). Before starting an EBSD mapping, an identification of the present crystal phases is required. If additional phases are identified during mapping, an offline rescan of the EBSD map with additional phases can be performed.

Working with pure synthetic phases, such as alite in the present analysis, limits the number of crystal phases that need to be considered. Because the alite sample was already characterized by high resolution X-ray diffraction analysis in combination with Rietveld analysis [23], the selection of possible crystal structures was straightforward.

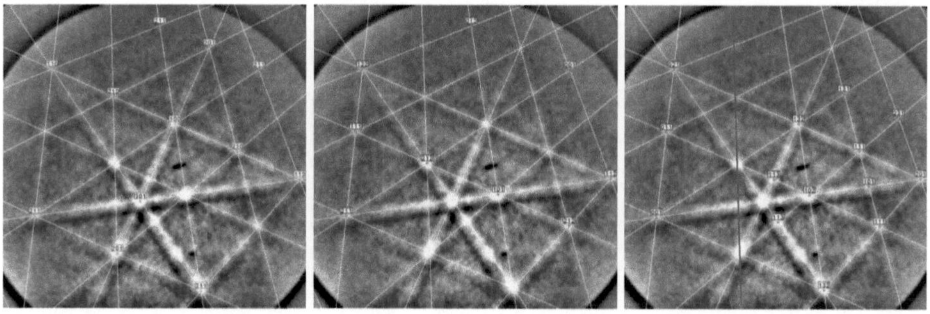

Fig. 13.16: Indexing of the recorded EBSD pattern of alite using monoclinic M_1 (left) [30], triclinic (middle) [31], and monoclinic M_3 (right) structure [32] for alite/C_3S. White lines correspond to calculated Kikuchi bands, red line is missing band in monoclinic M_3 structure.

According to the synthesis conditions and XRD phase identification, the monoclinic structure of alite [32] was proposed to be suitable for this material. The software package used for recording EBSD mappings allows one to directly import crystal structure data from X-ray diffraction analysis. Thereby, the structure factors of X-ray reflections are recalculated to match the requirements for electron diffraction [33]. In the present case, the Rietveld refined structure of monoclinic alite (i.e. M_3 structure) [32] was used for EBSD analysis. Thereby, it was compared whether monoclinic M_1 [30] or M_3 structures [32] or triclinic C_3S [31] would fit the observed EBSD pattern best. The automatic ranking of pattern indexing showed a very good fit (< 1) for all structures. The indexing criteria for monoclinic M_1 and triclinic polymorphs are very similar and fit the measured pattern best (Fig. 13.16, left and middle). The close similarity is not surprising because the monoclinic M_1 structure [30] is based on the triclinic structure proposed by Golovastikov [31] and refined by De La Torre [32]. Interestingly, the monoclinic M_3 structure [32] always shows the lowest agreement but the fit is still sufficient to use this structure for indexing. Thus, on the basis of EBSD phase identification alone a clear identification of polymorphs is not possible. By using additional information from synthesis and EDX analysis, one can exclude the triclinic polymorph. In conclusion, on the basis of EBSD phase identification, the monoclinic M_1 polymorph would fit the observed patterns best. Therefore, in the following EBSD mappings the crystal structure for the M_1 polymorph of alite is used.

Recording EBSD mappings requires similar presets as the EDX mappings mentioned above (i.e. step size of single analysis, time to record pattern resp. spectrum). At a user defined step size, EBSD patterns are recorded and analyzed. So called crystal orientation images are obtained (Fig. 13.15 and 13.17). One way of showing crystal orientations is to plot inverse pole figure (IPF) maps, as shown in Fig. 13.15 and 13.17 (legend see Fig. 13.17, right). Thereby, different colors reflect crystal directions in a stereographic projection [34]. The raw data obtained by EBSD mapping is shown in Fig. 13.17 (left). To get an orientation image as shown in Fig. 13.15, so called data clean up procedures have been applied.

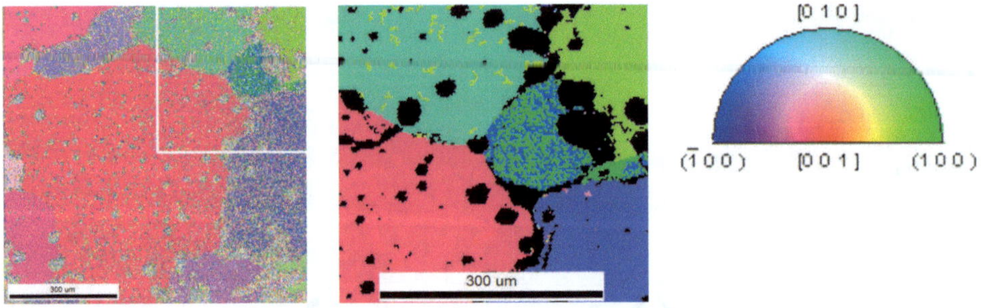

Fig. 13.17: Left: raw data of alite crystal orientation (IPF map) deduced from EBSD mapping. Middle: enlarged area of marked left image after first data clean up, pseudo-symmetric misindexing is still seen in spotty green/blue as well as cyan/yellow color grains. Right: Color legend of IPF maps shown in Fig. 13.15 and 13.17.

These allow misindexed points to be corrected and are indicated in detail in Section 13.2. Careful application of data clean up procedures is important because otherwise oversimplifications of the results may occur. One challenge of crystal orientation analysis, especially of low symmetric crystals, is that pseudo-symmetric misindexing might occur [27, 35]. This is a typical ambiguity in the identification of crystal orientation. An irregular patchy crystal orientation distribution (Fig. 13.17, middle) that does not follow any contrast indicated by electron imaging is one possible indication for pseudo-symmetric misorientation. The pseudo-symmetric relation needs to be identified and then the majority of the orientation solution is taken as valid. One pseudo-symmetry relation for crystal structure of monoclinic alite that needs to be considered has a rotation angle of 120° at crystal direction [301]. After the application of this clean up function, the spotty areas in grains in Fig. 13.17 (middle) vanish (compare with Fig. 13.15, right).

13.4.1.2.3 General approach of combined EDX-EBSD analysis of cement clinker

EBSD pattern can be used to identify crystalline phases and to record crystal orientation maps. The efficient identification of clinker phases on the basis of EBSD requires crystallographic data as a proper input and also some knowledge of the sample composition. Knowing the approximate chemical/crystallographic composition of the investigated phases reduces the number of phases considered for EBSD analysis. Therefore, phase identification by EBSD is always most efficient if in parallel the chemical composition of phases is determined by EDX spectroscopy [36]. Then EBSD mappings can be used for crystal orientation imaging, i.e. to analyze crystal orientations, grain boundaries, grain size distributions, strain, etc.

Generally, compared to most metals, the EBSD pattern contrast of silicates and aluminates contained in cement clinker is relatively low (compare Fig. 13.18). In addition, many possible crystal structures of calcium silicates possess a low symmetry

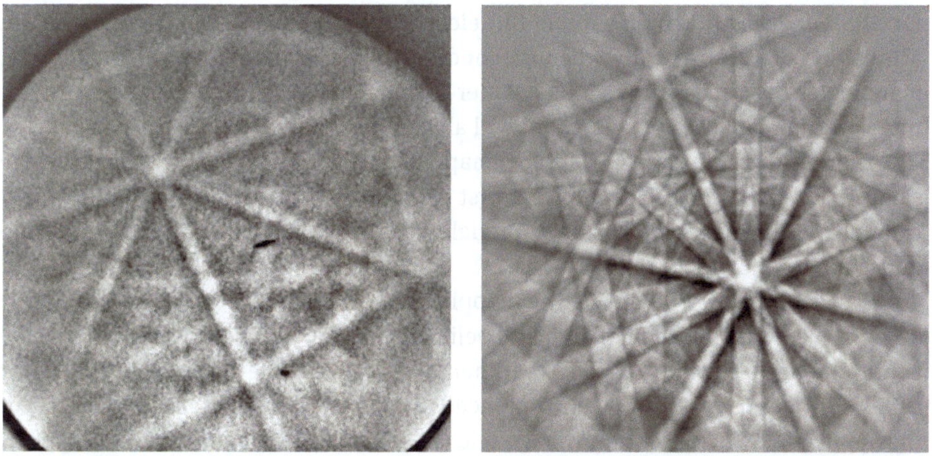

Fig. 13.18: EBSD pattern of tricalcium aluminate (left) and nickel (right) (20 keV, 16 nA). Pattern contrast, number of bands, and band width are significantly increased for metals such as nickel.

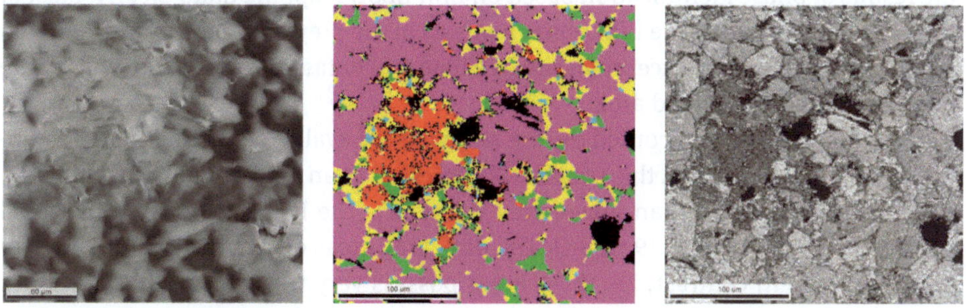

Fig. 13.19: Left: SE image of polished cement clinker (70° tilted). Middle: phase map deduced from combined EDX-EBSD map; magenta: alite, red: belite, yellow: aluminate, green: ferrite, cyan: Mg-rich aluminate, blue: periclase, black: pores. Right: IQ map indicating EBSD patterns contrast; light grey indicates high and dark grey low EBSD pattern contrast, black: no EBSD pattern.

(i.e. rhombohedral, triclinic, or monoclinic crystal structure). These facts pose major challenges for EBSD orientation imaging.

Combined EDX-EBSD mappings require that SEM parameters are set up to allow for EDX based phase identification and for optimum EBSD pattern quality. Clearly, this is a search for a compromise, i.e. a high enough resolution of EBSD and EDX measurements and an acquisition time that is as low as possible. The EDX parameters used for the results presented here allow at least a safe phase discrimination between aluminate, ferrite, potassium sulphate, free lime, and magnesium oxide, as well as alite and belite. Combined EDX-EBSD mappings do not deliver nearly the quality obtained for EDX mappings and phase analysis as described in the previous chapter. This is not achievable because of the inaccuracy introduced by the prerequisites necessary for EBSD analysis (i.e. high sample tilt, electrically non-conductive sample surface, acquisition of spectra at low vacuum condition). For proper EDX phase analysis, a mapping as described in previous chapter (using a carbon-coated sample in an untilted position and at a high vacuum) will always be required.

The starting point of an EDX-EBSD mapping is always an SE or BSE image of the sample (Fig. 13.19, left). This allows a first overview of microstructural features. For example, the SE image allows one to roughly distinguish between aluminate/ferrite and calcium silicates.

Based on the SE/BSE image, the mapping area is selected. For proper EDX phase analysis (i.e. differentiation of alite and belite, Fig. 13.19), a sufficient number of X-ray counts are needed. Thus, for the experimental setup, the time-determining factor for EDX-EBSD mapping is the dwell time needed to excite sufficient X-ray counts. For the applied 20 kV accelerating voltage and the relatively high beam current (approx. 16 nA), the dwell time for a single EDX measurement was set to approx. 100 ms. Using a step size of 1.4 µm for the full EDX-EBSD mapping, the total time needed to collect the area imaged in Fig. 13.19 and 13.21 was maximum 2 h. Images in Fig. 13.19 show the results of phase identification based on EDX-EBSD mapping data. Clearly, most of the phases described in the EDX phase analysis in the previous chapter are identified (Fig. 13.19, middle). The exceptions are free lime and potassium sulphate, which were not present in the analyzed area.

The EBSD patterns recorded during mapping are evaluated by their pattern contrast. In the software used, this contrast is expressed as an image quality (IQ) number, which is transferred into an image grey value [37]. The IQ map of the investigated clinker is shown in Fig. 13.19 (right). The dark grey areas of the image correlate with low EBSD pattern contrast, i.e. these areas possess very weak or no EBSD patterns. Normally, these are grain boundaries, empty, or resin filled pores, areas where topography is high and where poorly crystalline or strained phases are analyzed [37].

Comparing IQ with phase map (Fig. 13.19, middle and left) reveals that, among alite grains, a slight variation in EBSD patterns contrast also occurs. This is a known fact that due to the different orientation of crystals the EBSD pattern contrast varies [37]. The more interesting fact for clinker characterization is that belite appears always with darker grey values in IQ maps (compare Fig. 13.19, middle and left). This fact is further highlighted by the results in Fig. 13.20, where diffraction pattern for alite and belite are compared (recorded at the same SEM parameters).

Clearly, the belite EBSD pattern contrast (i.e. IQ grey value) is low compared to alite (Fig. 13.20). Obviously, not only large belite clusters (Fig. 13.19) but also belite inclusions in alite posses a low EBSD pattern contrast (Fig. 13.20). In fact, often it is very difficult to record belite EBSD patterns at all. One reason for this low EBSD pattern contrast of belite might be that, compared to alite, the topography is too rough. This is especially true for large belite crystals, that are known to appear with a rough lamella structure [20]. But the belite inclusions in alite are smooth and still possess a low EBSD pattern contrast (Fig. 13.20). Therefore, the other and more likely reason is that belite

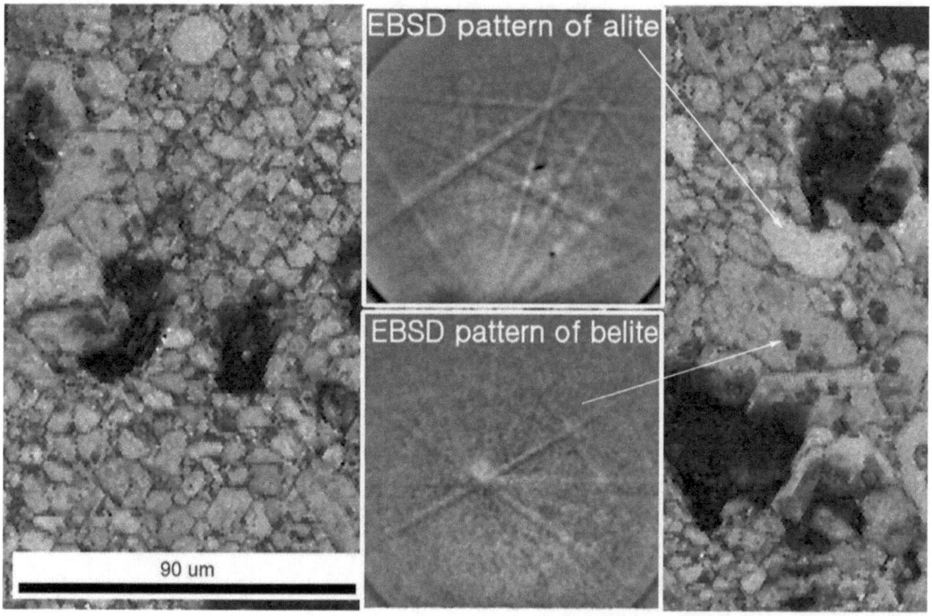

Fig. 13.20: IQ map of clinker and EBSD pattern of alite and belite, clearly EBSD pattern of belite always possess a low contrast, i.e. very faint diffraction pattern only, often with too few reflectors for proper indexing. Alite shows variations in EBSD pattern quality, but it is always sufficient for indexing.

crystallinity is effectively poorer or the crystallite size is very small compared to other phases contained in the investigated clinker. This finding also correlates with observations from X-ray diffraction, where belite is often observed with broader intensities. In X-ray diffraction analysis, this peak broadening is often attributed to low crystallite size and or intracrystalline strain [38]. Similar observations on the micro crystallinity of belite are described by optical light microscopy [20]. With some precautions, IQ maps can be used to evaluate crystal lattice distortion [39].

Besides phase analysis and phase distribution maps, combined EDX-EBSD analysis on cement clinkers allows for a characterization of crystal sizes, orientations, and grain boundaries. The results in Fig. 13.21 show the SE image and phase map of cement clinker as well as an IPF map of identified alite crystals. The IPF map for alite shown in Fig. 13.21 (left) reveals, as expected, that no preferred crystal orientation is obvious and that the crystal size of alite ranges from a few up to approximately 100 micrometers. Measuring the misorientation between grains (expressed as rotation angles) reveals the distribution of grain boundaries, as shown in Fig. 13.22 (left). Blue grain boundaries possess a misorientation angle larger than 15°, i.e. are regarded as regular, high angle grain boundaries [40]. Low angle (or sub-)grain boundaries [40] are depicted in yellow (5–15° rotation angles) and red (3–5° rotation angles). Obviously, most of the

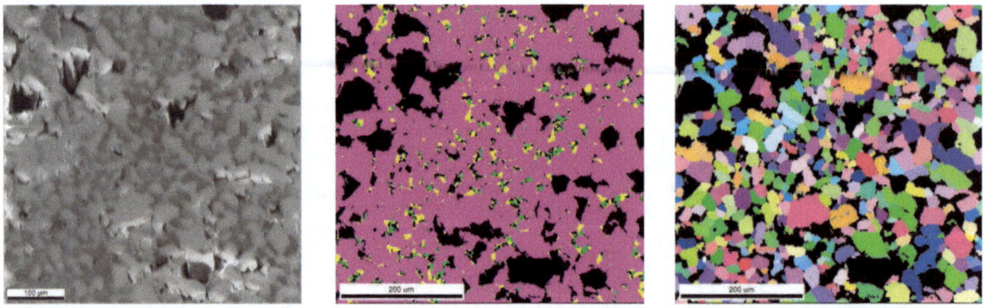

Fig. 13.21: Left: SE image of polished cement clinker (70° tilted); dark to black: pores, dark grey: aluminate, ferrite, light grey: calcium silicates (alite, belite). Middle: phase map deduced from combined EDX-EBSD map; magenta: alite, yellow: aluminate, green: ferrite, black: pores. Right: IPF map of monoclinic alite, legend see Fig. 13.17 (right).

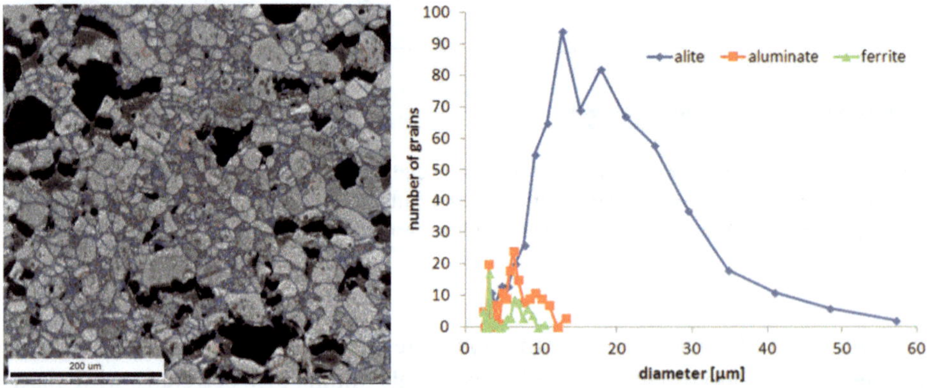

Fig. 13.22: Left: IQ map of alite overlaid with grain boundaries; blue: high angle grain boundaries (misorientation between grains > 15°), yellow and red: small angle grain boundaries (5–15° resp. 3–5° misorientation). Right: number of grains and crystal grain size of alite, aluminate, and ferrite as deduced from the mapping in Fig. 13.21.

detected alite grain boundaries are characterized as regular grain boundaries. Low angle grain boundaries can be taken as an indication for crystal distortion [40]. A safe estimation of alite crystal distortion on the basis of grain boundary analysis requires higher resolution EBSD mapping data than presented here (work is in progress).

The number of alite grains identified in Fig. 13.21 and 13.22 are more than 600, thus a good statistical base for grain size analysis is given. Grain size distribution of alite is shown in Fig. 13.22 (left). Additionally, grain size analysis of ferrite (70 grains) and cubic aluminate (160 grains) are shown in Fig. 13.22 (left). As expected, the grain size of ferrite and aluminate is much smaller than that of alite.

13.4.2 Characterization of hydrated cements

EDX phase mapping has been shown to be a valuable tool to identify and characterize phases of unhydrated cement clinkers. In the same manner, EDX phase mapping can be applied to hydrated cementitious materials. In the following, a short overview of the advantages and challenges of this method for characterizing the microstructure of hydrated cements will be given. Combined EDX-EBSD analysis on hydrated cementitious materials is not in the scope of the study. The reason is that it is very doubtful if the EBSD patterns of nano crystalline cement hydrates such as calcium-silicate-hydrates (C-S-H) can ever be obtained. The main reason is that these cement hydrates are very sensitive to electron beam damage [41]. Systematic investigations into this issue are needed.

Characterizing the microstructure of hydrated cements by SEM is still a challenge. The reason lies in the nanoscaled, porous structure and the beam and preparative sensitivity of the material. Recent developments in SEM, such as low vacuum and low voltage microscopy, allow for high resolution imaging of the microstructure (Fig. 13.23–13.28). During the last decade, this resulted in a strong improvement in the qualitative characterization of hydrated cement microstructure. For example, the intense intergrowth of nanoscaled hydrate phases in the capillary pore space is now clearly detectable (Fig. 13.24). Knowing this nanoscaled intergrowth, a clear explanation for the large scatter of results of EDX analysis is given, i.e. the volume where X-rays are excited is much larger (for C-S-H approx. 1.6 µm diameter at 10 kV acceleration voltage) than the individual hydrate phases (larger than area imaged in Fig. 13.23). Thus, all quantifications of EDX spectra carried out on hydrates in the capillary pore space must be regarded as analyses on several different hydrates. Fig. 13.24 shows the dense hydration rim around alite which is described as so-called “inner product” (Ip) [42] calcium-silicate-hydrate (C-S-H). Obviously, this Ip C-S-H appears to be a more homogeneous phase with nanoscaled pores. But a full proof of this assumption is missing. The SE image shown in Fig. 13.24 furthermore depicts that the alite surface is strongly corroded by dissolution structures.

In Fig. 13.25, a typical phase distribution of 28 d hydrated CEM I 42.5 R is given. For comparison, the SE image in Fig. 13.26 shows the phase distribution in GGBFS containing cement (CEM III/B) at the same hydration stage. Obviously, compared to alite in CEM I, the dense hydrate layers around slag particles are missing. Also, the porosity in CEM III/B is more homogenously distributed than in hydrated CEM I. In detail, Fig. 13.27 and 13.28 show the variation in morphology of C-S-H when slag is added to cement. Fig. 13.27 depicts the more fibrous C-S-H of CEM I 42.5 R, whereas Fig. 13.28 shows the foil-like C-A-S-H phases of CEM III hydration. This variation is typically known for GGBFS containing cements that possess a decreased calcium to silicon ratio and dissolve more aluminium [42]. Decades ago, TEM investigations had already shown that that a decreased calcium to silicon ratio leads to the formation of foil-like C-S-H [43]. Nowadays, the improvement resolution paved the way to image these variations in the morphology in the SEM.

Fig. 13.23: Hydrates such as ettringite, C-S-H fibers, and portlandite (CH) are growing in the capillary pore space of CEM I 42.5 R. In the following, these are called hydrate matrix.

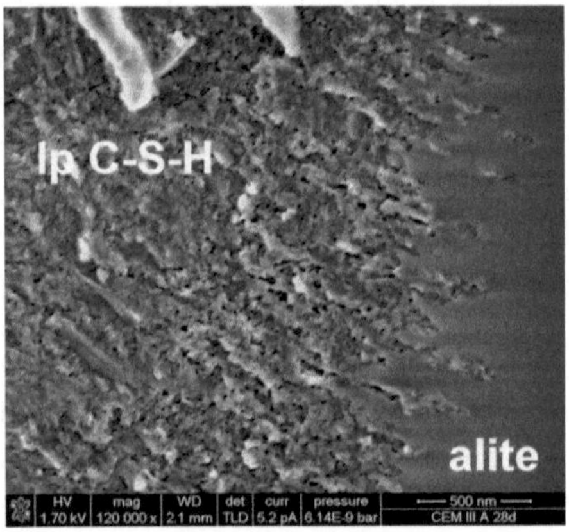

Fig. 13.24: Dense Ip C-S-H rim around alite with strong dissolution structures.

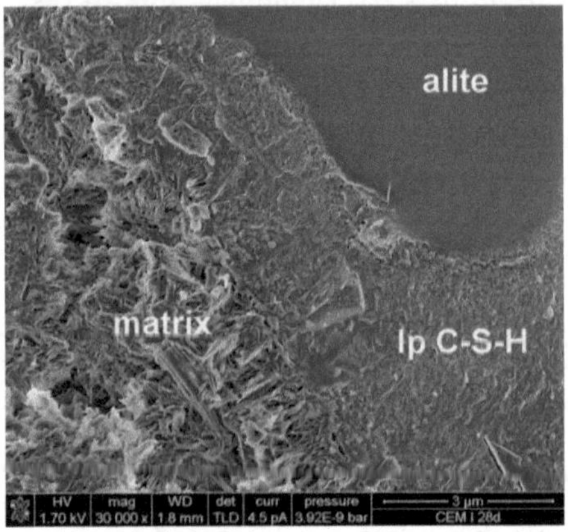

Fig. 13.25: Dense Ip C-S-H, alite, and matrix as normally observed in CEM I 42.5 R.

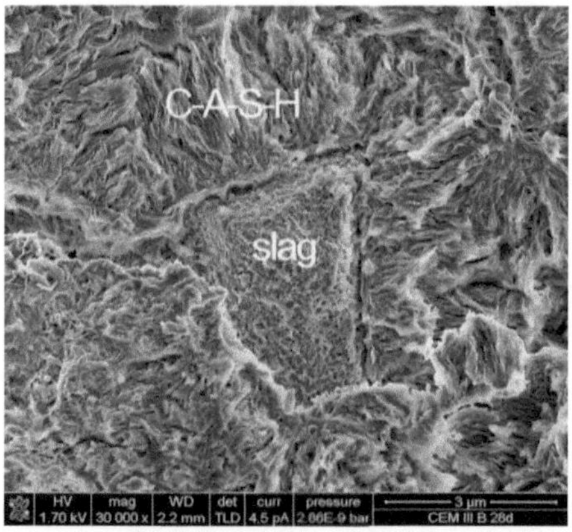

Fig. 13.26: C-A-S-H phases and slag particles in hydrated CEM III/B.

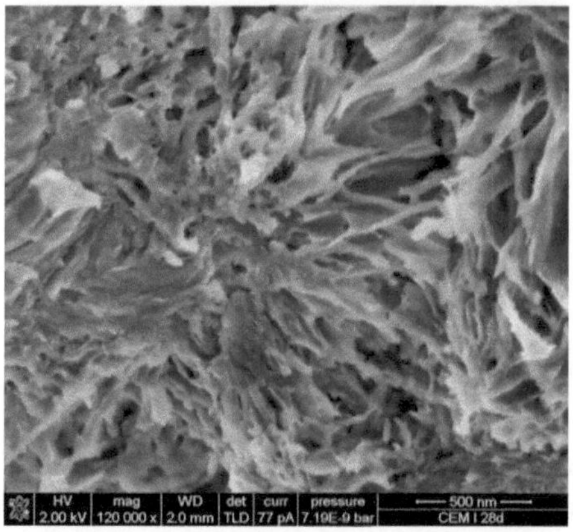

Fig. 13.27: Radial growing fibrous C-S-H in CEM I 42.5 R.

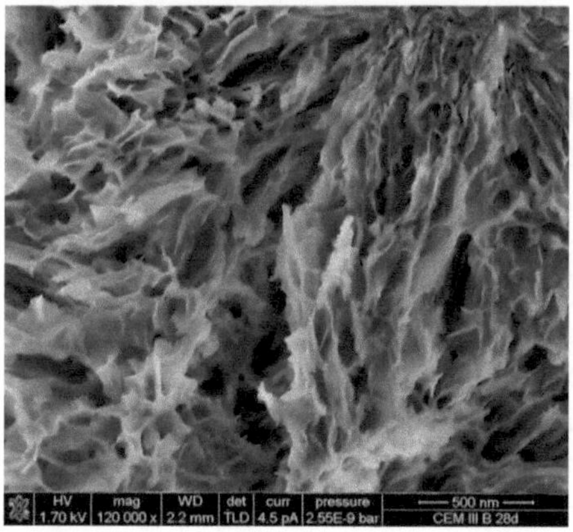

Fig. 13.28: Foil-like C-A-S-H phases in CEM III/B.

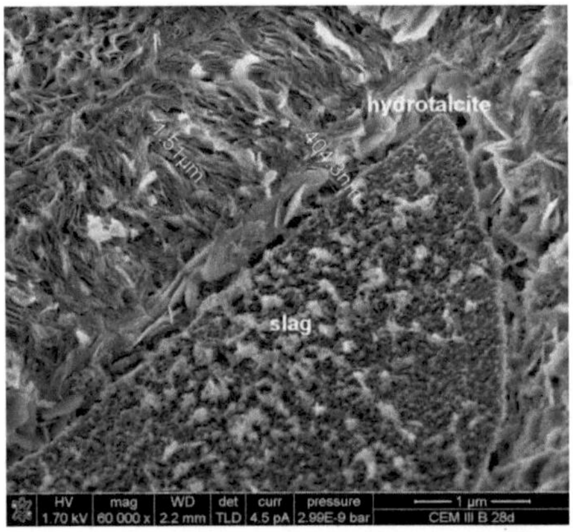

Fig. 13.29: GGBS particle surrounded by a gap partly filled with hydrate phases (CEM III/B).

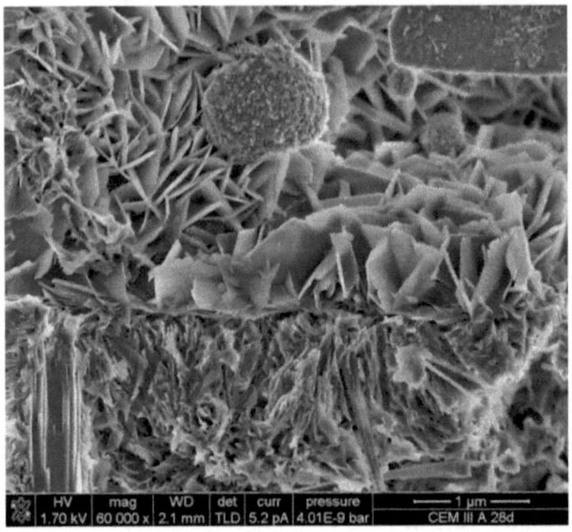

Fig. 13.30: Enlarged image of gap around GGBFS particle (top right corner) filled with platy, prismatic hydrates, and speroidal phase (CEM III/A).

Clearly, between the surface of slag particle that is covered by a few hydrates only (max. 1 µm layer) and the foil-like C-A-S-H phases, a gap is observable (Fig. 13.26, 13.29, and 13.30). Often, this gap is partly filled with platy prismatic hydrate phases (Fig. 13.29 and 13.30). These are known to be rich in aluminium and magnesium and are discussed as belonging to the hydrotalcite mineral group [17, 44–46]. According to previous studies [17, 44–46], these crystals are described as hydrotalcite-like phase or as Mg-Al layered double hydroxide (LDH) [17]. This phase is also discussed as being similar to AFm structure [47].

The SE images shown in Fig. 13.23–13.30 describe the hydrated cementitious matrix quite well. But for quantification of phase content and composition, the close-to native, fractured, and uncoated samples used for SE imaging are unsuitable. For quantification, resin embedded, polished sections coated with a few nanometres of carbon are required. Then, BSE images and or EDX phase analysis (as described in Section 13.4.1.1) can be used for quantitative analysis. But as will be shown in the following, for a targeted phase analysis of these polished sections, descriptions of nanostructure, as collected above, are definitely needed.

Similar to investigations on cement clinker (Section 13.4.1), BSE images and EDX phase maps of 28 d hydrated cements have been collected. For EDX phase mapping, a phase library with expected phases has been used and, similarly to investigations on cement clinkers, it was checked whether additional phases are present (see Sections 13.3.2.2 and 13.4.1.1 for detailed description).

EDX phase maps of 28 d hydrated reference cement (CEM I 42.5 R) are shown in Fig. 13.31 and 13.33. For comparison, BSE images of the respective phase maps are given in Fig. 13.32 and 13.34. The displayed phase maps of hydrated CEM I 42.5 R allow one to make a clear distinction between four hydrate phases: i.e. Ip C-S-H, CH, AFm (Al_2O_3–Fe_2O_3-mono phase of the Ca–Al–S–O–H system), ettringite, and hydrate matrix. Alite, belite, ferrite, and aluminate clinker phases are also identified (Fig. 13.31 and 13.33). Due to software limitations for available colors, either aluminate/ferrite (Fig. 13.31) or alite/belite (Fig. 13.32) have been merged to one phase. At increased magnification/EDX mapping resolution (Fig. 13.33), a better discrimination of small phases such as aluminate, ferrite clinker and ettringite, AFm hydrates is possible.

Color phase maps depict the spatial distribution of identified phases (Fig. 13.31 and 13.33). Clearly, detectable areas of C-S-H phases are related to surfaces of partly or fully hydrated silicate clinkers. In contrast, hydrates such as CH, ettringite, and AFm are also deposited in the pore space between clinkers. Due to the above-mentioned physical limitations of EDX analysis (excitation volume of X-rays) for nanoscaled cement hydrates, the matrix phase cannot be further differentiated and thus contains various hydrates (ettringite, AFm, fibrous C-S-H, CH) together with pore space (grey areas in Fig. 13.31 and 13.33).

As for investigations with cement clinkers (Section 13.4.1.1), EDX phase maps of hydrated cements can be used to generate EDX sum spectra for identified phases. These sum spectra have been quantified by applying standardless EDX analysis to characterize the hydrate phases of cements. Results are shown in Tab. 13.5.

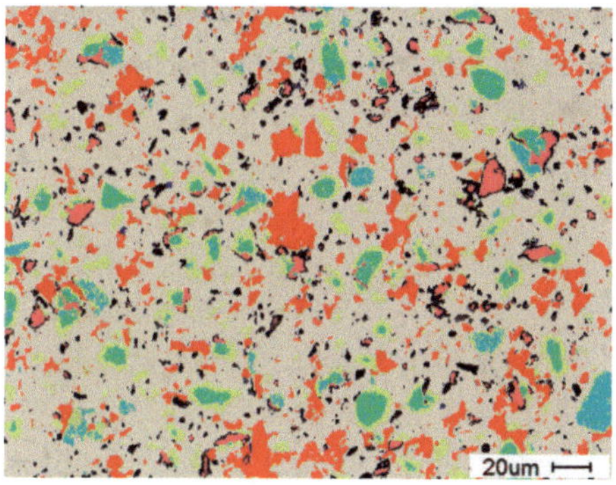

Fig. 13.31: EDX phase map of 28d hydrated CEM I; green: alite, cyan: belite, red: CH, yellow: C-S-H, grey: matrix, magenta: aluminate/ferrite, blue: ettringite, black: AFm.

Fig. 13.32: SEM-BSE image of respective area shown in Fig. 13.31.

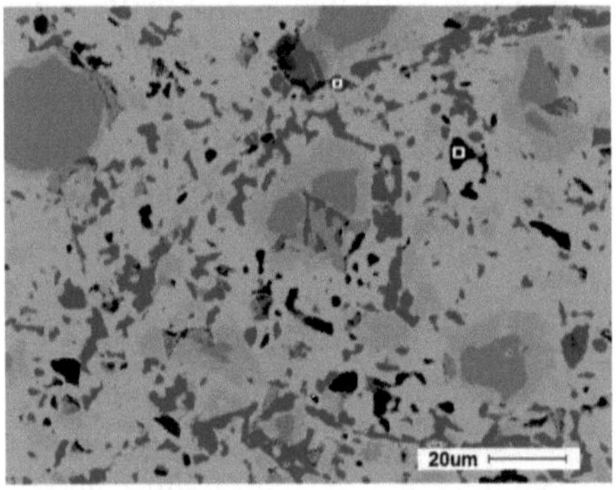

Fig. 13.33: Enlarged EDX phase map of CEM I. 42.5 R (28d); green: alite/belite, yellow: C-S-H, grey: matrix, magenta: aluminate, cyan: ferrite, blue: ettringite, black: AFm.

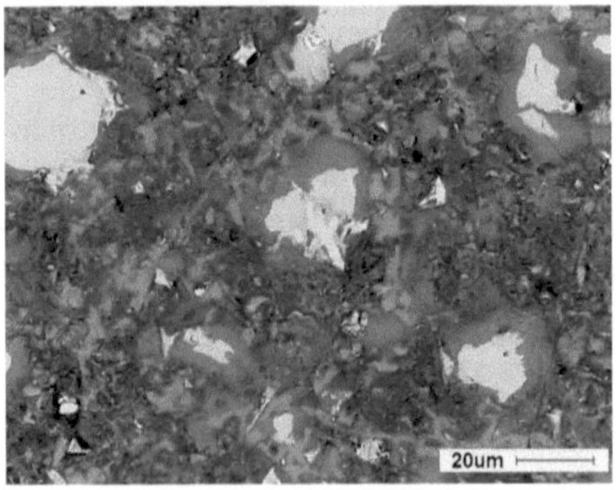

Fig. 13.34: SEM-BSE image of respective area shown in Fig. 13.33.

Tab. 13.5: Composition of selected cement phases contained in 28 d hydrated CEM I 42. 5 R, CEM III/A and CEM III/B (given in wt%). Phase composition is determined by standardless EDX analysis (details see Section 13.4.1.1).

		Na_2O	MgO	Al_2O_3	SiO_2	P_2O_5	SO_3	K_2O	CaO	TiO_2	Cr_2O_3	Fe_2O_3	ZnO	MnO	CaO/SiO_2 [SO_3/Al_2O_3]
Compositions of phases determined from quantification of EDX sum spectra identified by phase mapping (details see Section 13.4.1.1) and EDX analysis of alite taken from Taylor [17]															
Ip C-S-H	CEM I 42.5 R	0.2	1.7	2.4	28.0	0.1	5.3	1.3	58.1	0.2	0.2	1.5	0.9	–	2.1
	(std deviation)	(0.0)	(0.1)	(0.2)	(0.7)	(0.0)	(0.3)	(0.1)	(0.9)	(0.0)	(0.0)	(0.1)	(0.1)		
	CEM III A	0.5	2.5	5.0	32.5	0.2	3.0	0.4	54.4	0.3	–	1.1	–	–	1.7
Matrix	CEM I 42.5 R	1.5	1.5	3.9	21.1	0.4	7.6	1.9	59.2	0.2	0.1	1.9	1.0	–	2.8 [1.9]
	(std deviation)	(0.2)	(0.1)	(0.2)	(0.9)	(0.1)	(0.6)	(0.2)	(0.8)	(0.0)	(0.0)	(0.1)	(0.1)		
	CEM III B	1.2	3.3	6.9	32.2	0.4	3.8	1.3	48.9	0.5	0.2	1.5	–	–	1.5 [0.6]
Alite	CEM I 42.5 R	0.3	1.1	1.1	25.1	0.1	0.6	0.4	69.9	0.2	0.1	0.5	0.0	–	2.8
	(std deviation)	(0.0)	(0.0)	(0.1)	(0.3)		(0.1)	(0.0)	(0.5)	(0.0)	(0.0)	(0.4)	(0.0)		
	CEM I, Taylor [17]	0.1	1.1	1.0	25.2	0.1	0.1	0.1	71.6	0.0	–	0.7	–	0.0	2.84
AFm	CEM I 42.5 R	0.8	3.2	11.8	12.2	0.2	7.5	1.2	55.7	0.4	–	7.0	–	–	[0.6]
	(std deviation)	(0.1)	(0.5)	(1.0)	(1.0)	(0.1)	(0.4)	(0.1)	(0.8)	(0.0)		(0.2)			
	CEM III B	0.4	0.7	19.4	9.9	0.3	10.6	0.5	56.5	0.3	–	1.4	–	–	[0.6]
GGBFS hydration rim	CEM III B	0.7	8.2	8.8	32.3	0.4	3.3	0.6	43.3	1.1	0.2	1.0	–	0.2	1.3 [0.4]
Phase compositions determined from EDX spot analyses on separated phases															
Hexagonal, platy phase		–	16.3	10.1	24.5	–	3.1	0.6	37.7	3.3	–	4.5	–	–	1.5 [0.3]
(std deviation)			(5.5)	(2.9)	(4.6)		(0.8)	(0.9)	(6.0)	(1.3)		(0.4)			
Foil-like C-A-S-H		0.7	0.9	4.9	35.2	–	1.7	2.4	52.8	0.6	–	0.8	–	–	1.5 [0.3]
(std deviation)		(0.6)	(0.8)	(0.3)	(2.0)		(0.6)	(1.6)	(1.4)	(0.8)		(2.0)			

Due to the chosen EDX mapping conditions, the X-ray count number for the EDX sum spectra is high. Thus, elements present in low concentration can be clearly identified as peaks in the spectrum (Fig. 13.5). If peaks were identified, minor and trace elements such as phosphorus, titanium, potassium etc. are also included in the quantification. As described in Section 13.4.1.1, the quantification of the EDX sum spectra of very small phase areas should be avoided because then interferences with neighboring phase are included (rim effect). This effect can in the future be overcome by assigning more phases (i.e. rims of phases are to be identified as separate phase).

Standard deviations of alite and hydrate phases' composition, obtained on EDX sum spectra on five separate mappings (each mapped area: 82 000 μm^2, EDX point grid: 0.6 µm/pixel) of sample CEM I 42.5 R (28 d), are given in Tab. 13.5 in brackets. As seen for alite, the determined chemical composition (Tab. 13.5) is very similar to the mean alite composition given by Taylor [17]. The determined standard deviations for alite composition are also acceptable (for alite is always lower than 0.5 wt%). Thus, similar to findings for unhydrated clinker described in Section 13.4.1.1, it is shown that standardless EDX quantification can be used to quantify EDX sum spectra obtained from phase maps of hydrated cements. Accuracy of quantitative EDX results should always be evaluated by analysing materials of a known composition under the same conditions [14]. To evaluate the accuracy of standardless EDX analysis on the sum and spot spectra of hydrated cements, the Ca/Si of precipitated alpha C_2SH was characterized (Tab. 13.6). Obviously, standardless EDX quantification of sum spectra and spot analyses deliver a Ca/Si of 2.0, which is the expected value. In Tab. 13.5, results for Ca/Si of C-S-H phases of alite hydration are depicted. Within previous studies, the scatter of Ca/Si values for C-S-H phases in hydrated CEM I is broad (Tab. 13.6). Most of the Ca/Si values from literature plot in the range between 1.7–1.8 [42, 47, 48]. Thus, the Ca/Si deduced from EDX sum spectra of the current study lies within the expected range.

Surprisingly, the Ca/Si of Ip C-S-H phases determined on 30 single EDX spot analyses is somewhat lower, i.e. Ca/Si of 1.6 (Tab. 13.6, spot spectra). But since the determined standard deviation of Ca/Si of spot analysis is also larger than those obtained on phase maps (due to the lower number of counts in spot analyses) it is assumed that the number of spot analyses or the number of counts per spot is still too low to predict an accurate value.

Tab. 13.6: Comparison of Ca/Si (at.) for synthetic alpha C_2SH and C-S-H phases in CEM I as determined in the present study and various results taken from literature. All results are deduced from analysis in atom percent in order to compare with previous studies.

	Present study		**Allen [48]**	**Taylor [47]**	**Richardson [42]**
	Sum spectra	**Spot spectra**			
alpha C_2SH	2.0	2.0 ± 0.1	–	–	–
C-S-H	1.8 ± 0.0	1.6 ± 0.1	1.7	1.8 ± 0.2	1.2–2.3

Correlating the EDX phase distribution maps of hydrated CEM I 42.5 R with high resolution SE imaging shows that, at the resolution of EDX analysis, the Ip C-S-H rim around alite particles is identified as a homogeneous phase (Fig. 13.31 and 13.33). This is in line with findings from Richardson's TEM investigations, which depict the Ip C-S-H as more homogeneous areas than Op C-S-H, with only a few inclusions of other phases [42]. Looking at the results in Tab. 13.5, it is clearly seen that these Ip C-S-H (as shown in Fig. 13.24) contain minor elements, for example magnesium, potassium, sulphur, and aluminium. The existence of some trace elements, such as sodium, phosphorus, and titanium, is also proven. Which of these elements are adsorbed is not revealed.

Quantification of EDX sum spectra of AFm phase in Fig. 13.31 and 13.33 reveals low sulphur content. Due to the small sizes of individual AFm phase areas, the sum spectra might include characteristic X-ray intensities from neighboring phases. Therefore, in addition, the EDX spectra of areas indicated by white rectangles (Fig. 13.33) have been quantified. Results in Tab. 13.7 show that, as assumed, the analysis of AFm given in Tab. 13.5 contains elements (such as Si) from neighboring phases. Nevertheless, the trend given by analysis in Tab. 13.5, i.e. the low sulphur content of AFm phases, is confirmed. In addition the potassium detected by EDX spot analysis in the ettringite area is assumed to origin from adsorbed solutes (Tab. 13.7). Thus, as always for analytical electron microscopy, it is demonstrated that critical verification of results is needed.

Tab. 13.7: Chemical composition (wt%) deduced from EDX area analyses of blue and black phase in phase map in Fig. 13.33.

	Al_2O_3	SO_3	CaO	SiO_2	K_2O	SO_3/Al_2O_3
AFm (black)	31	3.2	65.8	–	–	0.1
Ettringite (blue)	15.9	33.2	47.3	1.5	2.2	2.1

By EDX phase mapping, besides portlandite (CH), C-S-H, and ettringite/AFm, a hydrate matrix is depicted in CEM I 42.5 R (28 d). As known from high resolution SEM-SE imaging, this matrix of consists of nanoscaled C-S-H, ettringite, AFm, portlandite, and capillary porosity (Fig. 13.23 and 13.25). Due to the intense intergrowth of the hydrates, they cannot be differentiated by EDX phase mapping even if the highest physically obtainable resolution is applied. Thus, the given composition of cement matrix (Tab. 13.5) is a combination of at least four phases; i.e. ettringite, AFm, portlandite, and C-S-H.

Comparing the microstructure of CEM I 42.5 R and CEM III/B matrixes (after 28 d of hydration), it was found that, due to the presence of GGBFS, the fibrous habit of C-S-H is changed to a foil-like habit (Fig. 13.26–13.28). This was already described in previous studies using transmission electron microscopy [45].

Comparison of compositions of Ip C-S-H of CEM I 42.5 R and CEM III/A hydration by quantification of EDX sum spectra are shown in Tab. 13.5. As expected, in the presence of GGBFS an increased Mg, Al, and Si content and a reduction in Ca concentration (Tab. 13.5) is observed. Thus, a reduction in the CaO/SiO_2 value of Ip C-S-H in CEM III/A is depicted. This is in accordance with findings from EDX point counting on hydrated slag cements [46]. EDX phase maps shown in Fig. 13.35 and 13.36 furthermore reveal that after 28 d only in CEM III/A is a significant amount of unhydrated alite found. CEM III/B at the same stage merely shows remnants of belite and very rarely alite (not shown by Fig. 13.36). In the microstructure of CEM III/B, after 28 d the identification of pure C-S-H is also difficult.

This can be taken as a clear indication that, by ongoing hydration in the presence of GGBFS, the C-S-H formed by alite hydration is transformed into foil-like hydrates of C-A-S-H type that possess a decreased Ca/Si but increased Al content.

If phase analysis is carried out for CEM III (containing GGBFS), an additional magnesium-enriched hydrate phase can be discriminated that envelops GGBFS particles (Fig. 13.35 and 13.36, blue color). If the information from phase maps is combined with high resolution SEM-SE images (Fig. 13.29 and 13.30) it becomes obvious that the blue phase (Fig. 13.35 and 13.36) consists of two different crystal habits, i.e. a platy, prismatic, and foil-like phase.

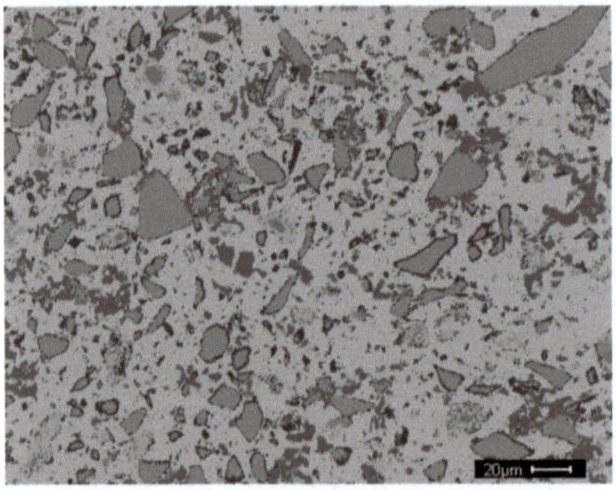

Fig. 13.35: Phase map of CEM III/A, 28 d; green: alite/belite, cyan: GGBFS, blue: GGBFS hydration rim, yellow: C-S-H, red: portlandite, magenta: AFm, grey: matrix.

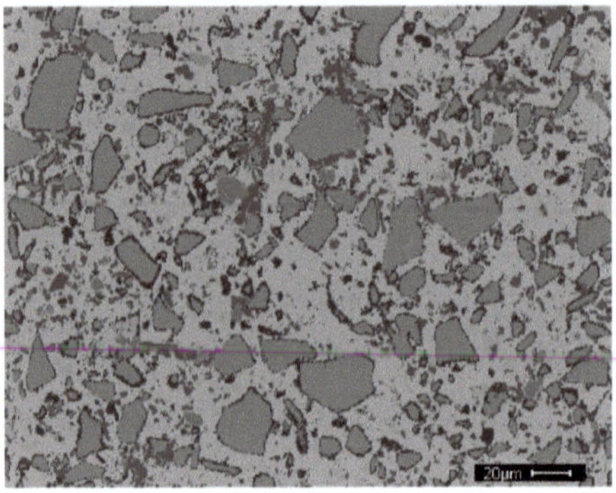

Fig. 13.36: Phase map of CEM III/B, 28 d. Color index: see Fig. 13.35, except green: only belite.

Thus, because of the small scale of the phases, in this case the quantification of EDX sum spectra delivers a mixed analysis of two phases (Tab. 13.5). Nevertheless, the results of EDX quantification clearly show that these phases contain significantly increased magnesium content. Because identification of platy prismatic hydrates is impossible on polished sections, EDX analysis of these phases can only be carried out on fractured surfaces or on separated phases. For both cases, accuracy and precision of EDX quantification is reduced [3, 49, 50]. Phase separation is achieved by:
(1) stopping hydration,
(2) gently grinding cement and dispersing it in alcohol [41],
(3) placing a droplet in suspension on a TEM sample holder (Fig. 13.37).

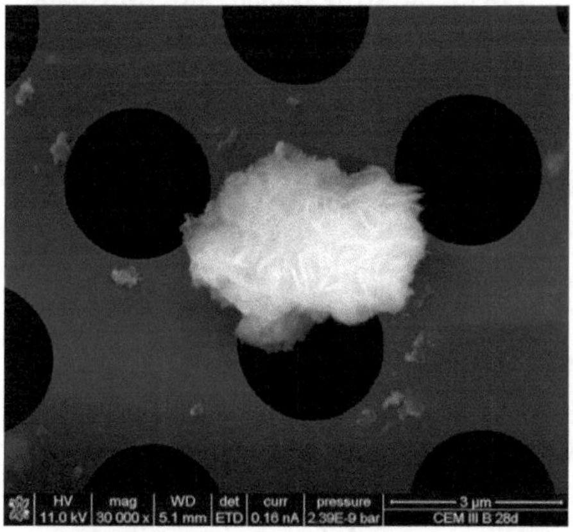

Fig. 13.37: Separated platy prismatic crystals (as depicted in Fig. 13.30) on a TEM sample holder (holey carbon foil).

The benefit of phase separation on TEM sample holders is that for very small sizes of analyzed phases, it is possible to avoid detecting X-rays that originate from neighboring phases [41]. The quantification of EDX spot analyses on separated platy, prismatic crystals and on foil-like phases is given in Tab. 13.5. These hydrates mainly consist of Ca, Si, Mg, Al, S, and Ti (in order of abundance). The relatively high standard deviation given for the determined chemical composition indicates that a separation of pure phase was probably not always achieved (i.e. intermixed nanoparticles or surface layers of evaporates from dried aqueous phases, etc.). Because these phases are also contained only in limited quantity it is very difficult and very time-consuming to detect many of them as separated single phases. Thus, TEM investigations were carried out to properly characterize these phases [51].

EDX analysis on separated foil-like C-A-S-H phases was also carried out to get an approximate chemical composition (Tab. 13.5). The determined composition is very similar to the Ip C-S-H composition in CEM III/A, except for potassium and magnesium content (Tab. 13.5). Surprisingly, the magnesium content of C-A-S-H is very low, i.e. it appears that magnesium from slag particles is precipitated only in close vicinity to slag particles (i.e. in the platy hydrotalcite like phase). If potassium detected in

C-A-S-H is only a nanoscaled adsorbed layer on the surface or if it is incorporated into the structure is unclear.

13.5 Conclusions

13.5.1 EDX phase mapping for characterization of cement clinker

The results shown above demonstrate that standards-based and standardless quantification of EDX spectra can be used to characterize the chemical composition of clinker phases. Using standardless EDX analysis, the accuracy of quantification might be significantly reduced for the elements present in an amount of less than one weight percent (trace elements). For the glass standard tested here, the standards-based EDX analysis delivered a maximum relative error of 3 %, whereas the standardless EDX analysis on sum spectra showed a maximum relative error of 4 %. This error is probably increased for elements present in low amounts (i.e. < 1 wt%, trace elements). The precision of the standardless quantification of EDX spectra is usually better than its accuracy. Thus, when working with standardless EDX quantification, the accuracy of the analysis should regularly be assessed by analysing homogeneous and defined standards of similar composition to the investigated materials [14].

EDX phase maps allow cement clinker phases to be identified by their chemical composition. The prerequisite is that the EDX mapping parameters are set up properly to ensure a proper number of X-ray counts and an adequate mapping resolution (see below). The EDX spectra of phases can be stored in a user-defined phase library and used for further analyses. Using a phase library that contains many realistic clinker phase compositions ensures that all phases that are to be discriminated chemically are found. These stored spectra are compared to those actually being measured and, if certain matching parameters (chi square fit) are fulfilled, the analyzed EDX pixel is assigned to a specific phase. Additionally, automated phase analysis and careful review of element maps (including RGB overlays) should be carried out to search for new phases that are not contained in the phase library. An application of this procedure revealed that, in addition to expected clinker phases (alite, belite, aluminate, ferrite, free lime, periclase, and arcanite), the investigated clinker contained a magnesium-rich aluminate phase.

EDX phase maps furthermore visualize the distribution of phases, i.e. as shown above, potassium sulphate forms inclusions in the aluminate and ferrite phases; belite partly forms inclusions in alite, etc.

As an indication for the presence or absence of trace elements (i.e. element present < 1 wt%) but also for getting approximate chemical composition of individual clinker phases, quantification of EDX sum spectra obtained from EDX phase maps is a helpful tool. For a precise analysis of the chemical composition of clinker phases, standards-based EDX quantification should be carried out. If the errors of standardless EDX analysis (see above) are acceptable, standardless analysis will yield faster results. In any case, the prerequisites for good precision and accuracy of results are proper sample preparation and, also, analyzed phases need to exceed a certain size, i.e. at least greater than 5–10 µm in diameter (for the given mapping parameters). For phases < 5–10 µm, the composition must be carefully validated by carrying out additional EDX spot or area analysis in the centre of the detected phases. Otherwise, the phase composition as determined from sum spectra might significantly deviate due to overlapping analysis of neighboring phases. The minimum phase area that is properly analyzed by EDX phase mapping can be lower if the total mapped area is reduced; respectively, the step size of mapping (resolution) is increased. Based on the chemical composition of cement clinker (the heaviest element of interest is usually iron), the minimum required acceleration voltage for proper EDX analysis is 12 keV. For calcium silicates, this leads to an area with approximately 1.8 µm diameter where X-rays are exited. Thus, for cement clinker, a better resolution of EDX mapping than 1.8 µm is only possible when no quantitative analysis of spectra is needed.

13.5.2 Combined EBSD-EDX analysis for characterization of cement clinker

The results shown here depict, for the first time, a strategy for how to apply combined EDX-EBSD analysis to characterize Portland cement clinker.

The first and most crucial step for EBSD analysis is the sample preparation of cement clinker. Conventional EBSD sample preparation methods used for metals and rocks fail for cement clinkers, i.e. no or very poor EBSD patterns are detected. The results of the present study clearly show that the resin embedded and mechanically pre-polished clinker surfaces need to be polished by using an argon BIB machine. Polishing should thereby remove all sample material that was distorted by mechanical sample preparation (sawing, polishing) and the topography of the sample surface should be kept as smooth as possible. Proper argon BIB milling parameters for cement clinker and the used polisher are indicated.

It is shown that for multiphase materials such as cement clinkers, a parallel acquisition of EBSD and EDX information is essential. The main reason for this approach is that EBSD analysis of cementitious silicates and aluminate is due to the low EBSD pattern contrast and crystal symmetry is quite challenging. In this way, EDX spectroscopy serves as a first step towards differentiating clinker phases by their chemical composition (including alite and belite). Before starting an EBSD mapping, a careful phase identification that selects the best fitting crystal structure model needs to be carried out. The number of crystal structures used for EBSD mapping should be limited. Thus, additional results from XRD phase analysis may be helpful to select proper crystal structures. In special cases, the XRD-Rietveld refined crystal structures can be used for EBSD analysis. Proper EBSD phase identification requires indexing manually selected patterns with various crystal structures. Only the best fitting/most probable structures should be used for EBSD mapping and automated indexing.

Recording combined EDX-EBSD mappings require optimized mapping parameters such as accelerating voltage, spot size, vacuum condition, step size, etc. For the SEM equipment used here, the EDX dwell time was the time-limiting factor for EDX-EBSD mapping. In general, EDX mapping parameters need to be set so that a clear differentiation of alite and belite and other clinker phases is possible.

EBSD mapping data of Portland cement clinker phases should be corrected for mis-indexed points. The data clean up functions applied are similar to those explained in previous studies. Additionally, clean up functions for pseudo-symmetric mis-indexing of monoclinic alite has been applied.

EBSD orientation imaging microscopy (OIM) can be used to evaluate grain boundaries of clinker phases. Results gathered for the investigated clinker show that for alite mainly high angle grain boundaries are detected. But some subgrain boundaries have also been found in the alite of technical clinker. These were not obvious for synthetic alite/C_3S. Since subgrain boundaries are indicators for crystal lattice defects, it would be interesting to correlate their occurrence with the hydraulic reactivity of clinker (work is in progress).

Analysing grain size distributions from EBSD-OIM allowed for comparing the grain sizes of alite, cubic aluminate, and ferrite. It is clearly shown that aluminate and ferrite possess a maximum grain size of 10–15 μm, whereas alite grains attain diameters up to 60 μm. Comparing with results from optical light microscopy, the measured mean alite grain size of 20 μm appears to be relatively small. Measuring alite crystallite sizes by optical light microscopy is very dependent on the operator and rather measures shape and not crystal properties. Thus, it is not surprising that by crystal orientation analysis a smaller mean grain size is detected than in optical light microscopy analysis. Measured crystal sizes for aluminate and ferrite are smaller than 15 μm, which is in accordance with expectations from optical light microscopy. For further investigations, it is planned to investigate if the crystal sizes of various clinker phases vary according to clinker burning conditions and raw meal composition. The dependence of clinker reactivity on the crystal sizes of clinker phases will also be investigated.

It is also shown that IQ quality maps, which reflect EBSD patterns' contrast, clearly indicate that, compared to alite, the crystallinity of belite is reduced. The origin of this reduced crystallinity might be the very small crystallite size of the partly amorphous structure of belite. This fact makes a grain size analysis of belite in the investigated clinker impossible. EBSD analysis on other clinkers types showed a similar low pattern contrast for belite.

13.5.3 EDX phase mapping and high resolution SE imaging for the characterization of hydrated cements

The results of the present investigation show that for the microstructural characterization of hydrated cementitious materials, various SEM imaging and analysis techniques need to be combined. The applied EDX phase analysis is shown to be a valuable tool for differentiating phases and their spatial distribution during cement hydration. In this way, hydrate phases such as C-S-H, C-A-S-H, CH, AFm, and ettringite can be differentiated and their spatial distribution revealed.

Furthermore, the chemical composition of identified phases can be revealed by quantification of EDX sum spectra obtained from mapping data. As for cement clinkers, this is a fast method for obtaining statistically relevant data on hydrate phase composition. Phase refers hereby to an area of similar chemical composition which might contain several crystallographically different components. Care should be taken when identified phases are of small size relatively to the mapping resolution. For the applied mapping conditions (i.e. acceleration voltage and density of hydrate phases), it was found that EDX sum spectra of areas with diameter smaller than 5–10 μm contained too many X-ray counts from neighboring phases. Thus, for an accurate analysis of phases smaller than 5–10 μm diameter, it is better to use selected area or spot analyses positioned in the middle of such phases. The approximate physical lower boundary of lateral and depth resolution of EDX spot analysis on hydrated cement based materials is approximately 2 μm (determined for 12 kV acceleration voltage and C-S-H phase composition). This value should be taken as the true resolution of EDX mapping (although the step size of mapping can be set a little lower, i.e. 0.6 μm in the present study). If the diameter of investigated homogeneous phases is smaller than 2 μm, EDX spot analysis on bulk polished sections will always lead to scatter induced by analysing more than one phase. Thus, it is never possible to get proper EDX quantification of hydrate phases (C-S-H etc.) that are intensely intergrown at the nanoscale in the cement matrix.

The present study showed that the analysis of small phases can alternatively be carried out after the mechanical separation of phases on a TEM sample holder. As shown for hydrotalcite-like phase and C-A-S-H phases, the standard deviation of these analyses is still high. This is mostly caused by the fact that EDX analyses of these small separated phases are based on a lower peak to background ratio of characteristic X-rays emitted and that only 5–10 phases have been found to be suitable for analysis. But EDX hardware and software (background and other corrections) for analysing nanoparticles are currently being developed and thus in the future may lead to more reliable results.

A good validation for the results of the applied standardless quantification of EDX sum spectra is that, for example, the determined alite composition matches very well with the alite composition determined with standards-based EDX analysis by Taylor [17]. The determined standard deviations of elemental concentrations are also within an acceptable range for most elements. For the Ip C-S-H that surrounds alite in CEM I 42.5 R, the determined Ca/Si is approximately 1.8 at.%, (see Tab. 13.6). This correlates very well with the mean value determined in previous studies [42, 47, 48]. Thus, for Ip C-S-H it is also shown that the quantification of phase composition by using EDX sum spectra obtained from mapping data delivers reliable results. The standard deviation of Ca/Si of C-S-H phases determined by phase analysis is also significantly lower than the standard deviation of EDX spot analyses (Tab. 13.6). Thus, it is shown that the applied EDX phase mapping not only identifies phases and reveals phase distribution images but also that te recalculation of phase composition from sum spectra reliably delivers the phase composition including minor phases (i.e. Ti, Zn, etc. at concentrations of minimum 0.1 at.%). In the future, automated multi-field mapping on large areas and following phase analysis will enable the quantification of phase concentration and composition contained in hydrated cements.

By comparing high resolution SEM-SE images with images of EDX phase distribution maps, it is revealed that after 28 d of hydration GGBFS particles (in hydrated CEM III) are surrounded by an approx. 500 nm thick layer of platy prismatic crystals followed by 1–2 µm thick layer of foil-like hydrates of C-A-S-H type. EDX analysis on separated single phases of platy prismatic crystal reveals that they contain a high Mg and Al content. Previous studies describe them as hydrotalcite-like [44, 52] or as Mg-Al layered double hydroxides (LDH) phase [53]. Since, as observed in the investigated CEM III (28 d), the quantity of these phases is very limited, it is questionable whether they have been properly identified by X-ray diffraction analysis. Recently by means of TEM-EDX and diffraction analysis on separated phases, it was proven that these platy minerals are hydrotalcite [51]. The hydration rim around GGBFS particles consists of two types of hydrates. The outer part of the slag hydration rim consists of the described C-A-S-H phases. These are also found together with other hydrates (nanoscaled ettringite, calcite, portlandite, etc.) in the CEM III matrix. They fully replace the C-S-H phases if the clinker content is as low as for the investigated CEM III/B.

References

[1] Scrivener KL, Snellings R, Lothenbach B, editors. A Practical Guide to Microstructural Analysis of Cementitious Materials. Boca Raton, USA: CRC Press, Taylor & Francis Group; 2016.

[2] Scrivener KL, Patel HH, Pratt PL, Parrott LJ. Analysis of phases in cement paste using backscattered electron images, methanol adsorption and thermogravimetric analysis. In: Struble LJ, Brown PW, editors. Materials Research Society Symposium. Cambridge University Press; 1987. 67–76.

[3] Goldstein J, Newbury DE, Echlin P, et al. Scanning electron microscopy and X-ray microanalysis, 3rd edition. New York, USA: Springer Science & Business Media; 2007.

[4] Stutzman PE, Bullard JW, Feng P. Quantitative Imaging of Clinker and Cement Microstructure. Technical Note 1877. Washington, DC: National Institute of Standards and Technology; 2015.

[5] Kocaba V, Gallucci E, Scrivener KL. Methods for determination of degree of reaction of slag in blended cement pastes. Cem Concr Res. 2012; 42: 511–525.
[6] Yio MHN, Phelan JC, Wong HS, Buenfeld NR. Determining the slag fraction, water/binder ratio and degree of hydration in hardened cement pastes. Cem Concr Res. 2014; 56: 171–181.
[7] Feng X, Garboczi EJ, Bentz DP, Stutzman PE, Mason TO. Estimation of the degree of hydration of blended cement pastes by a scanning electron microscope point-counting procedure. Cem Concr Res. 2004; 34: 1787–1793.
[8] Durdziński PT, Dunant CF, Haha MB, Scrivener KL. A new quantification method based on SEM-EDS to assess fly ash composition and study the reaction of its individual components in hydrating cement paste. Cem Concr Res. 2015; 73: 111–122.
[9] Lemmens HJ, Butcher AR, Botha PWSK. FIB/SEM and SEM/EDX: A new dawn for the sem in the Core Lab? Petrophysics. 2011; 52: 452–456.
[10] Pirrie D, Butcher AR, Power MR, Gottlieb P, Miller GL. Rapid quantitative mineral and phase analysis using automated scanning electron microscopy (QemSCAN); potential applications in forensic geoscience. Geological Society Special Publication. 2004: 123–136.
[11] Friel JJ. X-ray and Image Analysis in Electron Microscopy. 2nd edition. Princeton, USA: PGT, Priceton Gamma Tech; 2003.
[12] Newbury DE, Ritchie NWM. Performing elemental microanalysis with high accuracy and high precision by scanning electron microscopy/silicon drift detector energy-dispersive X-ray spectrometry (SEM/SDD-EDS). Journal of Materials Science. 2014; 50: 493–518.
[13] Newbury DE, Ritchie NWM. Is scanning electron microscopy/energy dispersive X-ray spectrometry (SEM/EDS) quantitative? Scanning. 2013; 35: 141–168.
[14] ASTM. E1508-2012a Standard Guide for Quantitative Analysis by Energy-Dispersive Spectroscopy. West Conshohocken, USA: American Society for Testing and Materials; 2012.
[15] Kikuchi S. Diffraction of cathode rays by mica. Proceedings of the Imperial Academy. 1928; 4: 271–274.
[16] Schwartz AJ, Kumar M, Adams BL, Field DP. Electron backscatter diffraction in materials science. New York, USA: Kluwer Academic/Plenum Publisher; 2000.
[17] Taylor HFW. Cement Chemistry, 2nd edition. London: Telford Publishing; 1997.
[18] Stutzman P, Heckert A, Tebbe A, Leigh S. Uncertainty in Bogue-calculated phase composition of hydraulic cements. Cem Concr Res. 2014; 61–62: 40–48.
[19] Stutzman P, Lespinesse G. Cementitious materials crystal structure database for X-ray powder diffraction analyses. International Cement Microscopy Association. 30th International Conference on Cement Microscopy 2008. 2008: 350–358.
[20] Campbell DH. Microscopical Examination and Interpretation of Portland Cement and Clinker. Portland Cement Association; 1999.
[21] Stutzman PE. Cement clinker characterization by scanning electron microscopy. Cement, Concrete and Aggregates. 1991; 13: 109–114.
[22] Wilson W, Krakowiak KJ, Ulm F-J. Simultaneous assessment of phase chemistry, phase abundance and bulk chemistry with statistical electron probe micro-analyses: Application to cement clinkers. Cem Concr Res. 2014; 55: 35–48.
[23] Naber C, Goetz-Neunhoeffer F, Göbbels M, Rößler C, Neubauer J. Synthesis of monocrystalline Ca3SiO5 using the optical floating zone method. Cem Concr Res. 2016; 85: 156–162.
[24] Link T, Bellmann F, Ludwig HM, Ben Haha M. Reactivity and phase composition of Ca2SiO4 binders made by annealing of alpha-dicalcium silicate hydrate. Cem Concr Res. 2015; 67: 131–137.
[25] Field DP. Recent advances in the application of orientation imaging. Ultramicroscopy. 1997; 67: 1–9.

[26] Humphreys FJ. Grain and subgrain characterisation by electron backscatter diffraction. Journal of Materials Science. 2001; 36: 3833–3854.
[27] Nowell MM, Wright SI. Orientation effects on indexing of electron backscatter diffraction patterns. Ultramicroscopy. 2005; 103: 41–58.
[28] Wright SI. Random thoughts on non-random misorientation distributions. Mater Sci Technol. 2006; 22: 1287–1296.
[29] Randle V, Caul M. Representation of electron backscatter diffraction data. Mater Sci Technol 1996,12,844-50.
[30] De Noirfontaine M-N, Courtial M, Dunstetter F, Gasecki G, Signes-Frehel M. Tricalcium silicate Ca3SiO5 superstructure analysis: a route towards the structure of the M1 polymorph. Zeitschrift für Kristallographie Crystalline Materials. 2012; 227: 102–112.
[31] Golovastikov NI. Crystal Structure of Tricalcium Silicate, 3CaOSiO2 = C3S. Sov Phys Crystallogr. 1975; 20: 441–445.
[32] De La Torre ÁG, Bruque S, Campo J, Aranda MAG. The superstructure of C3S from synchrotron and neutron powder diffraction and its role in quantitative phase analyses. Cem Concr Res. 2002; 32: 1347–1356.
[33] Wright SI. Fundamentals of automated EBSD. In: Schwartz A, Kumar JM, Adams BL, editors. Electron Backscatter Diffraction in Materials Science. Springer; 2000. 51–64.
[34] Rajan K. Representations of Texture in Orientation Space. In: Schwartz AJ, Kumar M, Adams BL, editors. Electron Backscatter Diffraction in Materials Science. Boston, MA: Springer US; 2000. 31–38.
[35] Nolze G, Winkelmann A, Boyle AP. Pattern matching approach to pseudosymmetry problems in electron backscatter diffraction. Ultramicroscopy. 2016; 160: 146–154.
[36] Nowell MM, Wright SI. Phase differentiation via combined EBSD and XEDS. Journal of Microscopy. 2004; 213: 296–305.
[37] Wright SI, Nowell MM. EBSD image quality mapping. Microsc Microanal. 2006; 12: 72–84.
[38] Klug HP, Alexander LE. X-ray Diffraction Procedures for Polycrystalline and Amorphous Materials. Wiley; 1974.
[39] Wright SI, Nowell MM, Field DP. A review of strain analysis using electron backscatter diffraction. Microsc Microanal. 2011; 17: 316–329.
[40] Passchier CW, Trouw RAJ. Microtectonics. Berlin-Heidelberg, Germany: Springer; 2005.
[41] Rößler C, Stark J, Steiniger F, Tichelaar W. Limited-Dose Electron Microscopy Reveals the Crystallinity of Fibrous C–S–H Phases. J Am Ceram Soc. 2006: 89; 627–632.
[42] Richardson IG. The nature of C-S-H in hardened cements. Cem Concr Res. 1999; 29: 1131–1147.
[43] Grudemo Å. The microstructures of cement gel phases. Stockholm, Sweden: Elanders Boktr. Aktiebolag; 1965.
[44] Wang SD, Scrivener KL. 29Si and 27Al NMR study of alkali-activated slag. Cem Concr Res. 2003; 33: 769–774.
[45] Richardson IG, Groves GW. Microstructure and microanalysis of hardened cement pastes involving ground granulated blast-furnace slag. Journal of Materials Science. 1992; 27: 6204–6212.
[46] Whittaker M, Zajac M, Ben Haha M, Bullerjahn F, Black L. The role of the alumina content of slag, plus the presence of additional sulfate on the hydration and microstructure of Portland cement-slag blends. Cem Concr Res. 2014; 66: 91–101.
[47] Taylor R, Richardson IG, Brydson RMD. Composition and microstructure of 20-year-old ordinary Portland cement-ground granulated blast-furnace slag blends containing 0 to 100 % slag. Cem Concr Res. 2010; 40: 971–983.
[48] Allen AJ, Thomas JJ, Jennings HM. Composition and density of nanoscale calcium-silicate-hydrate in cement. Nature Materials. 2007; 6: 311–316.

[49] Newbury DE, Ritchie NWM. Rigorous quantitative elemental microanalysis by scanning electron microscopy/energy dispersive x-ray spectrometry (SEM/EDS) with spectrum processing by NIST DTSA-II; 2014. 92360H-H-17.
[50] Newbury DE, Ritchie NWM. Is scanning electron microscopy/energy dispersive X-ray spectrometry (SEM/EDS) quantitative? Effects of specimen shape. Proceedings of SPIE – The International Society for Optical Engineering; 2011.
[51] Rößler C, Steiniger F, Ludwig HM. Characterization of C–S–H and C–A–S–H phases by electron microscopy imaging, diffraction, and energy dispersive X-ray spectroscopy. J Am Ceram Soc. 2017; 100: 1733–1742.
[52] Richardson IG, Groves GW. Microstructure and microanalysis of hardened ordinary Portland cement pastes. Journal of Materials Science. 1993; 28: 265–277.
[53] Richardson IG, Skibsted J, Black L, Kirkpatrick RJ. Character of cement hydrate phases by TEM, NMR and Raman spectroscopy. Advances in Cement Research. 2010; 22: 233–248.

Torsten Westphal* and Thomas A. Bier

14 Correlating XRD data with technological properties

Abstract: Several methods are described for linking XRD data with technological properties. Different methods are outlined for preparing raw XRD data for correlation analyses. Specific problems of diffraction measurements with respect to technological properties of cementitious materials are discussed. Concepts for correlation analyses are presented. Examples are reported where links between XRD data and technological properties were determined experimentally.

Keywords: cementitious materials, correlation analysis, powder diffraction, technological properties, Rietveld analysis, multivariate calibration, principal component regression, partial least squares regression, strength, heat flow, rheology, dimensional change

14.1 Introduction

XRD has become a fast, standard method for characterizing cements. Determining the technological properties of cements can be very time consuming, for instance in the case of the 28 days strength. Correlation analyses help to develop models that can predict such technological properties. There is clearly an economic incentive to find correlations between XRD data and the technological properties of cement-based materials.

Modern cement-based building materials are complexly composed. They can include a broad variety of binders and aggregates. Often, the building materials are tailor-made for specific purposes. Tailoring requires a profound understanding of how to influence the desired properties. Such an understanding is difficult to acquire but is important, and not just for tailoring specific solutions. Phase developments are fairly well understood. Correlating phase characteristics with technological properties helps improving our understanding of these materials.

Chemical reactions with corresponding changes to mineral phases are the driving force behind the development of technological properties. Such properties might be targeted properties like strength or just parameters to be considered like heat generation. Phase composition changes are complex processes and interrelated with material structures and changes thereof. Phase composition changes are therefore not easy

***Corresponding author: Torsten Westphal,** PST mbH, Bernburg, Germany,
torsten.westphal@pstbernburg.de
Thomas A. Bier, TU Bergakademie Freiberg, IKGB, Freiberg, Germany

DOI 10.1515/9783110473728-015

objects to study. Nevertheless, in order to purposefully modify technical properties, the relations between phase characteristics and technological properties need to be studied. Acquiring knowledge about phase characteristics and their consequences for technological properties helps not just to improve the properties of cement-based materials. In the long run, it also helps to reduce the costs of the production and application of such materials.

XRD has developed into a versatile, reliable, and easy-to-use method. It has become a preferred method for examining phase characteristics in cement-based materials. It has also developed into a powerful practical tool for cement production [1]. In industrial environments, powder XRD analyses tend to replace classical chemical analyses [2, 3].

For correlation attempts, quantifiable characteristics are required. Powder XRD can in principle be used to quantify many characteristics such as crystallite sizes, crystallographic lattice parameters, micro strain, phase amounts, or textures (preferential orientations). Practical experience shows that the demand of phase amount determination outweighs by far the demand for determining other phase characteristics.

14.2 Obtaining values from XRD patterns for correlation analysis

14.2.1 Quantitative analysis of XRD patterns

Basic considerations for the quantitative phase analysis of powder mixtures are provided by Klug and Alexander (1974) [4]. The intensity of a certain phase's diffraction signal is proportional to the amount of this phase. This is expressed by equation (14.1):

$$I_p = I_{p0} \cdot \frac{\mu_p}{\mu} \cdot x_p \tag{14.1}$$

x_p weight fraction of phase p
I_p diffraction intensity of phase p in the mixture
I_{p0} diffraction intensity of the pure phase p
μ_p mass absorption coefficient of phase p
μ mass absorption coefficient of the mixture

One then has the option either of using diffraction intensities directly or of translating diffraction intensities into phase amounts. Such translation can be done either by calibration methods or by pattern-fitting methods.

14.2.1.1 Classical calibration

Calibration is a fundamental method for quantifying properties. It is therefore mentioned in many textbooks about metrology [5, 6]. Classical calibration of diffraction data relates the height or the area of an undisturbed diffraction peak of a particular phase with the amount of this phase. The advantage of such calibration methods lies in its simplicity, which makes it easy to understand and to use. For Portland cement, the diffraction peaks of the main phases overlap heavily. This makes classical calibrations' reliance on undisturbed peaks a severe limitation in the field of cement-based materials.

14.2.1.2 Pattern fitting

Pattern fitting methods attempt to describe all diffraction peaks of a phase by mathematical functions. A measured diffraction pattern can be fitted by such mathematical descriptions and phase amounts can be obtained. In the context of quantitative phase analysis, the term "Rietveld Method" has become a general denomination for this approach. Rietveld published an algorithm to calculate a phase structure from its powders neutron diffraction pattern [7, 8]. This algorithm was adapted for analysing X-ray diffraction patterns. It can be used to determine phase amounts if the structures of all phases in a mixture are known. The pattern fitting approach is very versatile. It works standardlessly and without calibration for samples composed of crystalline phases with known crystallographic structures [9]. Detailed explanations of the Rietveld method can be found in textbooks [10, 11]. The limitations to the method are the need for a model to be refined and the need for specific software. The Rietveld method has become a standard method for the quantitative phase analysis of cement-based materials. This is also a topic discussed in the first two chapters of this book [12, 13]. In fact, cement scientists were among the driving forces behind the success story of the Rietveld method.

Besides the Rietveld method, the Debye scattering equation allows to fit diffractograms [14]. Although it is promising, in particular for disordered structures, this approach is actually not in use for cementitious materials.

14.2.1.3 Whole pattern calibration

Another, at present rarely used, approach is whole pattern calibration. This approach applies multivariate calibrations to diffraction patterns. Multivariate calibrations are known from spectral analyses [15, 16]. Spectra and diffractograms have similar data structures. Methods to quantitatively evaluate spectra are therefore applicable for diffraction data too. Whole pattern calibration does not require undisturbed peaks and can therefore be used when classical calibration fails. However, it relies on calibration and therefore requires reference samples. Multivariate calibrations are rarely discussed with respect to diffraction analysis. A brief description of the principle is given elsewhere [17]. More detailed information can be obtained for instance from Martins and Naes (1991) [18].

Whole pattern calibration can be considered a brute force approach. The complete raw diffraction data are used. All systematic variations between the diffraction patterns of the calibration samples are examined by statistical methods. These systematic variations are then related to phase amounts by regressions. An early standardless method for quantitative phase analysis that uses multivariate methods was published in 1984 [19]. This method did combine chemical and diffraction data. Soon after, methods relying only on diffraction data were introduced. The use of partial least squares regression (PLSR) has been demonstrated [20], and the use of principal component regression (PCR) for quantitative phase analysis has also been outlined [21]. With respect to cementitious materials, PLSR has been used to analyze blended Portland cement [22]. Similar analyses have been done with PCR [23].

Both PLSR and PCR are factor analysis methods. Factor analysis includes methods to evaluate datasets for causes of data variations. Factors in the statistical sense are latent variables underlying observed, non-random data variations. These statistical factors can represent physical sample characteristics such as phase amounts. PLSR and PCR are different ways of using such factors for calibration purposes.

PLSR is a stepwise fitting method that requires a pre-determined number of factors. In a first step, a least squares fit of the predictor variables (which are actually the intensities at all angular positions of the diffractograms) with the response variable (that is actually phase amounts) is computed. In a next step, the residuals of the fitted predictor and response variables are fitted again in a least squares process. This fitting of residuals is repeated until the number of fitting steps reaches the number of factors (Fig. 14.1). This procedure is intended to give a best possible fit between predictor and response variables.

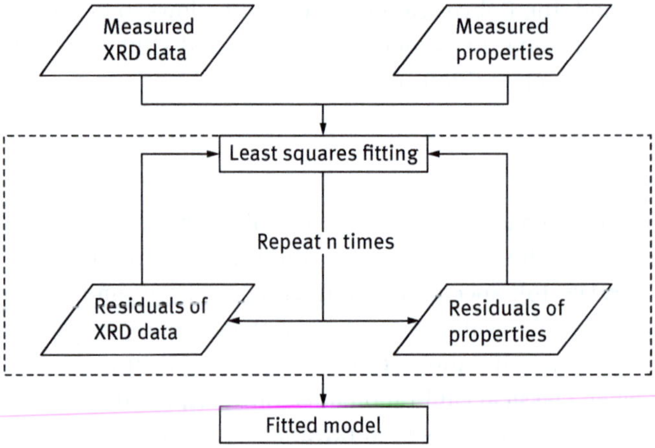

Fig. 14.1: Outline of the partial least squares regression process.

PCR is a fitting method that includes the determination of factors. PCR starts with the calculation of the principal components of a set of the predictor variables. Then the principal components are examined and all principal components are sorted out that represent random data variations. The principal components are considered as factors that represent non-random data variations. Multivariate regression is then performed with these factors and the response variable (Fig. 14.2). This procedure gives the user a greater influence on the fitting process.

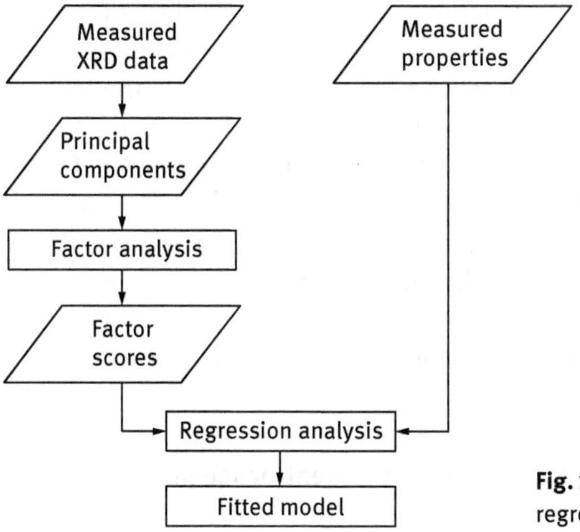

Fig. 14.2: Outline of the principal component regression process.

Apparently, the success of the Rietveld method since the 1990s has prevented the use of multivariate calibration methods for powder XRD. However, pattern fitting methods have become increasingly complicated to cope with the increasingly complex materials to be analyzed. Multivariate methods are intended to handle huge amounts of complex data. It seems therefore reasonable to consider whole pattern calibration as an alternative to the Rietveld method.

14.2.2 Numeric but non-quantitative phase analysis of XRD patterns

Since the weight fraction of a phase is proportional to its diffraction intensity, peak heights can be taken as numeric representations for phase amounts without any calibration. Peak heights are susceptible for biases and errors such as peak broadening. Peak areas therefore give more reliable results than simple peak heights. Consequently, intensities from a range around a peak can be summed up instead of using a simple peak height. This represents a peak area without peak fitting. However, special attention is required because variances in peak heights or areas can also be caused by a preferential orientation of crystals in the sample. Depending on the actual experiment, the preferential orientation of crystals can be valuable information about the sample structure or it can be just a bias. A preferential orientation is not a serious problem if it is equal for all samples of a series. This is often the case if preferential orientation results from the sample preparation and the sample preparation procedure is kept constant for all samples. Anyway, an experimenter should examine diffractograms for indications on preferential orientation.

For larger numbers of patterns, a statistical method, exploratory factor analysis, can be applied [24]. As mentioned earlier, the Factors in the statistical sense are latent variables considered to cause systematic variations in a data set. In sets of diffraction patterns, varying peak intensities can be observed. Varying phase amounts, for instance, can cause systematic variations in a set of diffraction patterns. In the sense of statistical factor analysis, the phase amounts are the latent variables causing systematic variations of the observed peak intensities. Therefore, phase amounts are factors in this statistical sense. A principal component analysis can be performed on a set of diffractograms. Thereby, different intensity variation patterns are separated. The patterns of systematic variations are the so-called factors. Factor scores represent the magnitudes of the underlying sample characteristics such as phase amounts. Factor scores can therefore be used instead of phase amounts or other phase characteristics.

14.2.3 Challenges of quantitative phase analyses of cement-based materials

A detailed explanation of powder XRD applied to cement with emphasis on the Rietveld method can be found elsewhere [12, 13, 25, 26].

A specific challenge is the large number of phase potentially present in cement-based materials. With alite, belite, ferrite, aluminate, free lime, portlandite, calcite, periclase, anhydrite, bassanite, gypsum, and quartz, there are already 12 phases to be considered in the case of dry ordinary Portland cement. This number increases for blended cements and can become larger than 30 phases when for instance fly ashes are constituents of such blends. Great care and experience is required when attempting to quantify phase amounts in such complex blends. A short estimation will point to the problem: Let's assume a number of 20 phases to be considered and a quantification uncertainty of ±1 weight percent for each phase. This would sum up in a worst case scenario to an overall quantification error of ±20 weight percent and thereby make quantification futile. Instead, in complex mixtures it may be advantageous not to quantify but to rely on diffraction intensities only.

Another challenge is the analysis of hydrate phases. Many hydrate phases have a known crystallographic structure. However, actual ambient conditions can cause structural variations. Structural variations can be considered with appropriate crystallographic and mineralogical knowledge. But sample preparation can cause structural changes for hydrate phases. Some hydrate phases have no known crystalline structure, such as C-S-H gel. Others are often found poorly crystallized like aluminium hydroxides. Quantifying phases with no or unknown structure is difficult in particular if more than one such phase is present in a sample.

Conventional and whole pattern calibration can be used to quantify phases with no or unknown structure if an appropriate calibration can be set up. The critical point for calibration is the reference samples. In the case of analysing dry mixtures such as blended cement, reference samples can be made from the blend's pure constituents. In the case of reaction products such as C-S-H gel, it would require synthesizing appropriate phases as they occur in the samples. This is usually impractical and often impossible.

There are various ways to handle phases with no or unknown structure. An internal standard can be used. This means adding a predefined amount of a phase with a known structure to a sample (spiking). By using an internal standard, a measure is provided to normalize to a phase with a known amount. Adding up all normalized phase amounts, a difference of up to 100 percent can occur. This difference represents all phases whose structures are not part of the refinement model, hence the phases with no or unknown structure.

Another option is using an external standard. Such an external standard is used to determine the instrumental influence on the X-ray intensity. Consequently, an absolute intensity scale is provided and the phase proportions obtained from pattern fitting can be corrected correspondingly. This so-called K-factor approach was introduced in 1988 [27]. It has been adapted to study early cement hydration [28].

A further option is describing the diffraction signal of a phase with no or unknown structure by a formal geometrical expression. Such a formal expression can be integrated into the Rietveld refinement. Given that the chemical composition of such a phase with no or unknown structure is known, a standardless quantitative phase analysis is possible [29]. Otherwise, the formal expression can be calibrated for quantitative phase analysis [30].

14.3 Technological properties

14.3.1 The assessment of technological properties

Technical properties are very divers. They include characteristics such as color, dimensional change, flow, freeze-thaw resistance, funnel time, heat generation, porosity, setting times, strength, sulphate resistance, and viscosity. Some properties such as strength are key properties, used to categorize building materials and to regulate their usage. Knowledge of other properties is required to plan a construction process accordingly. Heat generation for instance is important for constructing massive buildings like dams.

The assessment of the generally important properties is defined by international and national institutions for standardization. Typically, the amounts of materials, equipments to be used, test procedures, and timings are well-defined. The assessment of other properties can be less well defined and can therefore require a mutual agreement between the testing institution and its client about the testing conditions.

There are two basic types of characteristics, stationary and non-stationary. A stationary characteristic would not change over time. The density of water, for instance, remains 1.00 g/cm^3 at 4 °C no matter when it is measured. The technical properties of cement based material are basically non-stationary. Dry cement ages and wet cement hydrates. Both processes can be considered an asymptotic approximation to a final state. Consequently, any measurement requires a defined timing. The characteristics can only be considered quasi-stationary at very old ages.

Properties can be characterized by specific values such as setting time or 28-days strength. Properties can also be characterized by a series of values (curves) such as heat flow or rheological flow curves.

14.3.2 The challenge of relating technological properties to XRD data

The properties of cement-based materials depend on the properties of their constituents. The materials structure determines how the individual constituents' properties collectively form the properties of the bulk material. This means that the material structure translates the properties of individual phases into the technical properties of the building material. Structural features such as textures or particle sizes have effects on diffractograms, whereas dimensional change can lead to a peak shift in diffraction measurements. Fig. 14.3 compares three diffractograms of quartz sand and quartz powder, respectively. Obviously, peak heights are influenced by particle sizes. Furthermore, peak shifts due to changed sample heights can be seen. Variations of particle sizes and sample heights (which corresponds to a change of sample volume) are structural features clearly influencing diffractograms. Thus, diffraction data can contain both information about phases in a sample and information about the sample structure. Therefore, relating XRD data and technological properties seems reasonable.

Any attempt to relate two different properties requires corresponding data. This means that both properties should be determined under comparable conditions. Unfortunately, structural effects bias the determination of phase characteristics. Reliable traditional evaluation requires optimized sample conditions that differ significantly from real mortars or concretes. For instance, a particle size between 1 µm and 10 µm is suggested as optimum for instance by Klug & Alexander (1974) [4], but the aggregates have particle sizes up to several centimetres. Optimum sample conditions for powder XRD require grinding for most cement-based materials. This means that optimized sample conditions for powder XRD require the destruction of the original sample structure. Consequently, the theoretical relation between XRD data and technical properties can be lost by sample preparation for XRD.

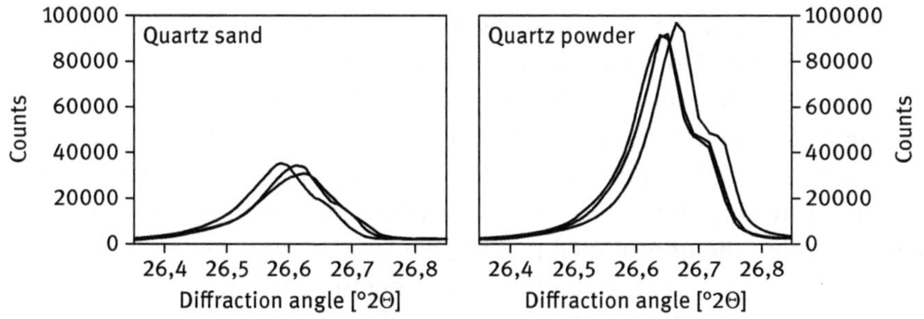

Fig. 14.3: Illustration of the influence of sample conditions on measurement results: The main quartz peak measured at different quartz sand and quartz powder samples under the same instrumental conditions.

When monitoring dynamically changing samples, the problem of measurement and sample conditions that are inappropriate for relating XRD and technological properties becomes even more important. Consecutive in situ XRD measurements became popular for monitoring the progress of a hydration. A mortar would have to be ground in order to obtain optimum XRD results. However, particle sizes have a profound influence on the hydration kinetics of the raw material phases such as C_3S [31]. This suggests that optimizing particle sizes for XRD can be a severe obstacle for attempts to relate XRD data with technological properties of real cement based products. Furthermore, the mixing regime has a significant impact on hydration kinetics and structure development [32]. The technical application of building materials requires mixing kilograms to tons. Technological properties are usually determined on samples in the size of kilograms. For in situ XRD, a few grams are sufficient. It is difficult to correctly downscale the mixing conditions of technical samples in a XRD lab. This means, in addition to fineness the mixing conditions are also crucial, and optimizing both for XRD can cut the link to technological properties.

Thus, if finding links to technical properties is the aim, an experimenter should use some of the material prepared for assessing technical properties. This is feasible for mortars with a maximum particle size still fitting in the sample holder for XRD. Coarser aggregates can be sorted out manually. But for such samples, pattern fitting methods cannot be used for quantitative phase analysis.

A specific challenge is linking technical properties to phase amounts. The increasing amounts of hydration products are driving forces for the developments of technological properties. In many cases, phase amounts alone are not sufficient to establish relations with technological properties. This is illustrated by the following thought experiment:

Ettringite crystals grow according to crystallographic principles in a specific, needle-like shape. It has preferential growth directions and it is therefore an anisotropic growth. To keep it simple, the ettringite is assumed to grow only in the most preferential direction and to grow perpendicularly on the host particle's surface.

An ettringite crystal can grow either unrestricted or confined by surrounding particles. Fig. 14.4 compares both scenarios. The ettringite amount doubles. Therefore, the ettringite crystals double their size. If the growth is unrestricted, the crystal growth simply consumes free space and eventually forms an interlocked structure (Fig. 14.4 (a)). During the unrestricted growth, porosity decreases as well as permeability. The interlocked structure can increase strength. If the growth is restricted, crystal growth pushes apart the confining particles. In Fig. 14.4 (b), doubling the crystal length also doubles the distance between the confining particles and consequently the length of the pores. In this scenario, porosity and permeability would increase as the crystals grow and with a less dense packing strength would eventually decrease. In summary, ettringite growth can trigger opposite effects on technological properties depending on whether it grows freely or is confined. Confined and unconfined growth can occur in the same material at the same time at different places.

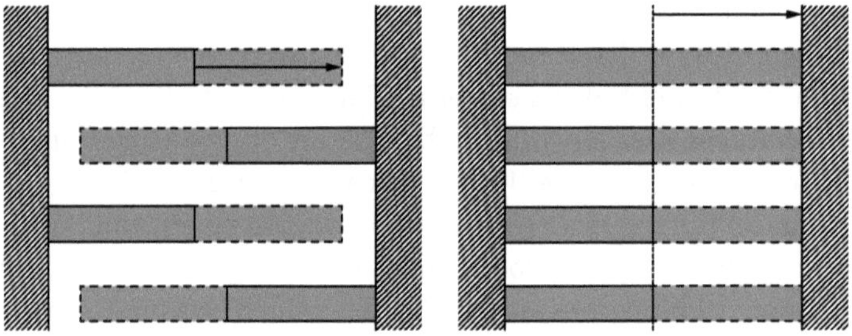

Fig. 14.4: Illustration for unconfined (a) and confined (b) growth. In both cases crystals double their size. In the unconfined case the gap between the walls is closing while in the confined case the walls are pushed apart.

The amount of ettringite can increase equally by a small number of large crystals growing or by a large number of small crystals growing. These scenarios are compared in Fig. 14.5. In both scenarios, the amount of ettringite is doubled by doubling the length of the crystals. This length doubling corresponds with the doubled distance between the confining particles. The distance increases more for a smaller number of large crystals (Fig. 14.5 (a)) than for a larger number of small crystals (Fig. 14.5 (b)). This means that the length change caused by ettringite growth under confined conditions depends also on the number of growing crystals.

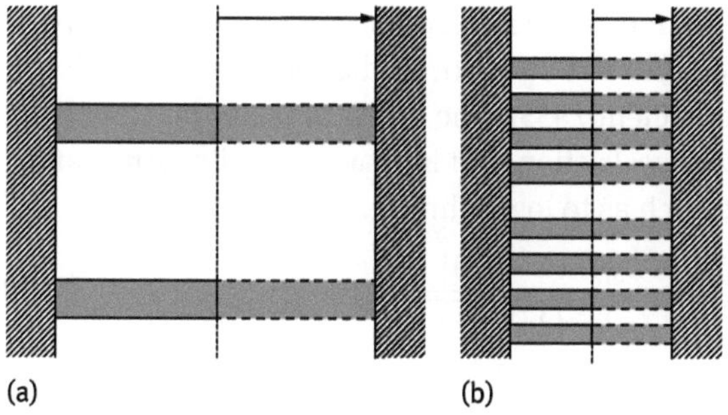

Fig. 14.5: Illustration of confined growth. In (a), a small number of large crystals are growing. In (a), a large number of small crystals are growing. In both cases the total phase amount is equal. The push-apart effect is stronger in (a).

Of course, the discussed scenarios are greatly simplified. But the scenarios point to a main insight: Phase development and technological properties are linked by complex relations. Structural features and their changes can be important influencing factors in those relations.

14.4 Methods to correlate XRD data and technological properties

14.4.1 Simple correlations analysis

For simple correlation analysis, the measured technical properties are directly compared with measured XRD data (Fig. 14.6). The kind of measured XRD data can range from peak intensities to phase amounts.

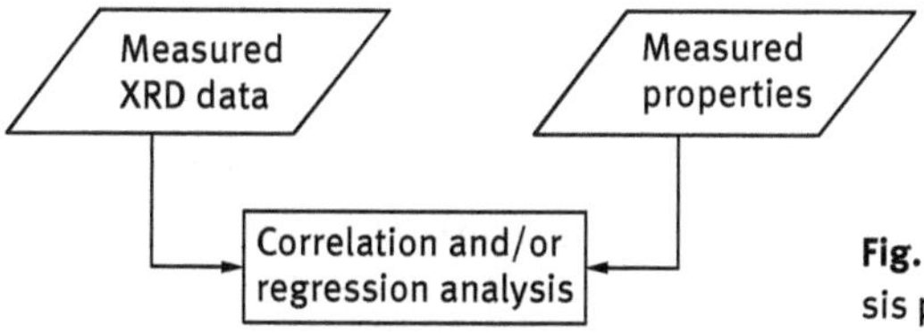

Fig. 14.6: Outline of a simple correlation analysis process.

The simplest way of performing a correlation analysis is to evaluate a scatter plot of two variables. If the variables are time series, both time series can be superposed in a diagram and compared for similar chronological patterns. Such simple and rather non-objective analyses can provide indications for correlations. But any such indication needs to be substantiated.

Correlation analysis evaluates relations between random variables. Classical correlation analysis implies linear relations. The strength of the relation is usually expressed by Pearson's correlation coefficient r [33]. The coefficient can be calculated by equation (14.2). Linearity is a severe restriction. But in many cases linearity can be obtained by simple transformations such as to logarithmize.

$$r = \frac{\sum[(x - x_{\mathrm{m}}) \cdot (y - y_{\mathrm{m}})]}{\sqrt{\sum(x - x_{\mathrm{m}})} \cdot \sqrt{\sum(y - y_{\mathrm{m}})}} \tag{14.2}$$

r Pearson's correlation coefficient
x actual value of the first variable
x_{m} average value of the first variable
y actual value of the second variable
y_{m} average value of the second variable

Calculating the correlation coefficient provides evidence about the strength of a relation and about the direction of the relation. Values of the coefficient range between −1 and 1. Zero indicates no correlation. The closer the values are to 1 or −1, respectively, the stronger is the correlation. A positive sign indicates that both variables vary in the same direction. A negative sign indicates that one variable increases while the other decreases.

Closely related to Pearson's correlation analysis is linear regression analysis. Regression analysis attempts to calculate a predictive model for related variables. In the case of linear regression, a linear relation is presupposed as given in equation (14.3).

$$y = c_0 + c_1 \cdot x + \varepsilon \tag{14.3}$$

y response variable
x predictor variable
c_0 intercept
c_1 regression coefficient
ε error term

With a sufficient number of measured xy pairs, the intercept and regression coefficient as well as the uncertainty (which is expressed by the error term) can be calculated. The number of required data pairs depends on the variances of these variables and on the desired level of significance. The quality of the regression is usually expressed by the coefficient of determination, which is Pearson's correlation coefficient squared.

14.4.2 Correlation analysis using known or presupposed models

For this kind of correlation, the measured and calculated technological properties are compared. The XRD data are used to calculate technological properties corresponding to a known model (Fig. 14.7).

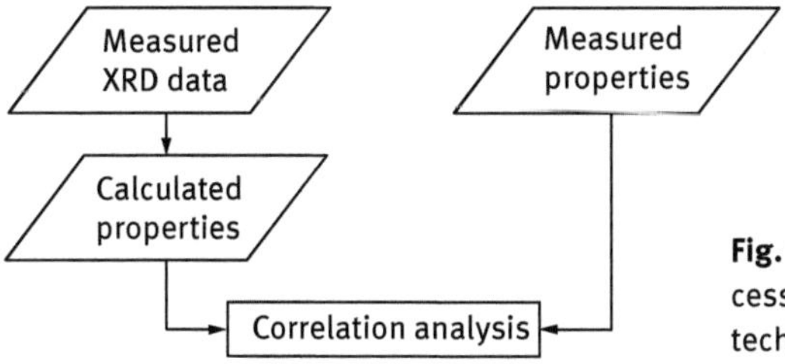

Fig. 14.7: Outline of a correlation analysis process with a presupposed model to calculate technological properties.

Such correlation analyses require formal relationships. Thermodynamic models are good examples. Phase amounts can be calculated from XRD patterns. Changing phase compositions represent chemical reactions. Chemical reactions and phase developments can be described by thermodynamic equations. With thermodynamic models, heat generation can be calculated from reaction enthalpies and chemical shrinkage can be calculated from net density changes.

Thermodynamic modelling of aqueous solutions, known from geochemistry, was introduced to cement science by Lothenbach & Winnefeld (2006) [34]. Ion solubility determines the dissolution and precipitation of solid phases. Consequently, the phase composition of hydrating cement based materials can be calculated time resolved and corresponding to mix design, provided the required thermodynamic data are available. Much of the data necessary is available from several databases or can be computed [35].

With known or presupposed formal relationships, properties can be calculated. The same properties can be determined in experiments. However complicated the model is, finally, measured and calculated properties are simply compared. The calculated data should equal the measured data. This means a simple linear relationship without offset is expected between measured and calculated data. Therefore, the agreement of calculated and measured properties can be tested easily by formal correlation or linear regression analysis as outlined in the previous section.

14.4.3 Correlation analysis if it is known which XRD characteristics are to be used

It can be that it is known that for instance amounts of certain phases should be related somehow to technological properties. But an actual model of this relation is not yet known. In this case, the required characteristics are calculated from the XRD data. Then regression analysis can be used to establish a relation between these calculated characteristics and the technological properties. A subsequent correlation analysis will verify or disprove the modelled relation (Fig. 14.8).

A typical scenario is that experiments have shown several phases and other characteristics as having significant influences on technical properties. But they are individually insufficient for reliable property prediction. In such cases, multivariate regression can help correlating phase and other sample characteristics with techno-

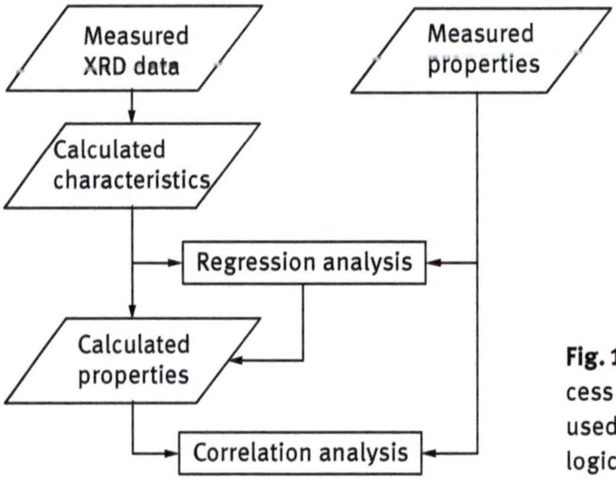

Fig. 14.8: Outline of a correlation analysis process with presupposed characteristics to be used but unknown model to calculate technological properties.

logical properties. Multivariate methods are essential in many fields. Consequently, there are many sources also for non-statisticians. A good introduction can be found for instance in Janert (2012) or Izenman (2013) [36, 37].

First, a general model needs to be set up for how the individual characteristics influences are interrelated with the technical properties. They could be for instance simply superposed, which would require just a multiple linear regression. There could be exponential or more complicated dose-response relationships and many more. Often, the data distribution indicates what kind of relationships can be expected or ruled out. These models are then refined by fitting processes. Also in such cases, measured data should equal the calculated data. Again, a simple linear relationship with no offset can be expected and the agreements of calculated and measured data can be tested by simple correlation analyses.

It requires nevertheless some trial and error runs to establish a reliable model.

14.4.4 Correlations when relations with technological properties are completely unknown

If nothing can be presupposed, a statistical approach is helpful. A diffractogram consists of several thousand data points which are the diffraction angles. All diffraction angles need to be examined. Individual evaluations of all these variables would require impractical efforts. Therefore, raw XRD data should be statistically examined for significant features in the dataset. Such significant features can be for instance principal components. The significant data features allow deducting XRD characteristics to be calculated and to be related to technological properties. These characteristics can be, for instance, peak or background heights, peak shapes, phase amounts, or just principal components scores. Regression analysis helps with establishing a formal model of the relation between these calculated characteristics and technological

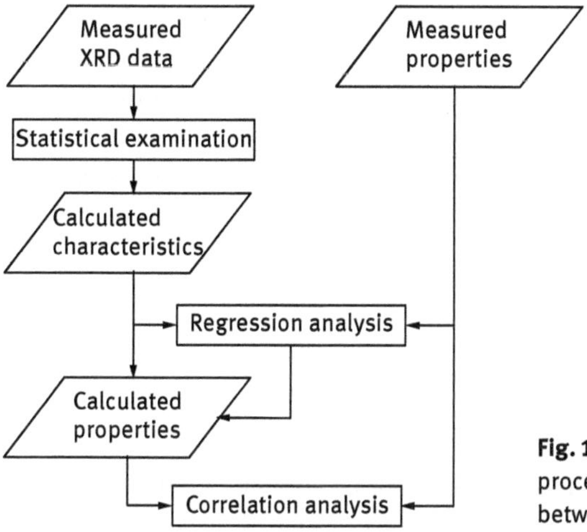

Fig. 14.9: Outline of a correlation analysis process with completely unknown relations between XRD data and technological properties.

properties. With this model, technological properties can be calculated. Finally, measured and calculated technological properties can be compared (Fig. 14.9).

In such cases, multivariate methods from the field of factor analysis can again be applied. In particular, partial least squares regression (PLSR) and principal component regression (PCR) are suitable [15, 16, 18–21]. These methods were already mentioned as methods for quantitative phase analysis in Section 14.2.1.3. But regression can be performed not just for phase amounts. It can similarly be used to establish models that quantitatively relate XRD data with technological properties.

PLSR is a regression algorithm that combines factoring and model fitting. It requires only the user to decide about the number of factors to be considered. A best possible fit between the set of predictor variables and the target variable is automatically calculated. Thus, PLSR is suitable for automated environments.

PCR is a three-step process. First, a principal component analysis is performed. In a second step, the factors must be identified. That means relevant principal components are selected. In a third step, multivariate regression is performed to fit the target property with the factor scores. This method requires two important user inputs: The decision of which principal components are considered factors and the decision of what regression model should be used.

PLSR is more user-friendly compared to PCR. However, the black-box-like regression approach has a significant draw back. The obtained regression model is not necessarily a correct physical model. This is no problem for quantitative phase analysis because a fundamental relationship between peak intensity and the amount of a corresponding phase exists (see equation 1). But in the case of technological properties and diffraction intensities, the nature of the relationship still has to be discovered, if there is any. Consequently, PLSR can be used to establish models that predict values for technological properties based on diffraction data. It does not necessarily describe the relation between diffraction data and technological properties in a physically meaningful way.

In the PCR approach, a user defines the regression models. The models can be formulated in physically meaningful ways. Compared to PLSR, PCR is more complicated to use. But PCR helps to potentially better understand the relationships between diffraction data and technological properties.

14.5 Case examples

14.5.1 Strength correlated with diffraction data

Strength is a key property of cement-based materials. Factors affecting strength are studied heavily. This includes articles about the contribution of phases, and clinker phases in particular, to the strength. Phase composition is nowadays determined by XRD. Consequently, any correlation that includes phase amounts can be seen as a correlation with diffraction data. Considering the titles of many articles and conference presentations, much effort is put into developing models predicting strength [38–43]. Consequently, correlation and regression analyses are common in articles about factors affecting strength. Various ways were found to correlate cement characteristics including the amounts of clinker phases with strength.

28 days strength is a key quality criterion for cements. Strength testing is time consuming. Strength prediction from characteristics of actually produced cement is therefore an effective quality control option for cement producers as well as for developers of concrete and mortar formulations. Actual cement quality improvement due to strength prediction implemented in the process control has been reported [44].

14.5.2 Heat flow correlated with diffraction data

The use of heat flow measurements has been well-established for characterizing the hydration process of cement-based materials. The heat flow is a thermodynamic consequence of chemical reactions and corresponding phase composition changes. Phase composition changes should therefore be correlated with heat flow. Such a correlation has been demonstrated elsewhere [45].

In this study, the hydration of an OPC paste was monitored with consecutive in situ XRD for the first 22 hours. Heat flow was measured on the same material and for the same time. The consecutively taken diffractograms were evaluated for phase amounts. The time-resolved phase composition changes were then used to calculate the generated heat. The calculated heat is then compared with the measured heat flow. The presented superposed graph of calculated and measured data is convincing, although a formal correlation analysis was not performed.

Thermodynamic modelling allowed heat flow peaks to be attributed to distinct chemical reactions and corresponding phase developments. Two main reactions were found with a significant influence on heat flow, silicate reaction (which is the alite consumption and subsequent C-S-H formation) and sulphate depletion (which is basically sulphate consumption by ettringite formation).

Besides these explanatory results, this study also has a more fundamental significance. It not only formally derives the fundamental relation between phase developments and heat generation, it also provides the first compelling experimental evidence that consecutive in situ XRD is a method capable of providing a thermodynamically complete representation of the hydration process of cement-based materials.

A predictive statistical model for 7-days heat of hydration of Portland cement has been attempted using quantitative powder XRD combined with non-linear multivariate data evaluation [46]. 53 different cements were analyzed for their heat of hydration, and their chemical and phase composition as well as for fineness, particle size distribution, setting times, and strength. The study did not provide a general model but it worked out parameters that should be included in actual models for heat of hydration. These include a structural mineralogical phase (preferably belite), a sulphate phase (preferably bassanite), a fineness parameter (preferably Blaine), and the content of ferrite or cubic aluminate.

14.5.3 Rheology correlated with diffraction data

The setting of cement paste is caused by hydrate phases growing into an interlocked structure. The growth of hydrate phases is a gradual process that increases viscosity until it sets. The growth of hydrate phases depends on the reactivity of the raw material phases. This reactivity is controlled by the solubility of the phases in water and the particle fineness. Consequently, the phase composition and fineness of cements should correlate with rheological data of the cement paste.

The characteristics of six Portland cements have previously been compared with the rheological behavior of these cements [47]. In this study, rheological measurements on the pastes of six Portland cements modified with three different super plasticizers were studied. The cements were characterized by chemical and phase composition as well as by fineness (Blaine). Phase amounts were obtained from Rietveld analyses of the dry cements. The rheological behavior of the cement pastes was compared with the individual characteristics of the cements. It was found that Blaine values (which represent particle fineness) as well as the amounts of alite and cubic aluminate are related with the area below the shear stress–shear rate curves. This area below the shear stress–shear rate curves was defined as "flow resistance". Therefore, a cement characteristic was defined with respect to flow resistance as a weighted sum of alite and cubic aluminate contents multiplied by the Blaine value (equation (14.4)).

$$\mathrm{CC} = B \cdot [d \cdot x_{\mathrm{aluminate}} + (1 - d) \cdot x_{\mathrm{alite}}] \tag{14.4}$$

CC	Vikan's cement characteristic
B	fineness in Blaine
$x_{\mathrm{aluminate}}$	amount of cubic aluminate
x_{alite}	amount of alite
d	weight parameter

$$\text{flow resistance} = a \cdot \mathrm{e}^{b \cdot \mathrm{CC}} \tag{14.5}$$

CC	Vikan's cement characteristic
a, b	regression coefficients

Regression analyses were performed with this cement characteristic as predictor and with flow resistance as response. Mostly exponential relationships corresponding to equation (14.5) were obtained. Only in one case was a linear regression sufficient to fit the data. The coefficients of determination were between 0.9641 and 0.9997. The actual fit and the regression coefficients a and b were found to depend on the kind and dosage of the superplasticizers.

Vikan's cement characteristic merges three of the most important cement properties into a single parameter. It can be an effective tool for relating cement reactivity to material properties other than flow resistance.

From a general point of view, it can be assumed that flow resistance and Vikan's cement characteristics are generally linked by an exponential relation. The linear relation can be seen as a special case for a specific material.

14.5.4 Dimensional change correlated with diffraction data

The shrinkage and expansion of building materials is an important topic. Dimensional change can cause cracking, flaking, spalling, and ultimately complete failure of a structure. Aggregates as well as cement paste can shrink or expand.

For aggregates, alkali-silica reaction (ASR) is usually considered the most important cause. ASR is the reaction of some quartz in the aggregates with the pore solution [48]. This creates a swelling gel. But others causes for expanding aggregates include the corrosion of pyrite and pyrrhotite [49, 50]. In adsition to chemical reactions, mineral phases can respond physically to changing ambient conditions such as humidity and temperature. Length change isotherms and thermal expansion coefficients of 12 different siliceous rocks have been measured elsewhere [51]. The results were compared with quantitative phase analyses by XRD. Significant correlations between chlorite content and drying shrinkage ($r^2 = 0.807$) as well as between quartz content and thermal expansion ($r^2 = 0.763$) were found.

There are also many potential causes for dimensional change of the cement paste. Such causes are, for example, delayed ettringite formation [52], carbonation, or sulphate attack [48] in the hardened paste. Prior to setting, a dimensional change can come from chemical shrinkage, settlement flow, sedimentation, and segregation. So there are mechanical transportation processes that change the material structure and thereby change the dimensions. This is superposed by a chemically induced density increase. Setting ends the mechanical transportation processes. After setting, chemical processes and the corresponding changes of phase characteristics become the main causes of dimensional change in cement paste.

The main hydration product of Portland cement is X-ray amorphous C-S-H gel. This makes XRD analyses of hydrating Portland cement paste and relating them to other sample characteristics specifically difficult. However, by using raw data of consecutive in situ XRD measurements such a relation has been reported [53]. A regression analysis of principal component scores and length change data has successfully been performed ($r^2 = 0.930$). Three distinguishable processes were identified in the XRD data that contribute to dimensional change: The most dominant process is the alite hydration associated with C-S-H formation. The second most dominant process is a microstructural development that includes the formation of a covering layer around quartz particles. The third process is gypsum consumption and corresponding ettringite formation. Interestingly, ettringite is involved in the last two processes. This study shows that the processes behind early age dimensional change affect diffraction pattern correspondingly. It is therefore possible to examine powder XRD data also for information about early age dimensional change.

In pastes with more aluminate phases, crystalline ettringite is a main hydration product, which is better observable than the C-S-H of the Portland cement pastes. Such systems are for instance ternary binders composed of Portland cement, calcium aluminate cement, and calcium sulphate. This ternary binder concept facilitates a broad variability in formulations. Consequently, hydration and dimensional change of such formulations vary considerably [54]. A link between ettringite formation and expansion of a hydrating paste of a ternary binder has been reported [55]. They compared the chronologies of dimension change and ettringite formation at early ages. The main expansion was observed when ettringite formation consumed temporarily formed gypsum. An influence of sulphate type on dimensional change and phase formation has also been observed [56]. It was observed that formulations with water-containing sulphates Bassanite or Gypsum expansion takes place while formulations with the water-free anhydrite did no expand. The onset of ettringite formation and dimensional change coincided. Thus, ettringite formation is related to dimensional change. But depending on the actual mix design, ettringite formation can be related with expansion as well as with shrinkage.

References

[1] Meier R, Anderson J, Verryn S. Industrial X-ray Diffraction Analysis of Building Materials. Reviews in Mineralogy & Geochemistry. 2012; 74: 147–165.

[2] Walenta G, Füllmann T. Advances in quantitative XRD analysis for clinker, cements, and cementitious additions. Powder Diffraction. 2004; 19(1): 40–44.

[3] Sorrentino F. Chemistry and engineering of the production process: State of the art. Cement and Concrete Research. 2011; 41: 616–623.

[4] Klug HP, Alexander LE. X-ray Diffraction Procedures For Polycrystalline and Amorphous Materials, 2nd edition. New York, Chichester, Brisbane, Toronto, Singapore: John Wiley & Sons; 1974.

[5] Fridman AE. The Quality of Measurements: A Metrological Reference. New York, Dordrecht, Heidelberg, London: Springer Science+Business Media; 2012.

[6] Gupta SV. Measurement Uncertainties Physical Parameters and Calibration of Instruments. Berlin Heidelberg: Springer-Verlag; 2012.

[7] Rietveld HM. Line profiles of neutron powder-diffraction peaks for structure refinement. Acta Crystallogr. 1967; 22: 151–152.

[8] Rietveld HM. A profile refinement method for nuclear and magnetic structures. J Appl Cryst. 1969; 2: 65–71.

[9] Bish DL. Howard SA. Quantitative Phase Analysis Using the Rietveld Method. J Appl Cryst. 1988; 21: 86–91.

[10] Taylor JC, Hinczak I. Rietveld made easy: a practical guide to the understanding of the method and successful phase quantifications. Canberra: Sietronics Pty Ltd; 2006.

[11] Young, RE, editor. The Rietveld Method. Oxford University Press; 1993.

[12] De la Torre AG, Santacruz I, León-Reina L, Cuesta A, Aranda MAG. Diffraction and crystallography applied to anhydrous cements. In: Pöllmann H, editor. Cementitious Materials Composition, Properties, Application. Berlin, Boston: De Gruyter; 2017.

[13] Aranda MAG, De la Torre AG. Diffraction and crystallography applied to hydrating cements. In: Pöllmann H, editor. Cementitious Materials Composition, Properties, Application. Berlin, Boston: De Gruyter; 2017.

[14] Thomas NW. A new approach to calculating powder diffraction patterns based on the Debye scattering equation. Acta Cryst. 2010; A66: 64–77.

[15] Gemperline PJ. Mixture analysis using factor analysis I: Calibration and quantification. J Chemometrics. 1989; 3: 549–568.

[16] Hamilton JC, Gemperline PJ. Mixture analysis using factor analysis II: Self-modelling curve resolution. J Chemometrics. 1990; 4: 1–13.

[17] Beebe KR, Kowalsky BR. An Introduction to Multivariate Calibration and Analysis. Analytical Chemistry. 1987; 59: 1007–1017.

[18] Martens H, Naes T. Multivariate Calibration. Chichester, New York, Brisbane, Toronto, Singapore: John Wiley & Sons; 1991.

[19] Starks TH, Fang JH, Zevin LS. A Standardless Method of Quantitative X-Ray Diffractometry Using Target-Transformation Factor Analysis. Mathematical Geology. 1984; 16: 351–367.

[20] Karstang TV, Eastgate RJ. Multivariate Calibration of an X-ray Diffractometer by Partial Least Squares Regression. Chemometrics and Intelligent Laboratory Systems. 1987; 2: 209–219.

[21] Mitsui T, Okuyama S, Fujimura Y. Determination of the Blend Composition Ratio of Cocaine to Hydrogencarbonate by X-Ray Diffraction Using Multivariate Analysis. Analytical Sciences. 1991; 7: 941–945.

[22] Fuellmann T, Witzke T. Partial least squares regression method for quantitative analysis of blended cements. In: Ludwig HM, Fischer HB, Bode KA, Beuthan C. Tagungsbericht Band 1 (proceedings part one) of 19. Internationale Baustofftagung ibausil 2015, Weimar. 513–519.
[23] Westphal T, Bier TA, Qoku E, Qorllari A. Examining large sets of XRD measurements with exploratory factor analysis. In: International Cement Microscopy Association, Proceedings of the Thirty-Eighth International Conference on Cement Microscopy 2016. Lyon: 16–30.
[24] Westphal T, Bier TA, Takahashi K, Wahab M. Using exploratory factor analysis to examine consecutive in-situ X-ray diffraction measurements. Powder Diffraction. 2015; 30 (4): 340–348.
[25] Aranda MAG, De la Torre AG, León-Reina L. Rietveld Quantitative Phase Analysis of OPC Clinkers. Reviews in Mineralogy & Geochemistry. 2012; 74: 169–209.
[26] Snellings R. X-ray powder diffraction applied to cement. In: Scrivener K, Snellings R, Lothenbach B, editors. A Practical Guide to Microstructural Analysis of Cementitious Materials. Boca Raton, London, New York: CRC Press; 2016.
[27] O'Connor, BH, Raven MD. Application of the Rietveld Refinement procedure in Assaying Powdered Mixtures. Powder Diffraction. 1988: 3; 2–6.
[28] Jansen D, Goetz-Neunhoeffer F, Stabler C, Neubauer J. A remastered external standard method applied to the quantification of early OPC hydration. Cement and Concrete Research. 2011; 41: 602–608.
[29] Riello P, Canton P, Fagherazzi G. Quantitative Phase Analysis in Semicrystalline Materials Using the Rietveld Method. J Appl Cryst. 1998; 31: 78–82.
[30] Scarlett, NVY, Madsen IC. Quantification of phases with partial or no known crystal structure. Powder Diffraction. 2006: 21: 278–284.
[31] Costoya Fernandez MM. Effect of Particle Size on the Hydration Kinetics and Microstructural Development of Tricalcium Silicate [dissertation]. École Polytechnique Fédérale de Lausanne; 2008. Thèse No 4102.
[32] Takahashi K, Bier TA, Westphal T. Effects of mixing energy on technological properties and hydration kinetics of grouting mortars. Cement and Concrete Research. 2011; 41: 1167–1176.
[33] Pearson K. Mathematical Contributions to the Theory of Evolution. III. Regression, Heredity, and Panmixia, Philosophical Transactions of the Royal Society. 1896; Ser. A 187: 253–318.
[34] Lothenbach B, Winnefeld F. Thermodynamic modelling of the hydration of Portland cement. Cement and Concrete Research. 2006; 36: 209–226.
[35] Damidot D, Lothenbach B, Herfort D, Glasser FP. Thermodynamics and cement science. Cement and Concrete Research. 2011; 41: 679–695.
[36] Janert PK. Data Analysis with Open Source Tools. Beijing, Cambridge, Farnham, Köln, Sebastopol, Tokyo: O'Reilly Media, Inc.; 2012.
[37] Izenman AJ. Modern Multivariate Statistical Techniques. New York, Heidelberg, Dordrecht, London: Springer; 2013.
[38] Brüggemann H, Bentrup L. Correlations between mineralogical clinker parameters and cement strength. In: International Cement Microscopy Association, Proceedings of the Eleventh International Conference on Cement Microscopy 1989; New Orleans. 226–245.
[39] Tsivilis S, Parissakis G. A mathematical model for the prediction of cement strength. Cement and Concrete Research. 1995; 25(1): 9–14.
[40] Kheder F, A1 Gabban AM, Abid SM. Mathematical model for the prediction of cement compressive strength at the ages of 7 and 28 days within 24 hours. Materials and Structures. 2003; 36: 693–701.
[41] García-Casillas PE, Martinez CA, Montes H, García-Luna A. Prediction of Portland Cement Strength Using Statistical Methods, Materials and Manufacturing Processes. 2007; 22: 333–336.

[42] Mechling JM, Lecomte A, Diliberto C. Relation between cement composition and compressive strength of pure pastes. Cement & Concrete Composites. 2009; 31: 255–262.
[43] Tsamatsoulis D. Optimizing the Cement Compressive Strength Prediction by Applying Coupled Linear Models. In: Matsorakis NE, Rudas IJ, editors. Advances in Electrical and Computer Engineering, Proceedings of the 17th International Conference on Automatic Control, Modelling & Simulation, Proceedings of the 14th International Conference on Artificial Intelligence, Knowledge Engineering and Data Bases, Proceedings of the 6th International Conference on Circuits, Systems, Control, Signals. 2015; Tenerife: 98–105.
[44] Tsamatsoulis D. Prediction of cement strength: analysis and implementation in process quality control. J Mech Behav Mater. 2012; 21(3/4): 81–93.
[45] Jansen D, Goetz-Neunhoeffer F, Lothenbach B, Neubauer J. The early hydration of Ordinary Portland Cement (OPC): An approach comparing measured heat flow with calculated heat flow from QXRD. Cement and Concrete Research. 2012; 42: 134–138.
[46] Azari H. Statistical Modeling of Cement Heat of Hydration Using Phase and Fineness Variables. NCHRP Web-Only Document 167, National Cooperative Highway Research Program, Transportation Research Board of the National Academies; 2010. [cited 2015 Apr 29]. Available from: www.trb.org/Main/Blurbs/164408.aspx.
[47] Vikan H, Justnes H, Winnefeld F, Figi R. Correlating cement characteristics with rheology of paste. Cement and Concrete Research. 2007; 37: 1502–1511.
[48] Taylor HFW. Cement Chemistry. London: Academic Press; 1990.
[49] Duchesne J, Fournier B. Petrography of Concrete Deteriorated by Weathering of Sulphide Minerals. In: International Cement Microscopy Association, Proceedings of the Thirty-Third International Conference on Cement Microscopy 2011; San Francisco.
[50] Duchesne, J, Fournier B. Deterioration of concrete by the oxidation of sulphide minerals in the aggregate. J Civ Eng Arch. 2013; 7(8): 922–931.
[51] Igarashi G, Maruyama I, Nishioka Y, Yoshida H. Influence of mineral composition of siliceous rock on its volume change. Construction and Building Materials. 2015; 94: 701–709.
[52] Taylor HFW, Famy C, Scrivener KL. Delayed Ettringite Formation. Cement and Concrete Research. 2001; 31: 683–693.
[53] Westphal T, Bier TA. About examination of in-situ XRD data by multivariate statistics and representation of heat flow and length change by these data. In: Ludwig HM, Fischer HB, Bode KA, Beuthan C, editors. Tagungsbericht Band 1 (proceedings part one) of 19. Internationale Baustofftagung ibausil 2015; Weimar. 361–371.
[54] Bier TA, Estienne F, Amathieu L. Shrinkage and Shrinkage Compensation in Binders Containing Calcium Aluminate Cement. In: Manghabai RJ, Glasser FP, editors. Calcium Aluminate Cements 2001, Proceedings of the International Conference on Calcium Aluminate Cements (CAC). Edinburgh: IOM Communications: 215–226.
[55] Evju C, Hansen S. Expansive properties of ettringite in a mixture of calcium aluminate cement, Portland cement and β-calcium sulfate hemihydrate. Cement and Concrete Research. 2001; 31: 257–261.
[56] Ohnishi K, Bier TA. Investigation into relations among technological properties, hydration kinetics and early age hydration of self-leveling underlayments. Cement and Concrete Research. 2010; 40: 1034–1040.

Johannes Södje

15 No cement production without refractories

Abstract: Cement manufacturing in huge amounts is still currently not possible without refractories. The development of the cement technology in rotary kiln systems started with the wet process kiln system with a production of ≈ 40 tons per day at the end of the 19th century. Currently, high efficient precalciner kilns with an average capacity of ≈ 3000 tons per day up to a maximum capacity of ≈ 12000 tons per day cover the world's cement demand. The refractory lining in the different rotary kiln systems has also undergone continuous modifications over the more than 100 year history of cement rotary kilns due to the changes in the operation of rotary kilns and burning technology. The application, the requirements, and the wear mechanism of refractories in the cement rotary kiln system are documented.

Keywords: refractories, cement kiln system, refractory wear mechanism

15.1 Introduction and application of refractories in cement rotary kilns (historical overview)

Goods like computer, car, air plane, steel, glass, cement, ceramic, etc. have become an integral part of the current technical world. In this context, three products are of particular importance, as they are basic materials and needed for many and partially essential intermediate and final goods:

(1) Cement,
(2) Steel,
(3) Refractories.

These materials represent a key position, because our current standard of living would be difficult to realize without them. Cement as binder and concrete is mainly used in huge amounts in the building-industry and steel in building, mechanical engineering, the automotive and ship-industry, etc., while refractories are necessary to produce steel, glass, cement, calcined lime, etc.

In general, cement and steel are well known, as everybody frequently comes into contact with them in their daily life. However, knowledge about refractories is usually limited to a small group of specialists. Nevertheless, the application of refractories for cement and steel manufacturing is indispensable, as steel is produced at temperatures up to ≈ 1800 °C and the main cement clinker production is manufactured at temperatures up to 1500 °C.

Johannes Södje, Refratechnik Cement GmbH, Göttingen, Germany, Johannes.soedje@refra.com

DOI 10.1515/9783110473728-016

The publication of Joseph Aspdin's patent "Improvement of the production of artificial stone" in 1824 is generally seen as the starting point of Portland Cement manufacturing [1, 2]. Ever since, the main demands for cement manufacturing were the improvement of cement quality and burning technology, and the application of suitable refractories. In the early years, the development and method of cement manufacturing can be characterized with pure empiricism.

Since the mid of the 19th century, commercial cement facilities were discontinuous shaft kilns, similar to the methods of lime burning and brick plants. The capacity of such discontinuous shaft kilns was between 1.7 and 3.4 tons of inhomogeneously burned cement clinker per week. The next improvement steps of cement burning technology were the introduction of the annular kiln (1864, developed by Friedrich Hoffmann) [1] and the development of the continuous shaft kiln by C. Dietzsch (1883) [1]. The capacity of the latter kiln type was about 20 tons cement clinker per day. Finally, the first rotary kiln for cement manufacture was introduced by Frederik Ransome, patented in England on May 2, 1885 [1, 3]. With the introduction of rotary kiln technology in the cement industry, the capacity of cement production massively increased, starting from about 40 tons per day in the beginning of the 20th century to maximum values of 10 000/12 000 tons per day currently.

Long wet rotary kilns were the standard rotary kiln type at the beginning of the 20th century, using slurry as raw feed. This kiln type is in operation in some countries of the world even today. Rotary kilns with grate preheater (so-called LEPOL-kiln) and long dry process rotary kilns were introduced in the twenties/thirties of the last century. For about 66 years, rotary kilns with cyclone preheaters (so-called DOPOL-kilns) were the next step toward increasing the cement capacity up to ≈ 800 tons per day. In the 1970s, precalciner kilns were introduced with an average capacity of about 3000 tons of cement clinker per day.

The specific energy usage to product 1t clinker could be reduced from about 6 GJ/t with the wet-kiln system to ≈ 3 GJ/t with the precalciner and multi-stage preheater dry kiln technology.

Early rotary kilns were gas fired, followed by oil fired kilns. Due to increases in the price of oil and gas, coal became the standard fuel in the late 70s and is still today. Mainly in the industrialized countries, coal is nowadays increasingly substituted by an increasing amount of petrol coke, tyres, plastics, materials with low calorific value, waste, etc.

The globalization of industrial markets – including the cement market – is additionally an incentive for cement plant operators to equip their plants with the latest technology or opt for new plants to maintain long-term competitiveness. In addition, sustainability is playing an increasingly important role in long-term success strategies. Cement plant operators must take several considerations into account to continue adapting to constantly-changing market pressures and regulatory requirements. These include:

- Capacity increases through the modernization of existing plants or the construction of greenfield/brownfield facilities
- Reduction in energy consumption using innovative technology
- Minimizing emissions by substituting fossil fuels with alternative fuels and materials
- Ensuring continuous production by using highly efficient refractory materials.

With the changes of rotary kiln operation and burning technology, the refractory lining also has undergone continuous modifications over the more than 100 year history of cement rotary kilns. Fig. 15.1 schematically illustrates the use of different brick grades in the main burning zone of rotary kilns since 1900.

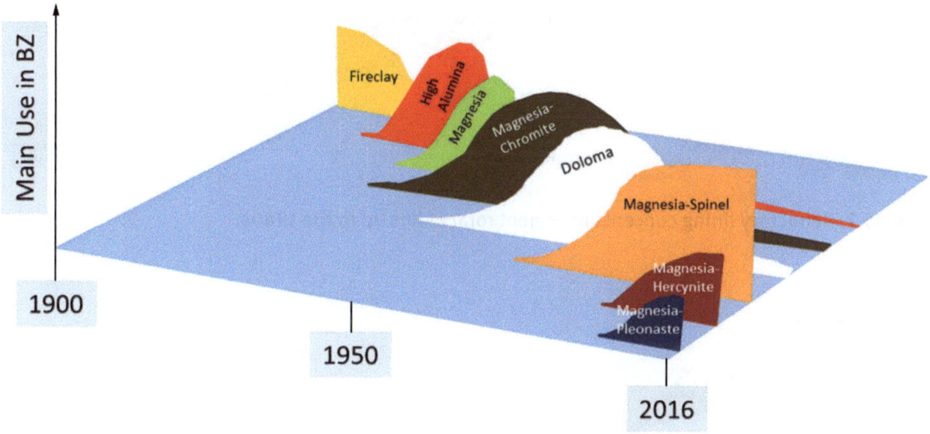

Fig. 15.1: Use of different brick grades in the main burning zone of rotary cement kiln since 1900 (schematic).

Until the 1940s, fireclay refractories and high alumina products were typically used throughout the entire kiln (Fig. 15.2).

Afterwards, basic brick grades were applied in the burning zone for the first time, Fig. 15.1. Trials were carried out with magnesia and magnesia chromite bricks during the mid-1930s and the 1940s. However, magnesia bricks could not compete with the performance of magnesia chromite bricks due to their poor structural flexibility. Dolomite bricks were installed in the main burning zone at the same time with similarly positive results as magnesia chromite bricks. But these brick grades exhibit sensibilities to CO_2, sulphur, and humidity.

The further development of high performance rotary kilns with increasing capacity and higher specific thermal cross sectional loads meant that magnesia chromite bricks now became essential in the hottest areas of the rotary kiln (burning zone, transition zones), Fig. 15.3. The peak level of magnesia chromite bricks usage was roughly between 1940 and 1990 (Fig. 15.1).

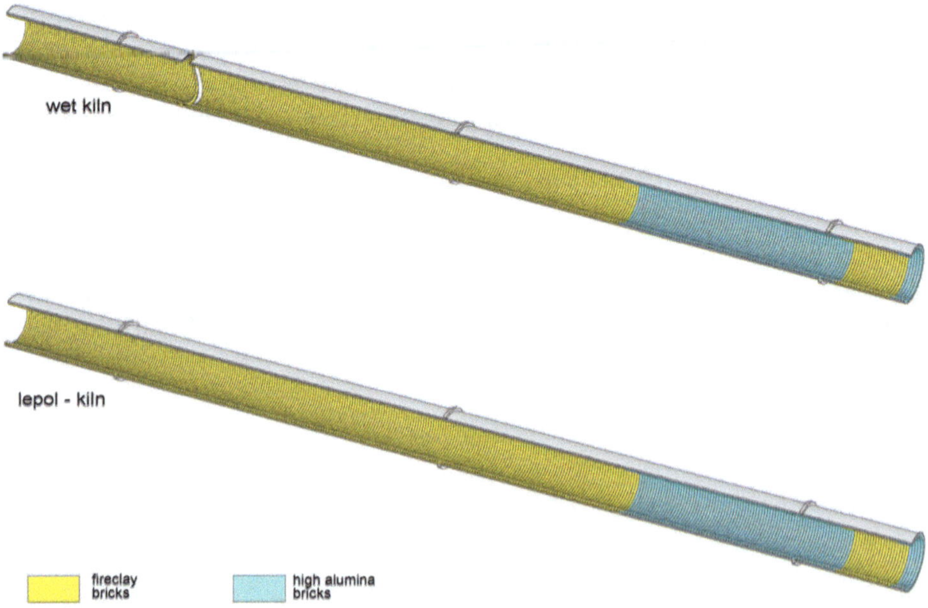

Fig. 15.2: Refractory lining concepts in cement rotary kilns up to the 1940s.

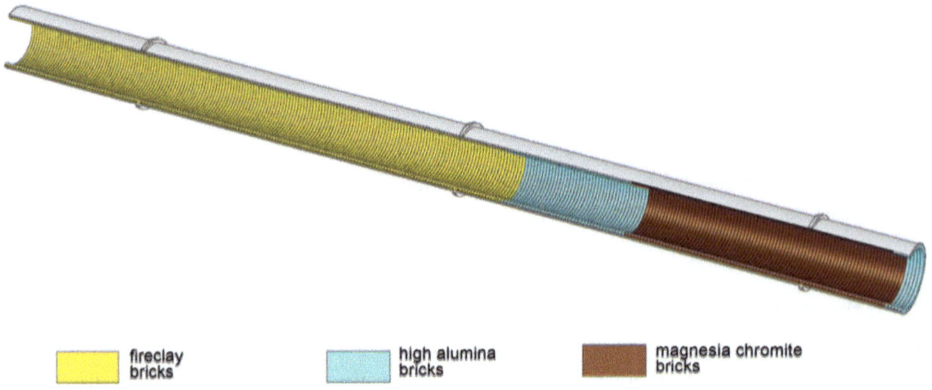

Fig. 15.3: Refractory lining concepts in cement rotary kilns up to the 1980s.

In the mid-seventies, chromite-free magnesia spinel brick grades were firstly tested in the coating-free transitions zones in Japanese cement rotary kilns (Fig. 15.1). The application of magnesia spinel bricks in Germany started in the early eighties. Fig. 15.4 broadly shows lining concepts for rotary kilns with the rising use of magnesia spinel brick grades between the mid-1980s and the 1990s.

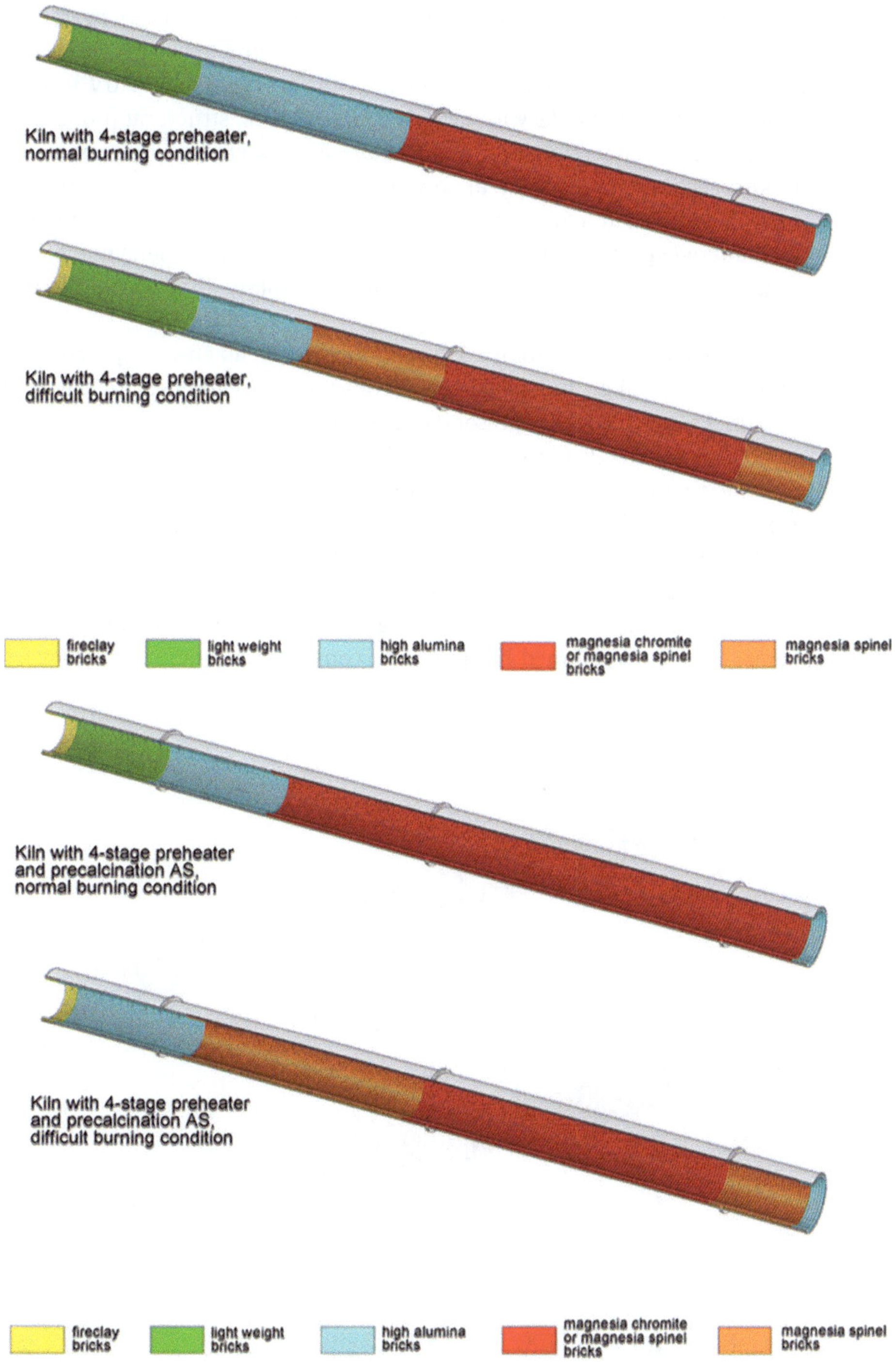

Fig. 15.4: Refractory lining concepts in cement rotary kilns between the mid-1980s and the 1990s. Top: rotary kiln with 4-stage preheater. Bottom: rotary kiln with 4-stage preheater and precalcination AS.

Due to the environmental impact of alkali chromate, which is caused by refractory products containing chromite, the demand for chromite-free basic bricks in the cement industry has increased. Consequently, the refractory industry has introduced new chromite-free bricks with modified spinel types, structural properties, and enhanced physical characteristics. Some of these brick grades were optimized for specific kiln sections, like tire area, severely thermally loaded zones, and the burning zone. In addition to magnesia spinel bricks with varying spinel content and spinel types (sintered spinel, in situ spinel, fused spinel), magnesia hercynite bricks and magnesia pleonaste bricks have been developed for rotary kilns within the last 20 years (Fig. 15.1).

Current lining concepts for rotary kilns mainly using magnesia spinel brick grades, magnesia hercynite, and magnesia pleonaste bricks besides fireclay and high alumina bricks/“SiC” bricks is presented in Fig. 15.5.

In the other thermochemically lower loaded kiln sections (outlet zone, preheating zone, safety zone, chain zone in long wet and dry kilns, inlet zone), lightweight refractory bricks, fireclay refractories, and high alumina brick grades are applied partially in combination with refractory concretes based on fireclay, andalusite, bauxite, and currently also SiC- and zircon-containing high alumina refractories. The same applies to the static areas of the kiln system (cyclone stages, calciner, riser duct, kiln hood, cooler systems), where a combined lining design consisting of brick and refractory concrete has proved to be the ideal solution.

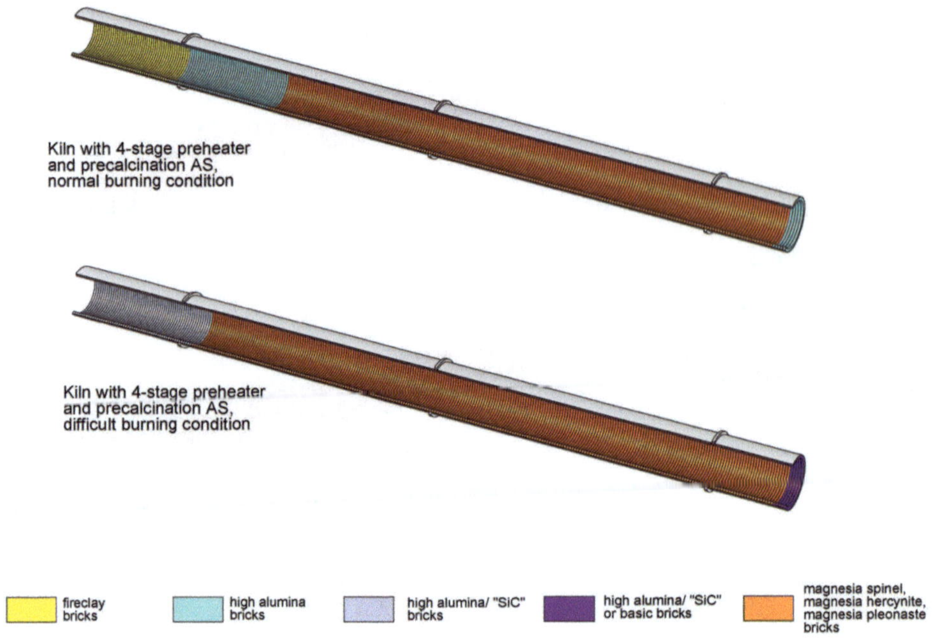

Fig. 15.5: Current refractory lining concepts in cement rotary kilns with precalcination.

The lining with refractory concretes was originally limited to a few zones in the cement kiln (kiln inlet and outlet, chain zone, burner pipe, some areas of the cooler, i.e. the bullnose, cooler shaft, elbow inlets of the satellite cooler, etc.). The introduction of the preheater and precalciner kilns has led to a significant increase in the use of unshaped refractory products since the 1970s. Fig. 15.6 shows a lining concept in static areas of the kiln system (calciner).

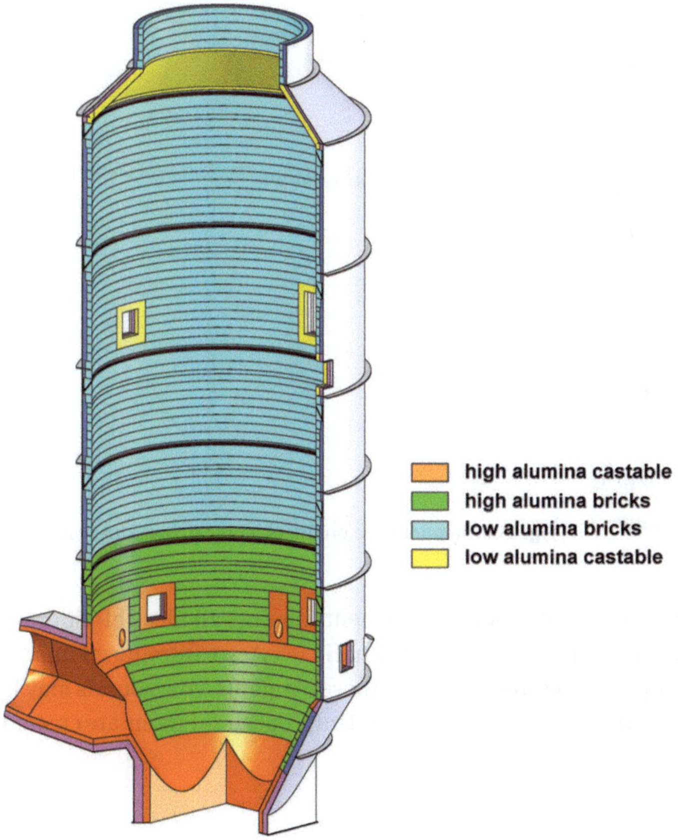

Fig. 15.6: Lining concept of a calciner type with refractory concretes and bricks.

15.2 Requirements for refractories in cement kiln systems

The cement clinker production from grinded and homogenized raw feed to clinker roughly includes the following process steps: preheating, calcination, sintering/ clinkerization, and cooling. Following these steps, increasing temperatures act on the refractory lining with kiln feed temperatures of ≈ 1500 °C in the hottest area of the rotary kiln under usual operating conditions, and finally the burned clinker cools down to ambient temperature within the cooler system (Fig. 15.7).

Fig. 15.7 additionally shows that the refractory lining has to withstand increased thermal and thermochemical loads in the hot zone of the kiln (transition zones, burning zone), and also in various levels of the calciner, especially in the area of additional burners, the riser duct, and the kiln hood.

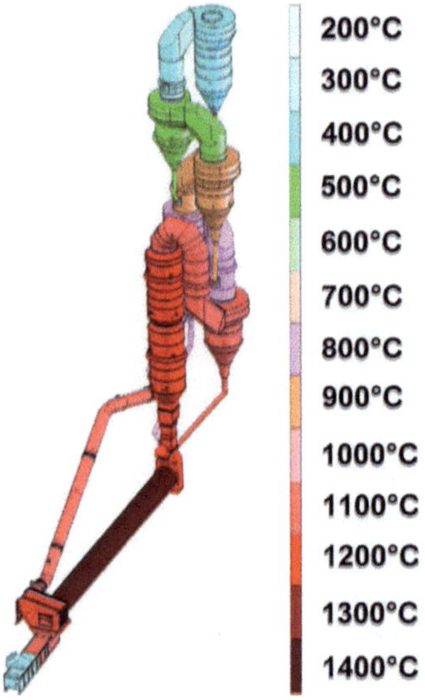

Fig. 15.7: Temperature profile of a precalciner kiln system.

With the fuel and kiln feed traces of volatile components, mainly alkalis, chloride, and sulphur compounds, as well as heavy metals (Pb, Zn, Hg, etc.), are introduced into the kiln system. Due to their properties and behavior, these elements/compounds may be enriched within different areas of the kiln system and internal circulations are created, as shown in Fig. 15.8.

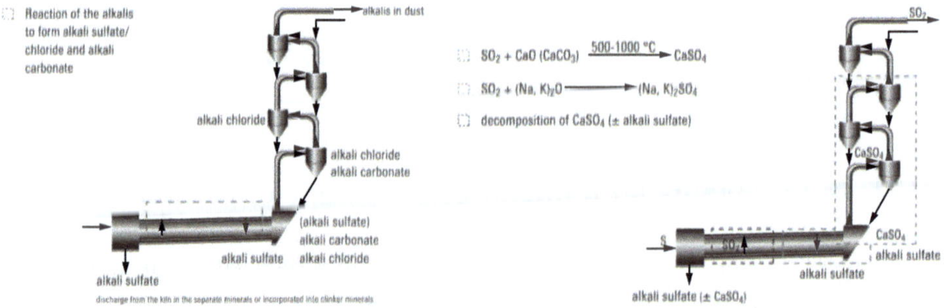

Fig. 15.8: Internal circulation of volatile components in the cement kiln system [4].

In the moving aggregate, the rotary kiln, additionally increased mechanical/thermo-mechanical loads may occur besides temperature and chemical attack, especially in the tire and deformed shell areas. In some parts of the kiln system, such as the raw meal ducts, tertiary air duct (TAD), and clinker beds of coolers, high abrasion resistance of the refractories is required. Tab. 15.1 roughly summarizes the operating conditions and refractory requirements in the different areas of a precalciner kiln system.

Tab. 15.1: Operating conditions and refractory requirements in different areas of a precalciner kiln system.

Area	Operating conditions	Requirements to refractories	
Preheater			
Cyclone stages #1–3	Gas temperature range: ≈ 250–650 °C, feed temperature range: ≈ 240–500 °C, high gas velocities, abrasion	Refractoriness	o
		Strength	++
		Abrasion resistance	+++
		Thermal shock resistance	+
		Elastification	o
		Volatile compound resistance	+
		Redox resistance	o
		Resistance to liquid phases	–
Cyclone stages #4, 5	Gas temperature range: ≈ 800–950 °C, feed temperature range: ≈ 700–900 °C, high gas velocities, abrasion, chemical attack, local reducing conditions (low NO_x)	Refractoriness	++
		Strength	++
		Abrasion resistance	+++
		Thermal shock resistance	+
		Elastification	o
		Volatile compound resistance	+++
		Redox resistance	++
		Resistance to liquid phases	+
Calciner			
Upstream	Gas temperature range: ≈ 1000–1200 °C, feed temperature range: ≈ 950–1100 °C, high gas velocities, abrasion, chemical attack and reaction, liquid phase attack	Refractoriness	+++
		Strength	++
		Abrasion resistance	+++
		Thermal shock resistance	++
		Elastification	+
		Volatile compound resistance	++
		Redox resistance	+
		Resistance to liquid phases	++
Down-stream	Gas temperature range: ≈ 800–950 °C, feed temperature range: ≈ 700–900 °C, high gas velocities, abrasion, chemical attack and reaction	Refractoriness	+++
		Strength	++
		Abrasion resistance	+++
		Thermal shock resistance	+
		Elastification	+
		Volatile compound resistance	+++
		Redox resistance	o
		Resistance to liquid phases	+

Tab. 15.1: (continued)

Area	Operating conditions	Requirements to refractories	
Riser duct	Gas temperature range: ≈ 900–1000 °C, feed temperature range: ≈ 850–950 °C, high gas velocities, abrasion, chemical attack and reaction, liquid phase attack	Refractoriness	+++
		Strength	++
		Abrasion resistance	+++
		Thermal shock resistance	++
		Elastification	+
		Volatile compound resistance	++
		Redox resistance	+
		Resistance to liquid phases	++
Rotary kiln			
Inlet zone	Gas temperature range: ≈ 900–1100 °C, feed temperature range: ≈ 850–1000 °C, chemical attack and reaction, liquid phase attack	Refractoriness	+++
		Strength	++
		Abrasion resistance	+
		Thermal shock resistance	++
		Elastification	+++
		Volatile compound resistance	+++
		Redox resistance	+
		Resistance to liquid phases	+
Calcining/ safety zone	Gas temperature range: ≈ 1000–1100 °C, feed temperature range: ≈ 950–1050 °C, abrasion, chemical attack and reaction	Refractoriness	+++
		Strength	++
		Abrasion resistance	+
		Thermal shock resistance	++
		Elastification	+++
		Volatile compound resistance	+++
		Redox resistance	+
		Resistance to liquid phases	+
Upper transition zone	Gas temperature range: ≈ 1200–1300 °C, feed temperature range: ≈ 1200–1300 °C, increased temperature, chemical attack and reaction, abrasion, unstable coating condition, mechanical loads from mid tire	Refractoriness	+++
		Strength	+++
		Abrasion resistance	+
		Thermal shock resistance	+++
		Elastification	+++
		Volatile compound resistance	+++
		Redox resistance	++
		Resistance to liquid phases	++
Burning zone	Gas temperature range: ≈ 1500–1700 °C, feed temperature range: ≈ 1400–1500 °C, high temperature, chemical attack and reaction, stable coating formation, redox condition	Refractoriness	+++
		Strength	+++
		Abrasion resistance	+
		Thermal shock resistance	+++
		Elastification	+++
		Volatile compound resistance	++
		Redox resistance	++
		Resistance to liquid phases	+++

Tab. 15.1: (continued)

Area	Operating conditions	Requirements to refractories	
Lower transition zone	Gas temperature range: ≈ 1200–1400 °C, feed temperature range:≈ 1150–1350 °C, high to severe temperature, unstable coating condition, redox condition, chemical and liquid phase attack	Refractoriness	+++
		Strength	+++
		Abrasion resistance	+
		Thermal shock resistance	+++
		Elastification	+++
		Volatile compound resistance	+++
		Redox resistance	++
		Resistance to liquid phases	+++
Outlet area	Gas temperature range:≈ 1000–1200 °C, feed temperature range: ≈ 1000–1200 °C, high to increased temperature, chemical attack and reaction, unstable coating condition, mechanical loads from outlet tire, high abrasion	Refractoriness	+++
		Strength	+++
		Abrasion resistance	++
		Thermal shock resistance	+++
		Elastification	+++
		Volatile compound resistance	+++
		Redox resistance	++
		Resistance to liquid phases	++
Burner pipe	Gas temperature range: 1500–2000 °C, high temperature, chemical attack and reaction, increased abrasion from high velocity gases/feed dust	Refractoriness	+++
		Strength	++
		Abrasion resistance	++
		Thermal shock resistance	+++
		Elastification	+++
		Volatile compound resistance	+++
		Redox resistance	++
		Resistance to liquid phases	++
Kiln hood	Gas temperature range: ≈ 1100–1200 °C, feed temperature range: ≈ 1050–1150 °C, increased temperature, chemical attack and reaction, abrasion	Refractoriness	+++
		Strength	+++
		Abrasion resistance	++
		Thermal shock resistance	++
		Elastification	++
		Volatile compound resistance	+++
		Redox resistance	++
		Resistance to liquid phases	+
Tertiary air duct (TAD)	Gas temperature range: ≈ 900–1000 °C, feed/dust temperature range: ≈ 850–900 °C, increased abrasion from high velocity gases/feed dust, possible chemical attack	Refractoriness	++
		Strength	+++
		Abrasion resistance	+++
		Thermal shock resistance	++
		Elastification	o
		Volatile compound resistance	++
		Redox resistance	o
		Resistance to liquid phases	+

Tab. 15.1: (continued)

Area	Operating conditions	Requirements to refractories	
Grate cooler			
Inlet zone	Gas temperature range: ≈ 1100–1200 °C, feed temperature range: ≈ 1050 °C, increased temperature, chemical attack and reaction, increased abrasion from clinker particles	Refractoriness	+++
		Strength	+++
		Abrasion resistance	+++
		Thermal shock resistance	+++
		Elastification	++
		Volatile compound resistance	+++
		Redox resistance	+
		Resistance to liquid phases	+
Bullnose	Gas temperature range: ≈ 1100–1200 °C, increased temperature, chemical attack and reaction	Refractoriness	+++
		Strength	+++
		Abrasion resistance	+++
		Thermal shock resistance	+++
		Elastification	++
		Volatile compound resistance	++
		Redox resistance	+
		Resistance to liquid phases	+
Cooler walls, Horse shoe	Gas temperature range: ≈ 1100–1200 °C, feed temperature range: ≈ 1100–1200 °C, increased temperature, chemical attack and reaction, severe abrasion from clinker particles	Refractoriness	+++
		Strength	+++
		Abrasion resistance	+++
		Thermal shock resistance	+++
		Elastification	++
		Volatile compound resistance	++
		Redox resistance	+
		Resistance to liquid phases	+
Grate level	Feed temperature range: ≈ 250–400 °C, high temperature, increased abrasion from clinker particles	Refractoriness	++
		Strength	+++
		Abrasion resistance	+++
		Thermal shock resistance	+++
		Elastification	++
		Volatile compound resistance	+
		Redox resistance	o
		Resistance to liquid phases	o

– low relevance, o relevant, + important, ++ very important, +++ significantly important

15.3 Wear mechanism of refractories in cement kiln system

Refractories installed in the cement kiln system are considered as working lining and are therefore subjected to continuous wear caused by thermal, chemical and mechanical loads (Fig. 15.9). Various loads of this kind are overlapping in the kiln system due to the interactions between refractory lining and kiln feed, flame, and kiln shell irregularities (deformation, ovality, etc.).

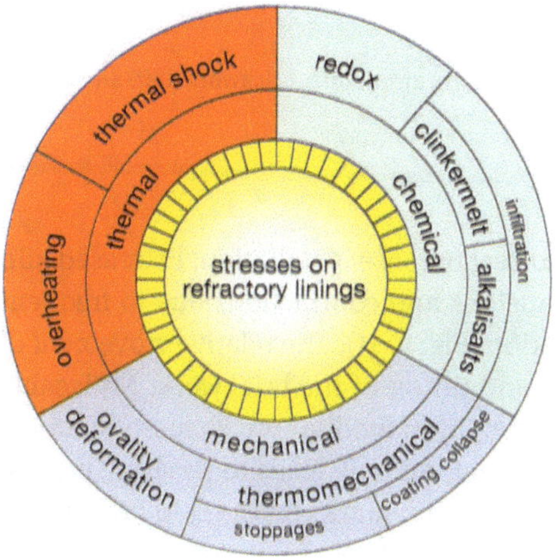

Fig. 15.9: Loads on refractory linings influenced by thermal, mechanical, and chemical loads [5].

Post mortem analyses of various refractory failures and corrective measures to overcome them are helpful tools for improving the kiln running time. The list of refractory wear is long and diverse. However, a selection of typical and extreme wear mechanisms are summarized in this section.

Typical kinds of lining wear can already be identified on site after kiln stoppage by comparing the present lining situation in the kiln with the following case studies.

15.3.1 Mechanical wear

15.3.1.1 Convex hot face spalling

Convex hot face spalling is mostly visible in the lower transition zone and kiln outlet area (Fig. 15.10), which is caused by excessive axial pressure on the lining or increased thrust within the last rings in the outlet area. The main reasons for this are insufficient clearance in the expansion joints, the installation of basic bricks without cardboard spacers, and frequent stoppages of operation after cardboard spacers are burnt off.

Fig. 15.10: Convex hot face spallings on the longitudinal joints are visible: brickwork (left) and single brick (right).

15.3.1.2 Displacement/spiralling

These phenomena are mostly caused by incorrect or loose installation, increased kiln shell ovality or deformation, repeated expansions and contractions due to frequent kiln stoppages, and changing coating formation (Fig. 15.11). The relative movement of brickwork occurs, finally leaving spiralling and tilting of the brickwork. Noticeable spiralling over large brickwork areas and tilting of individual bricks more often occur in large rotary kilns. The brickwork movement not only leads to shearing cracks, spallings, and broken corners and edges, but abrasion marks on the cold face side of bricks are additional signs of such influences (Fig. 15.12).

Fig. 15.11: Displacement and starting spiralling (left) and severe loosening of brickwork parts (right).

Fig. 15.12: Abrasion marks on the cold face side of a used brick.

15.3.1.3 Ovality

This lining appearance is caused by increased ovality in tire areas (Fig. 15.13). E.g. consumed tire shoes can increase the clearance and lead to excessive ovality. The ovalation (Ω) of a kiln shell should not exceed 1/10 of kiln diameter (Ω_{max} [%] $< 1/10D$ [m]), which can be seen as a usual practice. In any case, a shorter lifetime of the brickwork can be expected if this value is exceeded, especially if the ovalation value is above 0.7 %. Increased pressure loads cause tensions within the brickwork, and the hot face of the bricks is more loaded, finally resulting in locally limited spallings in the lining circumference.

Fig. 15.13: Deep spallings of some bricks in between completely perfect brick sections are visible in the circumference, indicating kiln shell deformation and/or increased ovality in the tire area.

15.3.1.4 Formation of grooves

Groove formation is a phenomenon which occurs occasionally (Fig. 15.14). If the groove formation occurs in the key area of the brickwork, the premature wear has to be seen in connection with the key shims. In most cases, brickwork rings were closed too tightly, or key bricks were damaged during installation by the use of a wrong hammer, or more than one iron plate to close the brickwork ring. Similar grooves have also been seen outside the key brick line. This is mostly caused by increased deformation of kiln shell, other irregularities of kiln shell, or locally incorrect brickwork installation.

15.3.1.5 Shear stress at the retaining ring

Increased thrust on the last brickwork ring towards the retainer may generate shearing cracks in the bricks and excessive movement and collapsing of brick in the retainer area (Fig. 15.15). Flexing kiln outlet, ovality, trumpet-like, or fatigued kiln outlet segments intensify such wear. Increased rotation speed additionally intensifies the brickwork movement especially in large rotary kilns.

Fig. 15.14: Groove formation is visible along the kiln axis in the lining, limited to 2–4 bricks wide.

Fig. 15.15: Collapsed bricks including shearing cracks at the cold face side due to severe shear stress at the retaining ring in the kiln outlet section: brickwork (left) and single brick (middle and right).

15.3.2 Thermal wear

15.3.2.1 Clinker melt infiltration

Clinker melt infiltration is mostly caused by increased thermal/thermochemical loads, including the formation of a high and low viscose melt phase portion, which infiltrates the bricks hot face, filling the pores and corroding the elastifier of basic bricks (magnesia spinel, chromium ore, hercynite, pleonaste, etc.), resulting in the formation of low melting mineral phases. These influences cause hot face densification and, finally, spalling of transformed hot faces if increased thermomechanical and/or mechanical loads occur (Fig. 15.16).

In the presence of chrome ore, the formation of C_2F and/or $C_4(A,Cr)F$ occurs (Fig. 15.17 (a)). Mayenite ($C_{12}A_7$), C_4AF/C_2F, and/or ye'elimite ($C_4A_3\bar{S}$), in the presence of sulphur, are the corrosion products in the case of MA-spinel (Fig. 15.17 (b)). Hercynite and pleonaste are degenerated to C_4AF/C_2F and mayenite (Fig. 15.17 (c)). Under extremely high temperature loads, the main component of basic bricks (magnesia = M)

Fig. 15.16: Densified hot faces of used bricks due to clinker melt infiltration, taken from the burning zone.

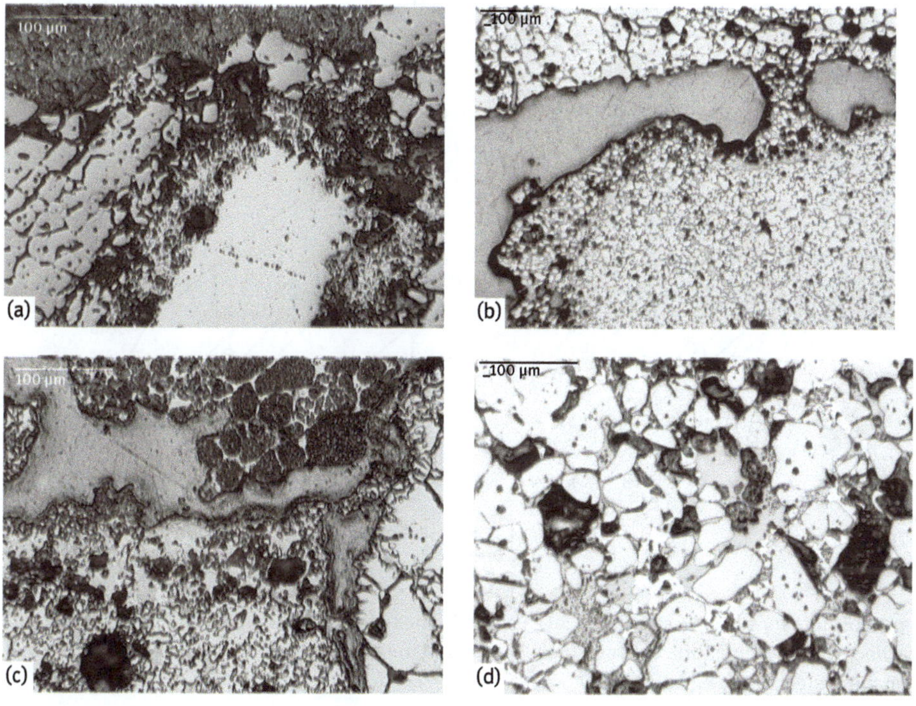

Fig. 15.17: Corrosion of different elastifiers and the main components of basic brick.

is additionally attacked and partially dissolved by the formation of melilite (C_4AMS_3) and C_3MA_2 (Fig. 15.17 (d)).

Similar effects occur if the chemical composition of the kiln feed is outside of typical cement clinker compositions.

In case of bricks and refractory concrete products from the system Al_2O_3–SiO_2, melts containing calcium aluminate silicates with an anorthite and/or gehlenite composition are formed at elevated temperature with relative low ternary eutectic melting points in the system CaO–Al_2O_3–SiO_2 (Fig. 15.18). The eutectic melting temperature is further reduced in the presence of volatile components (alkalis, halogenides, sulphur). Therefore, initial melting of such refractory linings can already occur at lower temperatures.

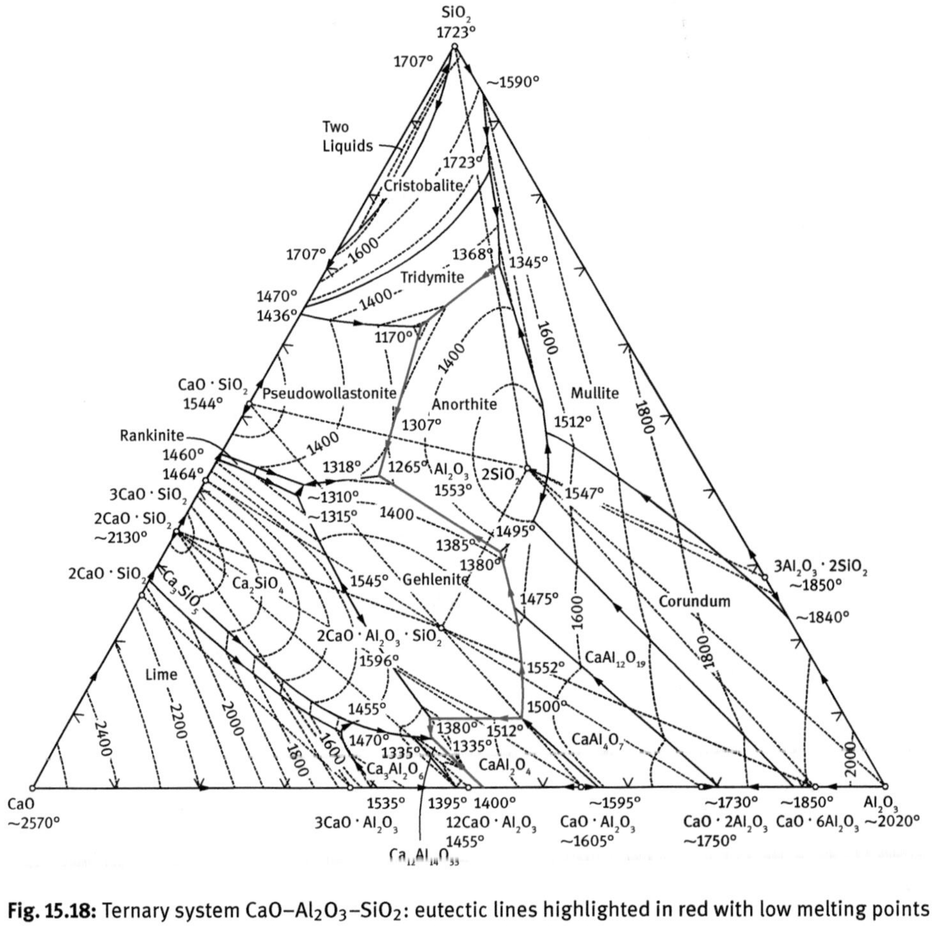

Fig. 15.18: Ternary system CaO–Al_2O_3–SiO_2: eutectic lines highlighted in red with low melting points reached by partial equilibria between the kiln feed and the brick and refractory concrete products from the system Al_2O_3–SiO_2 [6].

15.3.2.2 Overheating/lava-like coating

A continuous increase of severe thermal loads leads to an overheating of the lining and to a softening of the clinker melt infiltrated hot face, which can easily be eroded by the crossing kiln feed/clinker material. Erosion of the refractory lining finally occurs. The typical visual appearances of this type of stress are concave, wavy, sloping, and smooth hot faces (Fig. 15.19 (a), (b)). Additionally, a lava-like coating is formed and sticks to the brickwork hot face, which can be spalled off together with brick heads, if increased mechanical/thermomechanical loads are present (Fig. 15.19 (c)).

(a) (b) (c)

Fig. 15.19: Concave, smooth hot face, or lava-like coatings (from left to right) are typical signs of an overheating of the affected lining, taken from the burning zone and lower transition zone.

15.3.2.3 Thermal shock loads

Thermal shock loads occur if too rapid temperature rises during the heating-up of the kiln takes place or if too fast cooling and coating loss during operation occur. Thermal shock leads to thin spallings if the structural strength of the bricks is exceeded (Fig. 15.20).

Fig. 15.20: Thin spallings (≈ 20–30 mm) are visible due to sudden changes in temperatures.

15.3.3 Chemical wear

15.3.3.1 Infiltration of salts and attack of volatile compounds

The influence and interaction of salt compounds with the refractory lining depends on the prevailing temperatures, the oxygen partial pressure, and especially the alkali/chlorine/sulphur ratio in the different zones of the kilns system. The alkali/chlorine/sulphur ratio can be determined using the alkali/chlorine/sulphate modulus (ASM), defined by HOLCIM, Switzerland (Fig. 15.21).

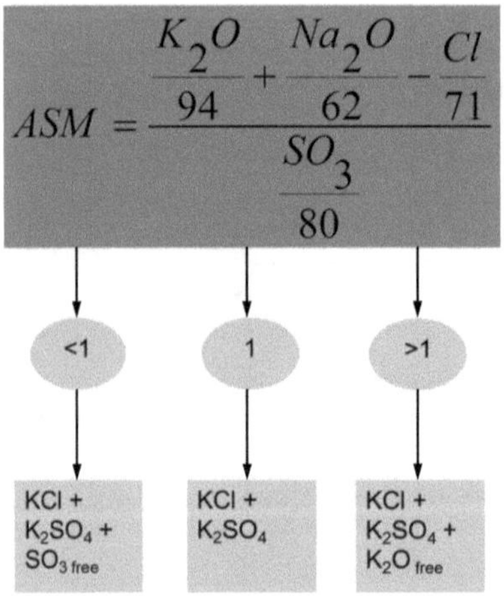

Fig. 15.21: Alkali/chlorine/sulphate modulus according to HOLCIM, Switzerland.

The following criteria have to be distinguished, which can result in various degradation effects in the refractory lining:

- Alkali salt condensation in a balanced ratio, i.e. ASM ≈ 1
- Excess of alkalis, i.e. ASM > 1
- SO_2/SO_3 excess, i.e. ASM < 1

The relationship between the degradation effects and the refractory lining is mentioned and in some cases discussed in detail elsewhere [7–9], providing a summary of the essential details of the three various cases.

15.3.3.1.1 Balanced ratio with alkali salts in kiln atmosphere (ASM ≈ 1)

In the case of a balanced alkali sulphate modulus, all alkalis, sulphates, and chlorides react in the kiln atmosphere to form alkali chloride and alkali sulphate salts, which are bonded in the kiln feed/clinker material or infiltrate into the brickwork. The salt compounds migrate in the brick structure until they reach their condensation temperatures (K_2SO_4: ≈ 1074 °C, KCl: ≈ 800 °C), Fig. 15.22. Sulphate salts are mostly detected in the lower transition zone and burning zone, together with alkali chloride salts in the upper transition zone. The crystallization of salts leads to densified horizons in the brick texture partially with frequently sharply limited infiltration horizons. The elasticity and the thermal shock resistance of the affected horizons are finally reduced. Changes in the properties of the refractory lining by chemical influences are rather unlikely in these cases.

Fig. 15.22: Used magnesia spinel bricks infiltrated with salts, leaving different densified horizons; left: arcanite (K_2SO_4); middle: sylvine (KCl); right: mixture of alkali chloride and sulphate salts.

During longer kiln stoppages and in the presence of increased humidity, a transformation of infiltrated salts by H_2O adsorption can occur. For example, the sulphate salt calcium langbeinite ($K_2SO_4 \cdot 2CaSO_4$) reacts to syngenite ($CaSO_4 \cdot K_2SO_4 \cdot H_2O$); K_2CO_3 forms $K_2CO_3 \cdot 1.5H_2O$. These reactions are associated with progressive structural weakening, which may lead to a more rapid degradation of the lining.

15.3.3.1.2 Excessive alkalis in kiln atmosphere (ASM > 1)

Under an oxidizing kiln atmosphere, an excess of alkalis leads to the oxidation of the chromium ore contained in magnesia chromite bricks, forming hexavalent alkali chromates and/or alkali chromate sulphates. Alkali chromate salts can already be macroscopically identified by their yellowish efflorescence (Fig. 15.23 (a)). This yellow efflorescence can be observed in the area of metallic anchoring of refractory concrete linings as well (Fig. 15.23 (b)). Here, the formation of alkali chromate is caused by alkali attack on the heat resistance anchors containing metallic chromium and nickel. The chromate formation can be described as follows (equations (15.1) and (15.2)):

$$Cr_2^{3+}O_3 + 2K_2O + \tfrac{3}{2}O_2 \rightarrow 2K_2Cr^{6+}O_4 \quad \text{(alkali chromate)} \qquad (15.1)$$

$$2K_2SO_4 + 2K_2O + Cr_2^{3+}O_3 + \tfrac{3}{2}O_2 \rightarrow 2(K_2SO_4/K_2Cr^{6+}O_4) \quad (15.2)$$

(alkali chromate sulphate)

In both cases, chromate formation may lead to premature wear of the affected brick or concrete lining, resulting in the spalling of embrittled brick hot faces or dropping of concrete fields due to the corrosion of the metal anchors.

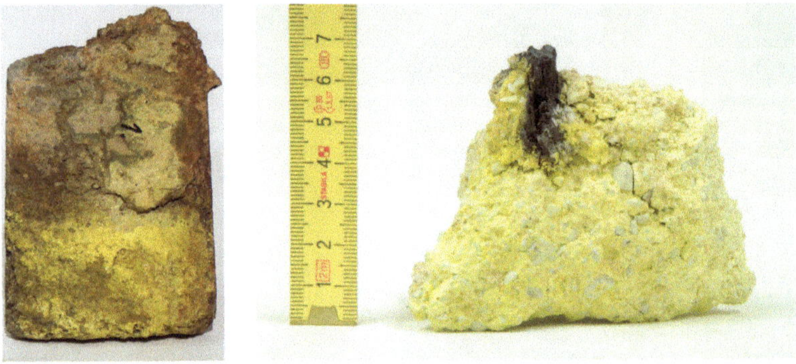

(a) (b)

Fig. 15.23: Yellowish efflorescences in case of magnesia chromite bricks (a), and in the vicinity of corroded anchors sticking to refractory concrete (b).

Refractories from the system Al_2O_3–SiO_2 react with excessive alkalis to form β-Al_2O_3, feldspar (sanidine: KAS_6), and feldspathoids (leucite: KAS_4, kaliophilite: KAS_4, kalsilite: KAS_2, KAS), depending on the Al_2O_3 content (equations (15.3)–(15.6)). The formation of these minerals is accompanied by a volume increase up to 30 %, which leads to a structural weakening of the affected refractory horizons. Finally, thermomechanical/mechanical loads cause spalling of altered horizons (Fig. 15.24).

$$A_3S_2 + 16S + 3K \rightarrow 3KAS_6 \quad \text{(sanidine)} \quad (15.3)$$

$$A_3S_2 + 10S + 3K \rightarrow 3KAS_4 \quad \text{(leucite)} \quad (15.4)$$

$$2A_3S_2 + 8S + 6K \rightarrow 6KAS_2 \quad \text{(kalsilite, kaliophilite)} \quad (15.5)$$

$$11A + K + N \rightarrow (K,N)A_{11} \quad (\beta\text{-}Al_2O_3) \quad (15.6)$$

Fig. 15.24: Alkali spalling effects: brick sample (left) and concrete sample (right).

15.3.3.1.3 Excessive sulphur oxides in the kiln atmosphere

If there is an excess of sulphates, alkali sulphate, and alkali chloride, salt condensation in the refractory lining may also occur, but the remaining free sulphate oxides can react either with the clinker melt or the components of the brick and concrete lining.

In the case of dolomite bricks, the CaO, one main component of this brick grade, is reformed into $CaSO_4$ or CaS. In accordance with the Boudouardian equilibrium, the organic impregnation substances in dolomite bricks can release carbon monoxide (CO) under operating conditions, which reacts with $CaSO_4$ to form CaS and CO_2. CO_2 can finally effect a re-carbonization of the CaO. These mineral formations are accompanied with a volume increase, leading to a weakening of the physical properties of affected brick horizons and finally to the destruction and spalling of the brick texture.

CaO-containing silicates (e.g. belite (C_2S)) occurring in low concentrations in basic brick grades react with the free SO_2/SO_3 to form lower melting silicates, such as merwinite (C_3MS_2) and monticellite (CMS). By this process, the magnesia of the brick is also corroded. Under the influence of continued SO_2/SO_3-attack, the formation of forsterite (M_2S) can occur. As a result, the sulphate salt anhydrite ($CaSO_4$) is formed, densifying the brick texture in deeper horizons together with other infiltrated salts. This so-called silicate corrosion leads to a weakening of the refractoriness and structural flexibility. Magnesia chromite bricks containing high levels of silicates are more susceptible than magnesia spinel bricks.

In cement-containing, and in particular in cement-rich refractory concretes, the formation of ye'elimite ($C_4A_3\bar{S}$) has been determined, accompanied by the corrosion of calcium aluminates from the cement phase. The ye'elimite formation leads to a loosening of the structure, which supports an easier and deeper migration of other volatile components in the refractory concrete lining, triggering further structural transformation.

The above described degradation mechanisms and their newly formed mineral phases are given by equations (15.7)–(15.14):

- dolomite bricks

$$4CaO + 4SO_2 \rightarrow 3CaSO_4 + CaS \tag{15.7}$$

$$CaSO_4 + 4CO \rightarrow CaS + 4CO_2 \tag{15.8}$$

(locally reducing atmosphere due to the tar residues in dolomite bricks)

$$CaO + CO_2 \rightarrow CaCO_3 \tag{15.9}$$

- basic bricks containing calcium silicates

$$2C_2S + MgO + SO_3 \rightarrow CaSO_4 + C_3MS_2 \quad \text{(merwinite, melting point: 1575 °C)} \tag{15.10}$$

$$C_3MS_2MgO + SO_3 \rightarrow CaSO_4 + 2CMS \quad \text{(monticellite, melting point: 1490 °C)} \tag{15.11}$$

$$CMS + MgO + SO_3 \rightarrow CaSO_4 + M_2S \quad \text{(forsterite, melting point: 1890 °C)} \tag{15.12}$$

- refractory concretes containing calcium aluminate cement

$$4CA_2 + SO_3 \rightarrow C_4A_3\bar{S} + 5A \tag{15.13}$$

$$4CA + SO_3 \rightarrow C_4A_3\bar{S} + A \tag{15.14}$$

15.3.3.2 Redox loads

Redox loads usually represent a low portion of the overall volume of wear cases, but can occur more frequently and in some cases determine the course of degradation caused by the inhomogeneous combustion of low grade fuels.

Basic brick grades with increased iron oxide contents in magnesia (alpine magnesia), e.g. magnesia chromite or magnesia hercynite bricks, have an increased susceptibility to these burning conditions. In a reducing atmosphere, the magnesioferrite ($MgFe_2^{3+}O_4$), contained in alpine magnesia is reduced to magnesiowustite ($(Mg,Fe^{2+})O$), which involves a significant reduction of volume (≈ 20 %). This influence is macroscopically indicated by the bleaching of the reduced hot face of the brick, Fig. 15.25. Fig. 15.25 (a) and (b) show the microstructure of unchanged high-iron magnesia with segregations of magnesioferrite and a high-iron magnesia after the influence of the reducing atmosphere.

If redox burning cycles occur, there are frequent changes between magnesioferrite and magnesiowustite, resulting in structural weakening of the affected horizons.

15.3.3.3 Corrosion of kiln shell

Occurrences of corrosion of the kiln shell are well-known in the cement industry and are undoubtedly related to the composition of the kiln gas atmosphere, Fig. 15.26 (a). Depending on the alkali/sulphate ratio and the oxygen partial pressure, reaction products, such as magnetite (Fe_3O_4), hematite (Fe_2O_3), pyrrhotite ($Fe_{1-x}S$), troilite (FeS), and pyrite (FeS_2) are formed under the prevailing temperature conditions, presented in the ternary system Fe–S–O, Fig. 15.26 (b).

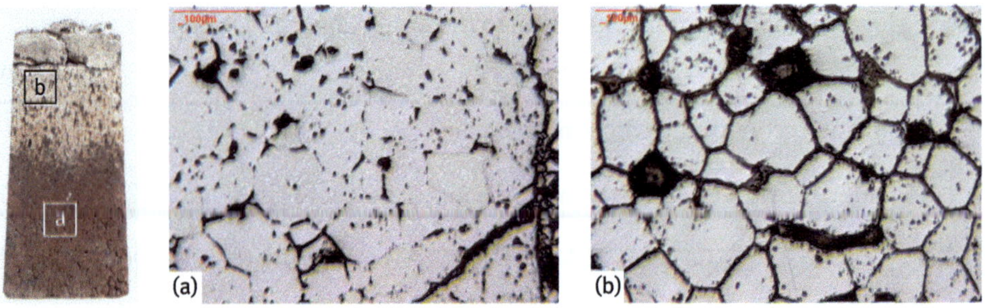

Fig. 15.25: Used magnesia chromite brick with bleached hot face (left), high-iron magnesia with segregations of magnesioferrite (middle), high-iron magnesia after the influence of reducing atmosphere (right).

(a)

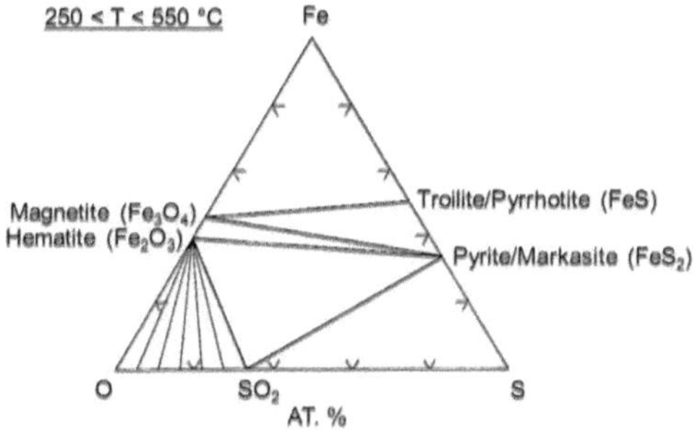

(b)

Fig. 15.26: Corroded kiln shell (a), and ternary system Fe–S–O with corresponding mineral phase formations (b).

15.3.4 Change of wear influences when using secondary fuels

Since the two major energy crises, primary high-grade fuels, such as fuel oil and gas, were replaced by cheaper and easily available fuels. As a first step, coal and lignite were utilized. Finally, the use of secondary fuels and waste has gained significantly in importance by today. For instance, in Germany the average content of secondary fuels used in the cement industry is currently about 60 % on average. The application of up to 100 % of secondary fuels in the cement industry is planned on the long-term. This is due not only to increased environmental awareness but also to the economic advantages of using them. However, secondary fuels are often very difficult to handle because of their inhomogeneity regarding calorific value, composition, particle size distribution, humidity content, availability, etc. Although recent years have seen the development of a particular market for secondary fuels of uniform quality to meet cement industry requirements, the process control of a cement plant will continue to remain rather undefined in the future due to changing fuel usage.

A comprehensive range of post-mortem studies on used refractory material from various areas of the cement kiln system provide sufficient evidence of changes in wear influences on the refractory, which occur in an undesired manner in many cases. The stresses by thermal overload, thermochemical load, redox loads, and attack by volatile components augmented with increasing usage of secondary fuels (Fig. 15.27).

The manifold changes of wear influences on the refractory lining can be summarized as followed:

- Risk of irregular temperature profile in the kiln (local thermal overload)
- Stronger chemical and thermochemical attack
- Risk of formation of local reducing atmosphere
- Stronger corrosion attack against the kiln/cyclone shell and the anchor system

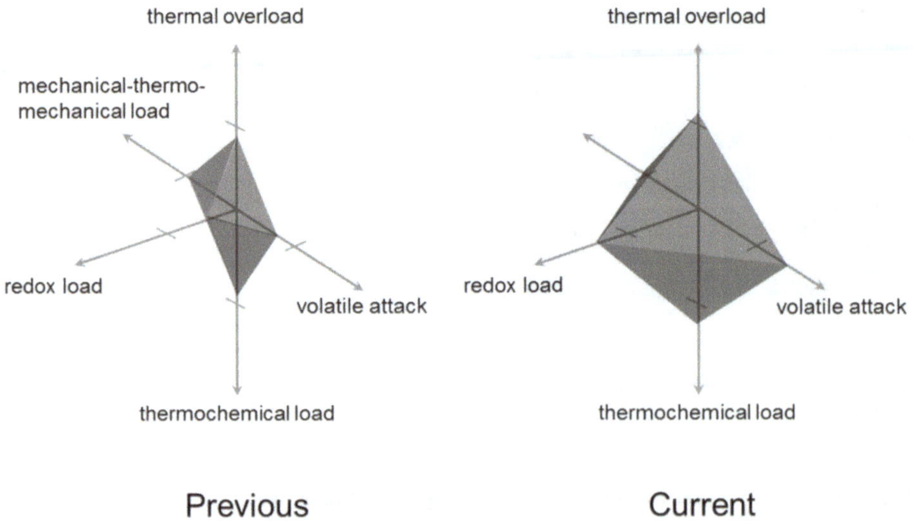

Fig. 15.27: Wear influences on refractory linings in the cement industry previous and currently.

The feeding of secondary fuels can lead to less optimum burning conditions in several areas in the kiln system (e.g. incorrect feeding, frequent changes of fuel types, particle size distribution, etc.) which can finally cause local overheating of the refractory lining. Wavy, concave surfaces or surfaces that appear to have melted and solidified are fundamental signs of thermal overload (Fig. 15.28).

Fig. 15.28: Thermal overloaded basic bricks from the lower burning zone and lower transition zone (left, in the middle), high alumina brick taken from the calciner (right).

The chemical and/or thermochemical wear of the refractory lining is traceable to the incorporation of volatile components into the system (alkalis, sulphur, chlorides, trace elements, etc.). A significant increase in these components, and especially heavy metals, in the system has been noted due to the increasing use of secondary fuels and raw material, and the associated processing difficulties. Depending on the quantity and atmospheric conditions, the volatile compounds and heavy metals have an affinity to form various salt compounds, such as alkali chlorides, alkali sulphates, and sulphides. Due to internal gas circulation processes, these compounds become significantly enriched in the system. While mainly individual salts, such as sylvine, arcanite, and anhydrite, could previously be detected in the structure of the lining, the condensation of mixed salts in the structure is currently determined to a greater degree. Calcium langbeinite, syngenite, aphthitalite, and langbeinite are the main mixed salts found in the refractory lining, in addition to alkali chlorides. These salt mixtures are liquid up to about 600 °C regardless of their composition (Fig. 15.29).

If potential flow paths are present, these salts also pass as far as the steel shell of the kiln system, where they cause a wide range of corrosion mechanisms such as hot gas corrosion, wet corrosion, and rusting effects on the steel during kiln shutdowns (Fig. 15.30).

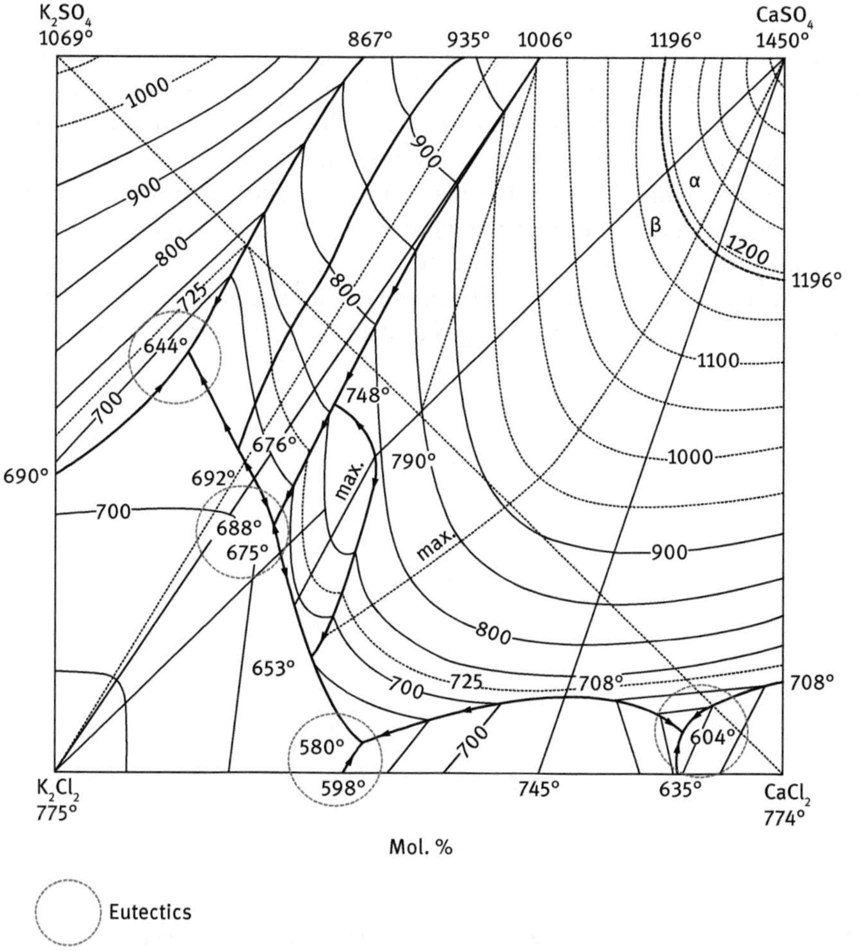

Fig. 15.29: Quaternary system K_2SO_4–$CaSO_4$–KCl–$CaCl_2$ with eutectic points (circles) [10].

Fig. 15.30: Strongly corroded effects on kiln shell (rotary kiln, left), and cyclone steel (right).

While in basic and chrome ore-free brick grades the open pores are generally filled by infiltration and condensation of the salts, leading to structural densification in various horizons (Fig. 15.31), additional intensive thermochemical reactions take place in dolomite bricks and non-basic linings. These include new mineral phase formations with obvious increases in volume, such as the formation of feldspar and feldspathoids in the case of high alumina linings. In the rotary kiln, this wear is manifested by thin spallings (so-called alkali spalling) (Fig. 15.32).

In the preheater, calciner, and kiln hood areas, an expansion of the non-basic lining on the hot side usually occurs before the alkali spalling effect takes place. Over

Fig. 15.31: Salt infiltration in different horizons of basic (left, in the middle) and non-basic brick grades (right).

Fig. 15.32: High alumina bricks showing alkali spalling (left), alkali spalling sample with white salt efflorescences (right).

time, this expansion exerts high pressure and traction forces on the entire refractory construction. If given expansion joints are then additionally filled with dust, salts, or kiln feed material and the compensators lose their ability to function due to the expansion of the lining underneath, then deformation or even tearing of the steel shell occurs. The lifting of partial areas of the calciner has also been observed (Fig. 15.33).

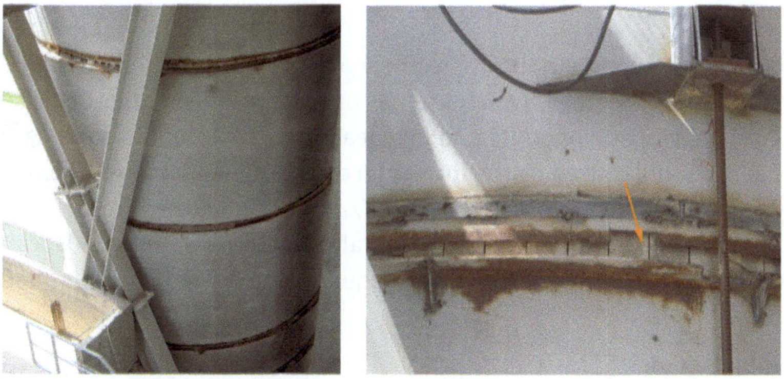

Fig. 15.33: Expansion of the refractory lining due to alkali attack with calciner shell lifted off.

Refractory linings have further been affected by the excessive presence of SO_2/SO_3 observed in the system in recent years. The use of sulphurous fuels, such as petrol coke, leads to an alkali sulphate modulus < 1. This means that free SO_2/SO_3 is widely available to thermochemically react with the refractory lining. In the case of basic bricks, a corrosion of the magnesia and calcium silicates, which are present in traces in the magnesia, may take place. For a long time this was only detectable mineralogically by X-ray analysis. Recent case studies on used bricks, however, show that this type of wear can even be detected macroscopically due to the high sulphur concentration in the system (Fig. 15.34). A texture with wavy pores is a typical sign of this wear, so-called silicate corrosion.

Fig. 15.34: Silicate and magnesia corrosion, surplus of SO_2/SO_3 in the kiln atmosphere (ASM < 1).

High sulphur compound contents in the atmosphere of a plant also foster the formation of coating rings and cloggings, which are not only found in the kiln but also increasingly occur in the upstream units and may lead to premature kiln stoppages (Fig. 15.35). This is primarily caused by the formation of spurrite minerals (carbonate spurrite ($2C_2S \cdot CaCO_3$), sulphate spurrite ($2C_2S \cdot CaSO_4$), and anhydrite ($CaSO_4$). The presence of salt melts, based on alkali chloride and alkali sulphate compounds, can further increase this effect, resulting in the more frequent removal of these rings and clogs.

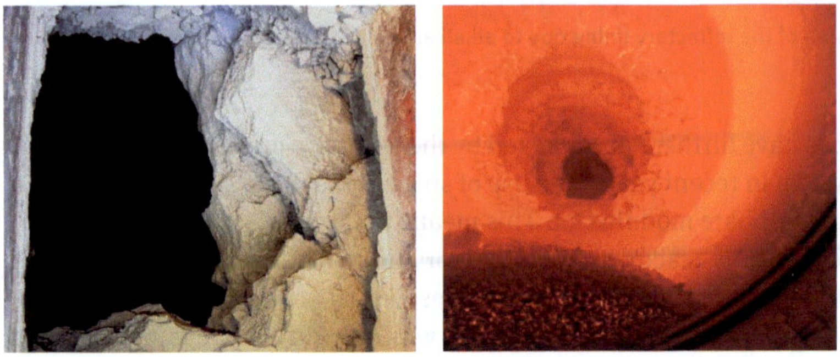

Fig. 15.35: Clogging in a preheater (left) and ring formation in a kiln (right).

(a) (b)

Fig. 15.36: Reducing atmosphere in the kiln system; bleached brick (left), with additional salt condensations (K_2S, K_2S_3, $KFeS_2$, K_2CO_3, right).

Clinker is normally burned under oxidizing atmospheres. The use of secondary fuels and the regulatory requirements for reducing NO_x emissions produce reducing atmosphere in some parts of the system. Unfortunately, these also influence the wear of the refractory material, which can be detected macroscopically on the hot side by strongly bleached horizons in the lining (Fig. 15.36 (a)). In the presence of salts, these conditions and/or redox burning conditions not only lead to the crystallization of alkali sulphates in the structure, but the condensation of alkali sulphite and alkali sulphides is again observed in the lining (Fig. 15.36 (b)). A typical H_2S smell is perceived during kiln stops, where the brickwork has been loaded in this way. The reason for this is the strong hygroscopy of the alkali sulphides, which draw moisture from the air and release H_2S gas (toxic in higher concentrations).

Reducing burning conditions also favor the Boudouard reaction (Fig. 15.37):

$$CO_2 + C \rightleftarrows 2CO.$$

Fig. 15.37: Reducing and redox burning conditions with carbon disintegration/condensation in the lining.

Due to the reducing atmosphere on the hot side, elemental carbon may deposit in lower horizons of the lining in a temperature range between ≈400–600 °C. Carbon deposits in the form of soot on the kiln shell are also observed. This effect is augmented catalytically by refractory products with higher contents of iron oxides with trivalent iron, such as magnesia chromite and magnesia hercynite bricks with alpine sinter magnesia. Serious consequential damage to the brickwork by excessive spalling is associated with this so-called carbon disintegration.

The increasing use of secondary raw material, such as waste materials from the non-iron industries, as an additive to the raw meal causes concentrations of heavy metals in the system. Most of all, sulphide compounds based on lead, cadmium, and bismuth could be found, especially in the lining in the upper transition zone (Fig. 15.38).

Fig. 15.38: Heavy metal concentration in the lining (left), massive PbS coating sample (right), detected in the upper transition zone.

Reducing burning conditions are essential for such condensations in the structure of the lining, which lead to embrittlement of the texture and spalling of affected horizons in the same manner as salt condensation.

Since the increasing installation of refractory concretes in cement plants, alkali chromate formation has again been increasingly found in used linings, especially in contact with corroded metal and heat resistant anchors (Fig. 15.23 (b)).

15.4 Conclusions

From the examples of wear given here, it is clear that the demands placed on the refractory lining have intensified. The main causes of this are increased plant capacities, additional combustion chamber modules, and an increasing use of secondary fuels and raw materials. Therefore, it is imperative for refractory companies to provide innovative solutions so cement producers can reduce and, indeed, prevent such wear mechanisms.

Under existing and future conditions, the following essential demands are placed on the refractory material and its installation.

- Minimizing infiltration and corrosion of refractory components
- Optimizing structural texture with higher flexibility and structural strength
- Renewable sealing of refractory hot face
- Optimization of installation technology, especially in the static of the cement kiln system, including a flexible and fast installation of high-grade refractory concretes and bricks
- Innovative insulation and protection options for the anchoring system and steel shell to reduce or even prevent different corrosion mechanisms of shell and anchors.

Therefore, from the refractory industry side, the optimization of the infiltration as well as the thermochemical and thermal resistance of refractory materials will continue to be the focus of major development work. Integrated concepts providing a systematic solution are at the forefront of this, which in addition to an adapted constructional brick lining as a wear layer, should also include a wide range of anchoring systems and insulating layer concepts.

References

[1] Stark J, Wicht B. Zum 100jährigen Jubiläum der Errichtung des ersten Zementdrehofens in Deutschland. ZKG International. 1997; 50: 407–416.

[2] English Patent no. 5022. Improvements in the Modes of Producing an Artificial Stone; 1824.

[3] English Patent no. 5442. Improvement in Manufacturing Cement etc.; 1885.

[4] Filippich M, Laschek D. Alkali-, sulphur- and chlorine circulations in cement kiln system. Refratechnik Report No. 62. Göttingen: Refratechnik Cement GmbH. 1–20.

[5] Künnecke M. Wear phenomena of refractory linings in the burning zone of rotary cement kilns. Refratechnik Report No. 55. Göttingen: Refratechnik Cement GmbH. 1–41.

[6] Bartha P, Hobrecht E, Klischat H-J, Weibel G. The use of basic and high alumina bricks in highly thermally and chemically stressed lime recovery kilns with special regards to thermal insulation considerations. Refratechnik Report No. 56. Göttingen: Refratechnik Cement GmbH. 1–35.

[7] Barthel H, Müller I. The influence of Alkali Oxide, Sulphur and Chlorine on the Wear of Magnesia Chrome Bricks in Rotary Cement Kilns. Interceram Special Issue. 1984: 18–21.

[8] Naefe H, Naziri M, Walk H. Optimum chrome-free burning zone lining for rotary cement kilns. ZKG International. 1993; 46: E159–E165.

[9] Kilb L, Södje J. Salt infiltrations in refractory linings due to the application of alternative fuels. Refra-Kolloquium 2000, Berlin. Göttingen: Refratechnik Cement GmbH. 186–198.

[10] The American Ceramic Society (sd). Phase Diagrams for Ceramics.

[11] Routschka G, Wuthnow H. Feuerfeste Werkstoffe: Praxishandbuch. Vulkan Verlag GmbH; 2011.

Tab. 15.2: Mineral data table. Salt compounds and newly-formed mineral phases determined by mineralogical analysis (XRD-analysis) in used refractory materials [11].

Compounds	Mineralogical Name	Chemical Formula	Melting Point (°C) [decomposition or transformation point]	Boiling Point (°C)	Density (g/cm³)
Alkali/alkaline earth salts	Sylvine	KCl	772	1500	1.98
	Halite	$NaCl$	801	1413	2.17
	Potash	K_2CO_3	897	decomposes	2.43
	Soda	Na_2CO_3	852	decomposes	2.55
	Buetschliite	$K_2Ca(CO_3)_2$			2.61
	Fairchildite	$K_2Ca(CO_3)_2$			2.44
	Kalicinite	$KHCO_3$			2.17
		$K_2CO_3 \cdot 1.5H_2O$			2.16
	Rinneite	$K_3NaFeCl_6$			2.55
	Erythrosiderite	$K_2FeCl_5 \cdot H_2O$			2.37
	Arcanite	K_2SO_4	1069	1689	2.66
		Na_2SO_4	884	1429	2.70
	Anhydrite	$CaSO_4$	1570	decomposes	2.96
	Aphthitalite	$3K_2SO_4 \cdot Na_2SO_4$	stable below 400/500 °C		2.70
		$K_2SO_4 \cdot Na_2SO_4$	867 (eutectic)		
	Calcium langbeinite	$K_2SO_4 \cdot 2CaSO_4$	936		2.68
	Glauberite	$Na_2SO_4 \cdot 2CaSO_4$			2.77
	Langbeinite	$K_2SO_4 \cdot 2MgSO_4$	930		2.83
	Syngenite	$CaSO_4 \cdot K_2SO_4 \cdot H_2O$	1004		2.57
		$K_2SO_3 \cdot 2H_2O$			
		$K_2(Cr^{6+},S)O_4$			
	Tarapazaite	$K_2Cr^{6+}O_4$	968		2.73
	Lopezite	$K_2Cr_2O_7$			2.68

Tab. 15.2: (continued)

Compounds	Mineralogical Name	Chemical Formula	Melting Point (°C) [decomposition or transformation point]	Boiling Point (°C)	Density (g/cm^3)
Alkali/alkaline earth sulphides		K_2S	840		1.81
		K_2S_2			1.95
		K_2S_3	252		2.14
		K_2S_6			2.21
		$KFeS_2$			2.56
	Oldhamite	CaS	2450	decomposes	2.56
Iron oxides/iron sulphides	Wustite	FeO	1369		5.70
	Magnetite	Fe_3O_4	1594		5.13
	Hematite	Fe_2O_3	1385		5.26
	Pyrrhotine	$Fe_{1-x}S$			4.60
	Troilite	FeS	1193–1199	decomposes	4.74
	Pyrite	FeS_2	1171		5.00
Heavy metal sulphides/ chlorides/oxides	Galenite	PbS	1114		7.50
	Perite	$PbBiO_2Cl$			8.16
	Greenockite	CdS			4.82
		$Cd_{9.5}Zn_{0.5}S$			
		$K_2Pb_2O_3$			5.98
Alkaline earth silicates	Merwinite	C_3MS_2	[1575]		3.15
	Monticellite	CMS	[1498]		3.20
	Forsterite	M_2S	1890		3.13

Tab. 15.2: (continued)

Compounds	Mineralogical Name	Chemical Formula	Melting Point (°C) [decomposition or transformation point]	Boiling Point (°C)	Density (g/cm³)
Alkaline earth aluminates/ alkaline earth aluminate silicates	Mayenite	$C_{12}A_7$	1455		2.69
	Melilite	C_4AMS_3	1390		2.95
		C_3MA_2			2.96
	Anorthite	CAS_2	1550		2.79
	Gehlenite	C_2AS	1590		3.04
	Ye'elimite	$C_4A_3\bar{S}$	1590		2.61
Alkali aluminate silicates/sulphates	Sanidine	KAS_6	[1150]		2.58
	Leucite	KAS_4	1150		2.51
	Kaliophilite	KAS_2	1686		2.47
	Kalsilite	KAS_2	923		2.59
	Diaoyudaoite	$(K,Na)Al_{11}O_{17}$	2000–2020		3.25–3.37
	Hauyne	$3NAS_2 \cdot CaSO_4$			2.50
	Nosean	$Na_8Al_6Si_6O_{24}SO_4$			2.30

Index

B

D

T

Bei Fragen zur Produktsicherheit wenden Sie sich bitte an:
If you have any questions regarding product safety,
please contact:

Walter de Gruyter GmbH
Genthiner Straße 13
10785 Berlin
productsafety@degruyterbrill.com